T0291563

Introduction to the Physical and Biological Oceanography of Shelf Seas

In this exciting and innovative textbook, two leading oceanographers bring together the fundamental physics and biology of the coastal ocean in a quantitative but accessible way for undergraduate and graduate students. Shelf sea processes are comprehensively explained from first principles using an integrated approach to oceanography – helping to build a clear understanding of how shelf sea physics underpins key biological processes in these environmentally sensitive and economically important regions. Using many observational and model examples, worked problems, and software tools, they explain the range of physical controls on primary biological production and shelf sea ecosystems.

Key features

- Opens with background chapters on the fundamentals of biology and physics needed to provide all students with a common, base-level understanding
- Develops the physical theory of each particular process in parallel with numerous data examples that describe the real-world impacts of physics on shelf sea biology
- Illustrates the success and failure of different model approaches to demonstrate their value as investigative research tools
- Boxes present extra detail and alternative explanations demonstrating the broader relevance of each topic
- Highlighted asides and anecdotes bring the reality and human aspects of ocean research work to life
- Physics sections include a set of non-mathematical summary points to help readers develop a qualitative understanding of the underlying processes
- Chapters end with summaries recapping key points to aid exam revision and problem sets that enable students to test their understanding

"This comprehensive and up-to-date book will be an ideal resource for both undergraduate and postgraduate students in pursuit of an all-round appreciation and understanding of the shelf seas. It really bridges a gap in the literature and the authors themselves pioneered much of the multidisciplinary research that has revealed a delicate interplay between the physical environment and life in the shelf seas."
Dr Robert Marsh *(University of Southampton)*

"Simpson and Sharples have combined courses in coastal physical dynamics and coastal biological oceanography to produce a textbook that is much greater than the sum of the individual disciplinary parts. Students and scientists alike will find the

discussions of sampling gear and deployment techniques an unusual and particularly useful aspect of this book. The authors are leaders in the study of the physics and biology of shelf seas and their experience and expertise are abundantly clear."
Professor Peter J.S. Franks *(Scripps Institution of Oceanography)*

"This text is a straightforward one-stop shop for students and professionals with a biological background who want to understand the basics of physical oceanography. It is very interesting and readable, and a great introduction the theoretical background a biologist needs to understand the large-scale physical dynamics of the world their organisms are inhabiting."
Professor Katherine Richardson *(Copenhagen University)*

"This book will prove to be a masterpiece with enduring value and fills a significant gap in physical oceanography textbooks by focusing on shallow seas. It reads well, is accessible to the intelligent, scientifically trained-non specialist and provides a solid foundation by which ecologists can learn much about the physical control of many ecological processes on shelf seas."
Distinguished Professor Malcolm Bowman *(State University of New York at Stony Brook)*

John Simpson leads a research group in the School of Ocean Sciences at Bangor University in Wales, which is developing new methods to observe and model turbulence and the mixing that plays a crucial role in biological production. He is a seagoing physical oceanographer with a broad interest in shelf seas and estuaries, and his research has focused on the physical mechanisms which control the environment of the shelf seas. He has taught Physics of the Ocean at Bangor and other universities worldwide for more than 40 years and was responsible for establishing the first Masters-level course in Physical Oceanography within the UK. In 2008, Professor Simpson was awarded the Fridtjof Nansen Medal of the European Geosciences Union for his outstanding contribution to understanding the physical processes of the shelf seas, and the Challenger Medal of the Challenger Society for his exceptional contribution to Marine Science.

Jonathan Sharples holds a joint chair at the University of Liverpool and the UK Natural Environment Research Council's National Oceanography Centre, and has taught courses in coastal and shelf oceanography at the universities of Southampton and Liverpool. He is an oceanographer whose research concentrates on the interface between shelf sea physics and biology. His work is primarily based upon observational studies at sea, combined with development of simple numerical models of coupled physics and biology. Professor Sharples has extensive seagoing experience off the NW European shelf and off New Zealand, having led several major interdisciplinary research cruises. His research has pioneered the use of fundamental measurements of turbulence in understanding limits to phytoplankton growth and controls on phytoplankton communities.

Introduction to the Physical and Biological Oceanography of Shelf Seas

JOHN H. SIMPSON
School of Ocean Sciences, Bangor University

JONATHAN SHARPLES
School of Environmental Sciences, University of Liverpool
and NERC National Oceanography Centre

CAMBRIDGE
UNIVERSITY PRESS

University Printing House, Cambridge CB2 8BS, United Kingdom

One Liberty Plaza, 20th Floor, New York, NY 10006, USA

477 Williamstown Road, Port Melbourne, VIC 3207, Australia

314-321, 3rd Floor, Plot 3, Splendor Forum, Jasola District Centre, New Delhi - 110025, India

79 Anson Road, #06-04/06, Singapore 079906

Cambridge University Press is part of the University of Cambridge.

It furthers the University's mission by disseminating knowledge in the pursuit of education, learning and research at the highest international levels of excellence.

www.cambridge.org
Information on this title: www.cambridge.org/9780521701488

First published 2012

A catalogue record for this publication is available from the British Library

Library of Congress Cataloging in Publication data

Simpson, John (John H.)
Introduction to the physical and biological oceanography of shelf seas / John H. Simpson, Jonathan Sharples.
p. cm.
Includes bibliographical references and index.
ISBN 978-0-521-87762-6 (Hardback) – ISBN 978-0-521-70148-8 (Paperback)
1. Oceanography. 2. Coasts. 3. Continental shelf. I. Sharples, Jonathan. II. Title.
GC28.S54 2012
551.46′18–dc23 2011030490

ISBN 978-0-521-87762-6 Hardback
ISBN 978-0-521-70148-8 Paperback

Additional resources for this publication at www.cambridge.org/shelfseas.

It can scarcely be denied that the supreme goal of all theory is to make the irreducible basic elements as simple and as few as possible without having to surrender the adequate representation of a single datum of experience.

Albert Einstein, 1933

I try not to think with my gut. If I'm serious about understanding the world, thinking with anything besides my brain, as tempting as that might be, is likely to get me into trouble.

Carl Sagan, 1995

CONTENTS

Colour plate section between pages 232 and 233.

PREFACE

The seas of the continental shelf where the depth is less than a few hundred metres experience a physical regime which is distinct from that of the abyssal ocean where depths are measured in kilometres. While the shelf seas make up only about 7% by area of the world ocean, they have a disproportionate importance, both for the functioning of the global ocean system and for the social and economic value which we derive from them. Approximately 40% of the human population lives within 100 km of the sea, and the coastal zones of the continents are host to much of our industrial activity. Biologically, the shelf seas are much more productive than the deep ocean; phytoplankton production is typically 3–5 times that of the open ocean, and globally, shelf seas provide more than 90% of the fish we eat. They also supply us with many other benefits ranging from aggregates for building to energy sources in the form of hydrocarbons and we use our coastal seas extensively for recreation and transport. The high biological production of the shelf seas also means that these areas are important sources of fixed carbon which may be carried to the shelf edge and form a significant component of the drawdown of atmospheric CO_2 into the deep ocean.

Understanding of the processes operating in shelf seas and their role in the global ocean has advanced rapidly in the last few decades. In particular, the principal processes involved in the workings of the physical system have been elucidated, and this new knowledge has been used to show how many features of shelf sea biological systems are underpinned and even controlled by physical processes. It is the aim of this book to present the essentials of current understanding in this interdisciplinary area and to explain to students from a variety of scientific backgrounds the ways in which the physics and biology relate in the shelf seas. Our motivation to write such a book came from our extensive experience of teaching undergraduate and post-graduate courses in physical oceanography and biological oceanography to students from diverse disciplinary backgrounds and the realisation that there was an unfulfilled need for a textbook to present the maturing subject of shelf sea oceanography combining the physical and biological aspects.

As far as possible, we have endeavoured to give the book an interdisciplinary structure and to make it accessible to a wide range of students from different disciplinary backgrounds. Some of the early chapters deal separately with the fundamental principles of physics and biology necessary to understand the later material. The later chapters are arranged along interdisciplinary lines to illustrate the impact of physical processes on the biological response from primary production up to higher trophic levels. A full understanding of the physics inevitably requires some use of mathematical notation and we have included this for students from physical science disciplines. At the same time, we have provided summaries of the 'essential physics' which allow shortcuts through the mathematical development and should help students coming from biological backgrounds with limited experience of physics and mathematics to grasp the key physical ideas and appreciate how they affect the biology. Understanding of new concepts and their application is facilitated by supporting material in the form of problem sets and numerical exercises, within the book text and also hosted on the book website at Cambridge University Press. The book should form a suitable course text for advanced undergraduate and postgraduate oceanography students, but we anticipate that it will also be appropriate to courses introducing physical and biological science students to oceanography.

Both of us are seagoing oceanographers who have studied diverse shelf sea systems in different parts of the world. Much of our understanding and insight into the way the shelf seas work, however, has come from extensive observational work during national and international campaigns in the tidally energetic shelf seas of northwestern Europe. Where possible we have used results from other shelf sea systems to illustrate parallels and differences between shelf sea systems but, inevitably, many of the examples we use are drawn from the European shelf which is now arguably the most intensively studied of all shelf systems in the global ocean. In this respect, we have not sought to produce a definitive volume on everything in shelf sea physical and biological oceanography. Rather, we have aimed to write a book that contains what we have found to be the key components of shelf sea physics and the way in which that physics impacts the biology in the European and other shelf sea systems. In doing so we have made extensive use of a variety of models, ranging from basic analytical constructs through to 1D turbulence closure models of vertical exchange to test simple and compound hypotheses about how the system works. By contrast, we have made rather little reference to large-scale 3D models which, while they are vital in applying understanding to the task of properly managing the shelf seas, have not yet contributed greatly to fundamental understanding of shelf sea processes.

Although shelf sea science has advanced rapidly in recent years, there are still many open questions about the processes involved, especially at the interfaces between physics, biochemistry and ecology. While a textbook is conventionally about established facts and well-supported theory, we have included some elements of conjecture and speculation in relation to the more interesting questions that remain, in the hope that they will stimulate further study and further refine our understanding of the shelf sea systems.

ACKNOWLEDGEMENTS

We would like to record our gratitude to the many individuals who have provided inspiration, advice and practical help in the preparation of this book. In particular we owe a debt to those (Joe Hatton, Ken Bowden) who inspired our interest in physics and physical processes in the ocean, and to those (Paul Tett, Patrick Holligan, Robin Pingree) who steered us towards interdisciplinary studies in shelf seas and whose work has helped to motivate us to write the book. Captain John Sharples and Eileen Ansbro Sharples had the courage to take their two kids on extended voyages aboard UK merchant vessels, which doubtless influenced the career path of one of us. Both of us have benefited greatly over the years from interacting with many able research students, too numerous to list, who have challenged our ideas and helped to refine them.

In the process of writing the book, we have received generous help and advice from many individuals, including Dave Bowers, Malcolm Bowman, Peter Franks, Mattias Green, Anna Hickman, Claire Mahaffey, Bob Marsh, Mark Moore, Kath Richardson, Tom Rippeth, Steve Thrope and Ric Williams. In several cases, their input has helped us to avoid mistakes in the text. However, the responsibility for any residual shortcomings rests squarely with us and we welcome notification by readers of any remaining errors.

We are also grateful to many colleagues and co-workers, including Gerben de Boer, Juan Brown, Byung Ho Choi, Mark Inall, Kevin Horsburgh, Jonah Steinbuck, David Townsend, Mike Behrenfeld, Clare Postlethwaite, Yueng-Djern Lenn, Flo Verspecht, Pat Hyder, Matthew Palmer, John Milliman, David Roberts, Oliver Ross and Alex Souza, for help in the acquisition and drawing of many of the figures, and Kay Lancaster for timely help in re-drafting and providing important finishing touches. Much of our use of satellite imagery comes courtesy of the UK Natural Environment Research Council's Earth Observation Data Acquisition and Analysis Service (NEODAAS) at the University of Dundee and at the Plymouth Marine Laboratory, with particular thanks to Peter Miller and Stelios Christodoulou.

Finally, we are pleased to acknowledge that all of our work is dependent on the ability to go to sea and make observations in often challenging conditions. This book would not have been possible without the professionalism and skills of the research vessel crews and technicians, on which we continue to rely.

GUIDE TO THE BOOK AND HOW TO MAKE THE BEST USE OF IT

We anticipate that readers of this interdisciplinary book will be a mixture of students and researchers who come to the subject of the shelf seas from a wide range of scientific backgrounds. At one extreme will be students of mathematics and physics who know little of biology and, at the other, students of biological subjects who have not pursued physical sciences beyond high school level. In between will be a broad group of students, including many who have already embarked on courses in marine science, who have some background in both physical and biological sciences.

In writing the book, we have endeavoured to cater to individuals from these diverse backgrounds without compromising the presentation of the science. In particular, we have structured the chapters to allow readers from a mainly biological background to appreciate the essence of the physical processes without having to follow the detail of the sometimes intricate mathematical arguments. Key processes are explained in more intuitive ways in box sections, many with illustrative diagrams. At the end of each chapter, the essential points are recapitulated in a chapter summary. There is also a selection of problems, of varying difficulty, and suggestions for further reading at the end of each chapter. In order to help students of all backgrounds familiarise themselves with key terminology, a full glossary is given at the back of the book.

The first chapter is a general introduction to the shelf seas, explaining their relation to the global ocean, their socioeconomic importance, the history of shelf sea investigations and the observational techniques now used in studying them. In Chapter 2 we explore the various physical forcing mechanisms which drive the shelf seas, determine their structure and supply the vital radiation input to drive photosynthesis. There follow three chapters concerned with the fundamental science which underpins our subsequent exploration of shelf sea processes: Chapters 3 and 4 focus on the basic physics of fluid motion, while Chapter 5 is concerned with the aspects of biogeochemistry and plankton survival involved in the shelf seas.

The book then moves to explore the main domains/regimes of the shelf seas in a series of five chapters. The cross-shelf schematic illustration in Fig. G1 provides us with a guide to where each chapter is focused. In Chapter 6 we consider the processes controlling thermal stratification, the partitioning of the shelf in stratified and mixed regimes and the controls exerted by stratification on the growth of plankton. The crucial role of low levels of internal mixing in supporting phytoplankton growth in the interior of the stratified regions is explored in Chapter 7, while Chapter 8 focuses on the physical nature and biological implications of the fronts produced by variations in tidal mixing. In Chapter 9, we consider the regions of the shelf where freshwater inputs from rivers play a major role, and in Chapter 10 we look at the

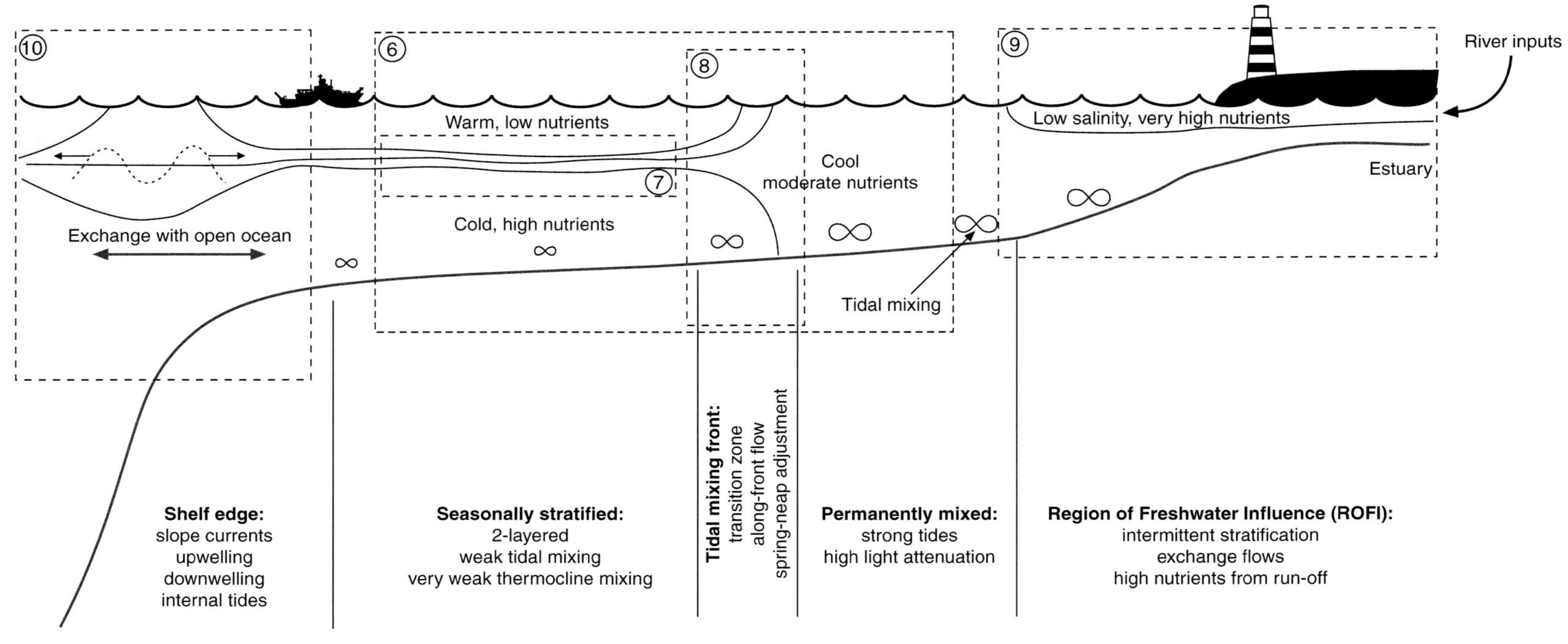

Figure G1 Schematic illustration of the shelf sea regimes. The dashed squares show the regions covered by individual chapters, with the relevant chapter number circled.

special physical processes of the shelf edge regime and their important biological consequences. The book concludes with an overview of progress and the remaining challenges in shelf seas, notably those in the Arctic and the tropics, and considers recent studies on the role of the shelf seas in relation to changes in the global ocean since the Pleistocene.

To help readers coming to the book from different disciplinary backgrounds, we offer a few suggestions on how best to approach it: Chapters 1 and 2 provide essential introductory background and should be readily accessible to all science students. For students who have already taken courses in physical oceanography, Chapters 3 and 4 may be largely revision but they will also be useful in applying knowledge of fluid physics to the shelf regime. Students without strong maths and physics may bypass some of the detailed argument here, certainly on a first reading, and make use of the boxes and summaries to pick up the essentials. Similarly, while much of Chapter 5 will already be familiar to students of biological oceanography, physics students will have a lot to learn here and may want to bypass some of the detail and, at least initially, rely on the summary. To help students identify the main points of the developing narrative, we have also put boxes around equations which are significant results to be applied in later sections or represent key stages in derivations. Students with previous training in both physical and biological marine science may want to bypass some of the tougher physics on first reading, returning later to follow the detail of the mathematics.

In addition to the problems for each chapter, many topics in the text are illustrated by visualisations and numerical models which are available on the book's website (http:www.cambridge.org/shelfseas). Much of the software is based on MATLAB and the programme scripts are available at the website for readers to copy, amend and use to explore their own ideas and understanding. The icon in the margin here is used throughout the book to indicate when there is relevant software on the book website.

SYMBOLS

Symbol	Name	Units
a	acceleration	m s^{-2}
a_e	Radius of the Earth	m
a_p	Phytoplankton cell radius	m
A	Albedo	none
A_0	Amplitude of oscillatory function	
A_n	Amplitude of tidal constituent	m
b	Buoyancy force per unit volume	N m^{-3}
$\boldsymbol{B}$	Buoyancy production of TKE	W kg^{-1}
B_G	Breadth of gulf	m
c	Phase velocity of waves	m s^{-1}
c_a	Specific heat of air	J kg^{-1} °C^{-1}
c_p	Specific heat of seawater	J kg^{-1} °C^{-1}
C	Conductivity	mS m^{-1}
C_d	Drag coefficient for wind stress	none
C_g	Geostrophic current speed	m s^{-1}
D	Ekman depth	m
e, e_s	Efficiency of mixing by tide and wind	none
E	Mass of the Earth	kg
E_d	Downward energy flux	W m^{-2}
E_k	Downward flux of PAR	W m^{-2}
Ek	Ekman number	
$E_s(k)$	Scalar spectrum for TKE	m^3 s^{-2}
E_T	Turbulent kinetic energy density	J kg^{-1}
E_v	Rate of evaporation	kg m^{-2} s^{-1}
$\boldsymbol{E_w}$	Energy density of waves	J m^{-2}
f	Coriolis parameter	s^{-1}
F	Force	N
$F(y)$	Function of variable y	
g	Acceleration due to gravity on Earth	m s^{-2}
g'	Reduced gravity	m s^{-2}
G	Gravitational constant	N m^2 kg^{-2}
$\underset{\sim}{G}$	Scalar flux vector	various
g_n	Phase lag of tidal constituent	degrees
g_p^b	Specific grazing rate	g C (g Chl) $^{-1}$ time^{-1}

Symbol	Name	Units
h	Water column depth	m
H_a	Attenuation factor for waves	none
H_n	Amplitude of a tidal constituent	m
H_T	Heat stored in water column	$J\ m^{-2}$
I	Total radiation energy flux	$W\ m^{-2}$
I_0	PAR flux incident at surface	$\mu E\ m^{-2}\ s^{-1}$ or $W\ m^{-2}$
I_K	Saturation light level in photosynthesis	$\mu E\ m^{-2}\ s^{-1}$ or $W\ m^{-2}$
$I_\lambda(\lambda)$	Spectral power density of radiation	$W\ m^{-2}\ nm^{-1}$
I_{PAR}	PAR flux at depth z	$\mu E\ m^{-2}\ s^{-1}$ or $W\ m^{-2}$
J	Flux of a scalar property	various
k	Wave number	m^{-1}
k_b	Bottom drag coefficient	none
k_b'	Constant in a linear bottom drag law	$m\ s^{-1}$
k_m	Molecular diffusivity	$m^2\ s^{-1}$
k_{NUT}	Half saturation concentration for nutrient uptake	$mmol\ m^{-3}$
k_s	Modified surface drag coefficient $= \gamma_s C_d$	none
K	Eddy diffusivity	$m^2\ s^{-1}$
K_{av}	Spectral average of K_d	m^{-1}
K_d	Diffuse attenuation coefficient	m^{-1}
K_{PAR}	Attenuation coefficient for PAR	m^{-1}
K_q	Eddy diffusivity for TKE	$m^2\ s^{-1}$
K_x, K_y	Horizontal Eddy diffusivity	$m^2\ s^{-1}$
K_z	Vertical eddy diffusivity	$m^2\ s^{-1}$
L	Turbulence length scale	m
L_H	Latent heat of evaporation	$J\ kg^{-1}$
L_O	Ozmidov length	m
L_S	Length scale for swimming plankton	m
m	Mass	kg
M	Mass of the moon	kg
M_s	Mass of a scalar substance	kg
N	Stability frequency	s^{-1}
N_z	Eddy viscosity	$m^2\ s^{-1}$
p	Pressure	Pa
p_0	Atmospheric pressure	Pa
$\mathbf{P}$	Shear production of TKE	$W\ kg^{-1}$
P_a	Power available to produce TKE	$W\ m^{-3}$
P_{Chl}	Phytoplankton Chl biomass concentration	$g\ Chl\ m^{-3}$
P^b_{dark}	Specific growth rate in the dark	$g\ C\ (g\ Chl)^{-1}\ s^{-1}$

Symbol	Name	Units
Pe	Peclet number	none
P_p	Rate of primary production	g C m^{-3} s^{-1}
$P_p{}^b$	P_p normalised by Chl biomass $= P_p/P_{Chl}$	g C (g Chl)$^{-1}$ s^{-1}
P^b_{max}	Maximum value of $P_p{}^b$	g C (g Chl)$^{-1}$ s^{-1}
P_r	Ratio of N_z/K_z	none
P_S, P_N	Stirring power at springs/neaps position of a TM front	W m^{-3}
P_T, P_W	Stirring power of tidal flow and wind	W m^{-2}
$\boldsymbol{P_w}$	Energy flux in waves	W m^{-1}
q	Turbulent eddy speed	m s^{-1}
q_a	Specific humidity of air	none
q_s	Specific humidity at saturation	none
Q_b	Back radiation from sea surface	W m^{-2}
Q_c	Heat loss by conduction	W m^{-2}
Q_e	Evaporative heat loss	W m^{-2}
Q_i	$Q_s (1-A) - Q_u$	W m^{-2}
Q^N	Phytoplankton cell nutrient quota	mmol N (mg C)$^{-1}$
$Q^N{}_{max}$	Maximum cell nutrient quota	mmol N (mg C)$^{-1}$
$Q^N{}_{min}$	Minimum cell nutrient quota (subsistence quota)	mmol N (mg C)$^{-1}$
Q_s	Solar energy input to sea surface	W m^{-2}
Q_{sed}	Heat exchange with sediments	W m^{-2}
Q_u	$Q_b + Q_b + Q_c$ = total heat loss through sea surface	W m^{-2}
Q_v	Heat gain from horizontal advection	W m^{-2}
ΔQ	Quantity of heat input per unit area	J m^{-2}
r^b_p	Phytoplankton-specific respiration rate	C (g Chl) $^{-1}$ time^{-1}
r_{SM}	Ratio of constituents S_2 to M_2	none
$R\ R'$	Moon and Earth orbital radii about centre of gravity of Earth – Moon system	m
R_d	River discharge	m^3 s^{-1}
Re	Reynolds number	none
Rf	Flux form of the Richardson number	none
Ri	Richardson number	none
RN	Rossby number	none
Ro, Ro'	Rossby radius (external and internal)	m
R_w	Freshwater inflow per unit width	m^2 s^{-1}
s	Concentration of a scalar	various
S	Salinity (or generic scalar)	none
SH	Simpson-Hunter stratification parameter	$\log_{10}$(m^{-2}s^3)
S_M, S_H	Stability functions	none
S_v	Velocity shear	s^{-1}

Symbol	Name	Units
t	Time	s
T	Temperature in Centigrade	°C
T_a	Air temperature	°C
T_b	Transport in bottom Ekman layer	$m^2\ s^{-1}$
T_c	Period of a tidal constituent	s
T_E	Period of Earth's rotation	s
T_I	Inertial period	s
T_K	Kelvin temperature	°K
T_m	Vertical mixing time	s
T_p	Wave period	s
T_{res}	Residence time	s
T_s	Temperature of the sea surface	°C
$\mathbf{T_w}$	Kinetic energy density of waves	$J\ m^{-2}$
$\hat{u}, \hat{v}$	Depth-mean velocity components in x, y	$m\ s^{-1}$
u,v,w	Velocity components in x,y,z	$m\ s^{-1}$
u',v',w'	Turbulent velocity components	$m\ s^{-1}$
u_b,v_b	Velocity at bottom boundary	$m\ s^{-1}$
u_g	Geostrophic velocity	$m\ s^{-1}$
u_{max}	Maximum nutrient uptake rate	$nmol\ l^{-1}\ h^{-1}$
$u^v{}_{max}$	Cell volume–specific uptake rate	$mmol\ m^{-3}\ s^{-1}$
u_{NUT}	Uptake rate of nutrient during incubation	$nmol\ l^{-1}\ h^{-1}$
u_s	Surface current speed	$m\ s^{-1}$
U, V	Depth-integrated transports in x, y	$m\ s^{-1}$
U,V,W	Time average velocity components (Chapter 4)	$m\ s^{-1}$
U_g	Group velocity of waves	$m\ s^{-1}$
$\mathbf{V_w}$	Potential energy density of waves	$J\ m^{-2}$
v_c	Phytoplankton swimming speed	$m\ s^{-1}$
x,y,z	Cartesian coordinates	m
W	Wind speed	$m\ s^{-1}$
z'	Fractional depth z/h	none
z_{cr}	Critical depth	m
α	Thermal volume expansion coefficient	$°C^{-1}$
α_n	Phase of tidal constituent in TGF	radians
α_q	Maximum light utilisation coefficient	$g\ C\ (g\ Chl)^{-1} m^2\ s\ \mu E^{-1}\ s^{-1}$
β	Salinity coefficient of density	none
γ	Compressibility of seawater	Pa^{-1}
γ_s	Ratio of surface current to wind speed	none
Γ	$R_f/(R_f + 1)$	none
δ	Angle	radians
ε	Rate of energy dissipation to heat	$W\ kg^{-1}$ or $W\ m^{-3}$

Symbol	Name	Units
ε_s	Emissivity	none
ζ	Vertical displacement	m
η	Surface elevation	m
η_v	Kolmogorov microscale	m
θ	Angle	radians
κ	von Kármán constant	none
λ	Wavelength	m
μ	molecular viscosity	N m^{-2} S
μ^N	Nutrient-dependent specific growth rate	time^{-1}
$\mu^N{}_{max}$	Maximum nutrient-dependent specific growth rate	time^{-1}
v	Kinematic viscosity	m^2 s^{-1}
ξ	Density gradient $= (1/\rho_0)\,\partial\rho/\partial x$	m^{-1}
ρ	Density of seawater	kg m^{-3}
$\hat{\rho}$	Depth-averaged density	kg m^{-3}
ρ_0	Reference density	kg m^{-3}
ρ_a	Density of air	kg m^{-3}
ρ_s	Surface density	kg m^{-3}
σ	Standard deviation of dispersion	m
σ_s	Stefan's constant	W m^{-2} K^{-4}
σ_{STP}	Density as ρ-1000	kg m^{-3}
σ_t	Density as ρ-1000 at zero pressure	kg m^{-3}
τ	Tangential stress	Pa
τ_b	Bottom stress	Pa
τ_W	Wind stress on sea surface	Pa
τ_x, τ_y	Horizontal stress components	Pa
ϕ	Velocity potential function	m^2 s^{-1}
ϕ_L	Latitude	degrees
Φ	Potential energy anomaly	J m^{-3}
χ	Surface slope $= \partial\eta/\partial x$	none
ψ	Angle	radians
$\Psi(\omega)$	Frequency spectrum of kinetic energy	m^{2}s^{-1}
ω	Angular frequency	s^{-1}
ω_a	Annual cycle angular frequency	s^{-1}
ω_n	Tidal constituent angular frequency	s^{-1}
Ω	Earth's angular speed of rotation	s^{-1}

Note: This list includes the principal symbols used in the book. You will find a number of additional symbols, which are used only locally and are defined at the point of use.

1 Introduction to the shelf seas

In this chapter we shall introduce the reader to the shelf seas, their extent and position in the global ocean and the motivation, both fundamental and applied, behind our efforts to understand and model the complex processes which control the shelf sea environment and ecosystem. We shall then briefly explain the historical development of shelf sea science and describe the technical tools which are now available and which have facilitated the relatively rapid advances of recent years. As well as discussing the principal observational techniques, in a final section we shall consider the role of numerical modelling and its potential contribution to developing understanding.

1.1 Definition and relation to the global ocean

Between the deep oceans and the continents lie the seas of the continental shelf. These shallow areas usually have rather flat seafloors and extend out to the shelf break, where the seabed inclination generally increases rapidly at the top of the continental slope leading down to the abyssal ocean. This abrupt change of slope is clear in the map of global bathymetry shown in Fig. 1.1a. It typically occurs at a depth of ~200 metres and a contour, or isobath, at this depth is often taken as defining the outer limit of the shelf seas. This choice is not critical, however, since the continental slope is so steep (~1:10); moving from the 200- to the 500-metre isobath involves little horizontal movement. Using the basis of a 500-metre definition, Fig. 1.2a shows that the shelf seas account for ~9% of the total area of the global ocean and less than 0.5% of the volume. The shelf seas have an influence and importance quite out of proportion to these numbers.

The shelf seas act to dissipate a high proportion of the mechanical energy input to the ocean. This is most obvious in the swell waves which are generated in the deep ocean and travel great distances before delivering large quantities of energy to be consumed in wave breaking and bottom friction on the shelf. Less obviously, perhaps, energy input in the deep ocean by the tidal forces also propagates onto the shelf in the form of very long waves which are dissipated by friction in the large tidal currents of the shelf seas.

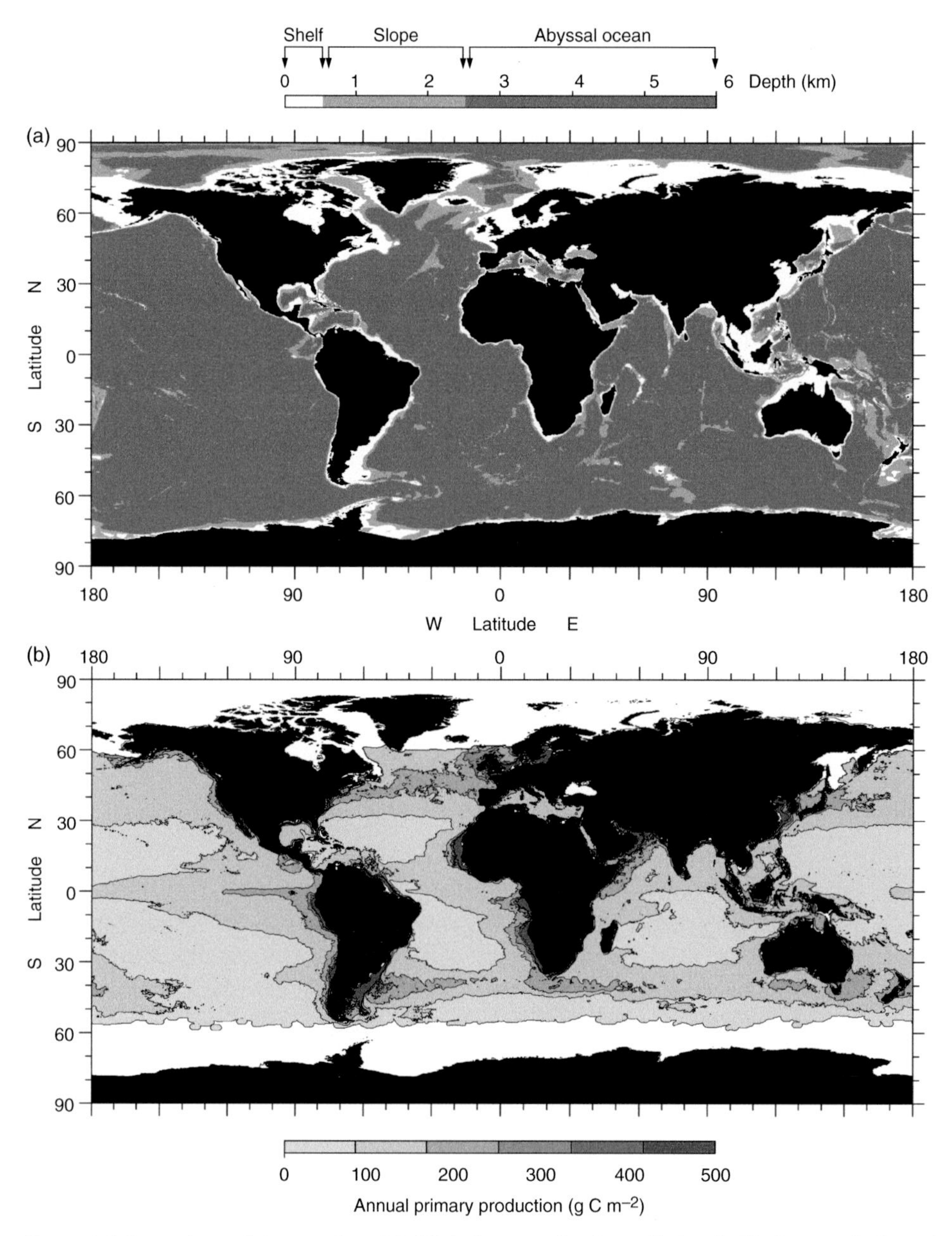

Figure 1.1 See colour plates version. (a) Global ocean depths, split by shelf, slope and abyssal ocean. Based on GEBCO bathymetry; (b) global annual primary production, using imagery from the MODIS satellite (Behrenfeld and Falkowski, 1997); data courtesy of Mike Behrenfeld, Oregon State University, USA: http://www.science.oregonstate.edu/ocean.productivity/index.php.

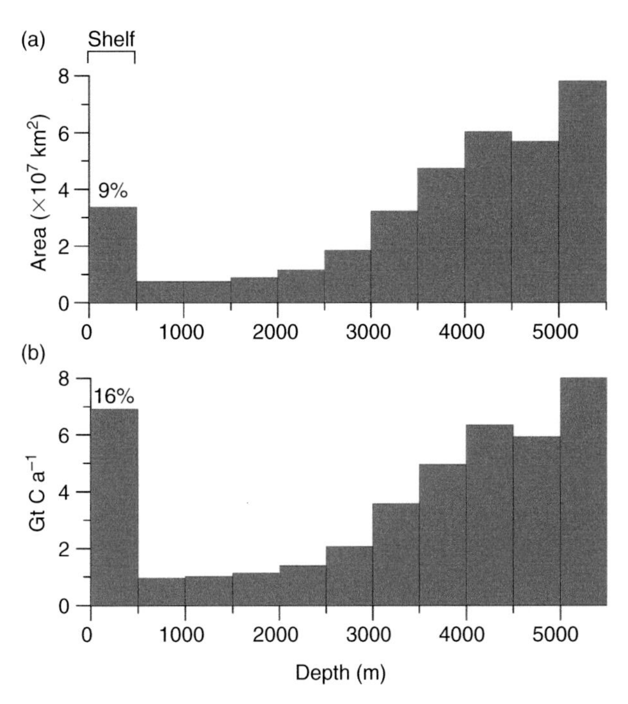

Figure 1.2 (a) The area of seabed in different depth ranges; if shelf seas are taken to be less that 500 metres depth they cover 9% of the ocean's total area. (b) Net primary production (gigatonnes of carbon per year) within different depth ranges; water shallower than 500 metres accounts for 16% of global production. The plots are based on the data in Fig. 1.1.

As a result of these external inputs and the direct action of the local wind at the surface, the shelf seas are physically energetic areas with vigorous stirring. They also receive large inputs per unit volume of solar energy which act to modify the density and thus create horizontal density gradients. Similarly, freshwater river discharge from the adjacent land, as well as rainfall, lower the density near the coast and thus contribute to density forcing of the shelf seas, which we will describe in Chapter 9. Because of the strength and variety of the forcing, the shelf seas are, in many ways, the most dynamic regions of the ocean and are host to much of the biogeochemical action. They play a major role in the growth of phytoplankton which constitutes the primary production of the oceans. A recent collation of available observations has suggested the annual primary production within the world's shelf seas is about 11 Gt C a^{-1} (Jahnke, 2010), which compares with a global total of between 45 and 60 Gt C a^{-1} (Longhurst *et al.*, 1995; Behrenfeld *et al.*, 2005). The distribution of net phytoplankton production in the ocean can be estimated from satellite imagery (Behrenfeld and Falkowski, 1997), though care needs to be taken in accounting for the contribution from subsurface phytoplankton growth (invisible to a satellite sensor) and the difficulty of linking ocean colour to chlorophyll concentrations in coastal regions where suspended sediments and dissolved organic material can influence the data (Longhurst *et al.*, 1995). Notice in the example of such a satellite-based estimate in Fig. 1.1b that the highest values of annual carbon fixation are located in narrow bands around the continents. Fig. 1.2b provides an analysis of that data showing that we can attribute ~16% of global marine primary production to the shelf seas. That amount of production occurs in 5% of the ocean surface sampled by the satellite (note in Fig. 1.1b that the satellite coverage misses the high latitudes, including substantial areas of shelf in the Arctic). A more detailed comparison between shelf and ocean production rates shows that the shelf seas have an average carbon fixation rate per unit area a factor of ~2.5 greater than in the deep ocean.

The abundant supply of phytoplankton provides the primary food source for the rich fisheries of the shelf seas. Globally the shelf seas are the dominant source of fish caught by humans; it has been estimated that over 90% of global fish catches come from the shelf seas and adjacent upwelling area over the continental slope (Pauly *et al.*, 2002). Also, primary production involves the drawdown of CO_2 from the atmosphere and the subsequent removal of some of the fixed carbon into the deep ocean. Again, despite their relatively small surface area, current best estimates are that the shelf seas are responsible for about 47% of the global annual export of particulate organic carbon (Jahnke, 2010). This high proportion of carbon export is partially a result of high primary productivity in the shelf seas, but it also arises because of the unique physical environment in shelf seas and at the shelf edge. The processes governing the shelf edge, and their biogeochemical and ecological consequences, are the subject of Chapter 10.

1.2 Economic value versus environmental health

We depend on the shelf seas not only for fisheries but for a wide range of benefits. The sediments of the shelf seas have been a major source of hydrocarbons, both oil and gas, and are also widely exploited as a source of aggregates for building. We use the shelf seas extensively for transport and for recreation, as many of us sail in them and spend our holidays on their shores. We place a high value on the coastal marine environment but, in many cases, compromise its health by using our shelf seas and estuaries as low-cost dumping grounds for our domestic and industrial wastes. The pressures on the shelf seas from human activity are the more acute because so many of us live close to the coast. Approximately 40% of the human population is located within 100 km of the sea, and coastal zones are host to much of our industrial activity. Even where direct disposal of wastes by dumping and through sewage pipelines has been stopped, rivers still carry large quantities of nutrients and pollutants from terrestrial agriculture and industry into the shelf seas where they have their first, and usually largest, impact. An excess of nutrients entering the shelf seas can have seriously adverse effects in producing nuisance blooms of phytoplankton which may themselves be toxic, or can lead to *hypoxia* (oxygen depletion) as they decay with consequent disruption of the ecosystem and the mass mortality of marine organisms. Heavy metals and organic pollutants tend to become concentrated as they progress up the food chain and may have toxic effects in higher trophic levels including humans.

In addition to these pressures through inputs from industry and agriculture, the welfare of the shelf seas is further threatened by the over-exploitation of fish stocks. It has been estimated that fishing pressures in the North Sea are such that ~25% of the total North Sea biomass is removed each year (McGlade, 2002), much of it by trawling which also ploughs up the seabed. Such intensive fishing can result in the collapse of economically important fisheries, as in the case of the cod on the Grand Banks (Hutchings, 1996). Recovery of fish stocks can take considerable time (Hutchings, 2000), and intense fishing can lead to radical changes to the ecosystem as a result of habitat disturbance and community shifts to non-commercial species (e.g. Johnson and Coletti, 2002).

In order to provide a basis for the rational management of the shelf seas, there is a clear need for a full scientific understanding of the way the shelf seas work. Improved knowledge of the processes involved is required to underpin the development of skilful numerical models which can be used to investigate the response of a given shelf sea to changes in the applied pressures, including the likely changes in sea level and atmospheric forcing which will soon arise through global warming. This is an ambitious but pressing requirement which provides strong motivation for the study of the science of the shelf seas.

1.3 The scientific challenge of the shelf seas

In addition to the important practical management requirements for improving knowledge of the shelf seas, their study is also motivated by the fundamental challenge to interdisciplinary science of understanding the diverse processes operating in the shelf seas and the way they interact. It has been increasingly recognized that this challenge is no less than that posed by the deep oceans with which it has much in common. A separate focus on the shelf seas, which has developed in recent decades, is appropriate because of the radically different regimes of the shelf and the deep ocean.

Understanding the shelf seas is thus a major interdisciplinary campaign of science; it calls on all the separate physical, chemical and biological aspects of oceanography to elucidate individual processes and aims to represent their interaction in conceptual and numerical models. This is an ambitious programme; until the latter part of the twentieth century there was only limited progress and the sole textbook on the shelf seas (Bowden, 1983) concentrated on the physical oceanography where significant progress was being made. Developments in the physics of the shelf seas during the last four decades have established a framework in which the principal physical processes are represented and which has formed the basis for interdisciplinary developments. To some extent, the problem of unravelling the complex mix of processes has been made easier by the fact that the physical system is, with a few exceptions, independent of the biogeochemistry and not subject to feedbacks from biological processes.

It is this emergence of a firm physical framework which has expedited progress in interdisciplinary areas and stimulated a number of major interdisciplinary campaigns and experiments (e.g. CalCOFI (CalCOFI, 2011), the North Sea Project (Charnock *et al.*, 1994), the PRISMA Project (Sundermann, 1997), the PISCO partnership (PISCO)). The progress resulting from such initiatives has stimulated interdisciplinary understanding and enabled the subject to advance to a point where we judged that the time was ripe for a text covering the science of the shelf seas from the physics through to the biology.

1.4 A brief history of scientific research of the shelf seas to 1960

While the scientific study of the deep ocean is generally considered to have started with the *Challenger* expedition in 1872–1876, the systematic investigation of the shallow seas did not develop until some time later. Some important early observations were made at coasts, notably of the tides which were first accurately recorded at the port of Brest in 1679. By the late nineteenth century, extensive measurements with tide gauges at coastal ports had allowed mapping of the tidal characteristics in many of the world's marginal seas and the development of reliable methods of tidal prediction (Cartwright, 1999). This early success based on measurements in shelf seas, arguably the first quantitative success of oceanography, was the exception, as the focus of marine studies was increasingly in deep water. Compared with the exciting challenge of exploring the vast interior depths of the deep ocean, the study of the more accessible shelf seas had less appeal. The *Challenger* expedition was followed by a number of similar major voyages of discovery by vessels sponsored by nations that were keen to share in the prestige of deep ocean exploration (Deacon, 1971). The shelf seas were largely neglected in these studies, which were much more concerned with discovering new forms of life in the ocean and less with mapping the physical environment and determining the processes which controlled it. A notable exception which did concentrate on the shelf seas was the German-sponsored study of the North Sea and the Baltic by the S.S. *Drache* in the period 1881–1884 which made some of the first large-scale surveys of temperature and salinity distributions in the summer regime of these regions. Victor Hensen, a German zoologist working in the North Sea and the NE Atlantic Ocean in the 1880s, is often credited as founding the discipline of biological oceanography. He recognised the fundamental importance of the plankton in supporting marine life, and coined the word 'plankton' (from the Greek 'planktos' meaning to wander or drift) that was formalised by Ernst Haeckel in 1890 to encompass all drifting organisms.

The burgeoning interest in marine science stimulated by the *Challenger* and other deep sea expeditions led to the establishment of a number of coastal marine stations. Amongst the first of these was Stazione Zoologica in Naples, Italy, established by the German zoologist Anton Dohrn in 1873 with support from several European countries to provide laboratory accommodation for marine scientists. Others included the Station Biologique Roscoff, France (1872), the Marine Biological Laboratory at Woods Hole in the United States (1885), and the Marine Biological Association Laboratory at Plymouth in the UK (1888). Initially, these marine stations were principally concerned with work in marine biology, but later they contributed to stimulating developments in other marine disciplines. Some marine stations also started long-term series of observations of physical variables, such as the remarkable record of temperature, salinity and nutrients carried out off Port Erin, Isle of Man, since 1904 (Allen *et al.*, 1998).

Early concerns about the possible effects of over-fishing led governments to set up agencies to promote scientific studies. In the United States, a Fish Commission was established in 1881, while in the UK, the Scottish Fisheries Board was given a research mandate in 1882 and many other countries established similar bodies to promote the study of marine science related to fisheries. Concerns about the impact of over-fishing were expressed at the International Fisheries Exhibition held in London in 1883 but were not shared by all. It was at that meeting that T. H. Huxley, one of the most eminent scientists of the day, famously discounted reports of scarcity of fish and asserted that

> any tendency to over-fishing will meet with its natural check in the diminution of supply, ... this check will come into operation long before anything like permanent exhaustion has occurred.

In hindsight, such reliance on the ability of the natural system to be able to resist the increasing pressures of industrial fishing appears naïve and unfounded. Fortunately such over-optimistic views, although influential at the time, did not deter governments from investing in fisheries research. In 1902, the national fisheries science organisations joined forces to establish the International Council for the Exploration of the Sea, ICES, to promote the study of the oceans in relation to fisheries. Scientific progress in relation to process understanding, however, remained limited as much of the fisheries research effort was concentrated on the practical problems of understanding the life cycles of commercial species and making stock assessments.

Nevertheless, important surveys of the physical properties and plankton distributions were accomplished through fisheries research. For instance, in the UK the early twentieth century saw work aimed at mapping the temperature and salinity distributions in shelf waters in an effort to understand fish stock distribution. At the Marine Biological Association in Plymouth, the first complete study of the seasonal changes in the physics, chemistry and plankton of a coastal water column was carried out in the late 1920s and early 1930s (Harvey *et al.*, 1935), and in an extensive series of cruises, Matthews (1913) documented the annual cycle of temperature, salinity and density variations over a large section of the north-west European shelf (the Celtic and Irish Seas). In the United States, similar pioneering studies of a large shelf sea area on the east coast, the Gulf of Maine, were undertaken by Henry Bigelow (Fig. 1.3). During the period 1912–1928, Bigelow studied the Gulf extensively and published three monographs on the physical oceanography, plankton and fishes. Bigelow was a strong advocate of an interdisciplinary approach to the then emerging science of oceanography and realised the need to move on from fact collecting to process understanding. In arguing the case for a new Oceanographic Laboratory (to be the Woods Hole Oceanographic Institute) he wrote:

> what is really interesting in sea science is the fitting of the facts together ... the time is ripe for a systematic attempt to lift the veil that obscures any real understanding of the cycle of events that takes place in the sea. It is this new point of view that is responsible for our new oceanographic institution.

This progression from fact collecting and mapping of the shelf seas to the testing of theories about the processes controlling the physical environment and the response

Figure 1.3 Left: Henry Bigelow at the wheel of the schooner *Grampus* in 1912 (courtesy Bigelow Laboratory). Right: Ken Bowden (courtesy University of Liverpool).

of the biological system developed gradually and only came fully to fruition in the second half of the twentieth century. A number of pioneering studies were undertaken before World War II. For example, Knudsen (1907) estimated residual flows in the North and Irish Seas using a continuity argument for mass and salt applied to the observed salinity distributions. The results revealed surprisingly small net flows; for the Irish Sea, Knudsen estimated a net northwards flow of $< 2\,\mathrm{cm\,s^{-1}}$ through the North Channel. In another important development, G. I. Taylor estimated the level of tidal energy dissipation in the Irish Sea (Taylor, 1922) and showed how it fitted into the global pattern of energy loss.

By the 1950s, the continuing accumulation of data sets, mainly through the work of the fisheries agencies, stimulated investigations of the processes controlling the distribution of water properties. Important contributions in the 1950s and 1960s were made by Kenneth Bowden (Fig. 1.3), who was amongst the first to investigate the mechanisms controlling the salt and heat content of the water column. By allowing for horizontal mixing in the salt balance, he developed an improved model of transport through the Irish Sea (Bowden, 1950) and showed that the net flow is $\sim 0.5\,\mathrm{cm\,s^{-1}}$, significantly lower than Knudsen's earlier estimate. This picture of weak residual flows was further supported by a study of the heat budget for the Irish Sea (Bowden, 1948), from which he concluded (correctly) that the seasonal changes in heat content can largely be accounted for by the transfer of heat through the sea surface, with the residual currents playing only a minor role. This understanding of the relative roles of advection and surface fluxes will be fundamental to our discussion of shelf sea stratification in Chapter 6.

Bowden also contributed to the understanding of tidal dynamics in shelf seas by determining the turbulent stresses in the water column which were bringing about the frictional dissipation previously estimated by G. I. Taylor. Bowden determined the stresses by inferences from the dynamical balance (Bowden and Fairbairn, 1952) and later went on to make the first direct measurements of stress in the ocean using the newly-developed electromagnetic flowmeter (Bowden and Fairbairn, 1956). With these and many other contributions, Bowden, a quiet and modest man, was responsible for stimulating progress in the physical oceanography of the shelf seas and for laying the foundation for many of the developments we shall be considering in later chapters.

At the same time, important progress was being made in understanding the biology of the shelf seas. Working over Georges Bank, Gordon Riley, along with Henry Stommel and Dean Bumpus, developed the fundamental understanding of how the spring bloom of phytoplankton is triggered by physical stability. From our twenty-first-century perspective, the idea of stability and light being key to rapid phytoplankton growth in the surface ocean can perhaps seem blindingly obvious, but in the 1930s and early 1940s there was some considerable effort aimed at demonstrating that the spring bloom was a product of grazing pressure, an idea related to a large degree to the understanding of terrestrial ecosystems. We will, however, see how grazing does play a more subtle role in determining the species that dominate the spring bloom later in Chapter 5. Riley applied statistical analyses to his careful sampling of phytoplankton biomass and growth over Georges Bank, leading to a key paper in biological oceanography (Riley, 1946) which included one of the first coupled theoretical models of physics and phytoplankton growth, the basics of which can still be seen in the codes of most coupled models today.

1.5 Instrumentation: 'Tools of the trade'

Progress in oceanography has been, and continues to be, closely linked to technical developments for making measurements in the ocean. This is true to some degree in most scientific disciplines, but in few areas is the constraint of technical limitation as severe as it is in our efforts to understand the workings of the ocean. The basic problem is that the ocean is largely impenetrable to electromagnetic radiation. Even in relatively clear water, light is absorbed on a scale of tens of metres, and at other wavelengths the absorption of energy is even more rapid. Only sound waves can travel relatively freely through the ocean, and even then, long-range propagation is restricted to low frequencies where the scope for the transmission of information is limited. So, unlike meteorologists who can watch the evolution of the atmosphere through the movement of clouds and measure velocities remotely with radar, oceanographers have to rely mainly on measurements from sensor systems lowered into the ocean or make the most of what can be learned from probing by acoustic methods. In this section, we shall consider the principal measurement tools and instrument platforms which are now available to determine the physical, chemical and biological characteristics through the water column.

Figure 1.4 A CTD being deployed over the side of the RRS *James Clark Ross*. The CTD rests within the lower part of the frame below the grey sample bottles. This allows the instrument to sample relatively undisturbed water as the package is lowered through the water column. (Photo by J. Sharples.)

1.5.1 The measurement of temperature, salinity and pressure (the CTD)

The density of seawater is a physical parameter which, as we shall see in Chapter 3, plays an important part in the dynamics of the flow in shelf seas. Until the 1960s, density determination in the ocean was largely based on a combination of measurements using specialised mercury-in-glass thermometers to measure temperature and the collection of water samples to allow the determination of the salinity (salt content) of the water by titration or laboratory measurements of conductivity. The methods were reasonably accurate, giving temperatures to $\sim\pm 0.01°C$ and salinity to ~ one part in 3000, but they were also rather complicated and labour intensive.

In modern practice these early methods have been almost entirely replaced by a profiling instrument package measuring conductivity, temperature and depth and referred to simply as a CTD, shown in Fig. 1.4. As the package is lowered on a cable through the water column, information from electrical sensors is transmitted to the surface via conductors in the cable. In most modern CTDs, temperature is measured by a high-quality platinum resistance thermometer while conductivity is sensed by a conductivity cell which may be directly coupled to the seawater via electrodes or indirectly through an inductively coupled system. Pressure, which is usually measured by a strain gauge sensor, is used to determine the depth of the CTD.

Salinity was originally defined as a mass ratio equivalent to the number of grams of salt in 1 kg of seawater. It is currently defined in terms of a near-equivalent ratio of the electrical conductivities of seawater and a standard solution of potassium chloride (KCl). Its value has to be inferred from the electrical conductivity (C) which varies with both salinity (S) and temperature (T) and to a lesser extent with pressure (p). Given C, T and p from measurement, S can be determined from accurately known polynomial functions $S = S(C, T, p)$. It is worth noting here that salinity, defined by a conductivity ratio, does not have any units. We should specify that we are using the *practical salinity scale* (referred to as PSS) when quoting such salinities. There is a tendency in the literature to use the *practical salinity unit* (psu), which is not strictly correct. Density, which varies with salinity, temperature and pressure, is then recovered from the equation of state $\rho = \rho(S, T, p)$ which is another complicated set of polynomial functions. Since the density of seawater varies by only a few percent over the whole ocean, the relatively small differences involved which are of great importance in relation to ocean dynamics are often represented in terms of the parameter σ_{STp} which is defined as:

$$\sigma_{STp} = (\rho(S, T, p) - 1000)\ [\text{kg m}^{-3}]. \tag{1.1}$$

For many purposes, in comparing the densities of water particles, it is sufficient to deal in terms of their density at the surface σ_{ST0} which is usually abbreviated to σ_T. A good quality CTD will be able to determine temperature with an accuracy of better than ~0.005 °C in temperature and ~0.01 in salinity. The pressure is also used to determine the depth of the CTD package from the relation $z = p/\hat{\rho}g$, with $\hat{\rho}$ the mean density of the water column above the CTD. There are several internet sources where you can find the polynomials used to calculate salinity and density, all using the algorithms in the UNESCO technical paper (Fofonoff and Millard, 1983). Recently there have been some important developments towards a thermodynamic equation of state for seawater (Millero, 2010).

The basic CTD sensor suite is normally attached to a frame along with a set of water bottles for taking water samples at selected depths (e.g. see Fig. 1.4). The bottles are arranged around the CTD frame in a 'rosette' configuration and can be triggered by electrical signals sent down the cable from the controlling computer on the research vessel. This arrangement is vital in physical oceanography to provide samples for salinity analyses and subsequent calibration of the CTD salinity. The rosette bottles are also necessary for a lot of interdisciplinary work as water samples are still needed for laboratory analyses and experiments. As well as vertical profiling from a stationary vessel, CTD sensor packages can also be carried by a towed undulating vehicle, two examples of which are shown in Fig. 1.5. These vehicles are able to porpoise up and down in a sawtooth pattern at speeds up to 8 knots (~4 m s^{-1}) behind a research vessel. This mode of operation allows for rapid sampling of the horizontal and vertical structure of the water column and is particularly useful in regions of high gradient which occur in oceanographic fronts. Oscillating to a depth of 100 metres, a horizontal resolution of ~500 metres can be achieved.

Figure 1.5 Two examples of towed, undulating vehicles that we have used in shelf sea research. Left: Seasoar, looking like a short, stubby-winged aeroplane. Right: Scanfish, a relatively simple wing-shaped vehicle that incorporates a downward-looking echosounder used for automated seabed avoidance – a very useful feature when working in shallow seas. (Photos by J. Sharples.)

1.5.2 Sensors for biogeochemistry and beyond

The basic physical measurements of pressure, temperature and electrical conductivity are, in principle, straightforward; the challenge lies largely with achieving the accuracy needed to quantify the small changes in density that drive mean flows in the ocean. When we want to make measurements of biogeochemical or ecological parameters, the challenge becomes far more extreme. There are still many measurements that require water to be collected and analysed using laboratory equipment on the ship, or even back ashore. We might need to transport samples from remote locations back to our institutes in such a way as to minimise the possibility of degradation, which could involve freezing to very low temperatures or chemical preservation. In addition to all of these practical issues, there is a key intellectual challenge in all interdisciplinary work: how can we make biogeochemical and ecological measurements that are compatible with the spatial and temporal patchiness that our physical measurements routinely show? This is an exciting and rapidly evolving field of ocean-observing technology.

The first, and by now most common, biogeochemical instrument that became available for routine use alongside temperature and conductivity was the chlorophyll fluorometer. The idea of measuring phytoplankton pigment as a proxy for phytoplankton biomass was pioneered by Gordon Riley in the 1940s, using methods developed in Plymouth by H. W. Harvey. The original colorimetric analyses of filtered seawater samples was replaced by a fluorometric technique; the plant pigment chlorophyll *a* fluoresces at a known, quite specific wavelength of light as a mechanism for dumping energy from photons that it cannot use in photosynthesis. We will describe fluorescence and photosynthesis in more detail in

Chapter 5. The development of the fluorescence technique opened the door to *in situ* chlorophyll fluorescence measurements alongside those of temperature and salinity. Practical *in situ* instruments first became available in the 1970s, and are now commonly used on lowered and towed/undulating CTDs, and, with the addition of batteries and some further control electronics, on moorings. By making changes to the optics (i.e. the wavelengths of the emitted and received light), fluorometers can also be used to observe dye releases in mixing experiments (e.g. Houghton and Ho, 2001), and can be made to measure pigments other than chlorophyll *a* (for instance phycoerythrin, a pigment associated with a cyanobacterium). Multi-spectral fluorometers measure several plant pigment concentrations simultaneously to build up a picture of the relative proportions of phytoplankton groups via knowledge of what pigments dominate in different phytoplankton. A more recent development in fluorometry is fast repetition rate fluorescence, which uses very rapid flashes of light to gradually (over hundreds of milliseconds) saturate the light-gathering capacity of phytoplankton cells and so measure the photosynthetic status of a population of phytoplankton as well as their biomass. We will look at this technique in more detail in Chapter 5. Measuring fluorescence is a convenient, but far from perfect, estimator of phytoplankton biomass. The fluorescence characteristics of chlorophyll vary depending on ambient light and also between species of phytoplankton, so calibrations will change during the day and between different phytoplankton communities. We also need to remember that a measure of chlorophyll, no matter how accurately you feel you have calibrated your fluorometer, is still only a proxy for phytoplankton biomass. A phytoplankton cell's carbon:chlorophyll ratio is not fixed. Ideally we really want to be able to measure phytoplankton carbon, but as yet there is no straightforward way to do that.

Chemistry has also seen technology developments that have led to some convergence with the way we make physical measurements. Two key parameters that can be measured routinely from a CTD package are dissolved oxygen (DO) and nitrate (NO_3^-). Measurement of dissolved oxygen involves the DO reacting with a cathode to set up a current in an electric circuit, with the seawater separated from the circuit by an oxygen-permeable membrane. DO sensors can be attached to moorings or interfaced with a CTD. With the latter it is important to note that DO sensors tend to have relatively slow response times (a few seconds) compared to conductivity and temperature sensors (both <0.1s in a good CTD) and so careful processing of the data is required to match the DO with the physical structure of the water column.

Instruments measuring all four macro nutrients (nitrate, phosphate, silicate and ammonium) in autonomous packages that can operate on moorings began to be commercially available in the 1990s; you will see an example of such an instrument's use in Chapter 6. Developing a technique with a response time suitable for use on a profiling CTD has been more challenging, but ultra-violet fluorescence-based instruments are now available for nitrate determination (e.g. Johnson and Coletti, 2002). Sensors are also available commercially for pH and for redox

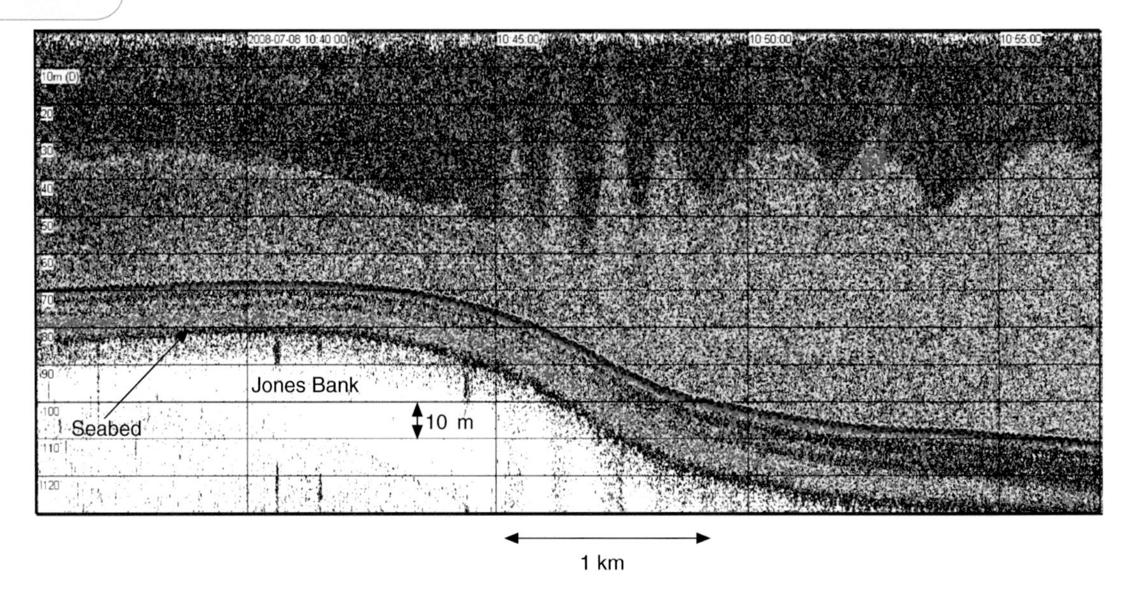

Figure 1.6 See colour plates version. A series of short internal waves on the edge of Jones Bank, Celtic Sea, as seen in the data from the 200 kHz transducer of an EK60 echosounder. Image courtesy of Clare Embling, University of Aberdeen, UK.

potential, and there are ongoing developments in the field of measuring some metals, such as copper and manganese, using sensors attached to CTDs.

There are several instruments that measure the size spectrum of particles in the ocean, such as the optical plankton counter (OPC) and the laser particle size analysers (LISST). The difficulty with data from these instruments is that they cannot identify the shape and nature of particles, a particular problem in shelf environments when plankton, detritus and suspended sediments are all present. Recent, exciting developments in the use of high-resolution digital imagery are changing this. For instance, a holographic technique using a laser and digital camera to image particle diffraction patterns and then de-convolving the patterns to generate images of the particles has recently become available (Graham and Nimmo-Smith, 2010).

For larger particles in the ocean, e.g. large zooplankton, fish larvae and fish, multi-frequency acoustic instruments are used to assess large-scale horizontal and vertical distributions. Considerable effort has been applied to the different acoustic signatures of key zooplankton and fish species (e.g. Brierley, *et al.*, 1998; Fablet, *et al.*, 2009), so that distributions of different species can be observed from vessels or from moorings. We often find that many of these particles, particularly the zooplankton, concentrate in gradients in the biogeochemical and physical structure of the water column, so that these acoustic techniques can provide revealing pictures of the physical environment. Figure 1.6 shows an example from some of our own work of acoustic backscatter data from an EK60 scientific echosounder common in fisheries studies, providing a startling image of internal waves over a bank in a shelf sea.

1.5.3 The measurement of the currents: the ADCP

While the need for direct measurement of ocean currents was evident from the start of oceanography, the emergence of effective technologies for current measurements is a comparatively recent development. Early current meters (e.g. the Ekman current meter, designed by Vagn Walfrid Ekman in 1903) were mechanical devices lowered from an anchored vessel; speed was measured from the revolutions of a propeller and direction from the deviation of a vane measured relative to an integral compass. Such devices were labour intensive and required a dedicated research vessel for the duration of the observations. Autonomous recording current meters deployed on a taught wire mooring, such as those shown in Fig. 1.7, did not appear until the 1960s. Even then, they were logistically demanding, expensive and vulnerable to damage by fishing vessels. Current meter moorings were, however, a great improvement on profiling with direct reading current meters from a research vessel, and were increasingly used in the shelf seas from the 1970s. As well as recording velocity data, recording current meters are frequently equipped with sensors for conductivity and temperature, so that salinity and density can be determined. The inclusion of a pressure sensor

Figure 1.7 A mooring ready for deployment in the eastern Irish Sea (depth about 35 metres). The instruments are Aanderaa recording current meters (RCMs), which use a rotor with the large vane aligning the instrument to the flow. The large spherical buoy (about 1 metre diameter) stretches the instrument line upward from an anchor on the seabed. Each RCM also had temperature and conductivity sensors to provide a coarse vertical profile of salinity. (Photo by J. Sharples.)

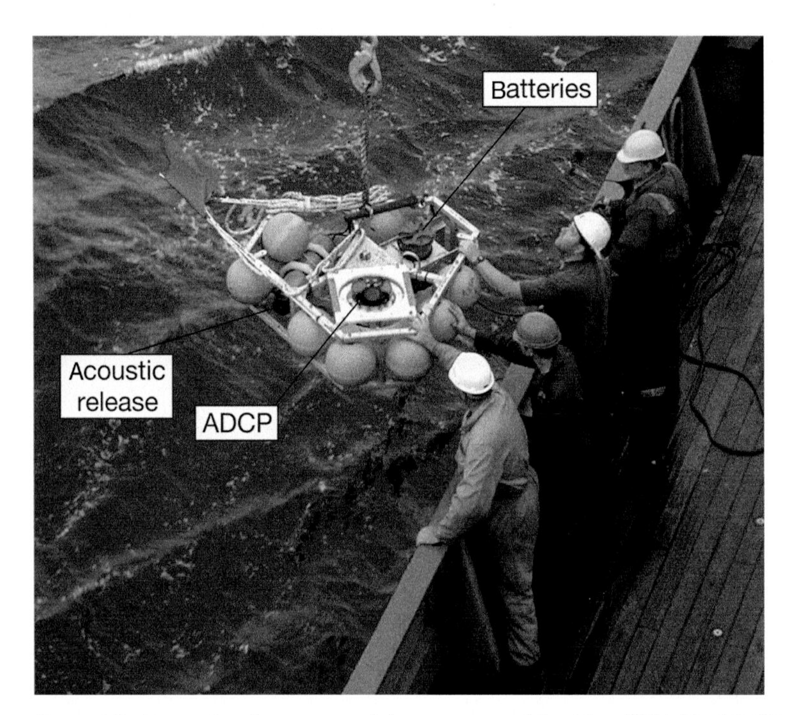

Figure 1.8 A mooring frame containing an upward-looking 300 kHz ADCP about to be deployed in the Celtic Sea. The orange buoys bring the frame back to the sea surface when, at the end of the deployment, the acoustic release is signalled to let go of a weight in the base of the frame. (Photo by J. Sharples.)

allows the depth of the measurements to be monitored and any set-down of the mooring in strong flows to be detected.

A major advance in current meter technology was the development of acoustic current meters based on the Doppler principle. Sound waves travelling away from a fixed source are reflected back by suspended particles in the water with a shift in frequency which is proportional to the water velocity in the beam direction. The idea of detecting this return signal and determining the velocity from the Doppler shift has long been known, but the signal strength of acoustic return is so weak ($\sim 10^{-9}$ of the transmitted power) that Doppler current meters for the ocean seemed to be technically unrealistic. It was only with significant advances in electronics and signal processing in the 1980s that practical Doppler instruments became available.

Modern Doppler instruments, such as the one shown in Fig. 1.8, typically have three or four acoustic beams oriented at an angle of 20° or 30° to the vertical and may be operated in upward- or downward-looking mode. The velocity at each level is found by determining the Doppler shift of a section of the return signal from an acoustic pulse sent out by the transducer. The three or four along beam velocities are then combined to evaluate the Cartesian current components at each level. In this way, a profile of the velocity in the water column is constructed. The instruments are widely known as *Acoustic Doppler Current Profilers* or simply ADCPs.

The range and resolution of these instruments is set by the acoustic frequency used and the typical densities of 'scatterers' (suspended sediments or zooplankton) in the water column. For the deepest parts of the shelf seas, a frequency of 150 kHz allows coverage of the full 200-metre water column but with a vertical resolution of ~4 metres. Higher resolution in shallower water can be achieved with instruments of higher frequencies but with range decreasing to ~20 metres at a frequency of 2 MHz. These instruments provide much fuller, non-invasive coverage of the velocity structure than could be achieved with a string of conventional current meters. When mounted on the seabed, they are also much less vulnerable to damage by fishing vessels than a conventional current meter mooring.

ADCPs are also used in a hull-mounted form on research vessels to monitor the flow field below the vessel. In the shelf seas a special 'bottom track' sound pulse returned from the seabed allows the determination of the ship's movement relative to the seabed. By subtracting this bottom track velocity from the water velocities relative to the ship, a vertical profile of velocity relative to the seabed can be recorded by the ADCP while the ship is engaged in survey work. This provides a useful supplement to measurements from bottom-mounted ADCPs and can be a valuable way of monitoring water movements during deployment and recovery operations.

In most situations, ADCPs are now the instrument of choice for velocity measurements and in the last three decades have helped to greatly advance oceanography. They are not, however, without disadvantages. Even when they have adequate range, the coverage of the water column by a bottom-mounted ADCP is restricted in two ways. First, near the transducers there is a 'blank' space ignored by the instrument because of the time taken to switch from transmit to receive modes. This can be up to several metres for low-frequency instruments. Second, there is a zone of no data at the water column boundary. For instance, in the case of an upward-looking ADCP no data can be retrieved from a thin layer at the sea surface. With a 20° beam angle, this layer has a thickness of 6% of the distance between the ADCP and the surface. A further limitation is that, while ADCP's can be equipped with temperature, conductivity and pressure sensors, the measurements of these parameters are restricted to the location of the instrument. In order to provide full water column temperature and salinity information to complement the velocity profile data, it is necessary to have a supplementary mooring equipped with a string of CTD sensors.

1.5.4 Drifters, gliders and AUVs

The alternative to measuring currents with instruments fixed relative to the seabed (the so-called *Eulerian current*) is to track drifting drogued buoys to determine the movement of water particles (the *Lagrangian current*). For a limited horizontal domain, such as a small estuary, surface currents can be determined simply by observing the movement of passive surface drifters or dye visually from an aircraft or a high vantage point on land. For larger scales the drifters can be tracked by relaying their GPS position by radio to a shore station or via a satellite link. To determine subsurface currents, it is necessary to use a large drogue located at a

Figure 1.9 Deploying a glider into the Irish Sea. The conductivity and temperature sensors are just visible underneath the left wing. The black tube on top is a turbulent dissipation instrument, with the shear sensors visible forward of the glider nose. (Photo courtesy of the National Oceanography Centre, UK.)

fixed depth and attached by a line to a small, low-drag surface unit which relays the drogue's position. This arrangement ensures that the measured current is that of the drogue and is not significantly influenced by variations of current with depth. A more radical solution, which avoids the surface connection, is to use neutrally buoyant floats which are tracked acoustically. Currents in the deep ocean have been determined in this way (Swallow, 1955; Dasaro *et al.*, 1996) but the requirement to set up a dedicated acoustic tracking system has limited applications and the method has been little used in the shelf seas. Most recent Lagrangian measurements in shelf seas have been based on satellite tracked drifters attached to large subsurface drogues. While they are used less frequently than Eulerian measurements, Lagrangian measurements provide an important complement to measurements from fixed moorings and are of particular value in relation to studies of dispersion.

Drifting floats with buoyancy control and equipped with CTD sensors, called Argo floats, are used for profiling in the deep ocean. These devices descend to a predetermined depth where they drift with the current for a period before returning to the surface. The profile data, along with position information, are then relayed to shore via a satellite link and the next descent begins. Argo floats are now present throughout much of the world ocean and are greatly enhancing the ocean database for the deep ocean. As presently configured, they are not suitable for shallow waters, but a further development of the profiling float in the form of a 'glider' seems likely to be a valuable tool for the shelf seas. Instead of descending vertically, the glider (Rudnick *et al.*, 2004) is equipped with wings and a rudder, as can be seen in Fig. 1.9, which enable it to glide down on a fixed compass bearing. After increasing its buoyancy at the end of the descent, the device glides upwards with a similar component of horizontal speed. Before launch, the glider's track is programmed to pass through a series of way points which, if required, can be updated during the periods when it is at the surface to download data. Gliders can carry a CTD plus additional sensors, such as fluorometers or turbulence probes (e.g. Fig. 1.9); the glider endurance is limited by the battery power available for sensor buoyancy changes, operation and communications. Initial trials indicate the considerable potential of such devices to undertake extensive and protracted surveys at modest cost.

Several powered autonomous underwater vehicles (AUVs) with the capability to undertake extended survey missions have been developed in recent years. Most have propellers driven by an electric motor and 'cruise' at speeds of ~1–2 m s^{-1}. They are controlled by onboard computers with instructions that can be updated when the vehicle surfaces. AUVs such as this have advantages over gliders in that they are faster (which is useful in tidally energetic environments) and can carry larger scientific payloads including one or more ADCPs. The technical development of AUVs has proved challenging and is continuing as engineers work to improve their range, which is again limited by battery capacity. There is also a need to reduce their size so that they can be launched from smaller vessels or even from the shoreline. At the time of writing, the contribution of AUVs to data gathering in the shelf seas is relatively small, but in the future, improved versions are likely to play an increasing role in oceanographic survey work.

1.5.5 Research vessels

While remote sensing and autonomous high-tech devices like gliders are increasingly contributing to the study of the ocean, research vessels still play an essential and central role. In particular, autonomous vehicles can only gather information on parameters that are measureable with electronic instrumentation, and therefore much of biogeochemistry and ecology still requires the ability to collect water samples. Ships with sufficient winches and lifting gear are also needed for the deployment of moorings, which enable the research vessel to get on with other activities and so greatly increase the cost-effectiveness of a research voyage.

Until recent years, specialist research vessels were not available and all manner of ships were adapted and pressed into service for surveying the ocean. Before the advent of modern instrumentation and mooring technology, almost any vessel equipped with a winch to deploy water bottles and reversing thermometers on a 4 mm wire could be used for station work. Prince Albert of Monaco, an enthusiastic marine biologist, famously had his royal yacht adapted for oceanographic work in the Mediterranean. At the other end of the scale, fishing vessels were a popular choice, with their holds converted to temporary laboratories although, in many cases, facilities for position fixing were lacking and accommodation for scientists was limited. The requirements for a modern vessel are centred on (i) the ability to deploy and recover heavy moorings using winches and a stern "A" frame, and (ii) a specialised winch with an armoured cable with electrical conductors for the deployment of a CTD/rosette system. The vessel would also carry a hull-mounted ADCP and one or more acoustic sounders for depth measurement and zooplankton/fish detection. To continuously monitor surface conditions, contemporary vessels are usually equipped with a flow-through system fed by seawater pumped from a subsurface inlet and measuring a suite of physical and biogeochemical parameters which are recorded in parallel with data from the ship's meteorological sensors. With the increasing move towards large interdisciplinary projects, the availability of berths for numerous scientists is vital.

1.5.6 Remote sensing of ocean properties

A limitation of the measurements made by floats, AUVs, and indeed research vessels is that they are not synoptic, i.e. the data are not sampled simultaneously at different locations, as is the case for an array of current meter moorings. Thus, it is not possible to construct true 'snapshot' pictures of the velocity and other parameters. The complex sampling scheme in space and time which results from the movement of the measurement vehicle presents a rather different perspective on the evolution of the flow and water column structure which, in many applications, presents a challenge in analysis.

Since the 1970s, the remote sensing of ocean properties has become an increasingly valuable supplement to direct observations by the techniques described in the preceding sections. The new information comes mainly from sensors carried on orbiting satellites, but aircraft-borne sensors are also used for the survey of surface properties. The extensive spatial coverage available through remote sensing has proved extremely valuable in interpolating and extrapolating limited data sets from *in situ* measurements.

Remote sensing can also be very cost-effective in comparison with ship observations, which are increasingly expensive. Depending on size, costs for a research vessel capable of 24-hour continuous operations could be somewhere between US$7000 and US$50 000 per day, and that is without allowance for days lost due to bad weather. So there is a strong motivation for oceanographers to use remote sensing data from satellites which is made available by agencies like the National Oceanic and Atmospheric Administration (NOAA) at modest cost.

An early success in remote sensing was the measurement of surface temperature by infra-red sensors on polar orbiting satellites which have been operated by NOAA since the mid-1970s. Temperature is now routinely measured with an accuracy of a few tenths of a degree Centigrade and the data coverage is global apart from small regions around the poles. The principles behind this technique will be explained in Chapter 2. There are two limitations: (i) infra-red sensing of temperature is only possible in cloud-free conditions when the satellite can 'see' the ocean surface, and (ii) the temperature measured is that of a very thin (<1 mm) near surface layer which may differ significantly from the average temperature of the surface water.

Differences in the colour of the surface ocean, as observed by sensors operating at optical wavelengths, have been used to deduce the concentration of the plant pigment chlorophyll and hence of phytoplankton. The interpretation of the radiation data here is not as straightforward as for infra-red measurements, but the distribution of chlorophyll has been successfully mapped for the deep ocean by optical scanners on satellites. In the shelf seas, the situation is complicated by the presence of other substances, notably suspended sediments and *gelbstoff* (coloured dissolved organic matter supplied to coastal waters mainly via rivers). These components complicate the accurate determination of chlorophyll. Intensive efforts to use optical data from several spectral bands to determine all three components have, as yet, been only partially successful. A recent and exciting development in the use of satellite sensors in biogeochemistry of the open ocean is the detection of natural fluorescence from surface chlorophyll, which can provide information on phytoplankton physiology, such as nutrient stress (Behrenfeld *et al.*, 2009).

The measurements of the sea surface elevation by microwave sensors (altimeters) on satellites have provided pictures of geostrophic currents and the mesoscale circulation in the deep ocean but works less well in shallow seas because of the large spatial scale of the 'footprint' of the microwave beam and interference from land at the coastal boundary. A more useful microwave technique for shelf seas is Synthetic Aperture Radar (SAR) which detects changes in surface roughness and thereby, in favourable conditions, allows the detection of internal wave motions. We will use some examples of this in Chapter 10.

1.6 The role of models – a philosophy of modelling

Numerical models are another important and powerful tool available to oceanographers. Increased computing power and improved analytical software such as MATLAB© have facilitated the construction of a great variety of numerical models of ocean systems. Such models can be used for many purposes, but from the scientific point of view, they play a crucial role in the testing of our conceptual models of how the ocean works. Here we set out the approach to the use of models in our own research which will be reflected in later chapters of the book.

As in other branches of science, oceanography advances as new observations stimulate hypotheses about how the system works. Predictions, based on one or more hypotheses, are then tested against further observations, eventually leading to improved understanding of the system. A major difficulty in oceanography is that we are rarely in a position to be able to carry out manipulative experiments on the ocean. Instead, we make observations of the real environment, and then have to try to tease apart the relative influences of the possible forcing mechanisms and interactions. In dealing with such complex systems in which many diverse processes are operating, the testing of hypotheses can often be undertaken only through the use of numerical models. We can think of the construction of the model as the assembling of a number of hypothetical ideas about how the system works which together amount to a 'compound hypothesis'. Comparing the results of the model with observational data then constitutes the required test of the compound hypothesis.

If the model compares well with observations, we conclude that the compound hypothesis is provisionally verified (i.e. not falsified) and our understanding is increased. We may then seek more exacting tests of the hypothesis in different scenarios and against new observations. We can also use the model to switch different forcing mechanisms on or off, which is a powerful way of gaining insight into how the real ocean operates. If the model fails, then we conclude that one or more aspects of the conceptual model need revision. Generally there is a presumption that we favour the simplest model (hypothesis); i.e. we follow the approach of Occam's razor, where we only trade model simplicity for a demonstrable increase in explanatory power.[1] Thus, only when the simplest model has been convincingly shown to fail do we proceed to introduce additional features to the model. Following this approach, models are built up in a logical step-by-step sequence, and by working with the simplest hypothesis, we keep to a minimum the number of adjustable parameters which are not well-constrained by theory or observations. Complex, coupled models of ocean physics, biogeochemistry and ecology can generate outputs that are very seductive. However, in our view, leaping straight into a complex model without either sufficient data to set the multitude of driving parameters, or sufficient observations with which to test the model outputs, is unlikely to lead to any reliable new insight into how the real world works.

Models of course have other uses. In principle, if well validated, they can be used to make predictions of, for instance, the response of the ocean to sea level rise or climate change. They can also be used to answer 'what if' questions such as those arising from, say, proposals to build a major tidal barrage and the need to assess near- and far-field changes to a region's tidal characteristics. In the physics domain such prognostic use of models can be undertaken with some confidence, but the ability of models to make useful predictions of biological changes in the shelf seas has not yet been established and remains as a major challenge for our science.

Throughout this book we use model-generated examples of the physics and coupled physics–biogeochemistry of the shelf seas. In all cases we have chosen models that satisfy our condition of simplicity, and that provide clear insight into the process(es) under consideration. We have also included a series of model systems on the book website (www.cambridge.org/shelfseas) that we have developed over the past 20 years or so in the context of our teaching and research; again, simplicity and insight are the characteristics we aim for with these models and the exercises based on them.

1.7 The future challenge and rewards of interdisciplinary studies

We have already indicated that the shelf seas present a major challenge to science requiring a high degree of interdisciplinary collaboration between physical, chemical and biological disciplines in order to elucidate the diverse processes of the shelf seas and their interaction. Given recent improvements in technology, which have greatly enhanced our capacity to undertake detailed process studies, and the strong foundation in physical understanding which has recently been established, we now see

[1] Father William d'Ockam was an English theologian in the fourteenth century. See http://en.wikipedia.org/wiki/Occam's_razor for a detailed account.

great potential for a further drawing together of the threads of physics, biogeochemistry and fisheries science to achieve a full synthesis of understanding of the shelf seas. Such a synthesis should illuminate the processes that govern the distribution and diversity of shelf sea ecosystems, clarify the role of the shelf seas in the Earth system and underpin predictions of how the shelf seas will respond to our shifting climate. The continuation and future development of this scientific campaign will require a cohort of scientists whose training enables them to straddle what, in the past, have seemed like disciplinary divides and approach the challenge of shelf sea science with a truly interdisciplinary perspective.

Summary

The shelf seas have an importance which is disproportionate to the relatively small fraction of the area of the global ocean which they occupy. Biologically they are among the most productive parts of the ocean and support most of the world's major fisheries.

Their importance to modern society is heightened by the fact that a high proportion of the human population lives close to the shores of the shelf seas and has become increasingly dependent on the resources available there; apart from fish and shellfish, they are also important sources of hydrocarbons and aggregates and we use them extensively for transport and recreation. Modern industrial society has tended to rely on the shelf seas as a convenient and economic way of disposing of toxic waste materials. Along with some of these waste materials, rivers also carry large quantities of nutrients leached from agricultural land. The proper management of the shelf seas in response to these many pressures provides strong motivation for understanding the workings of the shelf sea system and the development of predictive models of how they will respond to future changes. Apart from these practical applications, the shelf seas present us with a major interdisciplinary scientific challenge in unraveling the many diverse processes involved in the working of the shelf sea regime. The history of shelf sea research has been rather short; little was done until the early years of the twentieth century and even then progress was limited, partly due to meagre resource allocation but also because of the limitations of the available techniques for measurement at sea. The last few decades have seen rapid progress and substantial improvements in technology along with developments in theory. These have enabled our understanding of much of the basic physics to have advanced to a point where it provides a sound basis for elucidating the way physical processes influence and control so much of the biogeochemistry and the structures of ecosystems. Biological and chemical measurement techniques are improving rapidly, providing new process insights and data on space and time scales to match measurements of physical parameters. Alongside *in situ* measurement techniques and the remote sensing of ocean properties, soundly based numerical models are useful tools for furthering understanding and testing ideas.

FURTHER READING

Discovering the Oceans from Space, by Ian S. Robinson, Springer, 2010, p. 638.
Understanding the North Sea System, eds. Charnock *et al.*, Chapman and Hall, 1994, p. 222.
Reminiscences of an Oceanographer, by Gordon A. Riley, Scripps Institution of Oceanography Library: http://scilib.ucsd.edu/sio/biogr/Riley_Reminiscences.pdf.
The Marine Biological Association 1884–1984: One hundred years of marine research, by A. J. Southward and E. K. Roberts. Report and Transactions of the Devonshire Association for the Advancement of Science vol. 116, pp. 155–99, 1984. Available at: http://www.mba.ac.uk/nmbl/publications/occpub/occasionalpub3.htm.
A short biography of Henry Bigelow can be found at: http://www.bigelow.org/news/about/history/henry_bigelow/.

2 Physical forcing of the shelf seas: what drives the motion of ocean?

In this chapter we consider the powerful forces that drive the shelf seas, and supply the large amounts of energy which are dissipated within them. We shall see that these forces act mainly through the transfer of properties (momentum, heat, freshwater, etc.) at the sea surface and through the lateral boundary where the shelf seas meet the deep ocean. Together the various forcing mechanisms produce an energetic regime which, in most shelf seas, maintains a high level of energy dissipation far greater than that of the deep ocean. We begin by identifying the principal energy and momentum sources and then consider, in turn, the forcing mechanisms involved and the extent to which the resultant inputs are known and can be related to measurable parameters.

2.1 Energy sources

Perhaps the most obvious and striking form of mechanical energy input to the sea arises from surface wind stresses and pressure gradients imposed by the atmosphere. These forces drive ocean currents and generate surface waves whose impact at the coast can be dramatic and is often seen as symbolic of the ocean's power. In many shelf seas, however, energy input through tidal forcing is a more consistent and more powerful source of mechanical energy. Most tidal energy is delivered to the shelf in the form of energy fluxes in tidal waves which originate in the deep ocean, although there is also a (usually small) contribution arising from tidal body forces acting directly on the waters of the shelf seas. Both winds and tides inject very large amounts of kinetic energy to the ocean as a whole; total inputs have been estimated recently as ~1 and ~3.5 TW for wind and tidal inputs respectively (Munk and Wunsch, 1998).

These large mechanical energy inputs are, however, small in relation to the very large seasonal exchange of heat energy through the sea surface. The average rate of mechanical energy input per unit area to the shelf seas is ~0.1 W m^{-2}. This is almost negligible in comparison with seasonal heating and cooling rates which in temperate latitudes have an amplitude of ~100 W m^{-2}. Heating and cooling do not, of course, inject momentum directly into the surface of the ocean. Their primary effect is to modify the density of the seawater making it more, or less, buoyant. In changing the buoyancy of water particles, heat exchange modifies the water column stratification. In addition, differential heating or cooling between different areas will set up horizontal density gradients with attendant buoyancy forces and pressure gradients which will drive circulations. These motions can be thought of as releasing some of the potential energy of the density field which is set up by the heating or cooling. Similarly the input of freshwater, which is lighter than seawater by $\Delta\rho \sim 26\,\mathrm{kg\,m^{-3}}$, at the ocean boundaries constitutes an important second source of buoyancy forcing which injects potential energy and again tends to induce circulations which convert the potential energy to kinetic energy.

2.2 The seasonal cycle of heating and cooling

We begin our survey of the forcing mechanisms with energy exchange at the sea surface, the components of which are shown schematically in Fig. 2.1. The primary driver here is Q_s, the input of short wavelength ($\lambda \sim 0.5\,\mu\mathrm{m}$) radiation from the Sun, much of which is absorbed in the ocean. It is not returned to the atmosphere immediately but heats up the ocean and is ultimately returned back through the surface either by long wave back radiation (Q_b) or as a combination of heat fluxes due to evaporation (Q_e) and sensible heat transfer by conduction (Q_c). We shall now consider the characteristics of each of these fluxes in turn.

2.2.1 Solar heating, Q_s

Outside the Earth's atmosphere, energy arrives at a rate of $\sim$1.37 kW m^{-2}. The total energy intercepted by the Earth's disc, area πr^2 with r the radius of the Earth, is distributed over a hemisphere of surface area of $4\pi r^2$, so the average energy input per square metre at the top of the atmosphere is $\sim$340 W m^{-2}. This incoming radiation has a spectral energy distribution closely similar to the Planck radiation law for a black body:

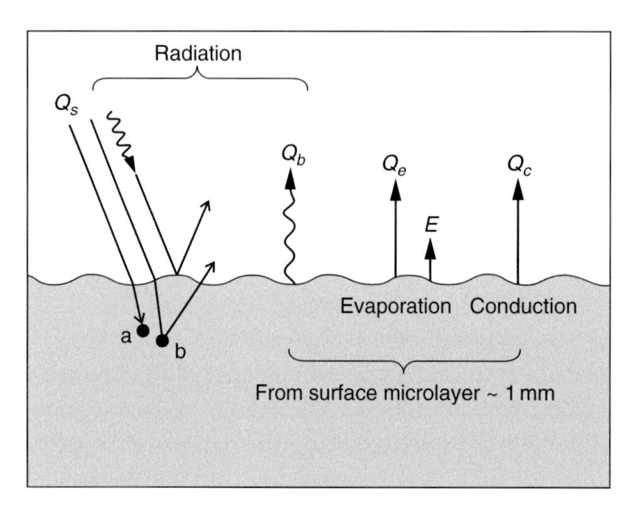

Figure 2.1 The heat fluxes across the air–sea boundary. a, b indicate the absorption and scattering of photons.

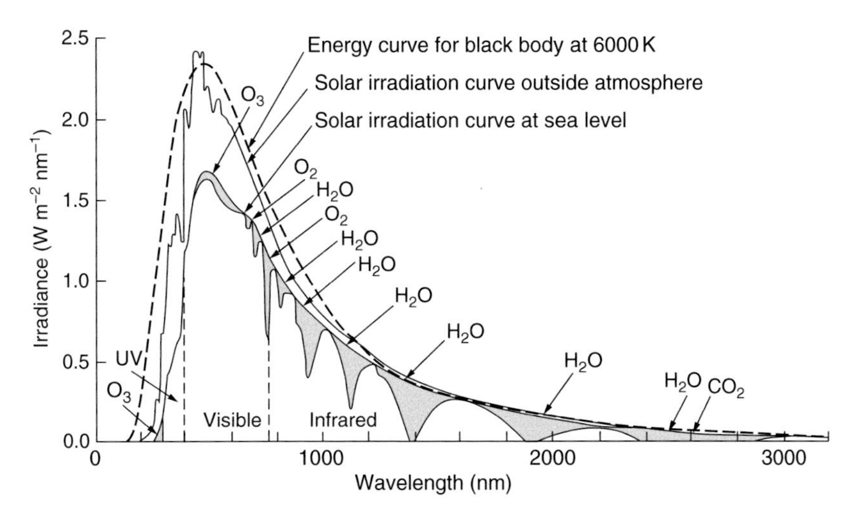

Figure 2.2 The solar spectrum, showing the main absorption bands associated with gases in the atmosphere. The wavelengths used in photosynthesis are between 400 nm and 700 nm. From (USAF, 1960); courtesy MacMillan.

$$I_\lambda(\lambda) = \frac{C_1}{\lambda^5 (e^{C_2/\lambda T_K} - 1)} \tag{2.1}$$

where T_K is the Sun's Kelvin temperature of 6000 K, λ is wavelength, and the constants are $C_1 = 3.74 \times 10^{-16}$ W m^2 and $C_2 = 1.44 \times 10^{-2}$m K. On its passage through the atmosphere, this distribution is modified by gaseous absorption, notably by O_2, O_3, H_2O and CO_2 whose absorption bands are clearly apparent in Fig. 2.2 as large 'bites' out of the spectrum which reaches the sea's surface. The spectral balance is also changed a little through scattering and absorption by the clouds.

On arrival at the sea surface, photons can be reflected or will pass into the sea to be either absorbed (a) or scattered back (b) into the atmosphere (e.g. Fig. 2.1). For overcast conditions in the atmosphere, only ~6% of the energy is returned by reflection. For clear skies a similarly low fraction of energy is returned except at times of low solar elevations (<20°) when surface reflection increases markedly.

Allowing for subsurface scattering, the total fraction of returned energy, termed the albedo A, is typically <0.08 and rarely exceeds 0.2 so that the great majority of the incident solar energy, $Q_s(1 - A)$, goes into the ocean. Almost all of it is converted to heat, though an important component in the visible band between 400 and 700 nm, termed *Photosynthetically Available Radiation* (PAR), is available for photosynthesis.

At wavelength λ the downward vertical flux of energy $E_d(\lambda)$ decays with depth due to absorption and scattering, according to:

$$\frac{dE_d}{dz} = K_d E_d \tag{2.2}$$

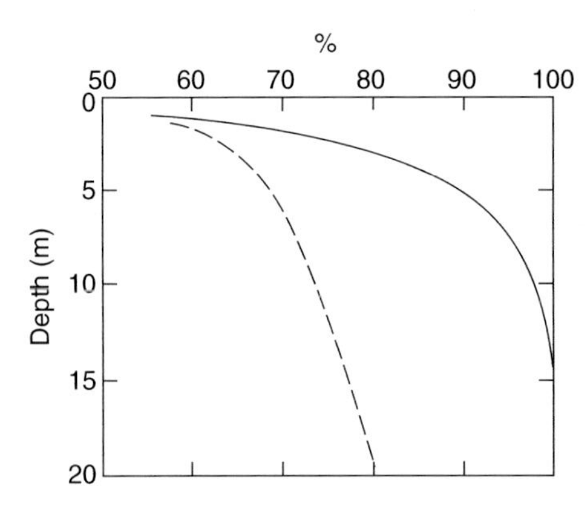

Figure 2.3 The fraction of energy absorbed by clear ocean water (dashed line) and shelf seawater (dash-dot line). Based on (Ivanoff, 1977) with permission from Pergamon Press.

where $K_d(\lambda)$ (m^{-1}) is the *diffuse attenuation coefficient* which varies with wavelength λ. When K_d does not vary with depth, the energy flux decays exponentially with increasing depth, i.e.

$$E_d(z) = E_0 e^{K_d z} \tag{2.3}$$

where $E_0(\lambda)$ is the energy flux at wavelength λ incident at the surface. Note that we define $z = 0$ at the sea surface, with z negative below the sea surface.

Most of the energy entering the ocean is absorbed or backscattered in the first few metres below the surface. The infra-red (IR) and ultra-violet (UV) components are attenuated particularly rapidly on length scales of a few mm or less. Only in the blue and green regions of the spectrum, which are included in the PAR wavelengths, does energy penetrate more than a few metres down the water column. Peak penetration in the clearest ocean water is at a wavelength of $\lambda = 0.45$ μm at which the diffuse attenuation coefficient $K_d \sim 0.03$ m^{-1}; this implies that ~5% of downward energy flux at this wavelength reaches a depth of 100 metres. In the shelf seas, attenuation values are usually considerably larger (0.1–0.4 m^{-1}) so that less than 5% of radiation in the blue-green penetrates beyond a depth of ~20 metres. We shall return to the biologically important question of the vertical distribution of PAR in Chapter 5, but for the moment we focus on the downward flux of the total radiant energy which is the primary source of heat for the ocean.

Measurements of total energy absorption (Ivanoff, 1977) in oceanic and shelf seawater types are shown in Fig. 2.3. For clear, open ocean waters, 55% of the incoming solar energy is absorbed in the first metre of the water column while more than 90% is absorbed in the first 15 metres. The higher turbidity levels typical of shelf seawaters reduce the 90% absorption level to ~5 metres or less. Since this is usually only a small fraction of the water column depth, we may often regard the heat input as being concentrated at the surface. A widely used and rather better approximation, of which we shall make use later, is to treat the heat input from a downward radiation flux of I W m^{-2} as a component of $0.55I$ input at the very surface of the water column, while the remaining $0.45I$ diminishes with depth as $e^{K_{av}z}$, where K_{av} depends on the water clarity and can be thought of as an average attenuation coefficient for the remaining components of the spectrum. In shelf seas, K_{av} is typically ~0.1–0.4 m^{-1},

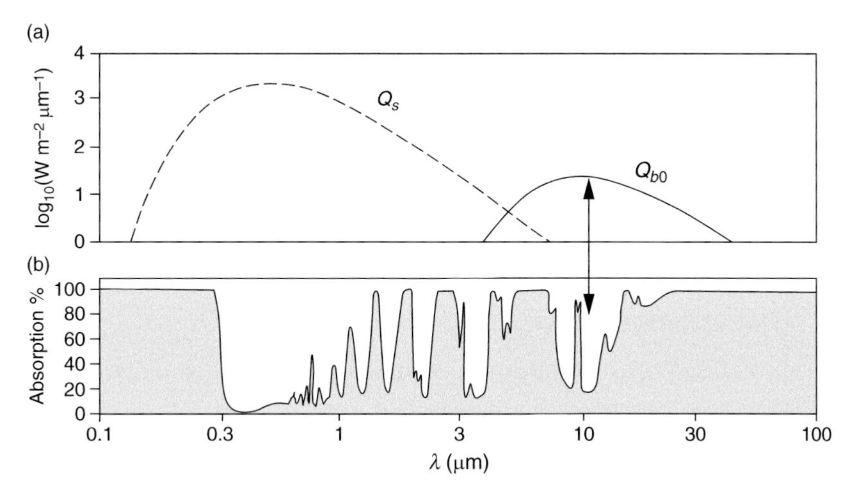

Figure 2.4 Long wave radiation and atmospheric transmission. (a) Spectral distribution of black body radiation from the sea surface, Q_{b0}, at $T_K = 283$ K (~10 °C) and for radiation from the Sun, Q_s, at $T_K = 6000$ K arriving at the top of the atmosphere; (b) An estimate of the fraction of energy absorbed during transmission through a cloud-free atmosphere showing the window for energy at the radiation peak for Q_b allowing energy to exit into space.

with the larger values being found in strongly mixed regions where there are high levels of suspended sediments.

2.2.2 Back radiation from the sea surface, Q_b

The sea surface acts as an emitter of radiation and closely approximates a black body with emissivity $\varepsilon_s = 0.985$ and a temperature ~270–310 K. The total amount of energy radiated upwards from the sea surface is given by Stefan's law: $Q_{b0} = \varepsilon_s \sigma_s T_K^4$ where T_K is the temperature in Kelvin and σ_s is Stefan's constant. Because of the fourth power law dependence on temperature, Q_{b0} from the sea surface ($T_K \sim$ 270–310 K) is much smaller than Q_s ($T_K \sim 6000$ K) as can be seen in Fig. 2.4a. Moreover, when conditions in the atmosphere are overcast, much of Q_{b0} is intercepted by clouds and re-radiated back downwards to the sea surface, so the net loss by long wave radiation Q_b is generally considerably less than Q_{b0}. Under cloud-free conditions, however, the situation is quite different. According to Wien's displacement law, peak emission in the Planck radiation spectrum at temperature T_K should occur at a wavelength of $\lambda = 2.897 \times 10^{-3}/T_K$ metres so that, for $T_K = 290$ K, the spectral peak will be at $\lambda \sim 10$ μm. Figure 2.4b shows that this peak in emission coincides with a minimum in atmospheric absorption by a cloud-free atmosphere which acts as a 'window' through which a high proportion of long wave energy can escape into space.

As well as being a significant contributor to the sea's heat budget, the back radiation is also important in that its intensity is dependent on sea surface temperature (SST). In cloud-free skies, satellite radiometers measure the long wave radiation in the 10–12 μm band. From the results the temperature may be found by inverting the Planck radiation law. Refinements of this procedure use a split spectral window technique to compensate for atmospheric absorption and achieve an accuracy of ~0.3 °C in the estimation of surface temperatures. Infra-red radiation is rapidly absorbed in seawater so that the outgoing radiation from the surface comes from a thin micro-layer of thickness ~1 mm whose temperature may differ from that of the surface mixed layer by up to 1°C (Schluessel *et al.*, 1990). As we shall see in Chapter 8, this use of satellite infra-red sensors in the mapping of SST has proved an invaluable tool for shelf sea oceanography.

2.2.3 Heat exchange by evaporation and conduction, Q_e and Q_c

In addition to the input and export of energy through short and long wave radiation, there are two other mechanisms of heat transfer across the ocean-atmosphere boundary. The most important of these, which frequently accounts for the major part of heat transfer back to the atmosphere, is the loss of heat through evaporation from the ocean surface. In this process, high velocity water molecules escape from the surface into the atmosphere, taking with them above average kinetic energy which constitutes a loss of latent heat from the ocean. This latent heat transfer amounts to $L_H \sim 2.5 \times 10^6$ J for each kilogram evaporated, so that for an evaporation rate E_v (kg m^{-2} s^{-1}), the evaporative heat loss $Q_e = E_v L_H$.

In contrast to the radiation terms Q_s and Q_b, both of which can be measured relatively easily with radiometers, evaporation is generally not easily measured even on land. Over the ocean, it often has to be estimated by a semi-empirical method which relates E_v to bulk parameters:

$$Q_e = E_v L_H = 1.5 \times 10^{-3} \rho_a W (q_s - q_a) L_H \tag{2.4}$$

where q_a and q_s are the specific humidity and its saturated value at SST respectively, W is the wind speed at anemometer height (usually 10 metres) and the constant 1.5×10^{-3} is the Dalton number.

The second non-radiative heat exchange mechanism is the direct transfer of heat by conduction Q_c in response to air-sea temperature differences, i.e. simply the transfer of heat between a relatively warm fluid to a relatively cool fluid. This heat transfer is referred to as 'sensible heat', which means that its effects are solely observed as a change in temperature (in contrast to latent heat where the heat flux is associated with a change in state). Heat loss via conduction is usually a much smaller term than the evaporative heat flux, but like the evaporative term it is difficult to determine directly and is usually estimated from bulk parameters according to the semi-empirical relation:

$$Q_c = 1.45 \times 10^{-3} c_a \rho_a W (T_s - T_a) \tag{2.5}$$

where T_s and T_a are the sea surface and air temperatures respectively, $c_a \approx 1000$ J kg^{-1} K^{-1} is the specific heat capacity of air, and $\rho_a \approx 1.3$ kg m^{-3} is the density of air. The coefficient 1.45×10^{-3} is termed the Stanton number.

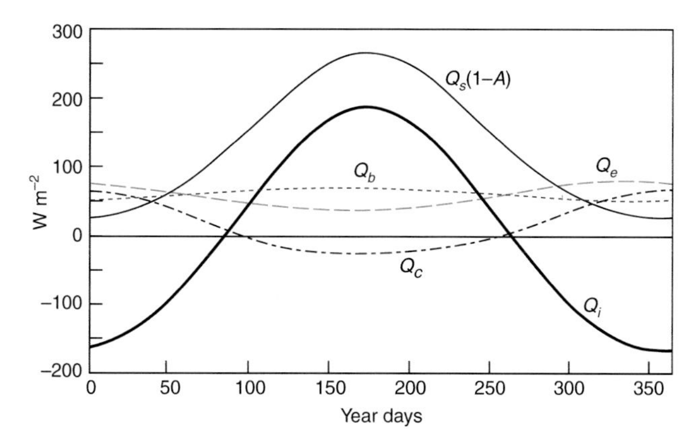

Figure 2.5 Typical pattern of seasonal variation of each of the heat flux terms in temperate latitudes.

In order to justify the representation of Q_e and Q_c by the bulk formulae of (2.4) and (2.5) and to determine the Dalton and Stanton numbers, it is necessary to make rigorous measurements of the two fluxes close to the sea surface in what is often a very energetic and challenging environment. Such measurements are best made a few metres above the sea surface and require the use of large, stable platforms, such as spar buoys, which are designed to minimise vertical displacement and maintain vertical alignment of the spar to which the instruments are attached (e.g. Graber *et al.*, 2000).

2.2.4 Seasonal progression of heat fluxes and the heat budget

The separate contributions to the vertical heat flux at the sea surface vary over the seasonal cycle and combine to determine the net heat flux and hence the heat budget of the ocean. Fig. 2.5 illustrates the typical pattern of seasonal variation of each of the heat flux terms in temperate latitudes. Notice that the heat input is dominated by the solar heating term Q_s, which is positive throughout the year with maximum and minimum inputs at the summer and winter solstices respectively. The net energy loss by long wave radiation, Q_b, has a weaker seasonal variation because of the relatively small proportionate change in the Kelvin temperature of the sea surface. The net heat flux $Q_i = Q_s\,(1 - \mathrm{A}) - Q_b - Q_e - Q_c$ is positive during spring and summer, and negative (net cooling) in autumn and winter

Apart from Q_b the other large component of heat loss through the sea surface comes from the evaporation term Q_e with a smaller contribution from the transfer of sensible heat Q_c. Looking at the example in Fig. 2.5, you can see that Q_c may change sign to become a heat gain for a period during the summer when air temperatures exceed SST. The two terms Q_e and Q_c tend to follow a similar seasonal pattern with larger values in the winter months when winds are generally stronger. This contrasts with Q_b which has a maximum in summer when SST is high and cloud cover is reduced. All three of the heat loss rates vary somewhat even in a particular shelf sea since they depend on SST, which as

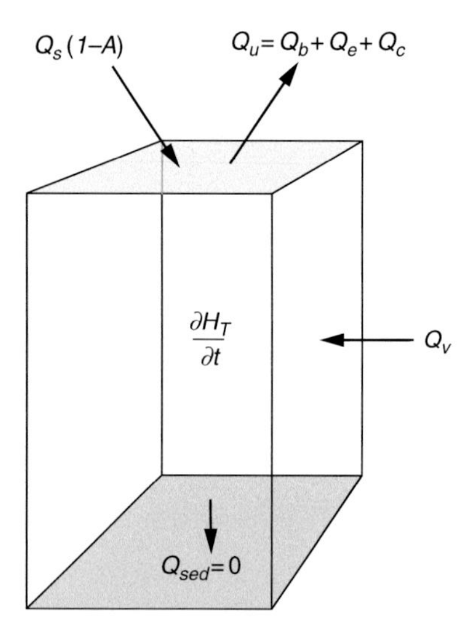

Figure 2.6 Heat fluxes affecting a water column.

we shall see in Chapter 6, has a seasonal cycle that is also controlled by a combination of depth and vertical mixing.

As with Q_b the back radiation term, Q_e and Q_c involve the transfer of heat from, or into, a thin surface microlayer of thickness ~1 mm. So all heat loss from the sea is drawn from this thin surface microlayer. This is in contrast to the radiational heat input Q_s which, while mostly absorbed in the first few metres, has components in the visible band which penetrate tens of metres below the surface.

The processes controlling the heat budget of the water column are shown schematically in Fig. 2.6. The total heat stored in the water column H_T (J m^{-2}) is defined as:

$$H_T = c_p \rho \int_{bottom}^{surface} T_K(z) dz \tag{2.6}$$

where c_p is the specific heat of seawater and $T_K(z)$ is the temperature of the water column at level z. The time rate of change of H_T is determined by the net exchange through the sea surface combined with Q_v, which is the gain or loss of heat due to the horizontal transport by the current. Note that Q_v is the *difference* between the heat entering and the heat leaving through the side walls of the column. Heat exchange between the water and sediments Q_{sed} at the seabed is small because of the low thermal conductivity of the sediments, which means that heat cannot penetrate far into the sediment bed. Over the seasonal cycle, the effect of heat exchange with the sediment has been shown to be roughly equivalent to adding an extra metre to the depth of the water column (Bowden, 1948), and may usually, therefore, be assumed negligible. With this assumption, the heat budget of the water column can be expressed as:

$$\frac{\partial H_T}{\partial t} = Q_s(1 - A) - Q_u + Q_v \quad \text{with } Q_u = Q_b + Q_e + Q_c. \tag{2.7}$$

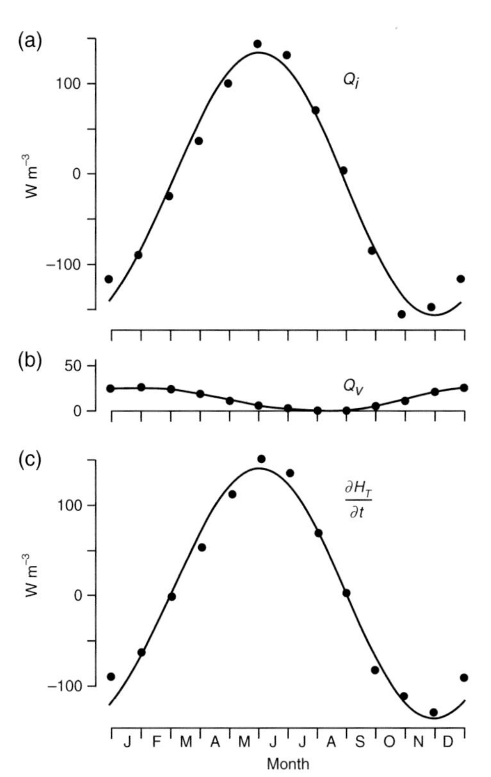

Figure 2.7 Annual cycles of the heat fluxes into the water column for a large area of the English Channel and southern North Sea based on 50 years' worth of data, after (Maddock and Pingree, 1982) courtesy of the Marine Biological Association, UK. (a) Net rate of heat input through the sea surface $Q_i = Q_s(1 - A) - Q_u$; (b) Rate of input of heat by advection Q_v; (c) The rate of change of heat content

$$\frac{\partial H_T}{\partial t} = Q_s(1 - A) - Q_u + Q_v$$

In many areas of the shelf seas, Q_v has been found to be small in relation to the vertical flux terms, so that H_T is controlled by $Q_s\,(1 - A)$ and the loss terms Q_u (e.g. Bowden, 1948; Dietrich, 1951; Maddock and Pingree, 1982). Figure 2.7 illustrates the average annual cycle of the terms in the heat budget based on an analysis of 50 years of observational data for a large, well mixed area of the English Channel and the southern North Sea (Maddock and Pingree, 1982). The net heat input through the surface $Q_i = Q_s\,(1 - A) - Q_u$ (Fig. 2.7a) is seen to be an order of magnitude greater than Q_v (Fig. 2.7b) and so dominates the seasonal cycle of $\partial H_T/\partial t$ (Fig. 2.7c).

The net change in heat storage, averaged over one or more seasonal cycles, should be close to zero. Hence, where Q_v is negligible, the total heat input by $Q_s\,(1 - A)$ must be balanced by the combined heat loss. This condition of no net change over the seasonal cycle can serve as a useful check on the consistency of flux estimates, providing, of course, that we can neglect the net changes due to longer term cycles or trends.

It is apparent from Fig. 2.5 and Fig. 2.7 that the time course of the net heat input $Q_i = Q_s\,(1 - A) - Q_u$ over the seasonal cycle may reasonably be approximated by the sinusoidal form:

$$\boxed{Q_i = A_0 \sin(\omega_a t + \delta)\ \mathrm{W\ m^{-2}}} \qquad (2.8)$$

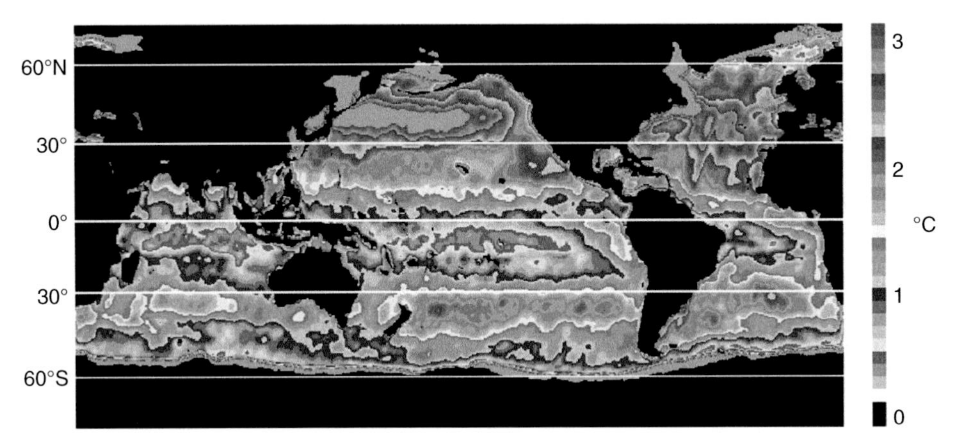

Figure 2.8 See colour plates version. The amplitude of SST (°C) over the seasonal cycle derived from the scanning multichannel microwave radiometer (SMMR) on board the NASA Nimbus 7 satellite from 1978 to 1987. From (Yu and P. Gloersen, 2005), courtesy of Taylor and Francis.

where A_0 is the seasonal amplitude, ω_a is the angular frequency of the annual cycle and ϕ is a phase angle. Since $Q_i = \partial H_T / \partial t$, the corresponding variation in heat storage is:

$$H_T = \overline{H}_T + H_{T0}\cos(\omega_a t + \delta) = \overline{H}_T - \frac{A_0}{\omega_a}\cos(\omega_a t + \delta) \tag{2.9}$$

where $\overline{H}_T$ is the mean heat content. By fitting the observed cycle of water column heat content H_T to a cosine wave function, H_{T0} and hence A_0 may be determined (Bowers and Simpson, 1990), providing that the Q_i dominates over Q_v. We shall return to the topic of surface heat exchange in a shelf sea in Chapter 6 and see how heat storage is influenced by water column depth and tidal stirring.

2.2.5 Variation of heat fluxes with latitude

Seasonal effects are strongest in the region extending from mid latitudes towards polar regions. In high latitudes, continuous ice cover acts to inhibit seasonal exchange. At low latitudes, seasonal influence diminishes until, in equatorial regions, there is generally rather little change in the water column heat storage over the annual cycle. The heat balance is then maintained on a shorter time scale with daily gains and losses almost equal and a minimal annual variation in water column heat storage.

An overview of seasonal thermal changes in the ocean is shown in Fig. 2.8 (see colour plates). The amplitude of the annual cycle of surface temperature has been estimated by fitting a sine wave function to ten years of SST data obtained from satellite microwave sensors (Yu and Gloersen, 2005). You can see that the amplitude exhibits a strong latitude dependence with pronounced maxima in mid latitudes in both hemispheres at ~35–45°N/S and a minimum in the tropics where it falls to ~0.5 °C or less. There is an interesting difference between the hemispheres with mid-latitude maxima exceeding 2.5 °C in the North Atlantic and North Pacific contrasting with weaker maxima ~2 °C in the southern hemisphere. This difference would seem to reflect the fact that the northern hemisphere has a higher proportion of land which possesses a much lower thermal capacity than the ocean. Land surface temperatures, therefore, exhibit a considerably larger range of variation which influences SST through heat transport by winds from land to ocean.

Although it serves well as a general indicator of seasonal heat exchange, the simple picture of SST variation in Fig. 2.8 (see colour plates) does not have sufficient horizontal resolution to separate out differences between the shelf seas and the open ocean and within the shelf seas. We should also remember that the picture in Fig. 2.8 represents only the surface temperature change and does not fully reflect changes in water column heat content, which also depends on water column depth and the strength of vertical mixing.

2.2.6 Thermal expansion and buoyancy changes

The transfer of heat in and out of the ocean does not directly drive currents in the ocean, but it has the important effect of changing the density of seawater and hence its buoyancy. As we saw in 1.5.1, the full variation of density (ρ) with temperature (T), salinity (S) and pressure (p) is expressed in the equation of state by a series of complicated polynomial functions (UNESCO, 1981). For our purposes, it is often sufficient to use a linearised equation of state:

$$\rho(T,S,p) = \rho_0(1 - \alpha(T - T_0) + \beta(S - S_0) + \gamma p) \tag{2.10}$$

which represents the changes of density over a restricted range of temperature salinity and pressure from a reference density ρ_0 (T_0, S_0, 0). The parameter $\alpha = -\frac{1}{\rho_0}\frac{\partial\rho}{\partial T}$ is the thermal expansion coefficient; $\beta = \frac{1}{\rho_0}\frac{\partial\rho}{\partial S}$ and $\gamma = \frac{1}{\rho_0}\frac{\partial\rho}{\partial p}$ are the equivalent parameters for salinity and pressure changes. A positive heat input of ΔQ (J m^{-3}) increases the temperature by:

$$\Delta T = \frac{\Delta Q}{c_p \rho}. \tag{2.11}$$

The increase in temperature causes a reduction in density $\Delta\rho$ which imposes a positive buoyancy force b (N m^{-3}) given by:

$$b = -g\Delta\rho = g\alpha\rho_0\Delta T = \frac{g\alpha\Delta Q}{C_p}. \tag{2.12}$$

Since this heating is concentrated near the surface, the upper water column develops vertical gradients of density with lower density water on top. If we wanted to re-distribute, or mix, the low density surface water throughout the water column, we would need to supply some energy. You can think of the mixing as acting to raise the centre of mass of the water column, which involves doing work; this is an idea that we will utilise further in Chapter 6. So the water column is made more stable by the surface heating. Conversely, surface heat loss increases the near-surface density, so we have a higher density surface layer sitting on top of lower density water. This is an unstable situation and the higher density layer will sink in a process called *convective overturning*. Thus heat loss acts to reduce water column stability. Where heating and cooling are not horizontally uniform, the result will be horizontal gradients of density, which, as we shall see in Chapter 3, will tend to drive water movements.

2.3 Freshwater exchange

While surface heat exchange is generally the main factor modifying the buoyancy of seawater in much of the shelf seas, river discharge may also constitute a significant buoyancy source. In addition, we need to take account of freshwater transfer at the sea surface through the process of evaporation (discussed previously in Section 2.2.3 in relation to heat exchange) and rainfall, both of which act to modify the salinity of surface waters.

2.3.1 Freshwater buoyancy inputs

Before being mixed into the deep ocean circulation, all freshwater from rivers must first pass through the shelf regime, where it is progressively diluted as it mixes with the ambient seawater. It is not surprising, therefore, to find that the impact of freshwater discharge is most evident in estuaries and in adjacent regions. In some cases this input may act as the dominant buoyancy source for all or part of the annual cycle. As an illustration, we consider the case of the River Rhine, the largest river discharging into the northwest European shelf seas, which has a mean annual discharge rate of $R_d \sim 2200\,\mathrm{m^3\,s^{-1}}$ with minimum flow $\sim 1\,000\,\mathrm{m^3\,s^{-1}}$ and maxima during floods $\sim 5000\,\mathrm{m^3\,s^{-1}}$. The density difference between freshwater and seawater is $\Delta\rho \simeq 26\,\mathrm{kgm^{-3}}$ so that the mean outflow of the Rhine represents a rate of buoyancy input to the North Sea of

$$R_d b = -R_d g \Delta\rho = 2200 \times 9.81 \times 26 = 5.61 \times 10^5\,\mathrm{N\,s^{-1}}. \tag{2.13}$$

This is equivalent to the buoyancy input by heating at $Q_{\max}$, the peak summer rate (see Fig. 2.5), over an area of:

$$\mathrm{Area} = \frac{-R_d \Delta\rho c_p}{\alpha Q_{\max}} \approx 10^4\,\mathrm{km}^2. \tag{2.14}$$

Within a region of this size, which represents a significant fraction ($\sim$10%) of the area of the southern bight of the North Sea, the Rhine river discharge exerts a stabilising influence comparable to that of surface heating in the summer. In such *Regions Of Freshwater Influence* (ROFIs), the freshwater buoyancy source maintains a distinctive regime, different from the rest of the shelf sea and sharing many of the characteristics of an estuary. These interesting regions and the processes operating in them will be the main subject of Chapter 9.

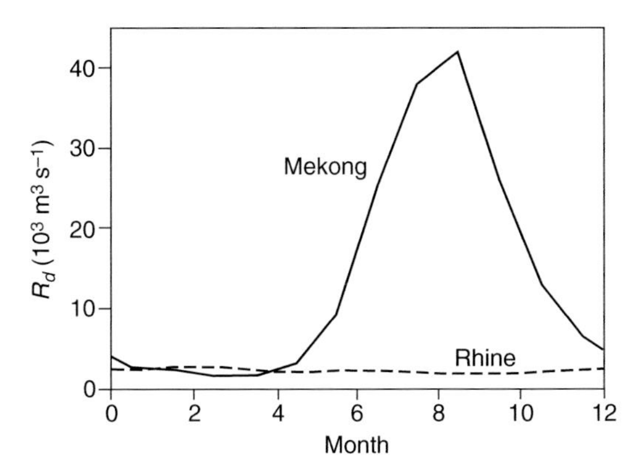

Figure 2.9 Average monthly discharge of the Mekong River based on data for 1933–53 and 1961–1966. The corresponding monthly discharge of the Rhine is shown as the dashed line. Data supplied by the Centre for Ecology and Hydrology, Wallingford, UK.

2.3.2 Seasonal cycles of freshwater input

Unlike the surface heat exchange, which follows rather predictable seasonal patterns depending on latitude, rivers vary widely in their annual discharge cycles. While many large rivers in temperate latitudes, such as the Rhine, maintain a substantial flow throughout the year, smaller rivers often have a highly variable discharge. Rivers in tropical and subtropical areas tend to exhibit particularly strong seasonal variation due to the monsoonal rain patterns. For example, Fig. 2.9 illustrates the seasonal cycle of the Mekong, a large tropical river in Southeast Asia, which discharges near the entrance to the Gulf of Thailand. The dry season flow of this river is of the same order as the mean flow rate of the River Rhine but increases ~40-fold during the peak runoff in the wet season.

2.3.3 Global distribution of freshwater input

From a global perspective, the tropical oceans receive a disproportionately high fraction of the total freshwater runoff entering the oceans. As can be seen in Fig. 2.10, the large discharges into the ocean from the tropical regions of South America, Africa and Southeast Asia greatly exceed those from the rivers in mid latitudes. Moreover, much of this freshwater buoyancy input to tropical seas is subject to strong seasonal variation, as in our example of the Mekong. This pronounced modulation of the buoyancy input as freshwater is in contrast to the buoyancy input by the heat flux in the tropics which, as we noted earlier (Section 2.2.5), tends to exhibit minimal seasonal variation in the tropics. Not all tropical rivers exhibit strong seasonal cycles; a notable exception is the Amazon, which carries a high proportion (~20%) of all the freshwater entering the oceans. Its discharge is relatively stable, varying by ~30% from the mean flow rate of 219 000 m^3 s^{-1}.

It is apparent from Fig. 2.10 that, after the tropics, the next largest discharges come from rivers debouching into the Arctic Ocean. Several very large rivers including the Yenisey, the Lena and the Ob discharging from the north coast of Eurasia combine with flow from the Mackenzie and other Canadian Arctic rivers to produce a combined mean input of ~150 000 m^3 s^{-1}, which is equivalent to more than two-thirds of

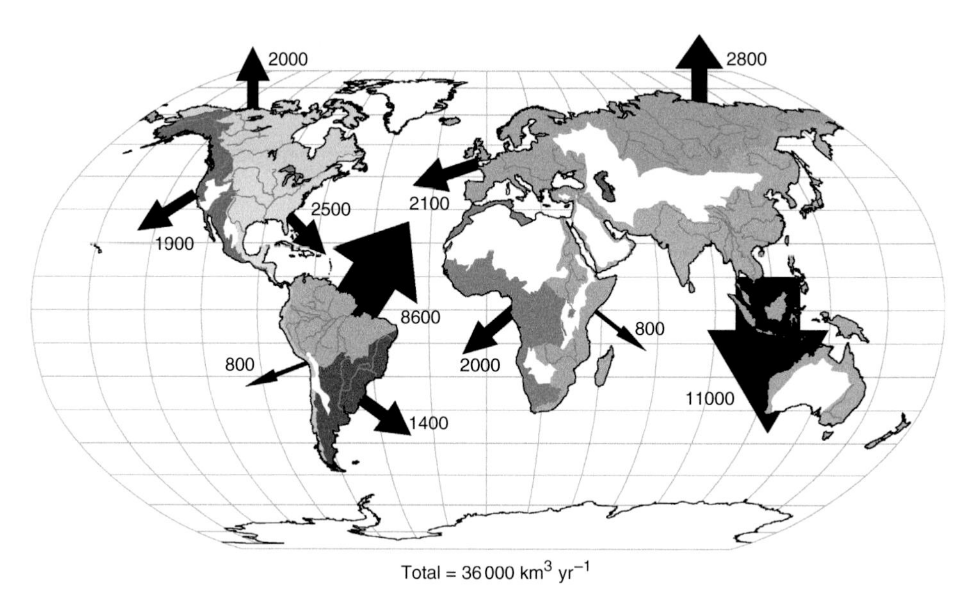

Figure 2.10 See colour plates version. Annual river discharge to the ocean from (Milliman and Farnsworth, 2011), courtesy Cambridge University Press. Figures represent freshwater runoff from coloured land areas in cubic kilometres per year (1 $km^3 yr^{-1} = 31.7 m^3 s^{-1}$).

the Amazon's discharge, or about 70 times the Rhine mean output. As well as being very large, these Arctic rivers are also subject to strong seasonal variation due to restriction of the flow during winter by ice formation and enhanced flow in spring and summer as terrestrial ice and snow melt. The result is that more than 90% of the annual delivery of freshwater to the Arctic Ocean occurs in the months of May, June and July (Dittmar and Kattner, 2003).

2.3.4 Surface fluxes of freshwater

Although the most conspicuous inputs of freshwater to shelf seas are those occurring through river discharge, direct sea surface exchange through evaporation and rainfall can also be important. For seas in arid regions such as the Arabian Gulf, annual water loss by evaporation can amount to as much as 4 metres. The loss of so much freshwater increases salinity and creates a hyper-saline regime with denser water forming, especially in shallow water. As an example, Fig. 2.11 illustrates the high salinities that occur in Spencer Gulf, a basin on the arid south coast of Australia with mimimal freshwater input. In response to evaporation, which exceeds precipitation for the whole year, salinity increases progressively from open ocean values of $S{\sim}36$ outside the gulf to $S{\sim}44$ at the head of the gulf, with the highest values ($S{\sim}48$) occurring late in the austral summer. The changes in density are correspondingly large and are responsible for driving a major component of the circulation in the Gulf (Nunes-Vaz *et al.*, 1990). At the other extreme, heavy precipitation in tropical seas can, at times, out-compete evaporation and lower

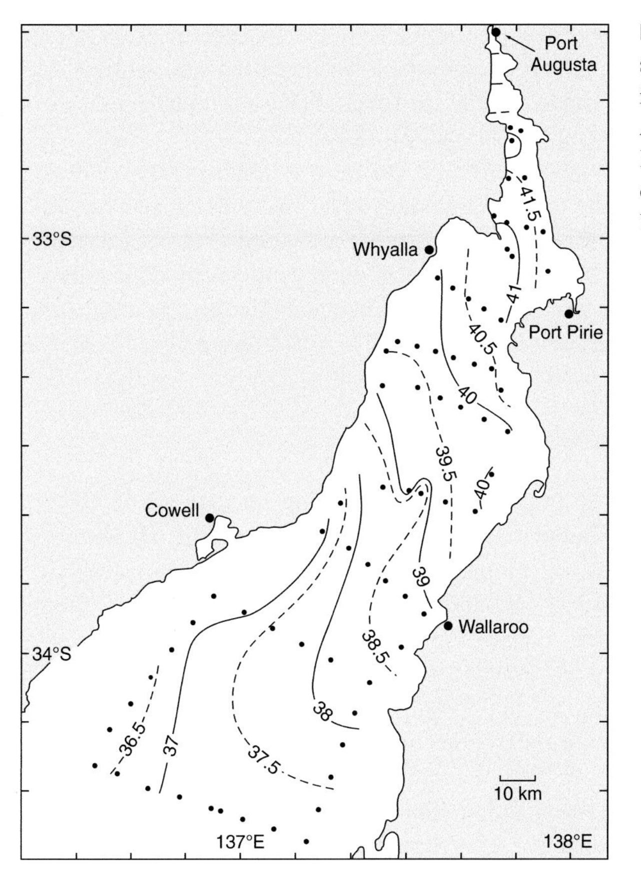

Figure 2.11 Depth-averaged salinity distribution in winter in Spencer Gulf, South Australia. From (Nunes and Lennon, 1987), courtesy of the American Geophysical Union.

surface salinity significantly, with a consequent reduction in density and the enhancement of water column stability.

2.4 Forcing by wind stress and pressure gradients

In addition to acting as a source and sink for heat and freshwater, the atmosphere also exerts a stress on the sea surface which involves the transfer of momentum from the wind to surface currents in the ocean. In principle this exchange of momentum can go either way but, since winds in the atmosphere tend to involve much higher velocities than surface ocean currents, we are mostly concerned with the downward transfer of momentum and energy to the ocean from the atmospheric boundary layer. The resultant stress τ_w on the sea surface is frequently represented in terms of a quadratic drag law

$$\tau_w = C_d \rho_a W(W - u_s) \cong C_d \rho_a W^2 \tag{2.15}$$

where ρ_a is the density of air, C_d is a drag coefficient, W is the wind speed and u_s is the component of surface current in the direction of the wind. In most cases $u_s \ll W$ so that u_s may be neglected. It should be noted that this relation has the same form as the bulk parameter representations of evaporation and sensible heat transfer rates with the relative momentum in the form of $(W - u_s)$ replacing the air-sea temperature difference, for example, in Equation (2.5).

Like the other coefficients, C_d has been determined by measurements of turbulent transfer in the marine boundary layer made from spar buoys. These measurements reveal that C_d is not strictly constant but varies with the degree of stratification in the atmosphere and with wind speed. Typically C_d is reduced by ~60% for stable conditions relative to the value for a neutral atmosphere. One of the simpler formulations linking C_d to wind speed is a linear trend (Smith and Banke, 1975):

$$C_d = (0.63 + 0.066W) \times 10^{-3}. \tag{2.16}$$

This result can be interpreted as indicating an increase in the roughness of the sea surface attributed to the increasing wave height. However, it is fair to say that the physics of momentum transfer from atmosphere to ocean is not yet well understood. In addition to the direct effect of the tangential wind stress on the sea surface, momentum is also transferred by surface waves which are driven by normal pressure forces. Large waves in a fully developed sea have considerable net forward momentum in the near surface layers, some of which is transferred to the mean flow when the waves break (Melville and Rapp, 1985).

As well as injecting momentum and energy into the ocean surface via surface stresses, the atmosphere also forces the ocean through the action of atmospheric pressure on the sea surface. The gradients of atmospheric pressure, which drive the winds, are also applied to the ocean and can induce significant flows as the ocean adjusts to the varying pressure field. Generally, however, the direct effect of pressure gradient is smaller by at least an order of magnitude than that of the corresponding wind stress.

2.5 Tidal forcing

In many shelf sea areas, the largest and most consistent mechanical forcing comes from the tides. The tide generation due to the Moon arises from an imbalance between the forces acting on a particle due to the gravitational attraction of the Moon and the centrifugal force due to the Earth's rotation about the centre of gravity of the Earth–Moon system. The balance between these forces is exact only at the centre of the Earth. At all points on the Earth surface, the small imbalance in these forces results in a *tide generating force*, which is $\sim 10^{-7}g$. The Sun exerts a similar force with a magnitude ~0.46 × that of the Moon. Together, these forces act on the waters of the ocean to drive the tides. Box 2.1 provides a more complete explanation of the tide generating forces.

Box 2.1 Tide generating forces

To understand how the forces generating the tides arise, we shall examine the tidal forces due to the Moon, which exerts more than twice the tidal influence of the Sun. Consider first the balance of forces which keeps the Moon in its almost circular orbit as it revolves around the mutual centre of gravity of the Earth–Moon system which is the point labelled C_g in Fig. B2.1.

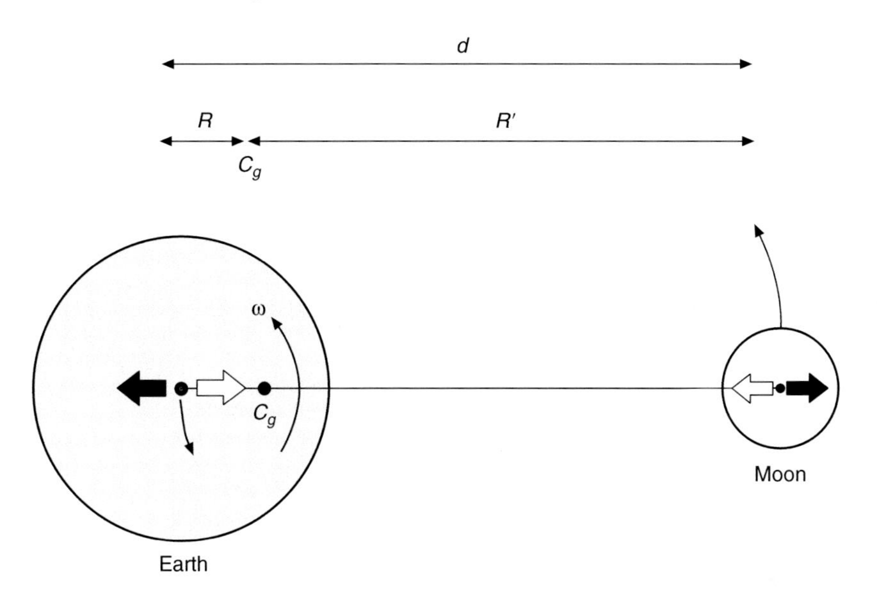

Figure B2.1 The Earth and Moon revolving about their mutual centre of gravity C_g. Thick white arrows indicate gravitational attraction; black arrows are centrifugal forces.

Because the mass of the Earth, E, is much greater than the mass of the Moon ($M \simeq E/81$), C_g is located within the Earth but it is important to remember that the Earth as well as the Moon revolves about C_g, albeit in a much smaller circle. The Moon remains in its orbit because of a balance between the mutual gravitational attraction of the Earth and the Moon on the one hand and the centrifugal force acting on the Moon on the other, a balance which can be expressed as:

$$\text{Gravitational attraction} = \frac{GME}{d^2} = MR'\omega^2 = \text{Centrifugal force} \qquad (2.17)$$

where $G = 6.674 \times 10^{-11}\ \text{m}^3\ \text{kg}^{-1}\ \text{s}^{-2}$ is the gravitational constant, $d \approx 384\,000$ km is the distance between the centres of the Earth and the Moon, R', is the distance between the centre of gravity C_g and the centre of the Moon and $\omega = 2.463 \times 10^{-6}\ \text{s}^{-1}$ is the angular speed of the Moon's orbit.

What is not quite so obvious is that there is an equivalent balance between the forces acting on the Earth, which can be written as

$$\frac{GME}{d^2} = ER\omega^2 \tag{2.18}$$

where R is the radius of the Earth's movement (see Fig. B2.1). This balance is exact only at the centre of the Earth. While the centrifugal force per unit mass $R\omega^2$ is the same for all points on the Earth, the gravitational pull of the Moon varies in magnitude and direction over the Earth's surface. It is this small difference between the gravitational and centrifugal forces which is responsible for generating the tides. To see how the magnitude and direction of the Tide Generating Force (TGF) can be calculated, look at Fig. B2.2, which represents the Earth–Moon system as viewed from above the plane of the lunar orbit.

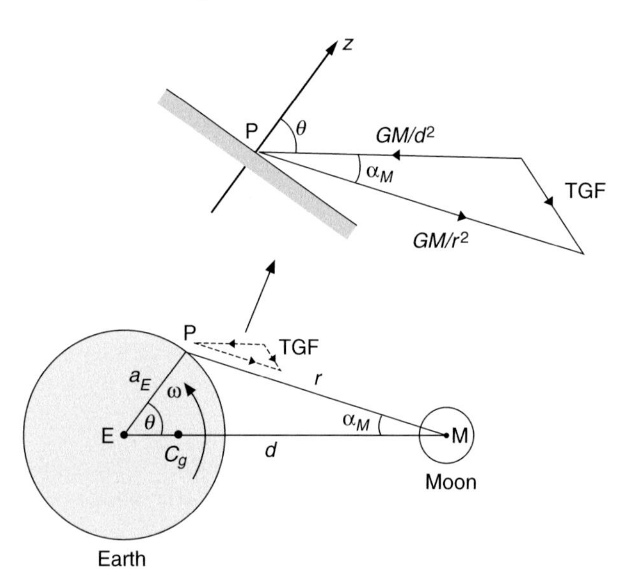

Figure B2.2 The Tide Generating Force (TGF) as the vector difference of attractive and centrifugal forces. The upper diagram is an expanded version of the vector triangle at P.

At an arbitrary point P, the attractive force of the Moon is different from that at the centre of the Earth; it is directed along the line PM which is at an angle α_M to the line of centres and it is also slightly larger because the distance PM is less than EM. The TGF at P can be calculated by taking the vector difference between the gravitational attraction at P and the centrifugal force. The resulting vector triangle is shown enlarged in Fig. B2.2; the magnitude of the centrifugal force, which is the same at all points on the Earth, is the same as the gravitational attraction at the centre of the Earth where the two forces are in exact balance. Taking the difference of the horizontal components of the forces in the vector triangle, and allowing for the fact that the radius of the Earth a is small compared with d, we find that the horizontal component of the TGF can be expressed as:

$$F_H = \frac{3}{2}\frac{M}{E}\left(\frac{a_E}{d}\right)^3 g \sin 2\theta \tag{2.19}$$

where θ is the angle between EP and the line of centres and $g = \frac{GE}{a_E^2}$ is the gravitational acceleration on the Earth. The magnitude of the horizontal TGF relative to g is just $\frac{|F_H|}{g} = \frac{M}{E}\left(\frac{a_E}{d}\right)^3 \sim 10^{-7}$. The corresponding vertical component has a similar magnitude but is negligible because it is in competition with g, whereas F_H, though small compared with g, is generally competing with the much smaller forces in the horizontal. The TGF varies over the globe according to $\sin 2\theta$ which means that it is zero at the sublunar point $\theta = 0$, on the great circle where $\theta = \pi/2$ and at the nadir where $\theta = \pi$. The force has a greatest magnitude at $\theta = \pi/4$ and at $\theta = 3\pi/4$ where it is directed towards the sublunar point and the nadir respectively, as you can see in Fig. B2.3.

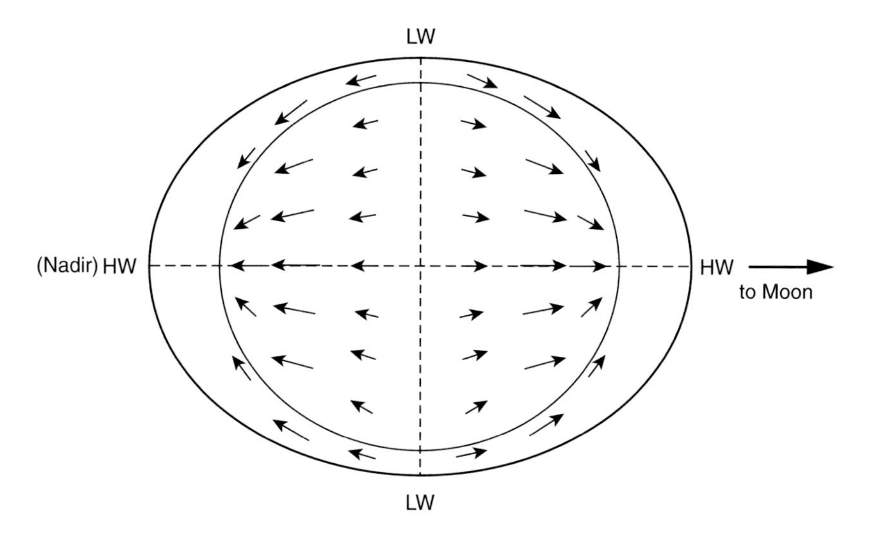

Figure B2.3 Distribution of the Tide Generating Force and the tidal ellipse which results from the Equilibrium model.

This pattern of forces, if acting on a uniform ocean covering the whole Earth, would tend to move the ocean waters towards the sublunar point and the nadir and away from the great circle where $\theta = 0$. We might expect this movement to set up slopes of the sea surface which would balance the TGF. This idea formed the basis of the *Equilibrium theory of the tides*, which was originally proposed by Newton who gave the first rational explanation of tidal forces in 1687. The water surface in the Equilibrium theory would be an ellipsoid of revolution with its major axis directed towards the Moon (Fig. B2.3) so that high water (HW) would occur at the sublunar point and at the nadir with low water (LW) on the great circle where $\theta = \pi/2$. In this theory, the maximum elevation at the HW points would be +35.4 cm and the minimum −17.7 cm at LW, giving a total range of only 53.1 cm. The tide due to the Sun would add almost half as much again. Such a range is considerably less than is observed in most parts of the shelf seas but is not so different from what we now know of the tides in the deep ocean, where a range of ~1 metre is typical.

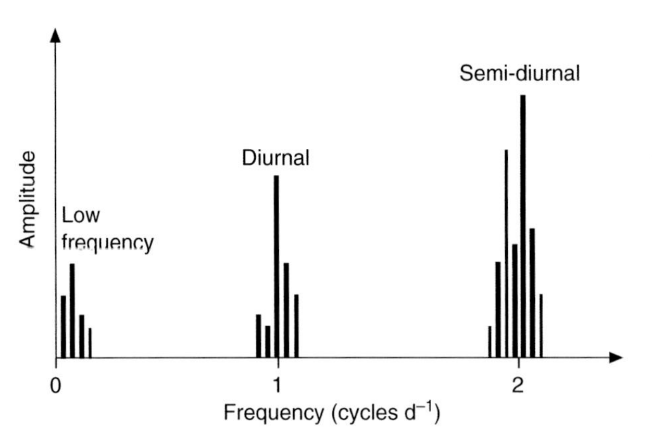

Figure 2.12 Schematic of the spectrum of constituents of the TGF.

2.5.1 Tidal constituents

As the Earth rotates, the tide generating forces of the Sun and the Moon vary in a regular fashion with periods set by the lunar and solar days. There are also more subtle variations at other frequencies due to the orbital motions of the Moon about the Earth and the Sun about the Earth. A rigorous mathematical analysis of the tide generating forces shows that the spectrum of the forcing, illustrated in Fig. 2.12, consists of a rather large number of spectral lines at specific frequencies. These lines are mainly concentrated in two spectral bands, termed *tidal species* or *bands*, which are centred on 1 cycle per day (the diurnal species) and 2 cycles per day (the semi-diurnal species) with some others at low frequencies (~1 cycle per month or less). Each separate frequency identifiable in the spectrum is referred to as a *tidal constituent*, and the TGF at each point on the Earth can, therefore, be written as a sum of these individual constituents:

$$\mathrm{TGF} = \sum_{n=1}^{N} A_n \cos(\omega_n t + \alpha_n) \tag{2.20}$$

where A_n and α_n are the constituent amplitudes and phases. The frequencies ω_n are known accurately from the astronomically determined motions of the Sun and Moon.

The response of the global ocean to this forcing can in theory be calculated without reference to observations if the bathymetry of the ocean is known. This prediction of the tides from first principles, originally proposed by Laplace, has proved very challenging and has only recently been accomplished using high resolution numerical models (Egbert *et al.*, 2004). The alternative, also anticipated by Laplace (Cartwright, 1999), is to treat the observed tide as a sum of harmonic terms at the same frequencies as the TGF but with different amplitudes and phases:

$$\eta(t) = \sum_{n=1}^{N} H_n \cos(\omega_n t + \alpha_n - g_n). \tag{2.21}$$

Table 2.1. Some of the more important tidal constituents.

Symbol	Name	Period (hours)
M_2	Principal lunar	12.42
S_2	Principal solar	12.00
N_2	Larger lunar elliptic	12.66
K_2	Luni-solar declinational	11.97
K_1	Luni-solar declinational	23.93
O_1	Larger lunar declinational	25.82
M_f	Lunar fortnightly	13.7 days

This idea, developed by Lord Kelvin, G. H. Darwin and A. T. Doodson amongst others, gave rise to the *harmonic method* of tidal analysis in which the tide is treated as a sum of independent, sinusoidal tidal constituents. The amplitudes H_n and phase adjustments g_n of the constituents are determined by analysis of the observed tidal elevation using least squares methods (see, for instance, Emery and Thomson, 2001 and book website). Although the number of terms N in the theoretical expansion of the TGF is large (~400), in practice the amplitudes of many of the constituents are small and the tide can be well represented by a limited number of constituents (typically ~20). A list of a few of the more important tidal constituents and their frequencies is given in Table 2.1. Constituents are identified by a symbol with a letter indicating something about the origin of the constituent and a subscript denoting the species (1 = diurnal, 2 = semi-diurnal). For example, the main semi-diurnal constituent driven by the Moon (frequently the largest of all constituents) has the symbol M_2. Next to this comes the main solar tide S_2 which, because its frequency is determined by the apparent movement of the Sun, has a period of exactly 12 hours.

In most parts of the ocean the semi-diurnal tides tend to predominate and the diurnal constituents are relatively small. There is, however, often a tendency for the two tides in the day to be unequal, an effect which arises from the fact that the Moon moves above and below the equator over the monthly cycle reaching a declination (angular height above the equator) which can be as large as 28.5°. When the Moon is above or below the equator, the axis of the ellipsoid moves with it and this gives rise to inequality between the two tides during a day which is represented, for example, in the diurnal constituent O_1. The Sun also moves above and below the equator (+/− 23.5°) so there are similar solar declinational constituents. In a few areas, the diurnal tendency takes an extreme form with large diurnal constituents resulting in only one tidal oscillation per day. There are also variations in the tide associated with the fact that the orbit of the Moon is an ellipse rather than a circle, which gives rise to 'elliptic' tidal constituents like N_2 which has a longer period than M_2 and beats with it to produce a monthly variation in tidal range.

As we have just noted, in many areas of the shelf seas, the main lunar and solar semi-diurnal tidal constituents M_2 and S_2 are the largest. These constituents have periods of 12.42 and 12 hours and their interaction produces a regular range cycle with a period of 14.79 days in which the tides alternate between the large *spring tides* and small *neap tides*. This fortnightly modulation in the strength of the tides is easily understood in terms of the positions of the Moon and Sun relative to the Earth. If the Earth, Moon and Sun are in line (at new Moon and full Moon) then the tidal ellipsoids generated by the Moon and the Sun are aligned. Thus the Moon HW and the Sun HW add together to produce a very high tide. Conversely, with the Moon–Earth–Sun forming a right angle, the Moon's HW coincides with the Sun's LW, resulting in a smaller HW. This *spring-neap cycle*, which can be extreme when the amplitudes of M_2 and S_2 are similar, has an important role in modulating mixing processes, as we shall see in Chapters 7, 8 and 9.

2.5.2 Tidal energy supply to the shelf seas

Most of the energy input to sustain the tides occurs in the deep ocean, but a large majority of the energy is dissipated in the shallower waters of the shelf seas where frictional forces are much greater. The tidal forces acting directly on the waters of the shelf seas also contribute but are responsible for only a small fraction of the energy input to the shelf seas. The majority of the energy input to the shelf seas comes from the deep ocean in the form of waves of tidal period. These waves cross the shelf break on to the shelf, where they are reflected and amplified before losing much of their energy in dissipation by frictional stresses acting at the seabed. We shall look more closely in Chapter 3 at the behaviour of these waves and the way that the shelf seas react to the tidal forcing which they bring, but for the moment we shall concentrate on the large scale budget of energy supply to the tides.

The Moon tides supply power to the Earth as a whole at the rate of ~3.2 TW (1 TW = 10^{12}W) with a further ~0.5 TW coming from the tides generated by the Sun. Tides in the solid earth (~0.2 TW) and in the atmosphere (~0.02 TW) make minimal contributions to energy dissipation, so that most of the total power supply (~3.5 TW) is accounted for by energy dissipation in the ocean. Until quite recently it was thought that almost all dissipation was concentrated in the shelf seas, but as knowledge of more remote seas improved, it became increasingly difficult to account for all the power input in terms of the observed and modelled tidal dissipation. It now appears that about 75% of the ocean's dissipation occurs in the shallow seas, with the rest (0.9 TW) being consumed in the deep ocean partly through the generation of internal tides and waves (Munk and Wunsch, 1998).

In Fig. 2.13 you can see that the distribution of the tidal energy input within the world's oceans and shelf seas is rather uneven, with a relatively small number of shelf sea areas accounting for a high proportion of the total. The large shelf region of Western Europe experiences particularly high dissipation, as do the comparable areas of the Yellow Sea and the large shelf area of northern Australia. In some cases, extremely high dissipation rates occur in relatively small regions, e.g. in the

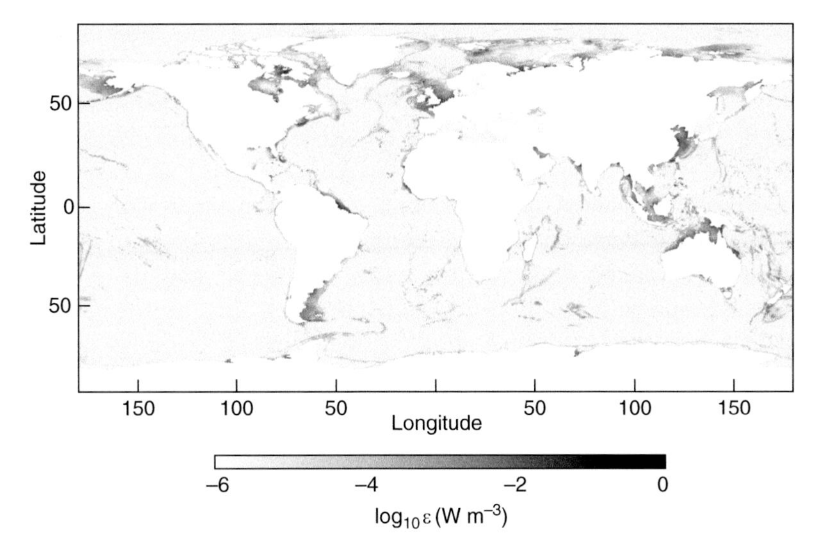

Figure 2.13 See colour plates version. Distribution of tidal dissipation in Wm^{-3} derived from a numerical model of the global ocean. Note the large contrast (~6 decades) in the magnitude of dissipation between low dissipation in much of the deep ocean and high values exceeding ~1 Wm^{-3} occurring mainly in the shelf seas. Figure from (Green, 2010), with permission from Springer.

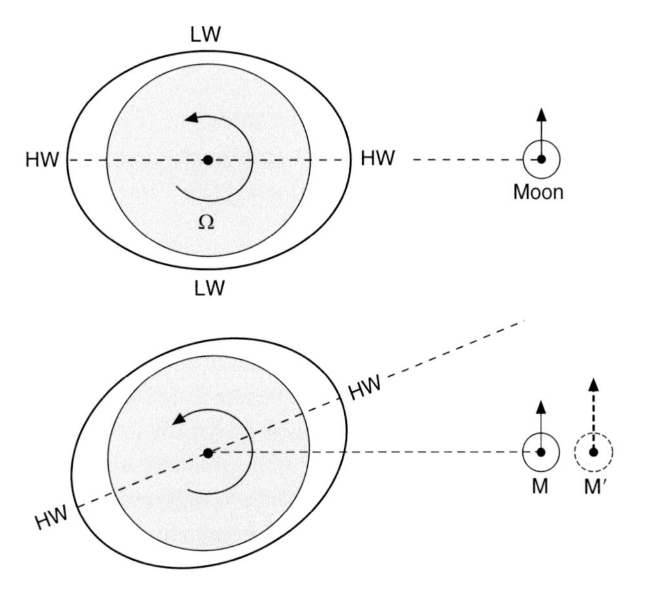

Figure 2.14 In a frictionless ocean, upper picture, the Earth and ocean rotate inside the tidal ellipsoid with the ellipsoid major axis directed along the line of centres towards the Moon. Consequently there are two tides per lunar day with the high water (HW) occurring when the Moon is overhead and at the nadir. In the lower picture, friction with the seabed pulls the ellipsoid round in the direction of the Earth's rotation, ahead of the Moon. This frictional stress between the ocean and the solid earth acts as a belt brake on the Earth's rotation, which is gradually slowed with a loss of kinetic energy of rotation and an increase in the length of the day.

Bay of Fundy which is host to the world's largest tides, and in the Cook Strait of New Zealand which alone dissipates more than 1% of the global total.

2.5.3 Source of energy dissipated by the tides

The energy dissipated by tidal friction in the ocean comes ultimately from the kinetic energy of the Earth's rotation. The frictional stresses acting in the ocean combine to exert a restraining torque on the Earth, tending to slow its motion and reduce its angular momentum. Figure 2.14 illustrates how friction with the seabed allows the daily spinning of the Earth to drag the tidal ellipsoid forward of the Moon's overhead position, which acts as a brake on the Earth's spin. Since the Earth-Moon system is not subject to external torques, the Earth and Moon must jointly conserve angular momentum.[1] The consequence is that as the Earth's rotation slows, the Moon must accelerate in its orbit. As it does so, it moves farther away from the Earth at a rate of ~2 cm per year, an adjustment that has been directly measured by laser ranging on a mirror left on the Moon by the Apollo astronauts (Dickey *et al.*, 1994). The acceleration of the Moon in its orbit can thus be determined with high precision and is the basis of our best estimates of the total tidal energy input to the ocean. It is interesting to note that tidal friction works both ways. Tides set up within the solid body of the Moon also generate frictional stresses, which have acted to slow down the Moon's spin to the extent that it now has one face tidally locked to the Earth (Gladman *et al.*, 1996).

Summary

The motions of the shelf seas are driven by a combination of buoyancy inputs from surface heat exchange and freshwater runoff and direct mechanical forcing by the atmosphere and by the tides.

In terms of energy, surface heat exchange in the seasonal cycle is the predominant buoyancy source for much of the shelf. Heat input is largely due to short wave radiation from the Sun, most of which is absorbed close to the surface. Heat loss occurs from a surface microlayer (< 1 mm thick) and is due to a mixture of long wave back radiation, latent heat loss due to evaporation and sensible heat transfer from ocean to atmosphere by conduction. These surface exchange processes generally dominate over horizontal heat fluxes and combine to control the seasonal cycle of heat storage in the water column. Buoyancy inputs from freshwater discharge may outcompete heat exchange as a buoyancy source, but usually only in estuaries and adjacent regions of freshwater influence (ROFIs). Particularly large inputs of freshwater occur in tropical shelf seas with, in monsoonal areas, a strong seasonal variation. The large shelf seas of the Arctic Ocean are also in receipt of freshwater inputs which greatly exceed those in mid latitudes.

[1] A fundamental law of physics requires that, in the absence of external forces, the angular momentum of a mechanical system must be conserved. A spinning top would maintain its spin forever were it not for weak damping due to air resistance and friction with the ground.

Changes in density caused by heat or freshwater may lead to vertical and horizontal gradients which contribute to water column stability and circulation respectively.

The mechanical forcing of the shelf seas, i.e. the direct input of momentum and kinetic energy, arises mainly from forces applied to the sea surface by the atmosphere and from the tidal energy input from the deep ocean. The principal atmospheric forcing is in the form of tangential stresses exerted by the wind on the sea surface. This stress is represented by a quadratic drag law with a drag coefficient which increases somewhat with wind speed.

The most regular and, in many cases, the dominant mechanical forcing of the shelf seas is the result of momentum and energy inputs from the tides. A tidal record of speed or sea level can be broken down into a sum of sinusoidal components, known as the tidal constituents. These constituents all arise from the orbital characteristics of the Moon about the Earth and the Earth-Moon system about the Sun. There are about 400 known constituents, though generally only a few dominate any particular region. The tide generating forces exerted by the Moon and the Sun inject energy mainly into the deep ocean. This energy takes the form of long waves which propagate on to the shelf and induce large tidal oscillations in which most of the incoming energy flux is dissipated by frictional effects. The energy consumed is taken from the kinetic energy of the Earth's rotation as the frictional stresses on the seabed act as a brake which is gradually slowing the rate of rotation. Most (~75%) of the tidal energy is dissipated in the shelf seas.

FURTHER READING

Changing Sea Levels: Effects of Tides, Weather and Climate, by David Pugh, Cambridge University Press, 2004.

Atmosphere-Ocean Dynamics, by Adrian E. Gill, Cambridge University Press, 1982. See particularly chapters 1 and 2.

Atmosphere and Ocean: A Physical Introduction, by Neil Wells, Wiley, 1997.

3 Response to forcing: the governing equations and some basic solutions

In responding to the forcing discussed in the previous chapter, the motions of the shelf seas are governed by fundamental physical laws. The principles involved, namely those of the dynamics expressed in Newton's laws of motion, and those of kinematics set by the geometrical rules of motion, are fundamentally the same as those which solid body movement must obey. However, to express these laws in an appropriate form for fluids is a little more difficult than for the solid body case. Starting from the basic principles we shall show in this chapter, without detailed derivations, how the equations of motion arise and illustrate the role played by each of the main terms in the equations. Along the way, we shall examine some simple force balances which describe particular forms of motion that are of special importance in the ocean generally and in the shelf seas in particular.

3.1 Kinematics: the rules of continuity

When water moves in the ocean, it has to obey some fundamental constraints which are independent of the forces driving the motion. These rules, termed kinematics, are mainly concerned with maintaining the *continuity* of the fluid and, in some cases, conserving its properties. Most fundamental and important is the rule that mass must be conserved. Since seawater is almost, but not quite, incompressible, we can often assert that volume as well as mass is conserved. When this is the case, we can readily derive a relation (see Box 3.1) between the velocity components u, v, w:

$$\frac{\partial u}{\partial x}+\frac{\partial v}{\partial y}+\frac{\partial w}{\partial z}=0. \tag{3.1}$$

This statement is the same for any Cartesian coordinate system, but we shall follow a general convention in oceanography and choose the z axis positive upwards, with $z = 0$ at the surface, and the x and y axes positive eastward and northward respectively. An alternative, and sometimes more convenient, form of the volume continuity statement comes from integrating Equation (3.1) between the bottom ($z = -h$) and the surface to give:

$$h\left(\frac{\partial \hat{u}}{\partial x}+\frac{\partial \hat{v}}{\partial y}\right)+\frac{\partial \eta}{\partial t}=0 \quad (3.2)$$

where $\hat{u}$ and $\hat{v}$ are the velocities averaged over the depth h and the vertical velocity w at the surface has been replaced by the rate of change of surface level $\partial\eta/\partial t$.

Box 3.1 Continuity statements

Consider the transport of fluid mass through a small fixed cuboid with dimensions δx, δy, δz which will act as a 'control volume', as illustrated below. We want to express the fact that what flows into the cuboid must come out, assuming that fluid is not destroyed or created within the volume of the cuboid. The fluid has density ρ and velocity components u, v, w (in Cartesian coordinates x, y, z) which generally will vary in space and time.

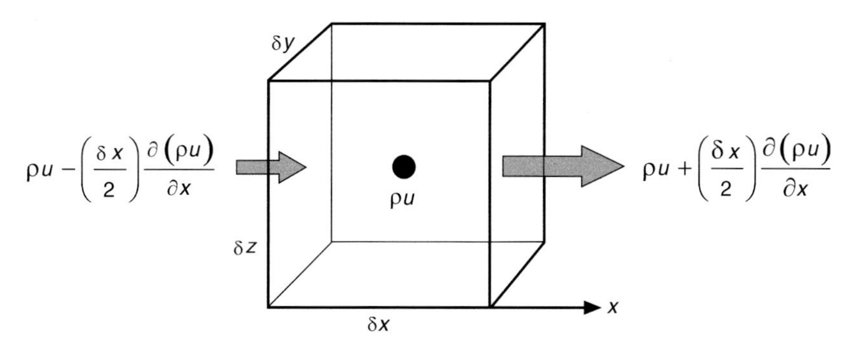

Figure B3.1

We start with flow in the x direction, with the mass flux, ρu, in the centre of the cuboid. The mass flow into the cuboid through the left face is the face area $\times$ velocity $\times$ density i.e. $(\delta y\delta z)\left(u\rho-\frac{\partial(u\rho)}{\partial x}\frac{\delta x}{2}\right)$, where the second term in the bracket accounts for any change in u and/or ρ in the x direction. For the right face, the mass flow will again be different due to the changes in u and ρ, and is given by $(\delta y\delta z)\left(u\rho+\frac{\partial(u\rho)}{\partial x}\frac{\delta x}{2}\right)$. The net gain of fluid mass due to flow in the x direction is the difference between what enters the box minus what leaves the box, which is just

$$-\,\delta x\delta y\delta z\frac{\partial(\rho u)}{\partial x}=-\text{cuboid volume}\times\frac{\partial(\rho u)}{\partial x}.$$

Combining with equivalent expressions for flow in the y and z directions, we see that the net rate of gain of mass in the control volume is:

$$-\,\delta x\delta y\delta z\left\{\frac{\partial(\rho u)}{\partial x}+\frac{\partial(\rho v)}{\partial y}+\frac{\partial(\rho w)}{\partial z}\right\}.$$

Setting this rate of increase equal to the cuboid volume × the rate of density increase, we have, after dividing both sides by $\delta x\ \delta y\ \delta z$, a statement of mass continuity:

$$-\left\{\frac{\partial(\rho u)}{\partial x}+\frac{\partial(\rho v)}{\partial y}+\frac{\partial(\rho w)}{\partial z}\right\}=\frac{\partial \rho}{\partial t}$$

If we can regard density as constant, then this relation simplifies to:

$$\boxed{\frac{\partial u}{\partial x}+\frac{\partial v}{\partial y}+\frac{\partial w}{\partial z}=0} \qquad (3.3)$$

which is a widely used statement of the continuity of volume. Equation (3.3) can be conveniently abbreviated to the statement that the divergence of the velocity is zero, i.e. $\nabla.\underset{\sim}{u}=0$ where $\underset{\sim}{u}=(u,v,w)$ is the vector representing the velocity components.

As well as insisting on the continuity of the fluid, we can often apply conservation rules to its properties. For example, changes in the total mass of salt in a region of the shelf seas must be matched by the net input or output at the boundaries. Such statements impose simple but powerful constraints which can greatly assist analysis. A further simplification which we can frequently exploit is that flow in the sea is mainly in the horizontal plane. This is because, in most parts of the ocean, density increases with depth so that a particle displaced upwards or downwards experiences a restoring force. Vertical movement is thus suppressed by stratification so that vertical velocities are very much less than those in the horizontal plane and the motion can often be regarded as two dimensional (2D). There are exceptions to this horizontal 2D pattern. For example, Fig. 3.1 shows two situations in which substantial vertical motions can occur in the ocean. In the first (Fig. 3.1a), local cooling of the sea surface can lead to the sinking of denser water in *convective chimneys* with surface flow converging at the top of the chimney and divergent flow away from the base of the chimney below. In the second example (Fig. 3.1b), horizontal flow at the surface converges along the line of a front, with vertical flows along the sloping density surfaces and a line divergence further down the water column. An example of such a line convergence, occurring along the plume front at the mouth of an estuary, can be seen in Fig. 3.1c. As we shall see later in this chapter, important exceptions to 2D flow also occur in the upwelling and downwelling motions at coastal boundaries.

3.2 Dynamics: applying Newton's Laws

The central principle of dynamics is Newton's second law of motion. This law is concerned with changes in the momentum of a body and can be expressed concisely in the statement that the acceleration of a body a is determined by the net force on the body F divided by its mass m, i.e. $a=F/m$. To express this relation for a fluid, we

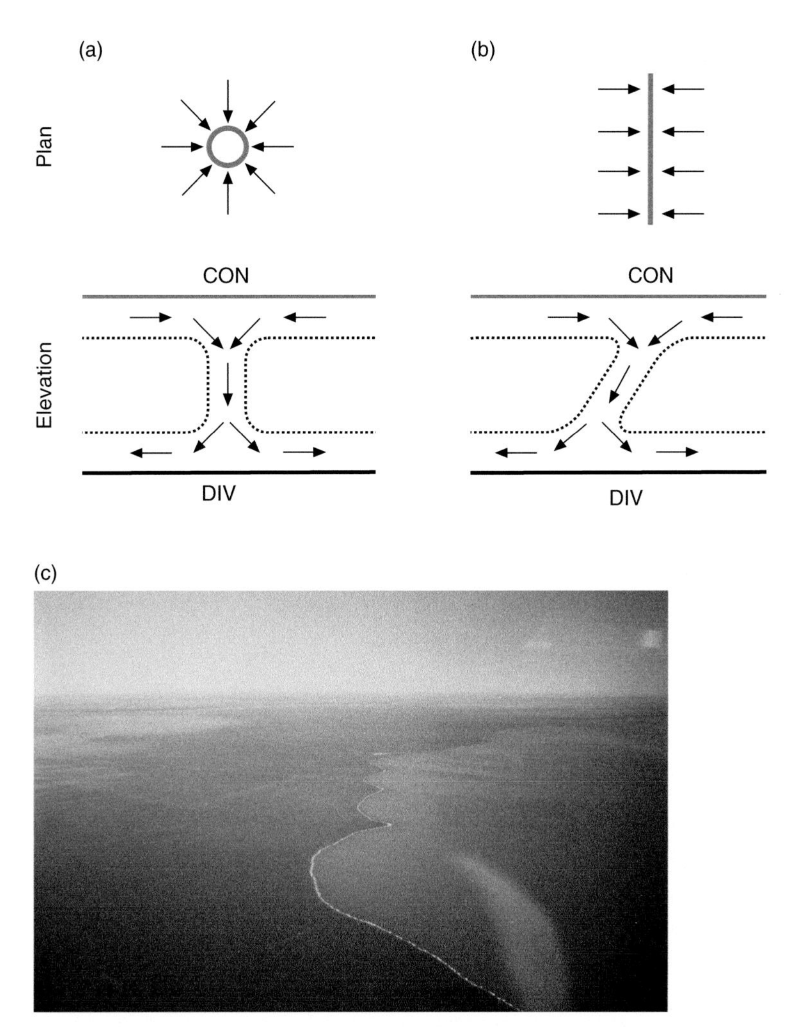

Figure 3.1 Examples of vertical flows with convergent (CON) and divergent (DIV) regions of horizontal currents. (a) Convergence towards a point at the sea surface; (b) Convergent flow to a line with sinking along a sloping interface; (c) An example of a frontal convergence along the Connecticut River Plume flowing into Long Island Sound. Brackish water from the estuary on the right of the picture is spreading over more saline water to the left, which is sinking under the estuarine water at the front. Convergence is indicated by the white line which is a strong accumulation of foam at the front. Picture taken by R. W. Garvine, April 26, 1972, from an altitude of 300 metres.

need to imagine the ocean as made up of a large number of relatively small particles, each of which must obey this rule. We can choose the volume and hence the mass of a particle if the density is known, but to apply the second law we need also to identify the forces acting on a particle and evaluate their sum to obtain the net force. At this stage we might anticipate that there will be force contributions from the pressure (F_p) and the frictional forces (F_f) acting within the fluid, and additional external force contributions (F_e). We might then express the particle acceleration in response to the net force F as:

$$a = F/m = (F_p + F_f + F_e)/m. \qquad (3.4)$$

There is, however, an important qualification to the use of Newton's second law if we want to apply it to the ocean or the atmosphere. Strictly, Newton's second law is a statement about the relation between acceleration and force in a non-accelerating reference frame. Since the Earth is rotating, our usual reference frame is accelerating and Equation (3.4) works only if we choose an external non-rotating reference frame, for example a frame related to the 'fixed' stars.

3.2.1 Coriolis force ($F = ma$ on a rotating Earth)

To overcome this reference frame difficulty, we divide the acceleration in a fixed reference frame into the acceleration as measured relative to coordinates fixed in the Earth (a_{rel}) and an additional component, termed the geostrophic acceleration (a_{geo}), which arises from the fact that the Earth is rotating about its axis with an angular velocity $\Omega = 2\pi$ radians/day:

$$a = a_{rel} + a_{geo}. \qquad (3.5)$$

The dynamical statement (3.4) can then be rewritten to read:

$$a_{rel} = (F - ma_{geo})/m = F'/m. \qquad (3.6)$$

So the simple proportionality of acceleration to net force can be retained for Earth coordinates if we add an additional 'force' term $F_c = -ma_{geo}$. This is the *Coriolis force*. It is strictly a mass × acceleration which has been moved from one side of the equation to the other, but there is no reason why we should not treat it as a force on equal terms with others (an example of a result in classical mechanics called D'Alembert's principle). To proceed further, we need to find out how to calculate the geostrophic acceleration. A rigorous derivation by a transformation of coordinates, given in many texts on geophysical fluid dynamics (e.g. Gill, 1982; Houghton, 2002), enables us to obtain all three components of the Coriolis force. In the ocean, vertical velocities are usually much less than horizontal velocities ($w \ll u, v$) so we only need to include the two horizontal components of the Coriolis force:

$$F_{cx} = 2\Omega v \sin\phi_L = fv; \quad F_{cy} = -2\Omega u \sin\phi_L = -fu \qquad (3.7)$$

where Ω is the angular velocity of the Earth's rotation[1] and ϕ_L is latitude. By convention we define the *Coriolis parameter* as $f = 2\Omega \sin\phi_L$. The Coriolis parameter determines the influence of latitude on the Coriolis force. The magnitude of f is greatest at the poles. It is positive in the northern hemisphere, decreases to zero at the equator, and is negative in the southern hemisphere. In the northern hemisphere, the Coriolis force acts to the right of the direction of motion; in the southern

[1] $\Omega = \frac{2\pi}{T_E}$ with T_E the rotational period of the Earth (24 hours), so $\Omega = \frac{2\pi}{24 \times 3600} = 7.27 \times 10^{-5}\ \mathrm{s}^{-1}$

hemisphere it acts to the left. In Box 3.2 we demonstrate how the Coriolis force arises in the particular case of a particle of water traveling due east.

The vertical component of the Coriolis force, $2\Omega u \cos \phi_L$, is generally very small in comparison with gravity ($\sim 10^{-5} g$) and other forces acting in the vertical and may usually be neglected. Why then do we have to include the horizontal components which are of the same magnitude as the vertical component? The answer is that the Earth's gravity has, by definition, no component in the horizontal and the other horizontal forces acting (e.g. pressure, frictional and tidal forces) are generally much smaller than g and of comparable magnitude to the Coriolis force.

Box 3.2 Coriolis force

Consider the acceleration of a particle moving due east on the Earth's surface at latitude ϕ_L as in the figure below. We wish to know its absolute acceleration; i.e. our viewpoint is that of an observer in a non-rotating frame of reference. The particle has an eastward velocity u relative to the Earth. Its velocity, as seen by the observer, is the sum of u and the velocity associated with Earth's rotation at the particle's position, which is just $u_{rot} = \Omega a_E \cos \phi_L$ where a_E is the radius of the Earth and Ω is the rate of angular rotation.

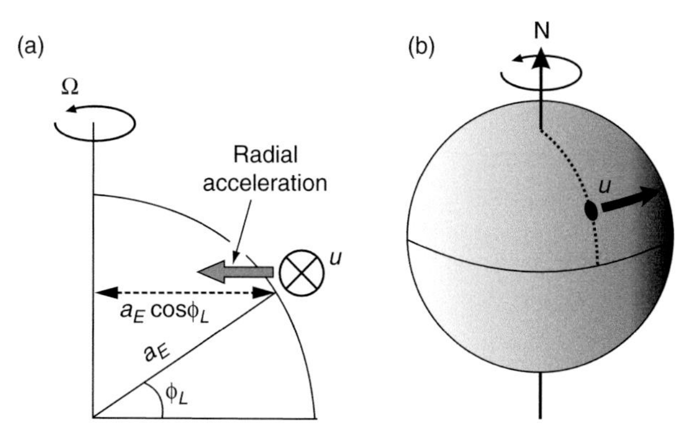

Figure B3.2

The radial acceleration of the particle moving in a circle in the non-rotating frame is $\frac{U^2}{r}$ where $U = u + \Omega a_E \cos \phi_L$ is the total speed of the particle and $r = a_E \cos \phi_L$ is the radius of the circular motion, i.e.

$$\frac{U^2}{r} = \frac{(\Omega a_E \cos \phi_L + u)^2}{a_E \cos \phi_L}.$$

Expanding the right side of the equation we see that the total acceleration consists of three components:

$$\text{total acceleration} = \Omega^2 a_E \cos \phi_L + \frac{u^2}{a_E \cos \phi_L} + 2\Omega u.$$

The first term is the centripetal acceleration of a fixed point at latitude ϕ_L as it rotates around the Earth's axis, while the second is the centripetal acceleration of a particle moving at speed u in a circular path on a non-rotating Earth. The final term $2\Omega u$ is the additional acceleration which results from the combination of rotation and the relative velocity. This acceleration acts at right angles to the Earth's axis of rotation and, therefore, has a component in the horizontal plane given by:

$$2\Omega u \cos(90 - \phi_L) = 2\Omega u \sin \phi_L = fu$$

which acts to the left of the direction of motion. The negative of this acceleration is the Coriolis force per unit mass which, therefore, acts to the right of the direction of motion. It is this term which we must include in the equations of motion if we are dealing with motion in a coordinate system rotating with the Earth.

3.2.2 The acceleration term: Eulerian versus Lagrangian velocities

There is a further complication in dealing with the acceleration term in Equation (3.6). Strictly, the second law is applicable to each particle of a fluid and relates the changes in momentum (mass × velocity) of the particle to the forces acting on it. The velocity here is that of the particle. In the ocean, velocities are usually measured, not for a particle, but at the particular point where a current measuring device is located; i.e. we make measurements of the velocities of lots of particles as they pass through a fixed point. Such velocities at a fixed point are termed *Eulerian* in contrast to *Lagrangian* velocity measurements which track an individual particle. In applying the second law, we need, therefore, to relate the acceleration of a particle of fluid in the x direction (written as Du/Dt) to the acceleration in a fixed Eulerian frame ($\partial u/\partial t$). This is done by noting that the velocity change experienced by a particle will be the sum of the local rate of change and the changes due to movement through the fluid where velocity will be changing with position.

Look at the illustration in Fig. 3.2. In a small time δt the displacement of the particle in the x direction will be $u\delta t$, which will result in a change in u of the particle

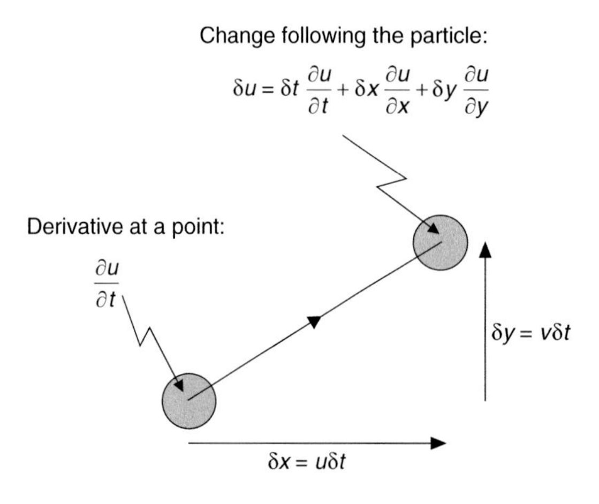

Figure 3.2 The circular particle moving in the x–y plane is displaced by $\delta x = u\delta t$ and $\delta y = v\delta t$ in a small time interval δt. If the velocity component u varies with x and y then the total change of u following a particle (i.e. in a Lagrangian frame of reference) will be the sum of the local change $\partial u/\partial t$ (i.e. in an Eulerian frame) and the additional changes due to the displacements $\delta x(\partial u/\partial x)$ and $\delta y(\partial u/\partial y)$. If the motion is in 3D, there will an additional term due to the displacement $\delta z = w\delta t$.

by $u\delta t\frac{\partial u}{\partial x}$, i.e. the distance travelled in the x direction multiplied by the rate of change of u in the same direction. Adding the equivalent contributions for movement in the y and z directions, the total change δu following a particle is just:

$$\delta u = \frac{\partial u}{\partial t}\delta t + u\delta t\frac{\partial u}{\partial x} + v\delta t\frac{\partial u}{\partial y} + w\delta t\frac{\partial u}{\partial z}. \tag{3.8}$$

Dividing by δt and then allowing time increment δt to tend to zero, we have for the total rate of acceleration following a particle:

$$\frac{Du}{Dt} = \frac{\partial u}{\partial t} + u\frac{\partial u}{\partial x} + v\frac{\partial u}{\partial y} + w\frac{\partial u}{\partial z}. \tag{3.9}$$

We can use the operator D/Dt, termed the *total* or *material derivative*, to express the rate of change for any property of a particle as it moves through the fluid. The last three terms on the right of Equation (3.9) are the *non-linear* terms. In many cases, when the spatial gradients are not too large, we can make the linearising approximation that $D/Dt = \partial/\partial t$ and neglect the non-linear terms.

With this form for the acceleration terms, we can now write the dynamical statements (Equation 3.5) for motion of a unit volume (we replace mass m with density ρ) in the x and y direction as:

$$\frac{Du}{Dt} = fv + \frac{F_x}{\rho}; \quad \frac{Dv}{Dt} = -fu + \frac{F_y}{\rho} \tag{3.10}$$

where we have included the Coriolis force components (Equation 3.7) and the terms F_x and F_y represent the net force (per unit volume) acting in the x and y directions respectively.

3.2.3 Internal forces: how do we include pressure and frictional forces?

In addition to the externally imposed forces such as wind stress acting at the sea surface, there are two important classes of force which operate in the interior of the fluid. Fluid particles exert normal forces on each other, where we are using 'normal' in the mathematical sense as acting perpendicular to a plane. All of the individual normal forces between fluid particles combine to make up the pressure which is defined as the normal force per unit area. At the same time, particles influence each other via frictional forces which act tangentially to exert stresses (force per unit area).[2] In the following paragraphs, we shall examine the way these two internal forces contribute to the dynamics, starting with the pressure field.

Pressure acts the same in all directions, i.e. the normal force on a small, solid plane located in the fluid would be the same whatever the plane's orientation. To determine

[2] For a pressure force, think of pushing down on the table. For the tangential shear force, think of the force you exert if you slide your hand along the table.

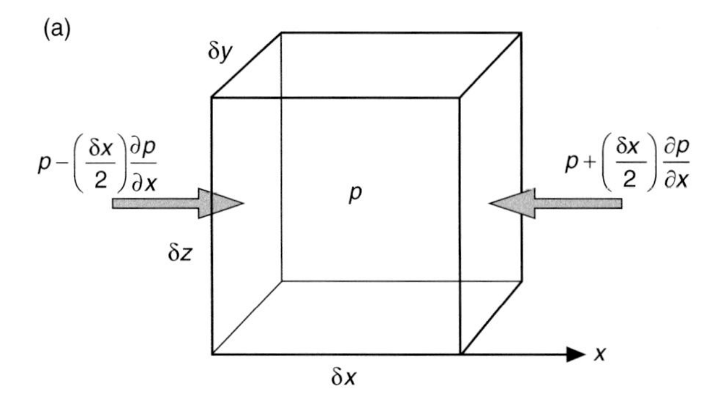

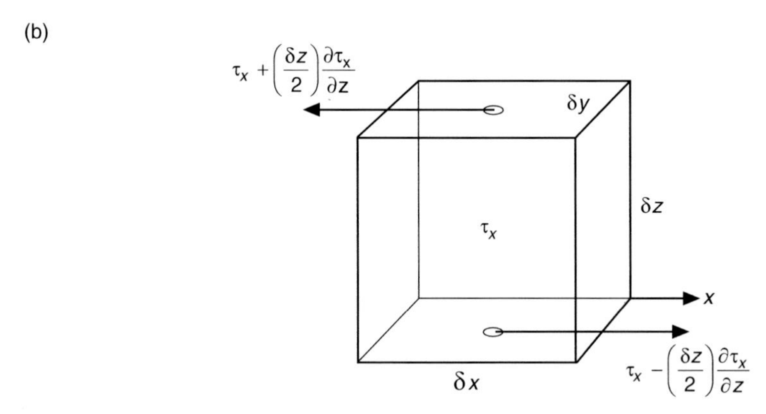

Figure 3.3 (a) Pressure and (b) shear forces acting on a cuboid in the x direction.

the net force on a particle of fluid, we need to know the gradients of pressure within the fluid. Figure 3.3a shows that the forces acting on a small cuboid of fluid in the x direction are $p_{LH}\ \delta y\delta z = (p - \delta x(\partial p/\partial x)/2)\ \delta y\delta z$ on the left-hand face and $p_{RH}\ \delta y\delta z = (p + \delta x(\partial p/\partial x)/2)\ \delta y\delta z$ on the right-hand face. These forces act in opposite directions, and if there is no gradient of pressure ($p_{LH} = p_{RH}$), there will be no pressure force in the x direction on the cuboid. In the presence of a gradient of pressure ($p_{LH} \neq p_{RH}$), the forces on the two faces will differ by:

$$(p_{LH} - p_{RH})\delta y\delta z = -\frac{\partial p}{\partial x}\delta x\delta y\delta z \equiv -\frac{\partial p}{\partial x} \text{ per unit volume.} \tag{3.11}$$

Similarly, the forces in y and z directions are determined by the y and z pressure gradients.

The effect of the frictional stresses on a fluid particle can be determined in a similar way. Consider again a small cuboid of fluid as shown in Fig. 3.3b. Think of this cuboid as situated in a horizontal flow in the x direction in which the velocity $u(z)$ varies with height (i.e. the flow is sheared in the z direction). A frictional stress between layers of fluid arises because they are moving relative to each other; a faster layer of water will tend to drag an adjacent slower layer along with it, transferring some of its momentum to the lower layer. In a laminar flow, the stress will be due simply to the molecular viscosity of the fluid. However, in the ocean, where the flow is often turbulent, much larger turbulent stresses are involved. For the moment we shall not distinguish between these forms of stress but represent the combined stress by $\tau_x(z)$ with the convention that this is the stress exerted by the fluid on the lower side of a plane on the fluid above it. With this definition, the bottom face of the cuboid in Fig. 3.3b experiences a force $(\tau_x - \delta z(\partial\tau_x/\partial z)/2)\delta x\delta y$ in the positive x direction. At the same time, the upper face of the cuboid exerts a force on the layer above it given by $(\tau_x + \delta z(\partial\tau_x/\partial z)/2)\delta x\delta y$. According to Newton's third law (action = reaction), the cuboid will experience an equal and opposite force in the negative x direction. The net force on the cuboid in the x direction will then be:

$$\left\{\left(\tau_x - \frac{\partial\tau_x}{\partial z}\delta z/2\right) - \left(\tau_x + \frac{\partial\tau_x}{\partial z}\delta z/2\right)\right\}\delta x\delta y = -\frac{\partial\tau_x}{\partial z}\delta z\delta x\delta y$$
$$\equiv -\frac{\partial\tau_x}{\partial z} \text{ per unit volume.} \tag{3.12}$$

Similarly, the force on a particle for flow in the y direction will be given by the analogous term $-\partial\tau_y/\partial z$.

3.2.4 The equations of motion (and hydrostatics)

We can now include both pressure and frictional stresses to state the equations of horizontal motion in x and y as:

$$\frac{Du}{Dt} = fv - \frac{1}{\rho}\left(\frac{\partial p}{\partial x} + \frac{\partial\tau_x}{\partial z}\right) + \frac{F_x}{\rho};$$
$$\frac{Dv}{Dt} = -fu - \frac{1}{\rho}\left(\frac{\partial p}{\partial y} + \frac{\partial\tau_y}{\partial z}\right) + \frac{F_y}{\rho} \tag{3.13}$$

where F_x and F_y now represent any additional forces which may be acting (e.g. tidal forces). We have included only the frictional stresses associated with vertical changes in the horizontal flow (termed vertical shear) and therefore neglected stresses due to horizontal shear in the flow. This is a reasonable simplification for much of the ocean, where conditions are laterally uniform, or nearly so, and vertical shear stresses dominate. In regions of rapid horizontal changes, additional stress terms will need to be included.

A dynamical statement analogous to the x and y equations is also available for the vertical dimension z, i.e.:

$$\frac{Dw}{Dt} = -\frac{1}{\rho}\frac{\partial p}{\partial z} - g. \tag{3.14}$$

It has a simpler form because, as explained above, the vertical component of the Coriolis force can be neglected and frictional forces acting in the vertical are usually small compared with g. In many situations, we can further simplify this equation by appealing to the fact that vertical velocities are small and consequently the vertical acceleration term Dw/Dt can be neglected in comparison with g. The only force in the vertical of comparable magnitude to g is then the pressure gradient term $\partial p/\partial z$ so that, to a good approximation, the force balance in the vertical is simply:

$$0 = -\frac{1}{\rho}\frac{\partial p}{\partial z} - g \text{ or } \frac{\partial p}{\partial z} = -\rho g. \tag{3.15}$$

This is the *hydrostatic approximation* in which the force of gravity on a particle is balanced by the Archimedean upthrust exerted on it by the pressure field. The equation is readily integrated to determine the pressure field of the ocean from the density field which is obtained from measurements of temperature and salinity profiles. In order to compute the pressure field accurately from density, we must take account of the smallest variations of density which we can resolve from temperature and salinity measurements. In practice, this means an accuracy in density of $\Delta\rho/\rho \sim 10^{-5}$, which requires temperature accurate to better than 0.0015 °C and salinity to better than 0.012 (PSS). This level of accuracy is not, however, necessary when we are dealing with density as part of the inertia term (i.e. mass × acceleration), in which case, it is usually good enough to use a fixed, reference density value[3] which we will denote by the symbol ρ_0. These different ways of treating density appear, at first sight, to be inconsistent but are readily justified and are widely used in what is referred to as the Boussinesq approximation (see Phillips, 1966). We shall frequently make use of this simplifying assumption in what follows.

The hydrostatic force balance (3.15), together with the x and y dynamical statements (Equation 3.13) and the continuity constraint (Equation 3.3) makes up the full set of four equations governing motion, termed the *equations of motion*. They have been derived under a particular set of assumptions (2D horizontal, incompressible, hydrostatic flow) and provide a suitable theoretical basis for most of the topics we shall discuss. However, you will appreciate that more general forms of the equations of motion may be needed when our assumptions are not applicable.

In general, the equations of motion give us four equations for the four unknowns (pressure and the three components of velocity) which can, in principle, be solved when friction can be neglected. If frictional stresses are important, then, to close the problem, we also need to know how to relate the turbulent stress components τ_x and τ_y to other properties of the flow. We shall return to the question of how to achieve this *turbulence closure* in Chapter 4 but, for now, we will proceed to examine some relatively simple solutions of the equations of motion in which the frictional stresses are neglected or specified in a very simple form.

3.3 Geostrophic flow

We start with the case of steady flow in which only the pressure gradient and the Coriolis force are involved and all other forces are excluded. The resulting balance leads to important relations between the velocity and density fields which are applicable to a wide range of situations in the ocean and the atmosphere.

[3] A commonly used value in shelf sea studies is $\rho_0 = 1026$ kg m^{-3}.

3.3.1 The dynamical balance

For a frictionless ocean ($\tau_x = \tau_y = 0$) in a steady state of motion ($\partial u/\partial t = \partial v/\partial t = 0$) with no external forces acting ($F_x = F_y = 0$), the linearised momentum equations (3.13) reduce to a balance between the pressure gradient and the Coriolis force:

$$fu = -\frac{1}{\rho_0}\frac{\partial p}{\partial y}; \quad -fv = -\frac{1}{\rho_0}\frac{\partial p}{\partial x} \tag{3.16}$$

where ρ_0 is the average density introduced in the last section. These are the equations of *geostrophic flow*. Squaring and adding the two relations gives:

$$f^2(u^2+v^2) = f^2C_g^2 = \frac{1}{\rho_0^2}\left(\left(\frac{\partial p}{\partial x}\right)^2 + \left(\frac{\partial p}{\partial y}\right)^2\right) = \frac{1}{\rho^2}\left(\left(\frac{\partial p}{\partial n}\right)^2\right) \tag{3.17}$$

which relates the current speed C_g to the magnitude of the pressure gradient $\partial p/\partial n$, i.e.:

$$C_g = \frac{1}{\rho_0 f}\frac{\partial p}{\partial n}. \tag{3.18}$$

The derivative $\partial p/\partial n$ here is taken in a direction n normal to the isobars which is the direction of maximum gradient, illustrated in Fig. 3.4. In much of the rest of this discussion we will use n as our general horizontal axis. The direction of the flow is at an angle ψ which can be found by dividing the x and y equations (3.16) to give:

$$\tan\psi = \frac{v}{u} = -\left(\frac{\partial p/\partial x}{\partial p/\partial y}\right). \tag{3.19}$$

This tells us something very important about the direction of a geostrophic flow. The direction of an isobar in the horizontal plane $(\partial y/\partial x)_p$ is set by the condition that

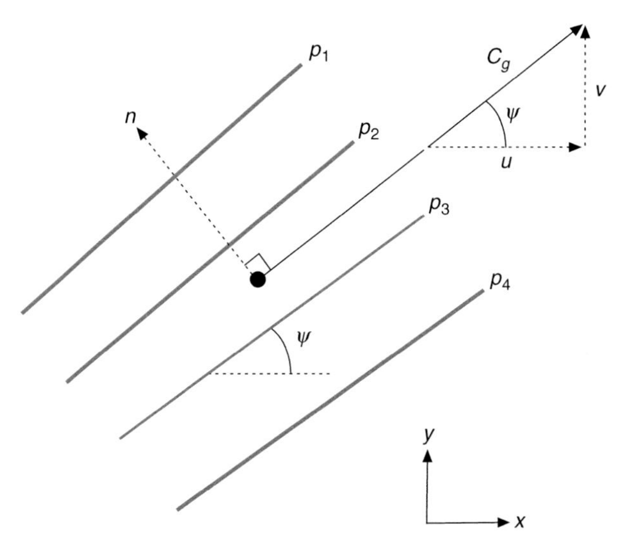

Figure 3.4 Geometry of isobars in the horizontal plane.

pressure is constant on an isobar, i.e. along an isobar changes in pressure in the x direction are balanced by changes in the y direction:

$$\delta p = \frac{\partial p}{\partial x}\delta x + \frac{\partial p}{\partial y}\delta y = 0$$
$$\text{so that } \left(\frac{dy}{dx}\right)_p = -\left(\frac{\partial p/\partial x}{\partial p/\partial y}\right) = \tan\psi. \tag{3.20}$$

Geostrophic flow is, therefore, parallel to the isobars, or perpendicular to the pressure gradient. The result is initially somewhat counterintuitive and in contrast to the situation in flows at the laboratory scale where fluid usually moves down the pressure gradient. Geostrophic flow, or a close approximation to it, occurs widely in the atmosphere and in the ocean. Equations (3.16) are extensively used in meteorology and oceanography to determine the velocity field from the pressure distribution. Before the advent of recording current meters, knowledge of the ocean currents was largely based on estimates of the geostrophic flow derived in this way from surveys of temperature and salinity.

3.3.2 The gradient equation

The pressure at a depth z, $p(z)$, arising from the vertical distribution of density through the overlying water column, is derived by integrating the hydrostatic relation:

$$p(z) - p_0 = \int_z^{\eta} \rho g dz = \int_z^{0} \rho g dz + \int_0^{\eta} \rho g dz$$
$$= \int_z^{0} \rho g dz + \rho_s g \eta \tag{3.21}$$

where p_0 is the atmospheric pressure acting at the surface, η is the surface elevation above mean sea level and ρ_s is the density at the surface which is assumed to be constant between $z = 0$ and $z = \eta$.

We next take the derivative to find the horizontal pressure gradient for two cases:

(i) The density ρ is constant or a function of z only

In this simple but important case, the density structure is the same everywhere in x and y and so the pressure gradient becomes:

$$\frac{\partial p}{\partial n} = \frac{\partial}{\partial n}\int_z^{0} \rho(z) g dz + g\rho_s \frac{\partial \eta}{\partial n} = g\rho_s \frac{\partial \eta}{\partial n}. \tag{3.22}$$

Note that the integral term is constant and has no variability in the horizontal, so it has a zero horizontal derivative. If we ignore any difference (typically $<1\%$) between

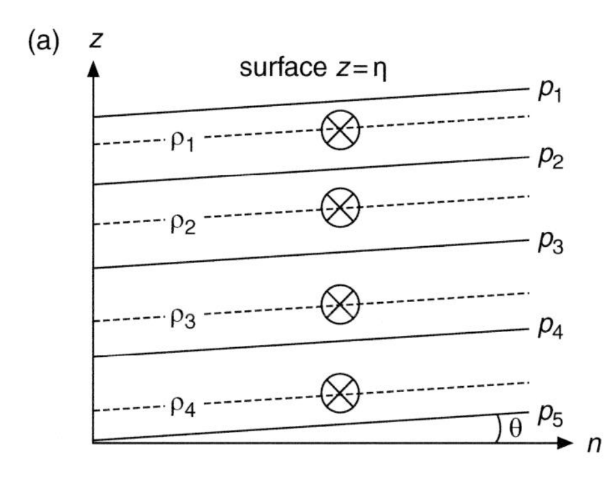

Figure 3.5 Isobars and isopycnals (dashed) in the vertical plane *n*-*z* where *n* is the direction of largest horizontal pressure gradient. In (a) density is constant or a function of *z* only so isobars, including that at the sea surface, are parallel to each other and to the isopycnals. This is the *barotropic* case in which flow is uniform in depth and directed into the paper (Northern Hemisphere).

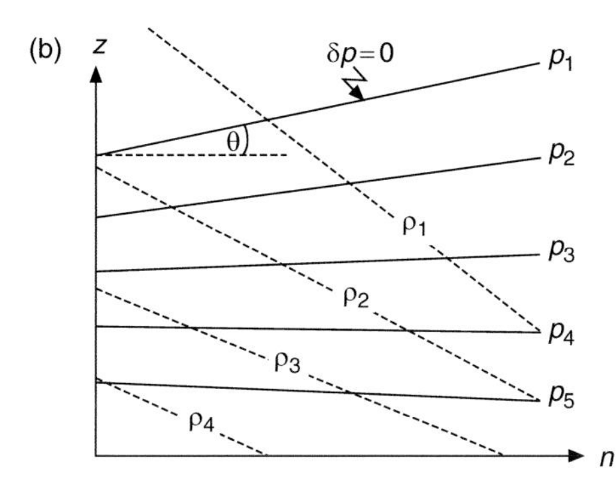

In (b) density is a function of horizontal position as well as depth. The isobars are inclined to the isopycnals and to each other in this *baroclinic* case.

the surface and the bulk water column density, i.e. set $\rho_s = \rho_0$, the geostrophic balance in Equation (3.18) then gives the current speed as:

$$C_g = \frac{1}{f\rho_0}\frac{\partial p}{\partial n} = \frac{g}{f}\frac{\partial \eta}{\partial n}. \tag{3.23}$$

We now see that C_g depends only on the surface slope and is the same at all depths. This situation is shown in Fig. 3.5a: when density is a function only of z, surfaces of equal pressure (isobars) are parallel to each other and parallel to surfaces of equal density (isopycnals). We refer to the flow as being *barotropic*.

(ii) The density ρ varies in x, y and z, i.e. $\rho = \rho(x,y,z)$

In this case, isobars may be inclined relative to each other and to the isopycnals, a situation of so-called *baroclinic* flow. Here it is most convenient to determine the pressure gradient by reference to the slope of isobars. In the vertical z-n plane shown in Fig. 3.5b, the slope of an isobar can be found from the conditions that the pressure is invariant on an isobar and obeys the hydrostatic condition. So as we move along a sloping isobar we can say that the pressure change will be:

$$\delta p = \frac{\partial p}{\partial n}\delta n + \frac{\partial p}{\partial z}\delta z = 0$$

$$\text{so that} \quad \left(\frac{dz}{dn}\right)_p = \tan\theta = -\frac{\partial p/\partial n}{\partial p/\partial z} = \frac{1}{\rho g}\frac{\partial p}{\partial n} \tag{3.24}$$

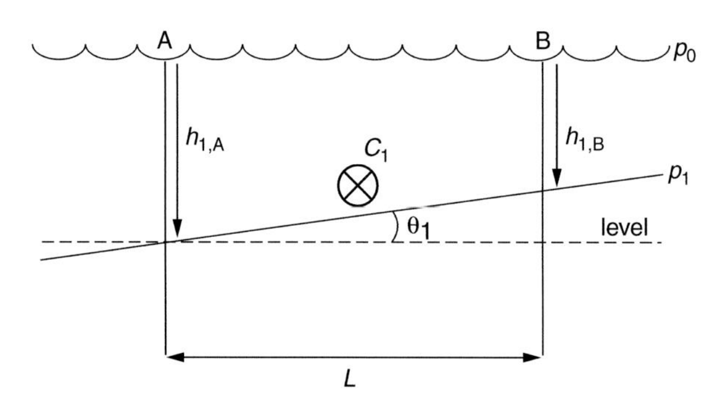

Figure 3.6 Application of the geostrophic method to calculate the relative velocity C_1 at the isobar p_1 between two stations A and B, with measured density profiles ρ_A and ρ_B.

where for the final step we have used the hydrostatic approximation (3.15). This means that the pressure gradient is $\rho g \times$ isobaric slope. Combining with the geostrophic relation (Equation 3.18) and setting $\rho = \rho_0$ (a fair approximation here), we have for the geostrophic current:

$$C_g = \frac{1}{f\rho_0}\frac{\partial p}{\partial n} = \frac{g}{f}\tan\theta. \tag{3.25}$$

So isobaric slopes are readily transformed into estimates of the geostrophic flow magnitude C_g by using Equation (3.25), which is known as the gradient equation. In the northern hemisphere (f positive), the flow is in a direction 90° anticlockwise from the pressure gradient vector; in the southern hemisphere, where f is negative, the flow direction is 90° clockwise from the pressure gradient. Note that, since the ocean surface is an isobar, Equation (3.23) for barotropic flow is a special case of the more general relation between velocity and isobaric slope represented by the gradient Equation (3.25).

Consider the practical application of the gradient equation, shown in the schematic of Fig. 3.6. If the density profile ρ has been measured at two stations A and B, the depth h_1 of an isobar p_1 at A and B can be determined by integrating the hydrostatic relation, e.g. at station A:

$$h_{1A} = \int_0^{p_1} \frac{dp}{\rho_A g} \tag{3.26}$$

and similarly for station B. If the stations are separated by a distance L, the slope of the p_1 isobar relative to the surface between A and B will be:

$$\tan\theta_1 = \frac{h_{1A} - h_{1B}}{L} \tag{3.27}$$

from which the velocity, relative to the surface and normal to the section A-B, can be derived using the gradient equation. Application of this approach to other isobars allows us to build up a profile of the geostrophic velocity relative to the surface and to obtain relative currents between any two isobars p_1 and p_2 from their relative slope θ_r as:

$$C_2 - C_1 = \frac{g}{f}(\tan\theta_2 - \tan\theta_1) \cong \frac{g}{f}\tan\theta_r \tag{3.28}$$

where the final approximation takes advantage of isobaric slope angles being generally $<<1°$. In order to complete knowledge of the geostrophic currents relative to the Earth, we need to know the slope of one isobar relative to the horizontal or the velocity on one isobar. In the past (before 1960), when direct measurements of current were rarely available, oceanographers relied heavily on geostrophic calculations to estimate ocean currents. In efforts to make geostrophic current estimates absolute rather than relative, it was often assumed, without proof, that there was a *level of no motion* at which the isobars were horizontal. In the shelf seas, comparison with current meter measurements indicates that assuming that there is no flow at the seabed is usually a reasonable approximation in determining geostrophic currents from the density field.

3.3.3 Thermal wind

An alternative (and somewhat simpler) approach to the application of the geostrophic balance is to calculate the so called *thermal wind*, a term which clearly comes from meteorology and indicates the origin of the method. The geostrophic balance Equations (3.16) are combined with the hydrostatic relation (Equation 3.15) to eliminate the pressure by cross-differentiating the equations. For example, using the x equation and the hydrostatic relation, we have:

$$\frac{\partial p}{\partial x} = \rho_0 f v \rightarrow \frac{\partial^2 p}{\partial z \partial x} = \rho_0 f \frac{\partial v}{\partial z}$$
$$\frac{\partial p}{\partial z} = -\rho g \rightarrow \frac{\partial^2 p}{\partial z \partial x} = -g\frac{\partial \rho}{\partial x}$$

so that:

$$\rho_0 f \frac{\partial v}{\partial z} = -g\frac{\partial \rho}{\partial x}. \tag{3.29}$$

Repeating the same procedure for the geostrophic balance in the y direction, the two components of the thermal wind shear can be written as:

$$\frac{\partial v}{\partial z} = -\frac{g}{\rho_0 f}\frac{\partial \rho}{\partial x}; \quad \frac{\partial u}{\partial z} = \frac{g}{\rho_0 f}\frac{\partial \rho}{\partial y}. \tag{3.30}$$

In order to deduce the velocity profile, these equations can be integrated upwards from the bed using the observed density gradients and the condition that $u = v = 0$ at the bottom ($z = -h$). As we shall see in Chapter 8, the thermal wind method provides us with an effective way of determining the mean baroclinic flow field in high gradient regions like fronts, even in the presence of oscillatory tidal currents.

3.4 Fundamental oscillatory motions: what a water particle does if you give it a push

We next examine two simple solutions of the dynamical equations which describe the motions that result when a particle is impulsively accelerated, i.e. when the particle is given a short, sharp push. This applies, for instance, to the pulse of wind stress felt by the sea surface as a storm passes. We need to consider separately the two cases of vertical and horizontal motion, both of which involve an oscillatory response at a fundamental frequency of the system.

3.4.1 Inertial oscillations

The ocean is again assumed to be frictionless, but this time we set the pressure gradient terms to zero and assume that we are dealing with a slab layer in which the motion does not vary with depth. The balance in the equations of motion is now between the acceleration of a particle and the Coriolis force acting on it:

$$\frac{Du}{Dt} = fv; \quad \frac{Dv}{Dt} = -fu \tag{3.31}$$

Differentiating the x equation with respect to time and substituting from the y equation gives:

$$\frac{D^2u}{Dt^2} = -f^2u. \tag{3.32}$$

This is an equation describing simple harmonic motion with angular frequency f. If the particle is given an initial velocity U along the x direction at $t = 0$, the subsequent motion is given by:

$$u = U_0 \sin ft; \quad v = \frac{1}{f}\frac{Du}{Dt} = U_0 \cos ft \tag{3.33}$$

which describes circular motion at a constant speed of U_0 and with an *inertial period*:

$$\boxed{T_I = \frac{2\pi}{f} = \frac{2\pi}{2\Omega \sin \phi_L} = \frac{12}{\sin \phi_L} \text{ hours}} \tag{3.34}$$

where Ω is the angular frequency of the Earth's rotation (see Section 3.2.1). So a particle, impulsively accelerated to a speed U, will describe circular motion with a period set by the latitude. Such circular motions are called *inertial oscillations*. The radius of the circular motion r_I is the perimeter of the circle divided by 2π:

$$r_I = \frac{U_0 T_I}{2\pi} = \frac{U_0}{f}. \tag{3.35}$$

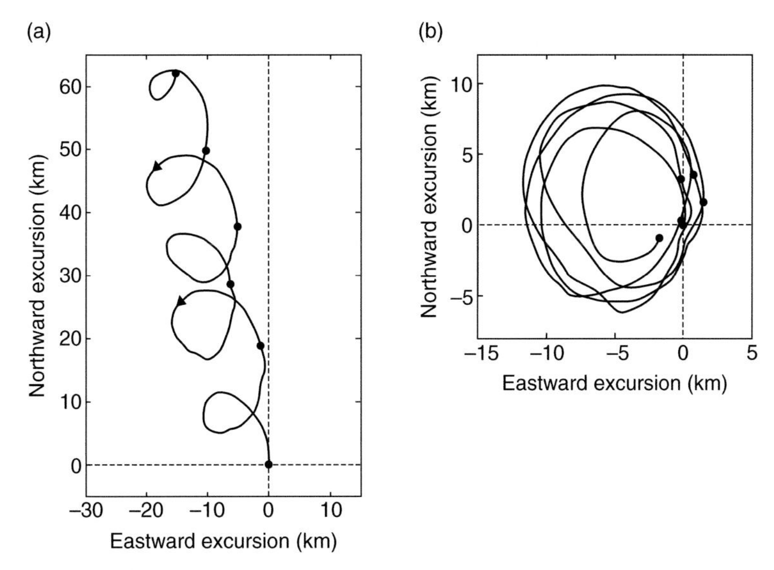

Figure 3.7 Oscillations at the inertial period. The track in (a) is a progressive vector diagram of the near-surface current over a five day period at a station on the Namibian shelf off SW Africa. The motion consists of anticlockwise rotating inertial oscillations superimposed on a mean flow. (b) is the vector diagram after the mean flow has been removed.

In order that the Coriolis shall balance the mass time acceleration in the circular motion, it must be directed inwards towards the centre of the circle. This means that inertial motions must be clockwise in the northern hemisphere and anticlockwise south of the equator. Circular motions at the local inertial frequency have been found to occur widely in the ocean in response to variations in wind forcing (Millot and Crepon, 1981; Schahinger, 1988). A short burst of wind stress will accelerate the surface mixed layer in the direction of the wind. When wind forcing ceases, the residual motion takes the form of inertial oscillations which, if frictional damping is weak, may continue for many cycles. Figure 3.7a shows an example from the southern hemisphere of inertial motions superimposed on a steady geostrophic flow. When the steady flow is subtracted, as shown in Fig. 3.7b, the anticlockwise circular motion with a diameter of ~10 km is clearly apparent. The large oscillations in this case apparently result from forcing by strong diurnal winds at a latitude close to 30° S where the inertial period is ~24 hours (Simpson *et al.*, 2002).

3.4.2 Water column stability and vertical oscillations

The last section demonstrates that if we give a water particle a push in the horizontal plane, it will oscillate at the inertial frequency f. An analogous natural frequency applies to vertical motions in a stratified water column. To see this, consider a water particle displaced upwards by an amount ζ in a stably stratified region of the ocean where the density gradient is $\partial\rho/\partial z$. Remember that the density gradient is a negative quantity for a stable structure in which density decreases upwards. The particle will have an excess density relative to its new surroundings of $\Delta\rho = -\zeta\,(\partial\rho/\partial z)$ and will experience a buoyancy force b on a unit volume of:

$$b = -g\Delta\rho = g\frac{\partial\rho}{\partial z}\zeta. \tag{3.36}$$

Imagine what happens to the water particle. It experiences this downward force, attempting to restore the particle to its stable position in the density profile. The acceleration will result in the particle achieving its maximum downward speed as it reaches this stable point, and so it overshoots deeper into the density profile. Now the particle is less dense than its surroundings, and so it is forced back upward in an attempt to reach its stable position. Again it overshoots, and the cycle repeats. Mathematically, if this is the only force acting, the motion of the particle, which has density ρ, will be described by:

$$\frac{Dw}{Dt} = \frac{D^2\zeta}{Dt^2} = \frac{g}{\rho}\frac{\partial\rho}{\partial z}\zeta = -N^2\zeta. \tag{3.37}$$

Again, this is the equation of simple harmonic motion, this time describing how the particle will oscillate vertically at the *buoyancy frequency* N given by:

$$\boxed{N = \sqrt{-\frac{g}{\rho}\frac{\partial\rho}{\partial z}}.} \tag{3.38}$$

Free, unforced vertical oscillations in the water column are restricted to frequencies lower than N and this sets an upper limit to the possible frequencies of internal waves, which are dealt with in Section 4.2.4. In the absence of friction the particle will simply keep oscillating; in reality we expect the oscillation to be damped as energy is lost to friction. The square of the buoyancy frequency N^2 is widely used as a measure of the stability of the water column, which, as we shall see in the next chapter, plays an important role in inhibiting turbulent motions.

3.5 Turbulent stresses and Ekman dynamics

Now we introduce friction, which plays a key role in the equations of motion. We will first approach this via a classical problem first solved by the Swedish oceanographer Vagn Walfrid Ekman in 1905 (Ekman, 1905). The problem arose from a question posed by Fridtjof Nansen who had observed that icebergs in the Arctic did not move in the same direction as the wind, but instead followed a path to the right of the wind direction.

3.5.1 Current structure and transport in the Ekman layer

Ekman asked what would be the form of the steady state current profile in an ocean forced by a steady wind if the only forces acting are frictional stresses between layers and the Coriolis force. In this scenario, the dynamical equations (3.13) simplify to:

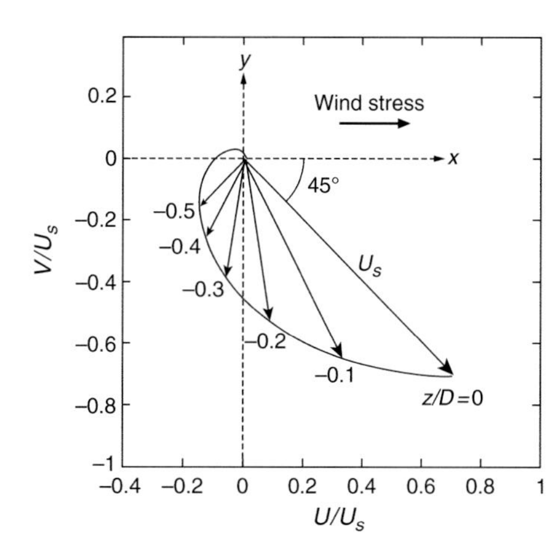

Figure 3.8 Velocity variation with depth due to a steady wind stress applied to an ocean remote from horizontal boundaries in the northern hemisphere. The hodograph shows the velocity vectors at different depths seen from above. The current decreases and rotates clockwise with increasing depth. Note that, at a depth of $z = -D$, where the current vector is reversed relative to the surface, the velocity and the frictional stress are reduced to $< 5\%$ of their values at the surface.

$$0 = fv - \frac{1}{\rho_0}\frac{\partial \tau_x}{\partial z}; \quad 0 = -fu - \frac{1}{\rho_0}\frac{\partial \tau_y}{\partial z}. \tag{3.39}$$

To make further progress we need to relate the frictional shear stresses τ_x and τ_y to other flow properties. Anticipating the discussion of frictional stresses in Section 4.3.4, we assume that the stresses are proportional to the velocity shear components, so that we can write:

$$\tau_x = -\rho_0 N_z \frac{\partial u}{\partial z}; \quad \tau_y = -\rho_0 N_z \frac{\partial v}{\partial z} \tag{3.40}$$

where N_z is known as the *eddy viscosity* and is here assumed to be constant. Substituting in Equation (3.39), we have:

$$fv = -N_z \frac{\partial^2 u}{\partial z^2}; \quad fu = N_z \frac{\partial^2 v}{\partial z^2}. \tag{3.41}$$

If there is a steady wind stress τ_w acting at the surface ($z = 0$) in the x direction, the solution of this pair of equations can be written as (Kundu and Cohen, 2008, see p. 617):

$$u = u_s e^{\pi z/D} \cos\left(\frac{\pi}{4} - \frac{\pi z}{D}\right); \quad v = -u_s e^{\pi z/D} \sin\left(\frac{\pi}{4} - \frac{\pi z}{D}\right)$$
$$\text{where } D = \pi\sqrt{\frac{2N_z}{f}}; \text{ and } u_s = \frac{\sqrt{2}\pi\tau_w}{\rho_0 f D}. \tag{3.42}$$

D is a length scale called the *Ekman depth*, which is used as an indication of the depth to which the influence of the surface stress penetrates. More specifically it is the depth at which the currents have diminished to $1/e$ of their surface magnitude given by u_s. The profile of current vectors in this near surface region, termed the *Ekman layer*, is illustrated in Fig. 3.8. You can see that as well as decreasing away from the surface, the current vector also rotates to the right (left) in the northern (southern) hemisphere so that at $z = -D$ the current is reversed in the direction relative to the surface. At the surface, the u and v components are equal and positive so the flow is at an angle of $\pi/4$ to the right (northern hemisphere) of the wind direction, a result that was in qualitative agreement with Nansen's iceberg observations.

Projecting the tip of the current vector at each level into the horizontal plane generates the famous *Ekman spiral* pattern, shown as the continuous curve in Fig. 3.8. This idealised form of flow has rarely been observed, partly because of the difficulty of measuring currents in the near-surface layers of the ocean, but also because the strong assumptions of the Ekman calculation are not usually satisfied. In particular the eddy viscosity is probably far from constant, especially if the ocean has varying stratification.

One very important result of Ekman's theory, which is independent of N_z, concerns the vertically integrated transport. If we return to Equations (3.39) and assume that $h >> D$ so that $\tau_x = \tau_y = 0$ at $z = -h$, then integration with respect to z gives:

$$\begin{aligned} f\int_{-h}^{0} v dz = fV = \frac{1}{\rho_0}\int_{-h}^{0}\frac{\partial \tau_x}{\partial z}dz = \frac{\tau_w}{\rho_0} \\ f\int_{-h}^{0} u dz = fU = \frac{1}{\rho_0}\int_{-h}^{0}\frac{\partial \tau_y}{\partial z}dz = 0. \end{aligned} \tag{3.43}$$

This means that the integrated volume transport is entirely in the y direction (i.e. perpendicular to the wind stress) and has magnitude, referred to as the *Ekman transport*, of $V = \tau_w/\rho_0 f$ (units: $m^2\ s^{-1}$). This fundamental result on wind driven transport expresses the fact that the wind stress at the surface is balanced by the Coriolis force summed over depth. In order that the resultant Coriolis force opposes the wind stress, the flow integrated over depth must be at right angles to the wind direction. This Ekman transport in response to wind forcing plays a major role in ocean circulation and underpins many conceptual models of the processes operating in the upper ocean.

3.5.2 The bottom Ekman layer

Currents near the seabed involve frictional stresses which result in a bottom layer structure closely analogous to that occurring in the surface layer. In this case the stresses are not forced externally but arise from the drag of the bottom boundary on the near-bed flow. As an example of the dynamics in the bottom layer, consider a steady flow u_g which, well away from the bottom boundary, is directed in the x direction and is in geostrophic balance, i.e.

$$fu_g = -\frac{1}{\rho_0}\frac{\partial p}{\partial y}. \tag{3.44}$$

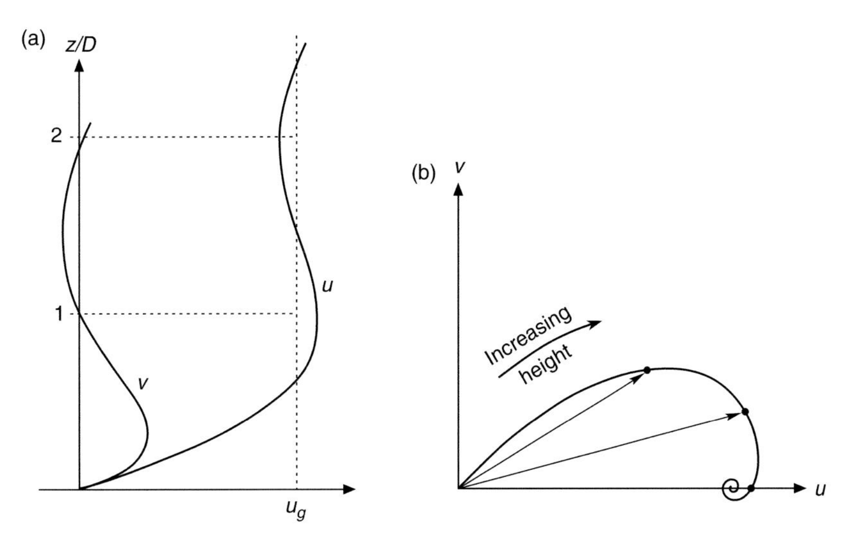

Figure 3.9 Ekman veering caused by bed friction acting on a geostrophic flow. (a) Vertical profiles of the u and v components of current; (b) the hodograph of the current vector. Adapted from (Kundu and Cohen, 2008), courtesy of Academic Press.

In the bottom layer, the x and y equations of motion for steady flow are just:

$$\begin{aligned} 0 &= fv - \frac{1}{\rho_0}\frac{\partial \tau_x}{\partial z} \\ 0 &= -fu - \frac{1}{\rho_0}\frac{\partial \tau_y}{\partial z} - \frac{1}{\rho_0}\frac{\partial p}{\partial y} = -fu - \frac{1}{\rho_0}\frac{\partial \tau_y}{\partial z} + fu_g \end{aligned} \tag{3.45}$$

where we have used Equation (3.44) to substitute for the pressure gradient. Expressing the stresses above in terms of the velocity shear and an eddy viscosity (Equation 3.40), we have:

$$\begin{aligned} fv + N_z\frac{\partial^2 u}{\partial z^2} &= 0 \\ -f(u - u_g) + N_z\frac{\partial^2 v}{\partial z^2} &= 0. \end{aligned} \tag{3.46}$$

For boundary conditions we have: at the bottom boundary $u = v = 0$ at $z = 0$. Far above the bottom, the motion is just the geostrophic flow $u = u_g$; $v = 0$ at $z = \infty$.

With these conditions, the solution to Equations (3.46) is (e.g. Kundu, 1990, see p. 622):

$$u = u_g\left(1 - e^{-\pi z/D}\cos(\pi z/D)\right); \quad v = u_g e^{-\pi z/D}\sin(\pi z/D) \tag{3.47}$$

which is in the form of the steady current plus a spiral similar to that which we saw at the surface. The spiral rotates clockwise with increasing height above bottom (northern hemisphere). A plot of the boundary currents (steady current + spiral) from Equation (3.47) is shown in Fig. 3.9. Close to the bed and continuing up to a height of $z = D$, the current is deflected to left of the upper layer flow. This deflection of the near-bed flow, known as *Ekman veering*, is widely observed in the ocean although it is generally less than indicated by our simple model with typical values 5~15° (Kundu, 1976; Weatherly and Martin, 1978). Veering also occurs in the atmosphere and is apparent in weather charts as a deflection of the surface wind relative to the isobars.

It is apparent in Fig. 3.9 that there is a component of transport in the positive y direction. Integrating Equation (3.47), we find this net transport to have the simple form:

$$\int_0^\infty v dz = \int_0^\infty u_g e^{-\pi z/D} \sin(\pi z/D) dz = \frac{u_g D}{2\pi}. \tag{3.48}$$

This flow perpendicular to the high level current can be thought of as the result of the weakening of the bottom current by boundary friction leaving an unbalanced component of the pressure gradient which drives the cross-stream flow. We shall see in Chapter 10 that this transverse flow in the bottom boundary layer can play a major part in downslope transport at the shelf edge where, outside the boundary layers, the flow is constrained to be parallel to the depth contours.

3.5.3 Response to the Ekman transport at a coastal boundary

Our derivation of the Ekman transport is based on the assumption that there are no horizontal pressure gradients, which is reasonable in the open ocean. Near to the coast, however, there can be no transport normal to the coastal boundary and so pressure gradients will develop in response to either onshore or offshore Ekman transport. Imagine a wind blowing parallel to the shore in the northern hemisphere with the land to the right of the wind direction. In this case the Ekman transport will be towards the coast and will act to pile up water against the coast, thus raising sea level and producing an opposing pressure gradient. As a result of this pressure gradient there will be a tendency for water to downwell at the coast and return seawards in the lower layers if the water is deeper than the thickness of the Ekman layer. Conversely, if the wind is in the opposite direction, surface water will move offshore, sea level will fall at the coast and the pressure gradient will act to promote onshore flow in the lower layers with upwelling of water near the coast. We shall examine the dynamics of response to wind forcing in these situations, starting with the simplest one-layer model.

One-layer model of the barotropic response

Consider the case where the transport is onshore as in Fig. 3.10a. We shall assume that the coast is straight, that the density ρ is constant and that conditions are uniform in the y (northward) direction. The flow is further assumed to be independent of depth, so we seek a solution of the linearised, vertically averaged equations of motion. Setting $Du/Dt = \partial u/\partial t$ etc. in Equations (3.13), integrating over depth and applying our assumptions, we have for the dynamical equations:

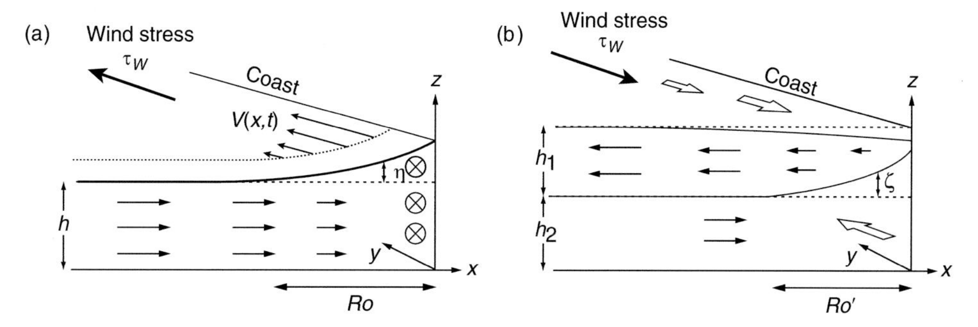

Figure 3.10 (a) Response of a homogeneous ocean in the northern hemisphere to a coast-parallel wind stress; (b) Upwelling, in a two-layer ocean in the northern hemisphere, forced by coast-parallel wind stress.

$$\frac{\partial \hat{u}}{\partial t} = f\hat{v} - g\frac{\partial \eta}{\partial x}; \quad \frac{\partial \hat{v}}{\partial t} = f\hat{u} + \frac{\tau_w}{\rho_0 h}; \tag{3.49}$$

where $\hat{u}, \hat{v}$ are the depth average velocities. The corresponding, vertically averaged version of the continuity Equation (3.3) is just:

$$\frac{\partial \eta}{\partial t} = -h\frac{\partial \hat{u}}{\partial x}. \tag{3.50}$$

For a wind stress τ_w switched on at $t = 0$, the solution of Equations (3.49) and (3.50) takes the form (e.g. see Gill, 1982, p. 396):

$$\begin{aligned} \hat{u} &= \frac{\tau_w}{f\rho_0 h}(1 - e^{x/Ro}); \quad Ro = \frac{\sqrt{gh}}{f} \\ \hat{v} &= \frac{\tau_w}{\rho_0 h} t e^{x/Ro}; \quad \eta = \frac{\tau_w}{\rho_0 c} t e^{x/Ro}. \end{aligned} \tag{3.51}$$

The Ekman transport, initially $h\hat{u} = \tau_w/(\rho_0 f)$ far from the coast, starts to decrease at an offshore distance determined by the parameter R_o and becomes zero at the coast ($x = 0$). Ro is an important length scale. It is the *Rossby radius of deformation*, and controls the extent of boundary influence. It can be thought of as the distance travelled in a time of $1/f = T_I/2\pi$ by a long wave at the phase speed of $c = \sqrt{gh}$. This makes intuitive sense, as the wave speed is the fastest that any signal, for instance a change in wind forcing, can be transmitted through the ocean. In mid-latitude shelf seas Ro is typically ~250 km.

While the shoreward transport is steady, the alongshore current and the surface elevation grow linearly in time to absorb the onshore flow which is converging at the coast. The pressure gradient due to the surface slope is also increasing with time and remains in geostrophic balance with the increasing alongshore flow.

This solution is a good model of what happens to the sea surface elevation after the onset of a strong wind stress and a storm surge develops at the coast. Of course, sea level does not go on rising indefinitely because the coastal current is eventually limited by frictional forces at the seabed which come into balance with the applied surface stress.

Two-layer model of upwelling and downwelling (the baroclinic response)

Our one-layer model assumes that the density is uniform and that the currents are barotropic. In order to see how the Ekman transport drives upwelling and downwelling motions near the coastal boundary, which will be important in addressing exchange between the ocean and the shelf in Chapter 10, we require a similar, slightly more elaborate model with two layers; a lighter warmer layer overlying a cold denser layer, as illustrated in Fig. 3.10b. Each layer has depth-uniform velocity and density and the interface between them is horizontal, at time $t = 0$, when the wind stress is switched on. The linearised dynamical equations for the top and bottom layers, again assuming the alongshore terms involving $\partial/\partial y$ are zero, take the form:

$$\begin{aligned} &\text{upper layer:} \quad \frac{\partial u_1}{\partial t} = fv_1 - g\frac{\partial \eta}{\partial x}; \quad \frac{\partial v_1}{\partial t} = -fu_1 + \frac{\tau_w}{\rho_0 h_1} \\ &\text{lower layer:} \quad \frac{\partial u_2}{\partial t} = fv_2 - g\frac{\partial \eta}{\partial x} - g'\frac{\partial \varsigma}{\partial x}; \quad \frac{\partial v_2}{\partial t} = -fu_2 \end{aligned} \tag{3.52}$$

where ς is the upward displacement of the interface between the layers and $g' = (\Delta\rho/\rho_0)g$ is the *reduced gravity* associated with the density difference $\Delta\rho$ between the layers. The surface pressure gradient can be eliminated by taking the difference between the corresponding equations for layers 1 and 2 to give:

$$\frac{\partial \tilde{u}}{\partial t} = f\tilde{v} + g'\frac{\partial \varsigma}{\partial x}; \quad \frac{\partial \tilde{v}}{\partial t} = -f\tilde{u} + \frac{\tau_w}{\rho_0 h_1} \tag{3.53}$$

where $\tilde{u} = u_1 - u_2$ and $\tilde{v} = v_1 - v_2$. The continuity equations for the two layers can be written as:

$$\begin{aligned} -\frac{\partial \varsigma}{\partial t} + h_1\frac{\partial u_1}{\partial x} = 0 \\ \frac{\partial \varsigma}{\partial t} + h_2\frac{\partial u_2}{\partial x} = 0 \end{aligned} \tag{3.54}$$

which can be combined to give:

$$-\left(\frac{h_1 + h_2}{h_1 h_2}\right)\frac{\partial \varsigma}{\partial t} + \frac{\partial \tilde{u}}{\partial x} = 0 \tag{3.55}$$

The solution to Equations (3.53) and (3.55) is then (see Gill, 1982, p. 404):

$$\begin{aligned} &\tilde{u} = \frac{\tau_w}{f\rho_0 h_1}\left(1 - e^{x/Ro'}\right); \quad Ro' = \frac{c'}{f} = \frac{1}{f}\sqrt{\frac{g' h_1 h_2}{h_1 + h_2}} \\ &\tilde{v} = \frac{\tau_w}{\rho_0 h_1} t e^{x/Ro'_0}; \quad \varsigma = \frac{\tau_w f Ro'}{\rho_0 g' h_1} t e^{x/R'_0} \end{aligned} \tag{3.56}$$

which has the same form as the barotropic case but with the important difference that the cross-shore length scale is now the *internal Rossby radius* $Ro' = c'/f$ in which $c' = \sqrt{g'h_1h_2/(h_1 + h_2)}$ is the speed of long internal waves. As $c'/c \ll 1$, Ro' is much smaller than Ro with values of order $Ro' \sim 10$ km.

With wind stress directed to the south (τ_w is negative, as in Fig. 3.10b), there is a steady offshore movement in the surface layer which results in an increase in the interface height ς at the coast and out to a distance of a few times Ro'. If the wind stress is maintained, this upwelling motion will continue, eventually exposing the cold lower layer at the surface and reducing the sea surface temperature. At the same time, nutrients, which are usually much more abundant in the lower layer, are brought to the surface with important implications for primary production which we will discuss in Chapter 10.

3.6 Long waves and tidal motions

As we noted in Section 2.5, most tidal energy input to the shelf seas does not come from the direct action of the tidal generating force but is delivered to the shelf from the deep ocean in the form of shallow water waves; i.e. waves with wavelengths large compared with the water depth. Also, we know that while the open ocean tidal range is ~1 metre, in shelf seas we see tidal ranges of up to an order of magnitude larger. Clearly something interesting happens to the tidal waves as they progress from the deep ocean into the shelf seas. Because of their importance in this context, we shall develop the theory of long waves, starting from the equations of motion for the case of waves in a non-rotating system. We shall then extend the theory to a rotating system and show how the shelf seas respond to the incoming tidal waves from the deep ocean. The introduction to the basic concepts of wave motions here should also provide a useful foundation for the fuller discussion of surface and internal wave motions which follows in Chapter 4.

3.6.1 Long waves without rotation

We start with the relatively simple case of waves of long period T_p and wavelength λ which is large compared with the depth. For such waves, the vertical acceleration Dw/Dt is small compared with g so that the pressure is given by the hydrostatic relation in Equation (3.15). We shall further assume:

(i) the density of the fluid is constant
(ii) the mean water depth h is constant
(iii) the waves propagate in the x direction so that $\partial/\partial y$ terms are zero
(iv) the motion is in a non-rotating frame (Coriolis terms can be neglected)
(v) the motions are small, so the equations can be linearised ($D/Dt \approx \partial/\partial t$)

With these assumptions, the x momentum Equation (3.13) and the continuity Equation (3.3) reduce to:

$$\frac{\partial u}{\partial t} = -g\frac{\partial \eta}{\partial x}; \quad \frac{\partial \eta}{\partial t} = -h\frac{\partial u}{\partial x}. \tag{3.57}$$

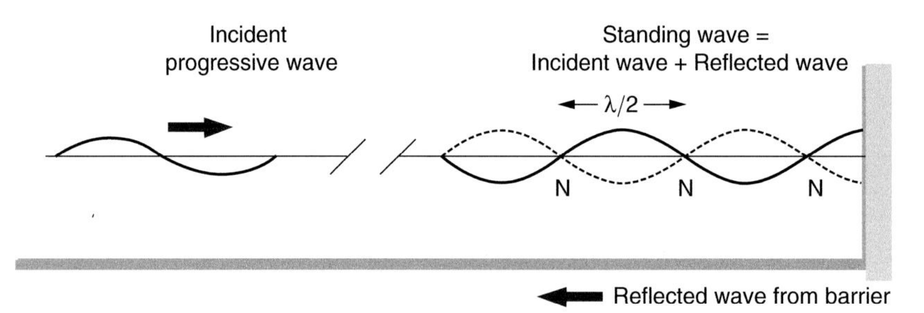

Figure 3.11 A progressive sine wave propagating from left to right is reflected at the barrier to produce a second wave of equal amplitude moving in the opposite direction. The two waves combine to produce a standing wave pattern with nodes (N) at intervals of $\lambda/2$. The nodes are points of zero surface displacement but maximum current amplitude. Notice that the first node is located at a distance $\lambda/4$ from the barrier.

Since η does not depend on z, it follows that the horizontal velocity u must be the same at all depths. Differentiating the first of these equations with respect to x and the second with respect to t, we can eliminate u to give the long wave equation:

$$\boxed{\frac{\partial^2 \eta}{\partial x^2} = \frac{1}{c^2}\frac{\partial^2 \eta}{\partial t^2}; \quad c^2 = gh.} \tag{3.58}$$

This equation describes waves travelling at a speed of $c = \sqrt{gh}$ which is independent of the wave frequency and wavelength. While the general solution of Equation (3.58) is any function varying as $(x \pm ct)$, we are mainly concerned with harmonic solutions such as:

$$\eta = A_0 \sin(kx \pm \omega t) \tag{3.59}$$

where the wave number $k = 2\pi/\lambda$, the angular frequency $\omega = 2\pi/T_p$, phase velocity $c = \pm\, \omega/k = \pm\, \lambda/T_p$ and A_0 is the wave amplitude. The $-$ and $+$ signs in (3.59) correspond to a sinusoidal wave travelling in the positive and negative x directions respectively. The particle velocity u is found by substituting for η in either of the Equations (3.57) and integrating to give:

$$\begin{aligned} u_f &= \frac{gk}{\omega} A_0 \sin(kx - \omega t) = \frac{g}{c} A_0 \sin(kx - \omega t) \\ u_b &= -\frac{g}{c} A_0 \sin(kx + \omega t) \end{aligned} \tag{3.60}$$

for the forward and backward travelling waves respectively. The general solution for waves travelling in the x direction along a channel of uniform depth is a combination of a forward and a backwards travelling wave with amplitudes A_f and A_b respectively:

$$u = u_f + u_b = \frac{g}{c}\{A_f \sin(kx - \omega t) - A_b \sin(kx + \omega t)\}. \tag{3.61}$$

As an example which will prove useful in our discussion of tidal waves in shelf seas, consider the case of a wave approaching a barrier as shown in Fig. 3.11. If the channel ends in a vertical barrier at $x = 0$, the velocity there must be zero at all times. Setting $u = 0$ at $x = 0$ in Equation (3.61) requires that $A_b = -A_f$ and the expressions for elevation and velocity can then be written as:

$$\begin{aligned} \eta &= A_f\{\sin(kx - \omega t) - \sin(kx + \omega t)\} = -2A_f \cos kx \sin \omega t \\ u &= A_f \frac{g}{c}\{\sin(kx - \omega t) + \sin(kx + \omega t)\} = 2A_f \frac{g}{c} \sin kx \cos \omega t. \end{aligned} \tag{3.62}$$

So the incident wave moving towards the barrier from the left in Fig. 3.11 is reflected back, with the reflected wave then interfering with the incident wave. The two *progressive waves* combine to form a *standing wave* with double the amplitude of the incident wave and nodes of elevation (points on the wave with zero range) for $\cos kx = 0$ (i.e. at $x = \lambda/4, 3\lambda/4, 5\lambda/4, \ldots$). The corresponding nodes of velocity (which are anti-nodes, or maxima, for η) occur for $\sin kx = 0$ ($x = 0, \lambda/2, \lambda, 3\lambda/2, \ldots$).

Waves of tidal period moving on to the shelf from the deep ocean travel at the long wave speed of $c = \sqrt{gh}$ which decreases from $c \sim 200\ \mathrm{m\ s^{-1}}$ in the deep ocean ($h \sim 4$ km) to $c \sim 30\ \mathrm{m\ s^{-1}}$ on the shelf ($h \sim 100$ m). Remember that a wave's wavelength, period and speed are related by $\lambda = cT$. The corresponding wavelengths at the main tidal period (12.42 hours) are 9000 km for the deep ocean and 1400 km for the shelf, so tidal waves certainly satisfy our initial assumption that $\lambda \ll h$. Tsunami waves, which are forced by movements of the seabed during earthquakes, also travel at $c = \sqrt{gh}$. In the deep ocean, with $h \sim 4000$ metres, the tsunami speed is $200\ \mathrm{m\ s^{-1}}$. These waves can, therefore, cross the ocean basins in less than a day.

3.6.2 Long waves with rotation (Kelvin waves)

The tide generating force acting over the large extent of the ocean basins drives the deep ocean tides which travel around the oceans in the form of long waves. The waves bringing tidal energy on to the shelf propagate to the coastal boundaries, where they are reflected and form standing wave patterns of the kind we have just discussed. There is, however, an important difference in the dynamics of tidal waves on the shelf from our simple non-rotating model. Because of the very large scales involved in the tidal long waves, we cannot ignore the influence of Coriolis force which modifies the structure of long waves and produces more subtle reflection patterns. Our non-rotating waves moved in the x direction and were uniform in the transverse y direction, so we needed only to consider the x momentum equation. When rotation becomes important, we need also to consider the dynamical balance in the y direction. Particle motion in the x direction involves a Coriolis force fu acting in the y direction. For a wave travelling in the x direction parallel to a coastal boundary, as shown in Fig. 3.12, this Coriolis force induces a balancing pressure gradient in the form of a sea surface slope normal to the coast, i.e.:

$$fu = -\frac{1}{\rho_0}\frac{\partial p}{\partial y} = -g\frac{\partial \eta}{\partial y}. \tag{3.63}$$

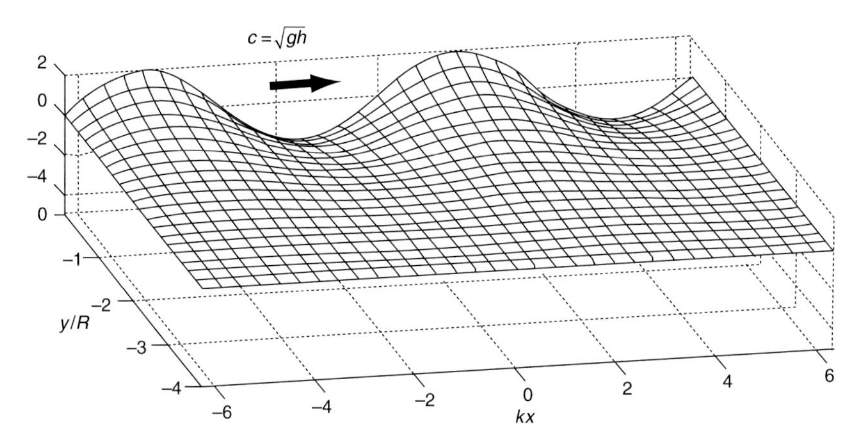

Figure 3.12 Schematic representation of a Kelvin wave travelling parallel to a vertical land boundary located along the axis $y = 0$. Note that, in this southern hemisphere case, the land is to the left of the direction of wave propagation. The amplitude of the waves decreases exponentially away from the boundary on a length scale set by the Rossby radius Ro so that at $y = -4Ro$ the wave amplitude is <2% of its value along the land boundary.

In the x direction, the dynamical balance remains unchanged from (3.58) so the waves again travel at the phase speed of $c = \sqrt{gh}$ but we now seek a solution for a progressive harmonic wave, which varies in y, of the form:

$$\eta = F(y)\sin(kx - \omega t) \tag{3.64}$$

where $F(y)$ is some function varying in the y direction only. The relation between u and η remains as in Equation (3.60) so that the particle velocity is given by

$$u = \frac{g}{c}\eta = \frac{g}{c}F(y)\sin(kx - \omega t). \tag{3.65}$$

Using Equations (3.64) and (3.65), the geostrophic balance (Equation 3.63) in the y direction then requires that:

$$\frac{\partial F(y)}{\partial y} = -\frac{f}{c}F(y). \tag{3.66}$$

$F(y)$ must then be of the form:

$$F(y) = e^{-fy/c} = e^{-y/Ro} \tag{3.67}$$

where $Ro = c/f$ is the external Rossby radius which we defined in Equation (3.51) during our discussion of coastal upwelling. The forward travelling wave is now described by:

$$\boxed{\eta = A_f e^{-y/Ro}\sin(kx - \omega t); \quad u = \frac{g}{c}A_f e^{-y/Ro}\sin(kx - \omega t).} \tag{3.68}$$

The resulting shape of this wave is shown in Fig. 3.12 for the case of the southern hemisphere. Notice for the southern hemisphere that the wave travels with the coast to its left. In the northern hemisphere, the wave direction would be reversed. The amplitudes of surface elevation and particle velocity of the wave both increase exponentially in the y direction. Waves of this kind are referred to as *Kelvin waves* and are widely used to represent tidal motions in shelf seas. The amplitude of the wave has its maximum value A_f at the boundary ($y = 0$) and decreases in the negative y direction to $A_f/e \simeq 0.37A_f$ at $y = -Ro = -\sqrt{gh}/f$, i.e. at a distance from the coast which is ~250 km for a typical shelf depth (h = 100 m) in midlatitudes. In the deep ocean, where h ~4000 m, the scale Ro is increased to ~1600 km.

3.6.3 Amplification and reflection of the tide

Tidal energy, input by the tide-generating forces acting on large basins of the deep ocean, travels around the ocean basins in the form of very long Kelvin waves (λ ~8000 km). At the shelf edge, it is transmitted on to the shelf, still in the form of Kelvin waves but travelling at a slower speed as h decreases. We shall see in Chapter 4 that the energy of these waves moves at the wave speed c, and that for a wave of amplitude A_0 the energy flux onto the shelf is given by $\boldsymbol{E}_w c$, where $\boldsymbol{E}_w = \frac{1}{2}\rho_0 g A_0^2$ is the energy density of the wave. Consider what happens to a wave travelling in the deep ocean with amplitude A_d, changing to amplitude A_s after crossing onto the shelf. As the waves moves into shallower water, c decreases, so, in order to maintain a steady energy flux, $\boldsymbol{E}_w$ must increase and with it the wave amplitude. The amplitude on the shelf will then be related to its value in deep water by:

$$\frac{A_s}{A_d} = \left(\frac{h_d}{h_s}\right)^{1/4}. \quad (3.69)$$

Taking h_s~100 metres and h_d ~4000 metres to be the corresponding depths, the amplitude of the tidal elevation therefore increases by a factor of ~2.5 as it crosses onto the shelf. The particle velocity amplitude is $|u| = A_0 g/c = A_0\sqrt{g/h}$ (Equation 3.68) so the ratio of shelf to deep water velocities is just

$$\frac{u_s}{u_d} = \left(\frac{h_d}{h_s}\right)^{1/2}\frac{A_s}{A_d} = \left(\frac{h_d}{h_s}\right)^{3/4}. \quad (3.70)$$

This implies a much larger amplification factor of ~16.

The incoming waves are also reflected at the coastline, and the combination of incident and reflected waves generates a standing wave component as in Equation (3.62) and Fig. 3.11 but with the added complication that we are dealing with Kelvin waves. Let us think about what happens to a Kelvin wave when it enters a rectangular gulf with width B_G and aligned in the x direction with $y = 0$ along the central axis. We can write a complete description of the motion by combining two Kelvin waves travelling in opposite directions:

$$\begin{aligned}\eta &= A_f e^{-y/Ro}\sin(kx - \omega t) + A_b e^{+y/Ro}\sin(kx + \omega t)\\ u &= \frac{g}{c}\left\{A_f e^{-y/Ro}\sin(kx - \omega t) - A_b e^{+y/Ro}\sin(kx + \omega t)\right\}.\end{aligned} \quad (3.71)$$

Notice that, if the gulf is narrow with $B_G \ll Ro$, then the exponential factors tend to unity ($e^{-y/Ro} \simeq e^{y/Ro} \simeq 1$) and (3.71) reverts to the non-rotating case (3.62).

3.6.4 Amphidromic systems

The combination of two Kelvin waves describes the motion well except in a small region close to the head of the gulf where the reflection of the incident wave occurs (a complication to which we shall return in a moment). Away from the head of the gulf, we can represent the tidal regime of a gulf simply in terms of the two Kelvin waves. The results are conveniently presented as a pattern of lines of equal amplitude (co-range lines) and lines of equal phase (co-tidal lines). Think of the co-tidal lines as representing all points at which each phase of the tide, e.g. high water (HW), will occur at the same time.

We can locate the co-tidal lines by choosing a particular condition for the phase of the tide in Equation (3.71). The condition for HW is that there is a maximum in η which means that $\partial\eta/\partial t = 0$. Applying this condition in Equation (3.71) for the case of complete reflection when $A_b = A_f$ leads to a relation between y and x for locations at which HW occurs at time $t = t_{HW}$:

$$\frac{\tan h(y/Ro)}{\tan kx} = \tan(\omega t_{HW}). \tag{3.72}$$

You can see in Fig. 3.13a the resulting pattern of co-tidal lines in the northern hemisphere. The lines (at intervals of 1 lunar hour) radiate outwards from central points termed *amphidromes* which are located where the nodal lines would be for a standing wave without rotation. As with the standing wave nodes, the amphidromic points are separated by $\lambda/2$. Notice that the time of high water (and any other phase of the tide) rotates around the amphidrome in an anticlockwise (clockwise) sense for the northern (southern) hemisphere.

The range of the tide at any position within the gulf can also be found by manipulating the expression for η in (3.71) and taking the modulus to obtain:

$$|\eta| = 2A_f\{\cos h^2(y/Ro)\sin kx + \sin h^2(y/Ro)\cos kx\}^{1/2}. \tag{3.73}$$

Contours of the range of the tide (the co-range lines) are shown in the lower panel of Fig. 3.13a. As there is no energy loss in this frictionless case, the amplitudes of incident and reflected waves are equal and the pattern is symmetrical about the centre of the channel with no tide ($\eta = 0$) at the amphidromic points to which the co-tidal lines converge. We can think of the Kelvin wave moving into the gulf hugging the coast to the right (northern hemisphere), reflecting off the head of the gulf, and then leaving the gulf still maintaining the coast on the right.

What happens if we make the gulf a little more realistic and allow friction to extract energy from the tidal wave? With increasing friction, energy is dissipated and the reflected wave is weaker than the incident wave so that $A_b < A_f$. This is the case

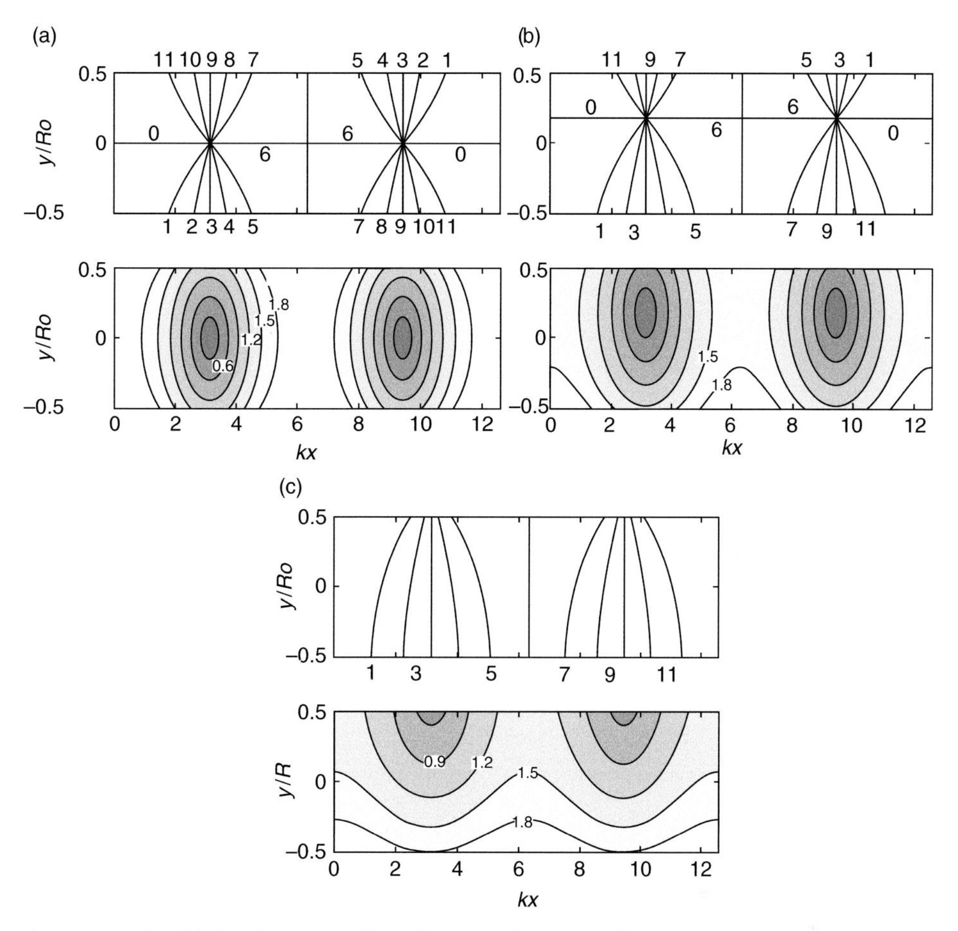

Figure 3.13 Co-tidal and co-range lines for a combination of two Kelvin waves travelling in opposite directions along a channel which has a width of one Rossby radius *Ro*. Top panels shows co-tidal lines along which phase is uniform, at intervals of $T_c/12$ where T_c is the period of the tidal constituent. Lower panels show the range of the tide relative to $2A_f$ the range of the incident Kelvin wave which is proceeding from left to right. In (a) the incident wave is fully reflected so that $A_b = A_f$; (b) is an example of partial reflection due to frictional losses of wave energy; (c) shows how the amphidrome becomes degenerate if frictional losses are high.

in Fig. 3.13b where you can see that the pattern becomes asymmetric for the case of $A_b/A_f = 0.7$ with the amphidrome being displaced to the left side of the gulf looking into the gulf in the northern hemisphere. In some situations, where frictional losses are large, the displacement may result in the amphidrome being located on land, in which case it is referred to as a *degenerate amphidrome*. Such an extreme displacement is illustrated in Fig. 3.13c for a case where $A_b/A_f = 0.25$.

Before we compare the Kelvin wave theory with examples of real co-tidal charts we should return to the question, which we deferred, of what happens at the landward end of the gulf where the Kelvin wave is reflected. Here account must be taken of the condition of no flow at the landward boundary. A famous analytical solution for the case of complete reflection in this region was obtained from some difficult mathematics by

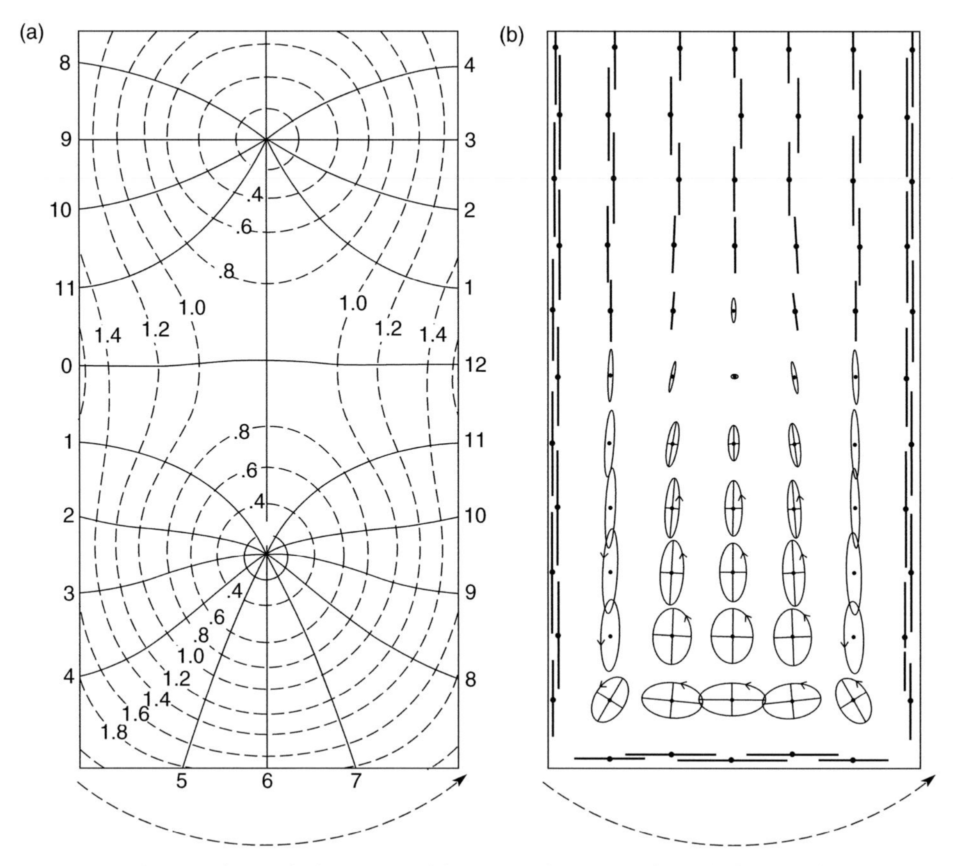

Figure 3.14 G. I. Taylor's solution to a Kelvin wave reflecting at the end of a gulf, from (Taylor, 1922), courtesy of the London Mathematical Society. (a) shows the co-phase (solid) and co-range (dashed) lines, (b) shows the tidal current ellipses. Numbers on the outside of (a) indicate the times of the co-phase lines in intervals $T/12$; co-range contours are labeled with a measure of the tidal range. The mouth of the gulf is at the top, and rotation is anticlockwise.

G. I. Taylor (Taylor, 1922). You can see the results in Fig. 3.14. There is some significant modification of the co-range lines extending to a distance of $\sim \lambda/2$ from the end of the gulf. Along the landward boundary the range increases by ~60% relative to the next antinode. The tidal currents in the first amphidromic system from the head of the gulf are also modified. In contrast to the simple Kelvin wave model, there are substantial cross-gulf flows as indicated by the more circular *tidal current ellipses* in this region.

An example of an amphidromic system in the real ocean is shown in Fig. 3.15 for the Yellow Sea, a large, shallow gulf between China and Korea driven by the ocean tide which enters from the East China Sea. This gulf can be regarded (with some imagination!) as being approximately rectangular with a length of ~1000 km. For the main lunar tidal constituent M_2, Fig. 3.15a shows that there are four clearly defined amphidromes within the gulf spaced at intervals of $\sim \lambda/2$, i.e. half a tidal wavelength. In each amphidromic system the tide rotates in an anticlockwise sense and, as in our Kelvin wave model, the phase changes 180° between adjacent systems, so that when it is HW on the coast of China at one amphidrome it is low

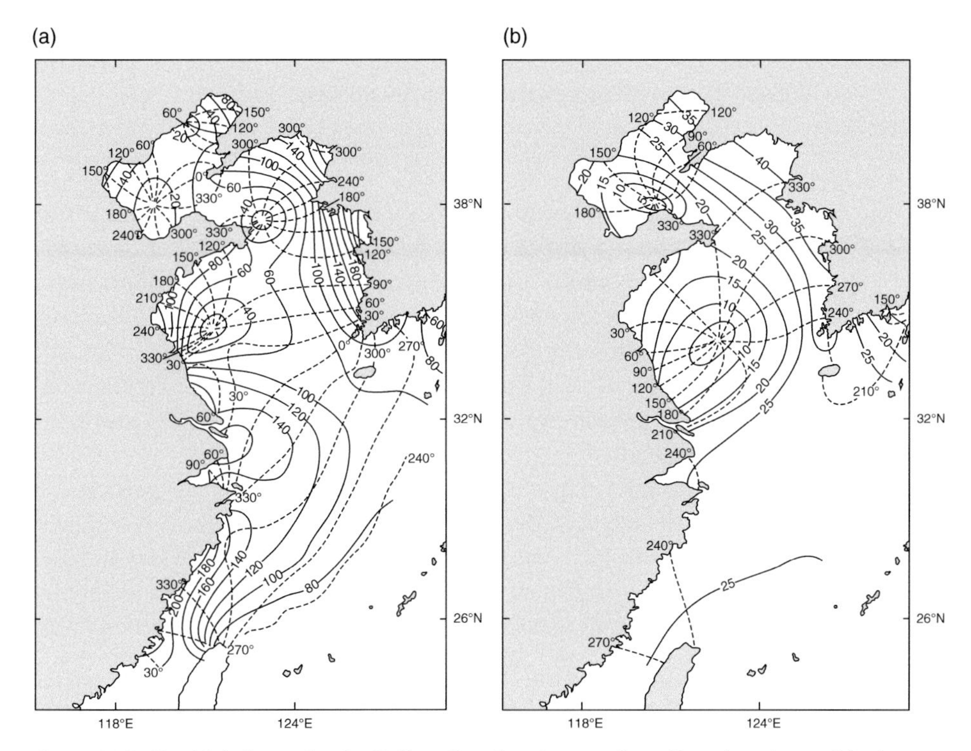

Figure 3.15 Co-tidal charts for the Yellow Sea showing co-phase lines (continuous) in degrees and co-range lines (dashed) in cm for (a) the M_2 semi-diurnal constituent and (b) for the K_1 diurnal constituent. Notice that the spacing of the amphidromes in (b) is approximately twice that in (a) because the period and wavelength of the diurnal waves are greater by a factor of ~2. From (Choi, 1980) with permission of the Korean Ocean Research and Development Institute.

water farther along the coast opposite the next. Note also that there is a marked asymmetry across the gulf with generally larger tides on the Korean coast because of amphidromes being displaced to the left when looking into the gulf (again in accord with the Kelvin wave model).

The pattern of the K_1 diurnal constituent is shown in Fig. 3.15b. The period and hence the wavelength of the tide is approximately twice that of semi-diurnal constituents and with it the spacing of the amphidromes. Consequently, the diurnal constituent exhibits only two amphidromes within the gulf. Again the response on the Korean coast is larger as the amphidromes are displaced over to the Chinese side of the gulf.

Box 3.3 Tidal current ellipses

Tidal currents generally involve two components of velocity u and v which combine together to move water particles around near-elliptical paths. For an individual constituent (say M_2 with angular frequency ω_{M2}), the components take the form of cosine functions of time t with different amplitudes (u_{M2} and v_{M2}) and phases (δ_u and δ_v):

$$u = u_{M2}\cos(\omega_{M2}t + \delta_u); \quad v = v_{M2}\cos(\omega_{M2}t + \delta_v)$$

The corresponding particle displacements are:

$$X = \frac{u_{M2}}{\omega_{M2}}\sin(\omega_{M2}t + \delta_u); \quad Y = \frac{v_{M2}}{\omega_{M2}}\sin(\omega_{M2}t + \delta_v).$$

Combining these sinusoidal displacements generates a particle trajectory which is an exact ellipse, as shown in Fig. B3.3. The values of the constituent amplitudes and phases control both the orientation θ and the shape of the ellipse which can vary from a circular motion (clockwise or anticlockwise) to oscillation back and forth along a straight line path (referred to as a 'degenerate ellipse'). The 'ellipticity' of the particle trajectory is the ratio of the minor axis OB to the major axis OA with the convention that positive/negative values correspond to anticlockwise/clockwise motion around the elliptical path.

The term 'tidal ellipse' can be used to denote either the particle orbit, as described above, or the locus of the tip of the velocity vector as at it rotates during the tidal cycle since both have the same shape (see Fig. B3.3). Tidal ellipses can be defined either by their constituent amplitudes and phases or alternatively by clockwise and anticlockwise rotating components. Relations between the two forms of representation are given in Pugh (1987).

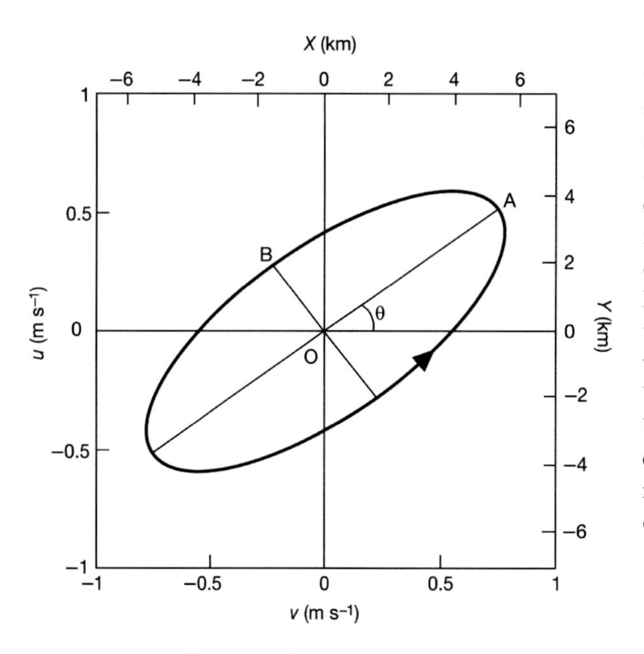

Figure B3.3 A tidal ellipse for the M_2 tidal constituent showing the major and minor axes (OA and OB respectively) and the orientation θ. The ellipse can be thought of as representing the displacement of a particle as shown on the km scales (right and top) or the track of the tip of velocity vector in the u-v plane as shown by the velocity scales (left and bottom). In this case the ellipse represents anticlockwise motion with an ellipticity of +0.4.

3.6.5 Tidal resonance

It should be clear now that the effect of rotation is to transform each node of the standing wave in the non-rotating case into an amphidrome. The amphidromic

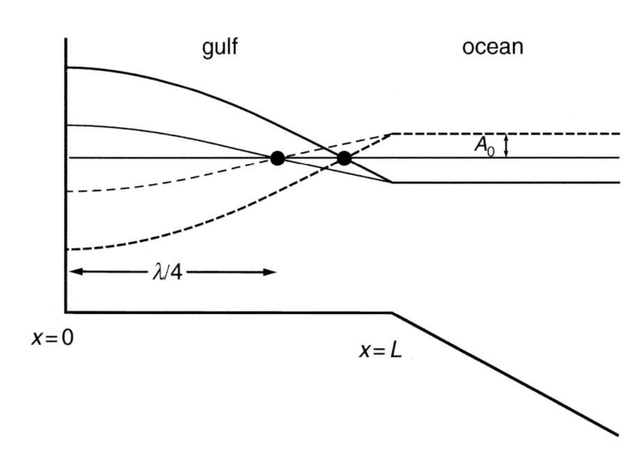

Figure 3.16 Tidal response of a narrow gulf showing the ocean tide being matched by a standing wave oscillation. A single node is located at $\lambda/4$ from the head of the gulf and initially the amplification of the tide is limited (grey curves). If the nodal point was located closer to the mouth, matching to the ocean tide would require a large increase in the standing wave response (black curves).

systems are spaced at intervals of $\lambda/2$ along the axis of the gulf, and the number of amphidromes within the gulf is dictated by its length and its depth (remember – depth controls the speed of the tidal wave, and so the tidal wavelength). At the open boundary of the gulf, the standing wave must match the ocean tide. We can demonstrate how this matching works for the case of a standing wave without rotation (Equation 3.62) which would apply to a narrow gulf. Figure 3.16 shows a section along the axis of such a narrow gulf which is long enough to contain one node of a standing wave. If the ocean tide is a sine wave of amplitude A_0, matching to a standing wave at the mouth of the gulf ($x = L$) requires that:

$$A_0 \sin \omega t = A_{sw} \cos kL \sin \omega t \tag{3.74}$$

so that the amplitude of the standing wave within the gulf will be:

$$A_{sw} = \frac{A_0}{\cos kL} = \frac{A_0}{\cos 2\pi L/\lambda}. \tag{3.75}$$

Equation (3.75) implies an interesting possibility. What if we change the length of the gulf so that it just happens that $L = \lambda/4$? The denominator in Equation (3.75) goes to zero, suggesting an infinite response of the wave amplitude inside the gulf. This is called *resonance*, and it suggests we might expect to see incredibly large tidal range and currents within the gulf. In practice the resonant motions would become limited by strong frictional effects and the ocean tide would be modified by the vigorous motion in the gulf (Arbic and Garrett, 2010). It also seems likely that any basin approaching resonance would experience very large sediment transport which would modify the bathymetry and move the system away from resonance. However, there are a number of tidal systems around the world that are close to resonance. A clear example is that of the Bay of Fundy, where the length of the gulf is $\sim\lambda/4$ for the main lunar tide so that there is a single node close to the ocean boundary. The result is a near-resonant response, which is responsible for the largest tidal range (~16 metres at spring tides) in the global ocean. Note also from Equation (3.75) that similar resonant responses will occur in gulfs whose length is close to $3\lambda/4$, $5\lambda/4$, $7\lambda/4$ etc.

At this point you can see why tides in shelf seas are often so much larger than the open ocean tide. The tidal amplification is the result of two processes: first, the very large increase in wave height and tidal currents, by factors of ~2.5 and ~16 respectively, as the tide crosses into the shallower water of the shelf while conserving energy; and second, the possibility of further amplification governed by the proximity of the geometry of a semi-enclosed shelf region to resonance with the tidal wave.

3.7 Tidally averaged residual circulation

Having looked at the basic responses of the shelf seas to forcing by wind, tides and density differences, we might ask at this point: what is the effect of the combined responses in terms of the net movement of water particles? In other words, what is the extent of the *residual circulation*? Most of the movement induced by tidal forcing is oscillatory in nature and does not result in a significant net displacement when the flow is averaged over a tidal cycle. Similarly, wind forcing is highly variable in direction and so the net effect of wind-induced movements tends to be greatly reduced by averaging over a period of months. On the other hand, horizontal density gradients, if sustained, can drive significant geostrophically balanced residual flows. In addition, energetic tidal motions can result in rectified currents through processes involving the non-linear terms in the equations of motion.

Few shelf sea areas have been subjected to the intensive long-term study with current meters needed to properly determine their residual circulation. Even where measurements have been made, there can be difficulties in resolving small residual flows in the presence of strong tidal flows. The alternative to direct current measurement is to determine the residual currents on the basis of tracer distributions. For example, Fig. 3.17 shows the distributions of Caesium Cs^{137}, a radionuclide with a half life of ~30 years which is released from the Sellafield nuclear processing plant on the northwest coast of England. The contours of this tracer indicate a residual flow northward from the Irish Sea and passing around the north of Scotland and on into the North Sea. This observed Cs^{137} distribution has been used in conjunction with a numerical model to infer the details of the residual flows (Prandle, 1984). With the exception of a few relatively energetic flows at the shelf edge and along some coasts where density gradients are large, the long-term average currents over much of this area are found to be weak, typically < 2 cm s^{-1}. Such small flows are consistent with the limited contribution of horizontal transport to the heat budget, which we found in Section 2.2.4. Together, these results provide a basis for the assumption, which we shall make in later chapters, that vertical exchanges predominate over horizontal fluxes in the control of water column structure. We will see in Chapter 8 that stronger residual flows can be associated with fronts in shelf seas; while still much weaker than the tidal currents, these consistent residual currents can play an important role in the transport of, for instance, fish larvae.

Consideration of these weak residual flows within the shelf sea interior also allows us to lay to rest a commonly held misconception about circulation through the seas off Western Europe. Low average flow rates imply long residence times for water particles. For example, the time for a water particle to pass through the North Sea is

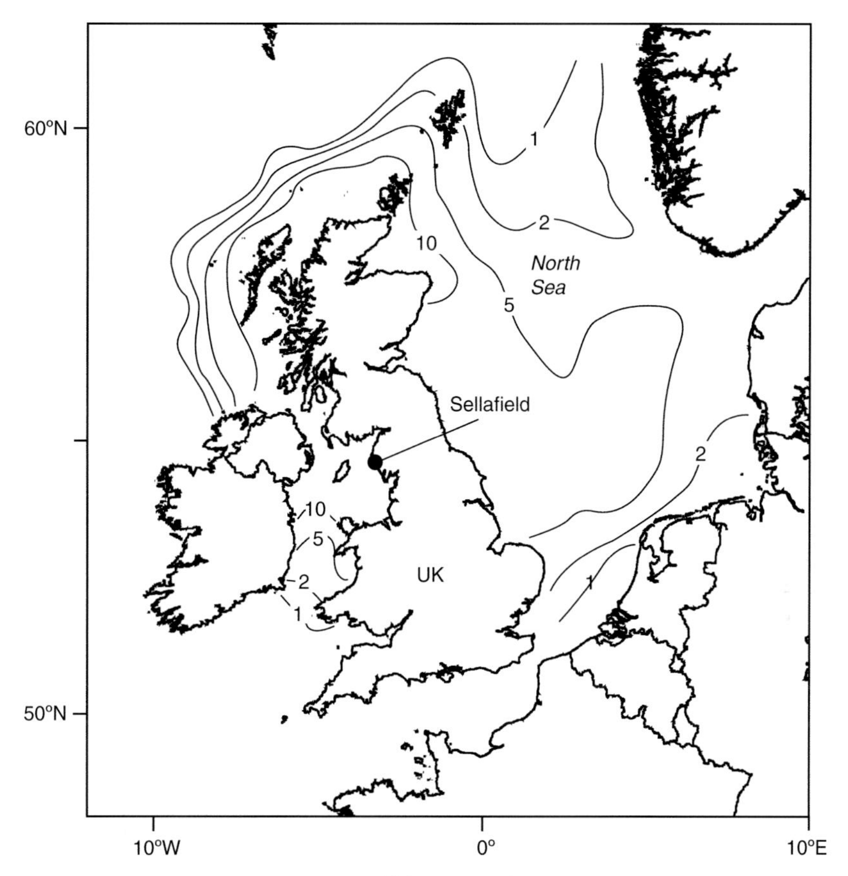

Figure 3.17 Observed distribution of ^{137}Cs (pCi l^{-1}) around the UK in August 1979. The source of the radioisotope is waste water from the Sellafield nuclear plant in NW England. After Prandle, 1984, courtesy of the Royal Society, London.

~3 years. For the Irish Sea the transit time is ~1–2 years, which corresponds to an average northward flow of <0.5 cm s^{-1}. Such weak net flow, originally inferred by Bowden (1950) and later confirmed by direct measurement (Brown and Gmitrowicz, 1995), is not consistent with the erroneous view that the Gulf Stream flows through the Irish Sea to directly warm the west coast of the UK.

Summary

In response to the forcing mechanisms, the shelf seas move in a way that is controlled by the fundamental laws of fluid motion which arise from the principles of continuity and Newton's laws of dynamics. The dynamics require that the equations be set up in a way that allows for the rotation of the Earth and enables us to work in coordinates which are fixed relative to the Earth. When this is done, an additional virtual force (the Coriolis force) is found to play a key role in all large-scale dynamics in the ocean and the atmosphere.

We next considered how the other internal (friction) and external (e.g. wind stress at the sea surface, horizontal pressure gradients) forces are included in the equations of motion and proceeded to examine a series of simplified force balances which illustrate some of the basic forms of motion observed in the ocean starting with geostrophic currents in which the forces can be simplified to a balance between horizontal pressure gradients and the Coriolis force. We looked next at the effect of a sudden horizontal push on the ocean, showing how oscillatory inertial currents are set up as the initial flow is diverted by the Earth's rotation. Oscillatory flows can also be set up vertically in a stratified water column when water is displaced away from its equilibrium position, with the rate of the oscillation being the buoyancy frequency.

Driving the sea surface with a steady wind leads to the classical Ekman spiral form of the vertical profile of currents. An important result of the theory is that the net transport is at right angles to the wind direction. Wind-driven dynamics were then extended to the case where the Ekman transport interacts with a coastal boundary to produce upwelling and downwelling motions; remember here that the 'up' and 'down' refer to the movement of the thermocline, which will have implications for nutrient and carbon fluxes to and from shelf seas. Ekman effects are also seen at the seabed boundary where they result in a deviation of the current vector to the left (northern hemisphere) and a net transport normal to the direction of the overlying geostrophic flow.

We then considered the question of why tides in shelf seas are so much larger than in the open ocean. The tidal wave is generated in the deep sea and increases in amplitude upon crossing into the shallow water on the shelf. If the wave encounters a gulf or semi-enclosed sea with the right geometry (length and depth), then the tides can resonate inside the gulf, leading to the exceptionally large tides seen in coastal regions such as the Bay of Fundy and over parts of northwest Europe. The Earth's rotation plays a role in how tides propagate around the shelf seas, leading to the observed amphidromic systems and, when frictional losses are incorporated into the analysis, an explanation for the differing tidal ranges on opposite sides of a gulf. Finally we considered what is known of the long-term average currents in the shelf seas and found that, at least for the Northwest European shelf seas, these flows are remarkably small so that the residence times for water particles in the shelf seas tend to be a year or more.

FURTHER READING

Gill, A. E., *Atmosphere-Ocean Dynamics*, Cambridge University Press.

Kundu, P. K., and I. M. Cohen, *Fluid Mechanics*, 4th edn, Academic Press. This edition includes some excellent video material of fluids processes.

Pugh, D. *Tides, Surges and Mean Sea Level*, John Wiley.

Problems

3.1. A cubic metre of near-surface water in the core of a steady current at latitude 33° N is moving to the northeast (045° T) at a speed of 1.75 ms^{-1}.
Determine:

(i) The Coriolis parameter in this latitude
(ii) The magnitude and direction of the Coriolis force acting on the water particle
(iii) The magnitude and direction of the slope of the sea surface assuming the current is in geostrophic balance.

3.2. Explain how the horizontal components of the Coriolis force are balanced in inertial oscillations and show that the period of these oscillations T_f is equal to $2\pi/f$ where f is the Coriolis parameter. Estimate T_f and the scale of the inertial circle at latitude 32° S if the current speed is 0.27 ms^{-1}.

3.3. (i) A steady wind blows over the open ocean remote from land boundaries at latitude 55° N. If the wind stress is 0.5 Pa and directed to the east, what will be the magnitude and direction of the vertically integrated transport? Explain which forces are involved and any assumptions you make.
(ii) If a wind of this magnitude is switched on abruptly, what will be the form of the motion in the approach to a steady state? What other 'forces' are involved in the transient motion?

(HINT: Check your answer to part (i) and investigate the transient response of part (ii) by using the Ekman3 option at the website.)

3.4. A rectangular gulf of length 900 km, breadth 150 km and depth 60 metres co-oscillates with the adjacent ocean. If the amplitude of the M_2 constituent at the ocean boundary is 1.2 metres, estimate:

(i) The range at the head of the gulf.
(ii) The position of the nodes in the gulf.
(iii) The maximum level difference across the gulf if it is situated in latitude 35° S.

Describe the tidal regime in the gulf and make a schematic plot of the co-tidal and co-range lines.

3.5. How does energy dissipation by bottom friction influence the position of amphidromes in the shelf seas? Estimate the ratio of reflected and incident amplitudes of the two Kelvin waves in a rectangular gulf of depth 46 metres at latitude 57° N if the single amphidrome in the gulf is displaced by 47 km from the axis of the gulf.

4 Waves, turbulent motions and mixing

In this chapter we shall look at waves and turbulence, two forms of motion which are of particular importance in the shelf seas because of their roles in bringing about the mixing which re-distributes properties such as heat, salt, momentum and substances dissolved or suspended in the water. There is a marked contrast in character between the two: waves are generally highly ordered motions which are amenable to precise mathematical description, while turbulence is chaotic in nature and it can usually only be represented in terms of its statistical properties. Both waves and turbulence are large scientific topics in their own right and are the subject of more than a few specialised textbooks. Here, we shall focus on those aspects of surface waves, internal waves and turbulence theory which are necessary to the understanding of processes in shelf seas, and we shall leave the more specialised aspects for the interested student to pursue from the further reading list.

4.1 Surface waves

We have already developed the theory of long waves in Chapter 3 from the basic equations and shown how such waves can help us to understand tidal motions in shelf seas. The more general theory of surface wave motions, in which there is no restriction on wavelength, is more involved and a full treatment is beyond the scope of this book. Here we shall simply present the assumptions and the principal results of the theory of infinitesimal waves and give a physical description of the motion involved. As well as being useful in themselves, the results of surface wave theory introduce us to many of the concepts relevant to the understanding of the more complicated motions involved in internal waves.

4.1.1 The first order velocity potential

The starting point is the recognition that the waves are irrotational, which means that, if we were able to freeze (i.e. make solid) a small water particle anywhere in the wave field, we would find that it was not rotating. In fluid mechanics, this essential

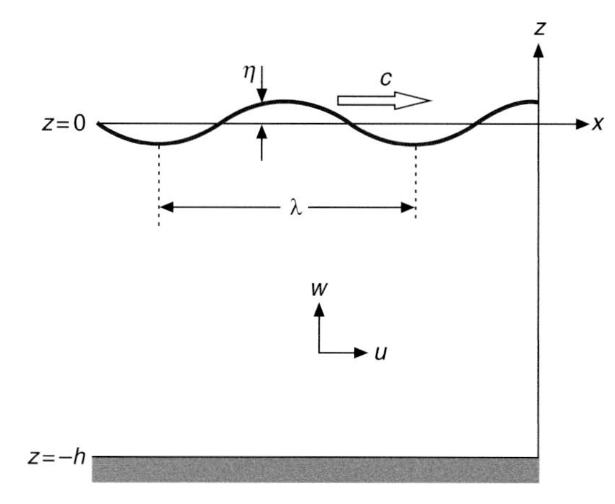

Figure 4.1 Definition of parameters for a surface wave.

fact is expressed by the statement that the *vorticity* of the fluid is zero. Vorticity is a vector quantity with components in x, y and z. For a wave travelling in the x direction, for example, the y component of vorticity is $\frac{\partial u}{\partial z} - \frac{\partial w}{\partial x} = 0$. The significance of the waves having zero vorticity is that the velocity field may then be expressed as the derivative of a potential function,[1] i.e. we can write:

$$\underset{\sim}{u} = (u, v, w) = \left(-\frac{\partial \phi}{\partial x}, -\frac{\partial \phi}{\partial y}, -\frac{\partial \phi}{\partial z}\right) = -\nabla \phi. \tag{4.1}$$

The analysis is then directed to find the form of ϕ which satisfies the continuity Equation (3.1) which, on substituting, becomes:

$$-\left(\frac{\partial^2 \phi}{\partial x^2} + \frac{\partial^2 \phi}{\partial y^2} + \frac{\partial^2 \phi}{\partial z^2}\right) = -\nabla^2 \phi = 0. \tag{4.2}$$

This is a well known equation in maths and physics called Laplace's equation. We can set up the problem as shown in Fig. 4.1. We choose the x axis to be the direction of propagation of a plane wave which is uniform in the y direction so that the $\partial/\partial y$ terms are zero. As well as Laplace's equation, the velocity potential ϕ must also satisfy the boundary conditions:

(i) at the bottom boundary ($z = -h$) flow cannot pass through the seabed:

$$w = -\frac{\partial \phi}{\partial z} = 0;$$

(ii) at the surface any vertical motion will be seen as changes in sea level:

$$w = -\left(\frac{\partial \phi}{\partial z}\right)_\eta = \frac{\partial \eta}{\partial t};$$

[1] A potential function is one from which a vector field (e.g. velocity) can be derived by taking the gradient of the function. If such a representation of the vector field is possible, then there is a mathematical requirement that vorticity or its equivalent is zero.

(iii) A third boundary condition arises from the dynamics in the form of the time dependent linearised Bernoulli equation for irrotational flow (e.g. Kundu and Cohen, 2008) which relates simultaneous changes in pressure, p, and the time derivative $\partial\phi/\partial t$ of the potential function:

$$\frac{p - p_0}{\rho} = -\frac{\partial \phi}{\partial t} - gz.$$

Requiring that the pressure at the surface should equal the (constant) atmospheric pressure (p_0), the Bernoulli integral boils down to

$$\eta = \frac{1}{g}\left(\frac{\partial \phi}{\partial t}\right)_{z=0}.$$

The latter two boundary conditions require that the waves are small (so called *infinitesimal waves*). The solution of this set of equations for a harmonic progressive wave travelling in the x direction with wavelength λ and amplitude a takes the form:

$$\phi = \frac{gA_0}{\omega}\frac{\cosh(k(z+h))}{\cosh kh}\cos(kx - \omega t) = \frac{gA_0}{\omega}H_a(z)\cos(kx - \omega t) \tag{4.3}$$

where ω is the wave's angular frequency, and $k = 2\pi/\lambda$ is the wave number. The factor $H_a(z) = \cosh(k(z+h))/\cosh kh$ controls the attenuation of wave motion with depth. The phase speed c of the waves is given by:

$$\boxed{c^2 = \frac{\omega^2}{k^2} = \frac{g}{k}\tanh kh.} \tag{4.4}$$

Velocity components for the water particles within the wave are obtained from (4.3) by differentiation:

$$\boxed{\begin{aligned} u &= -\frac{\partial \phi}{\partial x} = \frac{gk}{\omega}A_0\frac{\cosh(k(z+h))}{\cosh kh}\sin(kx - \omega t) \\ w &= -\frac{\partial \phi}{\partial z} = -\frac{gk}{\omega}A_0\frac{\sinh(k(z+h))}{\cosh kh}\cos(kx - \omega t). \end{aligned}} \tag{4.5}$$

The surface elevation η and the pressure variation p' induced by the wave at depth z are then:

$$\boxed{\begin{aligned} \eta &= \frac{1}{g}\frac{\partial \phi}{\partial t}\bigg|_{z=0} = A_0\sin(kx - \omega t) \\ p'(z) &= \rho\frac{\partial \phi}{\partial t} = \rho g A_0 H_a(z)\sin(kx - \omega t). \end{aligned}} \tag{4.6}$$

For long waves ($\lambda \gg h$), $kh \ll 1$ so that $\cosh kh \to 1$, $\tanh kh \to kh$ and $H_a(z) = 1$. The pressure is then hydrostatic and the phase velocity becomes $c = \sqrt{gh}$. These results are thus consistent with our analysis of long waves in 3.6.1.

For waves with $\lambda \ll h$, termed *deep water waves*, $kh \gg 1$ and $\cosh kh \rightarrow e^{kh}/2$ so that the attenuation factor simplifies to $H_a(z) = e^{kz}$. At the same time, $\tanh kh \rightarrow 1$ so that the phase velocity in Equation (4.4) becomes:

$$c^2 = \frac{\omega^2}{k^2} = \frac{g}{k} = \frac{g\lambda}{2\pi}. \tag{4.7}$$

Since we must also have $c = \lambda/T_p$, we can substitute for λ and obtain the useful relations:

$$\boxed{c = \frac{gT_p}{2\pi}; \quad \lambda = \frac{gT_p^2}{2\pi}} \tag{4.8}$$

giving the phase velocity and wavelength of waves in deep water in terms of the wave period T_p.

In many situations where the depth is ~100 metres or greater, locally generated wind waves in the shelf seas satisfy the condition $\lambda < h$, which means that they are not substantially influenced by the presence of the bottom. They can reasonably be treated as deep water waves conforming to Equation (4.7) and decaying with depth according to $H_a(z) = e^{kz}$. We shall next look at the particle motions for these waves and the way in which they are modified when the waves encounter shallow water.

4.1.2 Orbital motions

From Equation (4.5), the particle velocities for deep water waves simplify to:

$$\begin{aligned} u &= -\frac{\partial \phi}{\partial x} = \omega A_0 e^{kz} \sin(kx - \omega t) \\ w &= -\frac{\partial \phi}{\partial z} = -\omega A_0 e^{kz} \cos(kx - \omega t). \end{aligned} \tag{4.9}$$

This means that the particles describe circular orbits with a radius A_0 at the surface and decreasing with depth as $A_0 e^{kz}$. These circular particle motions, which are in-phase at all depths, are illustrated in Fig 4.2a; the particles move forward (in the direction of wave propagation) under the wave crests and backwards under the troughs. As λ/h increases, the flow extends to the seabed where the condition that flow cannot penetrate the seabed becomes important. With only horizontal flow allowed along the bottom boundary, the circular orbits give way to ellipses which flatten towards the bed as shown in Fig. 4.2b. For very long waves ($\lambda \gg h$), like those discussed in Section 3.6, the ellipses become extremely flattened (see Fig. 4.2c); the amplitude ratio of w to u is $k(z + h)$ which is zero at the bed ($z = -h$) and takes a value of $kh = 2\pi h/\lambda$ at the surface ($z = 0$). The motion is, therefore, practically rectilinear at all depths; an obvious example of this in the ocean is the particle motion in tidal flow.

4.1.3 Waves of finite amplitude

Of course real waves are not infinitesimally small and, if we want to be rigorous, the above first order theory has to be refined when we are dealing with steep waves. In

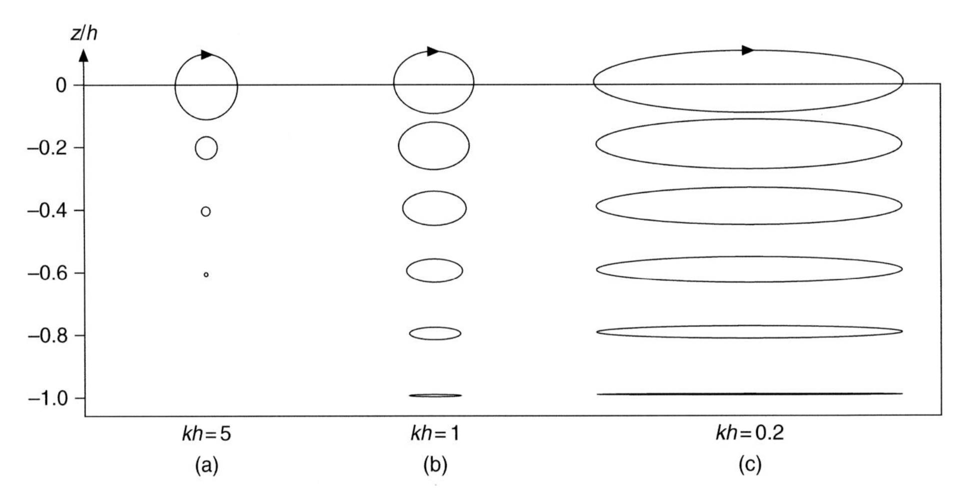

Figure 4.2 Particle orbits for surface waves. (a) $kh = 5$; $\lambda = 1.25h$; short waves with circular orbits decreasing exponentially with depth. (b) $kh = 1$; $\lambda = 6.28h$; intermediate case with elliptical orbits decreasing with depth. (c) $kh = 0.2$; $\lambda = 31.4h$; long waves with elliptical orbits.

higher order approximations for waves of finite height, like the Stokes wave (see Chapter 5 of Kinsman, 1965), the wave profile becomes asymmetric with sharper crests and flatter troughs, as you may have observed is the case for real steep waves approaching a beach. Steep waves also travel a little faster than those of small steepness. For example, a wave with steepness of $ka = 2\pi a/\lambda = 0.1$ has a phase speed in deep water which is ~1% greater than that given by Equation (4.7). Waves with steepness $ka \geq 0.14$ become unstable and break. Even for waves close to this maximum steepness, the correction to the phase speed does not exceed 10%. Generally, for waves of small and moderate steepness, the first order velocity potential theory does not differ greatly from the finite amplitude theory. It therefore serves as a reasonable model for real surface waves and is widely used as such. It is also usually much easier to apply than the higher order theories, which are mathematically complicated.

One important finite amplitude effect in waves is the residual transport which they induce. For strictly infinitesimal waves, the orbits are closed and there is no net drift of particles due to the wave motion. As the wave amplitude increases, the speed of a particle changes slightly as it moves around its orbit; for example the speed will be greater at the highest point in the orbit than at the lowest point because of the exponential variation in velocity with depth. These changes combine with similar effects due to the horizontal motion to give an 'unclosed' orbit and a net forward motion illustrated in Fig 4.3. We can deduce the magnitude of the forward drift for waves in deep water using the Eulerian velocities specified in Equation (4.9) and allowing for a finite orbital amplitude. The resulting residual current, called the *Stokes drift*, is given by

$$\bar{u}_{St} = k^2 A_0^2 c e^{2kz} \tag{4.10}$$

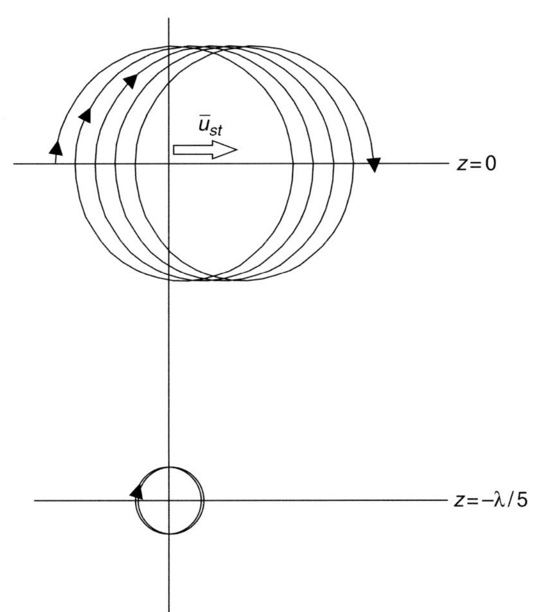

Figure 4.3 Particle motions at the surface ($z = 0$) and at $z = -\lambda/5$ in a surface wave with amplitude $A_0 = 0.25$ metres, period $T_p = 6$ s, wavelength $\lambda = 56.2$ metres. The forward drift results from the particle's (Lagrangian) velocity differing slightly from the (Eulerian) velocity at the centre point of the orbit. The Stokes drift velocity at the surface based on Equation (4.10) is $\overline{u}_{St} = 0.73$ cm s^{-1} which may be compared with the particle's instantaneous orbital speed of ~ 26 cm s^{-1}.

which is a net flow in the direction of propagation. It has a maximum of $\overline{u}_{St\,\max} = k^2 A_0^2 c$ at the surface and falls off rapidly with depth, being < 5% of its surface value at $z = -\lambda/4$. The depth-integrated, time-averaged transport by the Stokes drift is just $\overline{U}_{St} = kA_0^2 c/2 = \pi A_0^2/T_p$, so for waves of period $T_p = 6$ s and amplitude $A_0 = 1$ metre, there would be a wave-induced flow of 0.52 m^2 s^{-1} with a not-insignificant mean surface current of 0.12 m s^{-1}.

4.1.4 Energy propagation

Waves have both kinetic energy due to their orbital motions and potential energy due to vertical displacement of particles. For deep water waves, it is readily shown using u and w from Equation (4.9) that the average kinetic energy per unit area is given by:

$$\boldsymbol{T_w} = \frac{1}{2}\rho \int_{-\infty}^{0} (u^2 + w^2)dz = \frac{\rho g A_0^2}{4} \tag{4.11}$$

while for the average potential energy we have

$$\boldsymbol{V_w} = \frac{1}{\lambda}\int_0^{\lambda} \frac{1}{2}\rho g \eta^2 dx = \frac{\rho g A_0^2}{4}. \tag{4.12}$$

Hence, in this case the energy of the waves is equally divided between the two forms and the total energy is just $\boldsymbol{E_w} = \boldsymbol{T_w} + \boldsymbol{V_w} = \rho g A_0^2/2$. This energy is transmitted by the pressure forces acting within the waves. The rate of working by the pressure force on a plane normal to the direction of wave in a depth element dz is just $pudz$ per unit width, so that the average total rate of working is given by the integral:

$$\boldsymbol{P}_w = \overline{\int_{-\infty}^{0} pudz} = \rho g\omega A_0^2 \overline{\sin^2(kx - \omega t)} \int_{-\infty}^{0} e^{2kz} dz = \frac{\rho g A_0^2}{2} \frac{c}{2} = \boldsymbol{E}_w U_g \tag{4.13}$$

where u and p are taken from Equations (4.6) and (4.9) and the overbar denotes the time average. Energy is, therefore, transmitted at a rate which is the product of the energy density $\boldsymbol{E}_w$ and a velocity $U_g = c/2$, termed the *group velocity* (because it is also the speed at which wave groups travel). The result that energy flux is the product of energy density and group velocity has been obtained here for a sinusoidal deep water wave. It is applicable more generally, to a wide range of wave motions, with the group velocity given by:

$$\boxed{U_g = \frac{\partial \omega}{\partial k}.} \tag{4.14}$$

For the surface waves being considered, we have from Equation (4.7) that:

$$\omega^2 = gk \quad \text{so that } U_g = \frac{\partial \omega}{\partial k} = \frac{g}{2\omega} = \frac{gT_p}{4\pi} = \frac{c}{2} \tag{4.15}$$

which is in accord with the result of (4.13).

4.1.5 Wave breaking and near-surface processes

Energy input to the wave field by the wind is propagated around the shelf seas in the way described in the last section. Some of it ends up being dissipated at the coast in wave breaking on beaches and through bottom friction in shallow water. A large fraction of the energy, however, is dissipated in wave breaking in deep water out in the open sea in a process termed 'whitecapping'. In, and close to, areas where waves are being actively generated by the wind, the waves are sufficiently steep so as to become unstable and break. Wave breaking may also be induced by the interaction between the local generated wind waves and long swell waves arriving from distant storms in the deep ocean.

As illustrated in the aerial photograph of the sea surface in Fig. 4.4, the whitecaps which result from wave breaking are highly turbulent patches. Within these patches, the energy of the breaking wave is transformed into turbulent motions which stir the near surface layers of the ocean and dissipate to heat. This energy input has an important role in driving near surface mixing and hence promoting exchange between the ocean and atmosphere. Its influence, however, is restricted to a relatively shallow layer just below the surface, much shallower than the depth to which significant orbital velocities extend ($\sim \lambda/2$). Most of the energy is thought to be dissipated within a length scale set by the wave height parameter $H_{1/3}$ which is the average height of the highest one third of the waves.

Breaking waves also act to enhance ocean–atmosphere exchange through the formation of bubbles in whitecaps. The motion in breaking waves is so violent that the integrity

Figure 4.4 Images of the sea surface taken from the bridges of research ships in the Celtic Sea (height above sea level approximately 25 metres) showing whitecaps for wind speeds of 25 m s^{-1} on the left and 15 m s^{-1} on the right. The whitecaps are growing or decaying areas of high turbulence produced by wave breaking. The fraction of the ocean surface occupied by whitecaps increases with wind speed, as does the energy transfer to turbulence. (Photos by J. Sharples.)

of the surface is disrupted and air is entrained into the sea surface layers as large numbers of small bubbles are formed. Because of their small size, the bubbles rise rather slowly and some are drawn downwards in the turbulent circulation which follows whitecapping. The high pressure within the bubbles accelerates the transfer of gaseous components into solution in the surface waters (see Thorpe, 2005, Chapter 9 for a detailed account).

Surface waves are important, not only because they transmit mechanical energy and stir the near surface layers, but also because they transfer momentum to the surface layers. As we have just seen, when waves break their energy is dissipated, but the wave momentum, associated with the Stokes drift (4.1.3), is transferred to the surface current system and adds to the momentum input directly by the wind stress.

The Stokes drift of surface waves is also involved in the generation of an important component of near surface flow, known as *Langmuir circulation*. This circulation is responsible for the formation of parallel lines of foam and other buoyant material, termed 'windrows', which are often apparent when the water surface in the ocean and in lakes is subjected to wind stress. The windrows are aligned roughly in the wind direction, as is evident in Fig. 4.5a where the surface wind-generated waves can been seen with wave fronts orthogonal to the lines of foam. Windrows are the result of surface convergence in the transverse circulation in the vertical plane normal to the wind stress as shown in the schematic of Fig. 4.5b.

The mechanism responsible for the transverse motions appears to involve a subtle interaction between surface waves and the shear flow. A small perturbation of an initially uniform downwind Stokes drift flow is amplified by transverse forces in an instability which leads to convergence and, hence, sinking of fluid where the flow perturbation is growing. Conversely, divergence and upwelling occur where the downwind flow is reduced (see Fig. 4.5b). The details of this mechanism have been formulated theoretically (Leibovich, 1998) and studied in numerical simulations (see Thorpe, 2005, chapter 9). In both observations and simulations, the cell structure is found to be

(a)

(b)

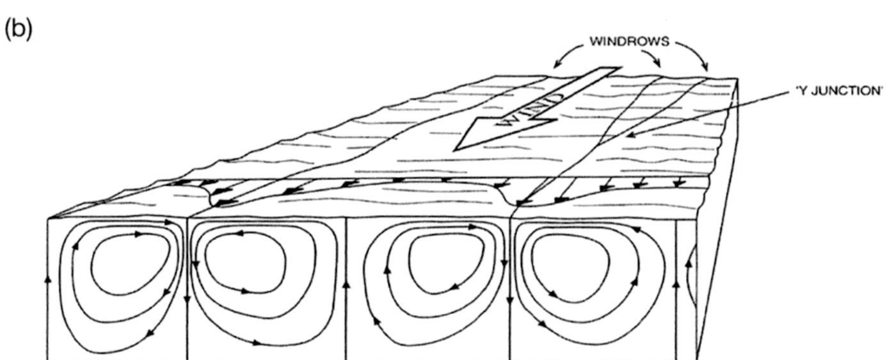

Figure 4.5 (a) Windrows (foam lines) generated by Langmuir circulation in a lake. The mean separation of the foam lines is ~8 metres. Reproduced from (Thorpe, 2007), courtesy of Cambridge University Press. (b) A schematic view of the circulation pattern. Note the cells of alternating clockwise and anticlockwise transverse circulation normal to the wind direction. The windrows are located in the convergent zones which occur at alternate cell boundaries. The pattern of cells is unsteady and continually evolving with new cells forming and others merging (as at the y junction shown). Reproduced from (Thorpe, 2005), courtesy of Cambridge University Press.

unsteady and is continuously evolving with the formation of new cells and the merging of others along with considerable variation in the spacing of the windrows. Langmuir circulation is, therefore, highly variable in character with a strong random element so that it is now often referred to as 'Langmuir turbulence'. Although flow speeds in the transverse cells are relatively small (typically a few cm s^{-1} at the surface), the persistence and widespread occurrence of this type of circulation means that it is now regarded as one of the major processes promoting mixing of the near-surface layers.

4.2 Internal waves

Internal waves are a feature of stratified fluids. They have many properties in common with surface waves but they are more diverse in character and have the reputation of being fiendishly difficult to fully comprehend. We cannot, however, ignore them as they are major players in the shelf sea system. Like surface waves, they transmit energy and provide a source of power for stirring but, unlike surface waves, internal waves can involve energetic motions in the interior of a fluid well away from the surface boundary and deliver power for internal mixing where there are no other available sources.

Here we shall develop a basic theory of internal waves for the relatively simple case of an ocean with two homogeneous layers. We shall make use of a similar approach to that which we applied to surface waves in order to illustrate the principal properties of this class of waves.

4.2.1 Velocity potential for waves on the interface between two layers

We will assume again that the motion is irrotational so that the velocity potential method can be applied. This assumption is valid in each of the two layers, although not in the interface between the two layers which is assumed to be very thin. We will use the system illustrated in Fig. 4.6. The upper layer has a depth h_1 and density ρ while the corresponding values for the lower layer are h_2 and $\rho + \Delta\rho$. If ζ is the vertical displacement of the interface, a sinusoidal progressive wave travelling

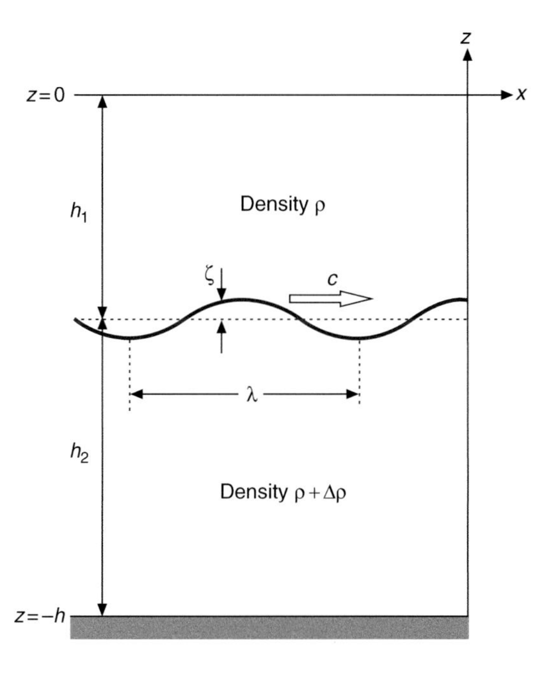

Figure 4.6 Definition of parameters for an internal wave.

along the interface in the positive x direction would have the same form as a surface wave, i.e.:

$$\zeta = A_0 \sin(kx - \omega t). \tag{4.16}$$

The corresponding velocity potential describing motion within the two layers may be obtained by essentially the same methods as used for the surface waves (see Phillips, 1966, p. 167) to give:

$$\begin{aligned} &\text{upper layer:} \quad \phi = -\frac{\omega A_0}{k}\frac{\cosh kz}{\sinh kh_1}\cos(kx - \omega t) \qquad 0 > z > -h_1 \\ &\text{lower layer:} \quad \phi = \frac{\omega A_0}{k}\frac{\cosh k(h+z)}{\sinh kh_2}\cos(kx - \omega t) \qquad -h_1 > z > -h_2 \end{aligned} \tag{4.17}$$

from which we can find expressions for the horizontal and vertical particle velocities as in the surface wave case. The phase speed c of the waves is given by:

$$\boxed{c^2 = \frac{\omega^2}{k^2} = \frac{g'}{k}(\coth kh_1 + \coth kh_2)^{-1}} \tag{4.18}$$

where $g' = g(\Delta\rho/\rho_0)$ is the reduced gravity.

In many situations in shelf seas, we are concerned with waves of long period (up to the semi-diurnal (~12 hours) and diurnal (~24 hours) periods). For such waves, the wavelength is long compared with the depth ($\lambda \gg h$) so that $kh \ll 1$ and we can approximate $\coth kh_2 \cong (kh_2)^{-1}$ etc. so Equation (4.18) then simplifies to:

$$c^2 = g'\frac{h_1 h_2}{h_1 + h_2} = g'\frac{h_1 h_2}{h} \tag{4.19}$$

where $h = h_1 + h_2$ is the water column depth. If, in addition, the upper layer thickness $h_1 \ll h$, the phase speed can be further approximated by $c = \sqrt{g'h_1}$. This is of the same form as the expression for the speed of long surface waves, but the speeds involved are much lower, mainly because $g'/g = \Delta\rho/\rho_0 \sim 10^{-3}$ which makes the internal waves travel ~30 times slower than their surface counterparts.

4.2.2 Particle motions in internal waves

The motion described by the velocity potential of Equation (4.17), termed the lowest mode or the first vertical mode, is an important reference solution which we will now explore in more detail. The particle velocities u and v are readily obtained by differentiation of the velocity potentials in the two layers. For example, in the bottom layer ($-h_2 < z < -h_1$) at a particular location ($x = 0$) we have:

$$\begin{aligned} u &= -\frac{\partial \phi}{\partial x} = -\omega A_0 \frac{\cosh k(h+z)}{\sinh kh_2}\sin \omega t \\ w &= -\frac{\partial \phi}{\partial z} = -\omega A_0 \frac{\sinh k(h+z)}{\sinh kh_2}\cos \omega t \end{aligned} \tag{4.20}$$

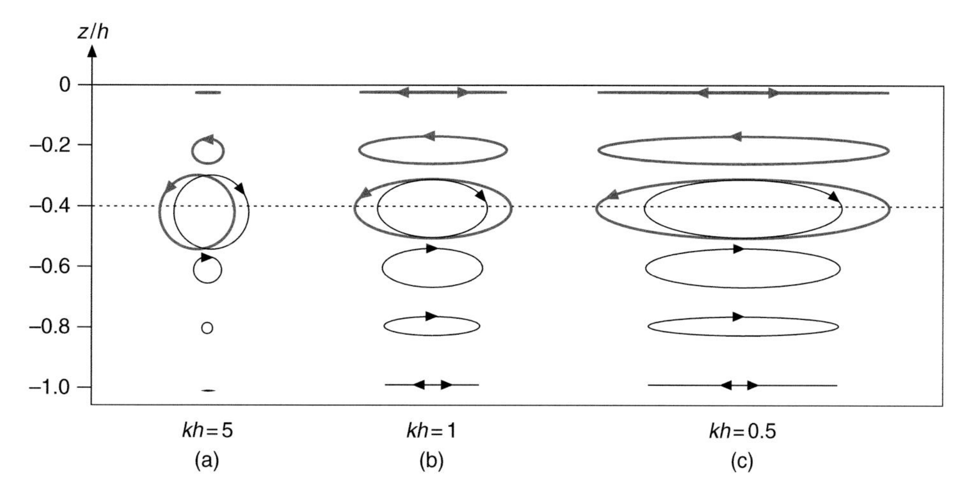

Figure 4.7 Particle orbits for sinusoidal progressive internal waves in a two-layer system derived from the velocity potential (Equation 4.20). (a) $kh = 5$; $\lambda = 1.25h$; short waves with circular orbits decreasing exponentially in both layers from a maximum at the interface. (b) $kh = 1$; $\lambda = 6.28h$; intermediate case with elliptical orbits becoming weaker and increasingly flat towards the boundaries. (c) $kh = 0.5$; $\lambda = 12.5h$; long waves with elliptical orbits; amplitude of horizontal movement is practically uniform in each of the two layers.

which describes elliptical motion with the particles revolving clockwise about their mean position. The ratio of the amplitudes of vertical to horizontal motion is therefore:

$$\frac{|w|}{|u|} = \tanh k(h + z) \tag{4.21}$$

which is the same as for the case of a surface wave in an un-stratified ocean.

If the depth is large relative to the wavelength ($kh \gg 1$), the ratio of w and u components tends to unity and the motion becomes circular in the interior of the layer and dies out towards the boundary, as shown in Fig. 4.7a. Notice here that the bottom layer in an internal wave is behaving in a similar manner to a surface wave. In the upper layer, the particles revolve in the opposite, anticlockwise, direction and again the motion diminishes in amplitude as we move from the interface towards the upper boundary. At the interface, where the sense of the orbits changes abruptly, there is node (a zero) in the horizontal velocity with a large velocity difference (shear) across the interface. This shear, which increases with the wave amplitude, can result in instability of the wave and the generation of turbulence, as we shall see in Section 4.4.

For longer waves ($kh \leq 1$) the particle orbits, shown in Fig. 4.7b, c, become flattened as they do for surface waves. The vertical velocity is zero at both top and bottom boundaries and increases linearly with distance from the boundaries to a maximum on the interface, while the horizontal velocity in each of the two layers becomes depth uniform with a 180° phase shift between the layers.

The particle orbits shown here are for small (strictly infinitesimal) amplitude waves for which the orbits are closed. For finite amplitude internal waves, just as for surface

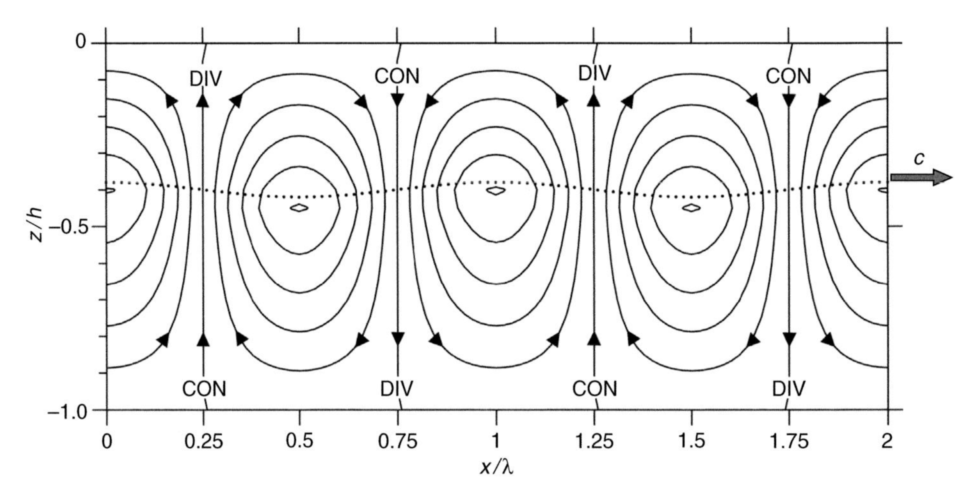

Figure 4.8 A snapshot of the instantaneous streamlines in a long internal wave with amplitude $a = 1$ metre travelling from left to right through a two-layer ocean. The wavelength is $\lambda = 3000$ metres, the water depth is $h = 100$ metres so that $kh = 2\pi h/\lambda = 0.21$ and the upper layer has a depth of $h_1 = 40$ metres. The thick grey line indicates the displacement of the interface. Notice the regions of convergence (CON) and divergence (DIV) in the surface flow. This streamline picture is obtained from the velocity potential of Equation (4.17).

waves (Section 4.1.3), water particles will sample the velocity along their orbital trajectory where the velocity differs from that at their mean position. As a result, the Lagrangian (particle) velocity will differ from the Eulerian (fixed point) velocity and the orbits will not be closed: the particles will, therefore, 'drift'. This rectified flow is analogous to the Stokes drift of surface waves and results in an integrated volume transport V_I in the direction of wave advance given by:

$$V_I = \frac{1}{2} k A_0^2 c (\coth kh_1 + \coth kh_2) \tag{4.22}$$

where the two terms represent transport in the upper and lower layer respectively. Note that for deep water ($kh \gg 1$), coth $kh_2 \to 1$ so that transport in the lower layer is the same as that for a surface wave (see Section 4.1.3).

At the surface ($z = 0$) the vertical velocity $w = 0$, so, we might deduce that there is no surface displacement by the wave. In fact, it can be shown (see Phillips, 1966, p. 169) that there is very small displacement of the surface which is out of phase with the interface movement and smaller by a factor of $\sim \Delta\rho/\rho_0 \approx 1/1000$ so that, in many situations, it can be neglected.

An interesting property of propagating internal waves is that the surface flow is convergent ($\partial u/\partial x < 0$) and divergent ($\partial u/\partial x > 0$) over the alternate nodes of ζ. You can see this pattern of alternate convergence and divergence in Fig. 4.8 which shows the instantaneous velocity field in a propagating internal wave as streamlines of the flow. Buoyant material at the surface tends to accumulate in the convergences, creating transverse bands at intervals of one wavelength along the direction of propagation. Such regular bands of surface material can be useful indicators of the presence of internal waves (Pingree and Mardell, 1985). The alternation of the surface flow direction may also cause the modulation of the roughness of the sea surface by changing the steepness of short surface waves. We shall see in Chapter 10 that, under favourable conditions, this effect may allow the detection of internal waves from satellites using synthetic aperture radar (SAR) or even by the standard marine radar aboard research vessels.

4.2.3 Energy and the group velocity of internal waves

Like surface waves, internal waves transport potential and kinetic energy as they propagate. The potential energy density of an internal wave is found by evaluating the average change in potential energy resulting from the vertical displacement of water particles within the wave. For our wave travelling on the interface between two layers, we have the potential density as:

$$\boldsymbol{V}_w = \frac{1}{2}\rho g' \overline{\zeta^2} = \frac{1}{4}\rho g' A_0^2 \tag{4.23}$$

which is the same result as we had for surface waves except that g is now replaced by the reduced gravity g'. As for surface waves, the kinetic and potential energy densities of this type of internal wave are equal (see Problem 4.1) so that the total energy is just:

$$\boldsymbol{E}_w = \boldsymbol{V}_w + \boldsymbol{T}_w = 2\boldsymbol{V}_w = \frac{1}{2}\rho g' A_0^2. \tag{4.24}$$

This energy moves at the group velocity which can be found from Equations (4.14) and (4.18) as:

$$U_g = \frac{\partial \omega}{\partial k} = c\left\{\frac{1}{2} + \frac{kh_1 \sinh^2 kh_2 + kh_2 \sinh^2 kh_1}{\sin 2kh_1 + \sinh 2kh_2}\right\} \tag{4.25}$$

where c is again the phase velocity. This rather unwieldy result has two simple limiting forms. For long waves ($kh \ll 1$), it becomes $U_g = c$, so that group and phase velocities are equal, as we found for long surface waves. For short waves ($kh \gg 1$) Equation (4.25) reduces to $U_g = c/2$ which is again the same result as for surface waves when the depth is large compared with the wavelength.

4.2.4 Continuous stratification and rotation

The above description of internal waves in an ocean with two homogeneous layers serves us well as a reference model of internal motions but, in the real ocean, the density structure is not usually so simple. Generally we are dealing with a water column which is continuously stratified with the degree of stability varying with depth. In this case, the motion is not restricted to the lowest mode and can take on more complicated forms of vertical structure in higher modes. The mth mode is characterised by having m nodes in the vertical structure of the horizontal velocity. Hence the simple wave we have described above is referred to as a *mode 1 internal wave*, with the node at the interface between the layers. Each mode has its own propagation speed which generally decreases with increasing mode number.

The form of all the modes and their speed of propagation can be found from the observed density structure which is used to calculate the buoyancy frequency (Equation 3.37). Normal mode analysis solves an eigenvalue problem for the modes using well-established methods (Phillips, 1966) which are now readily applied using freely available MATLAB codes (Klink, 1999). In many cases in the ocean, it seems that the lower modes are dominant with much of the energy contained in the lowest mode. It is, therefore, usually unnecessary to consider more than the first few modes in analysing observations of internal waves.

The possible frequencies of internal waves are limited in two ways. No solution is possible for an internal mode with a frequency $\omega > N_{max}$ where N_{max} is the largest value of the buoyancy frequency in the water column density profile (see 3.4.2). At the same time, there is a low frequency limit which is set by the Earth's rotation, whose influence we have not yet considered. When Coriolis forces are included in the equations of motion, it can be shown (Kundu, 1990) that wave type solutions are only possible for frequencies greater than the inertial frequency f. Internal waves on a rotating Earth are, therefore, limited to the frequency range $N_{max} > \omega > f$ with a corresponding wavelength set by the phase velocity of the mode involved. Rotation also affects energy distribution in internal waves. When we include the Coriolis forces in the dynamical equations (Gill, 1982), the ratio of potential and kinetic becomes

$$\frac{V_w}{T_w} = \frac{\omega^2 - f^2}{\omega^2 + f^2}. \tag{4.26}$$

For $\omega \gg f$ the ratio is unity as we found for non-rotating waves, but as $\omega \to f$ it diminishes and the potential energy becomes zero at $\omega = f$. In this limit vertical motion is suppressed and all energy is in the horizontal motion which has the form of inertial oscillations.

4.2.5 The importance of internal waves

Generally the energy content of internal waves is small compared with that of surface waves and the energy is propagated around the shelf seas much less rapidly. Internal wave energy fluxes are, therefore, very much less than those involved in surface waves and it is reasonable to ask whether these motions need to be considered as an important component of the shelf sea system. The answer, as we shall see in Chapters 7 and 10, is that they play a significant role in the interior of the stratified shelf seas, a region which is remote from both the surface and bottom boundaries where almost all of the mechanical energy is input to the water column. Consequently, very little energy reaches from the boundaries to the stable, relatively tranquil interior and the flux of energy from internal waves, generated mainly by stratified flow over topography, assumes considerable importance as one of the few sources of energy available to drive the vertical mixing and maintain biogeochemical fluxes through the thermocline.

4.3 Turbulence and mixing

We turn now from the well-developed theory of wave motions to a less well understood class of motions which play a crucial role in our later discussion of processes in the shelf seas. Turbulence is the random fluctuation of flow and properties which is widely observed to occur in moving fluids. We experience it in the gustiness of the wind and see it in the eddies which are apparent in the swift flow of a river. It is present for much of the time almost everywhere in the shelf seas and plays a key role in the dynamics and in the mixing of water properties. It has long been recognized that turbulence has its origin in the instability of larger scale motions which arise through the non-linear nature of the equations of motion, but detailed understanding of the processes of turbulence remains limited, with much of the theory being of a semi-empirical nature.

In this section, we shall describe the nature of turbulence, its generation by stirring, its dissipation through frictional stresses and the way it brings about dispersion and mixing. The subject of turbulence is extensive and our coverage will, necessarily, be selective, with a focus on those aspects which are relevant in the discussion of processes in the shelf seas in later chapters. For a fuller account of some of the topics discussed below and a more general view of turbulent processes in the ocean, the reader is referred to the excellent volume by Thorpe (Thorpe, 2005).

4.3.1 The nature of turbulence and its relation to mixing

Marine turbulence is sometimes referred to as the 'disorganized motion of the ocean'. Turbulence has a chaotic aspect, first described in the late nineteenth century by the physicist Osborne Reynolds on the basis of studying the flow of water in pipes. At sufficiently low flow speed U_0 in pipes of small diameter d, Reynolds found that dyed fluid particles followed straight line paths parallel to the pipe wall in what is referred to as 'laminar flow'. But as U_0 or d increased there came a point of abrupt transition to a turbulent flow regime in which fluid particles no longer stayed on wall-parallel streamlines but acquired random components of velocity both across and along the pipe axis. Reynolds deduced that the switch between laminar and turbulent conditions was controlled by the relative magnitude of inertial and viscous forces, which he expressed in a non-dimensional number $Re = U_0 d/\nu$ where ν is the kinematic viscosity of the fluid (units $m^2\ s^{-1}$). For values of this *Reynolds number* of greater than about 2000 the flow was turbulent and dye, introduced at a point into the flow, was rapidly diffused across the pipe cross-section by turbulent flow components.

Here we are interested in the high Reynolds number flows we often find in the shelf seas, and we want to be able to describe how the associated turbulence behaves, how it mixes water properties and how we can measure and quantify it. The Reynolds number is also a vital part of understanding the mechanics of plankton life. In the Reynolds number definition above, we can replace the current speed with a plankton sinking or swimming speed, and the relevant lengthscale is the size of the plankton. In anticipation of the discussion of the Reynolds number in Chapter 5, let us assume a plankton cell diameter of 10 μm and a swimming speed of 0.1 mm s^{-1}. Setting the kinematic viscosity of seawater at 1.1×10^{-6} m^2 s^{-1} yields a Reynolds number of 10^{-3}. Hence life for the plankton is dominated by viscous forces; we will look at some of the consequences of this later in Chapter 5.

The random motions of turbulence may be thought of as a combination of eddies of different scales, and it is the combined effect of these eddies which is responsible for the dispersive tendency of turbulence which brings about mixing of fluid properties. If we label two particles in a turbulent fluid and follow their subsequent motion, we will see that their separation, while fluctuating, tends to increase with time. The rapid dispersion which results from turbulence contrasts sharply with the laminar flow situation where the transfer of fluid properties is limited to molecular diffusion, which is generally a very slow process. Molecular diffusion is governed by *Fick's law* in which the flux J of a fluid property is related to the gradient of its concentration $\partial c/\partial n$ by:

$$J = -k_m \partial c/\partial n \tag{4.27}$$

where k_m is the molecular diffusivity which is generally a small quantity. For example, the diffusion of heat in water has a molecular diffusivity $k_m = 1.4 \times 10^{-7}$ m^2 s^{-1}. If there were no turbulence in a shelf sea of depth $h = 100$ metres, and all heat exchange was by molecular diffusion, it would take a time $\sim h^2/k_m \sim 2260$ years to achieve equilibrium. In reality, when turbulence is involved, vertical mixing times are usually much less than one year.

It is important to distinguish between stirring and mixing and the way that they are linked by turbulence, a relationship often represented in the process of mixing milk into a cup of coffee. Stirring is the action of introducing mechanical energy into the flow (the movement of the spoon), while mixing is the irreversible inter-mingling of the milk and the coffee fluid at the molecular level. The two are linked by the suite of turbulent eddies which act to stretch out the interface between milk and coffee fluids while maintaining large gradients at the interface. In this way, turbulence creates a large interface area and high gradients which accelerates mixing by molecular diffusion.

As well as being dispersive, turbulence is also necessarily dissipative, i.e. it converts kinetic energy of motion into heat. This happens as energy is progressively transferred from larger to smaller and smaller scales, ending up at the smallest scales (typically millimetres) where molecular viscosity becomes important. In the smallest eddies, there are relatively large velocity gradients, which result in correspondingly large stresses between fluid particles given by the stress law for molecular viscosity:

$$\tau_m = -\rho \nu \frac{\partial u}{\partial n} \tag{4.28}$$

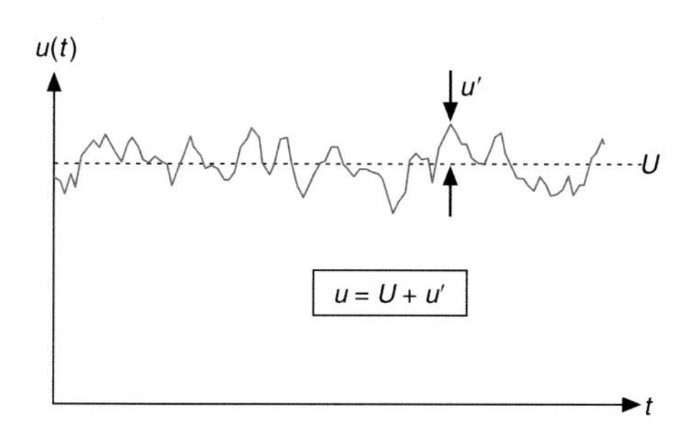

Figure 4.9 Plot of velocity against time (or x) showing the definition of the turbulent component u' as the difference between the instantaneous velocity u and its mean value U. Notice that with this definition the average $\overline{u'} = 0$.

where the stress τ_m is parallel to the local flow direction and $\partial u/\partial n$ is the velocity gradient normal to the flow. These frictional stresses convert the kinetic energy of the smallest eddies into heat and prevent energy being transferred to even smaller scales. Energy may also be consumed by turbulence in mixing a stratified fluid, i.e. by doing work in moving fluid along vertical density gradients. If the energy supply is insufficient to supply these demands, turbulence cannot be maintained and will die out. We will return to this important issue of the competition between mixing and stratification in Section 4.4 and again in Chapter 6.

In summary then, turbulence is random in character; it is necessarily dispersive and brings about the mixing of fluid properties; it is also inherently dissipative and requires an input of energy to sustain it.

4.3.2 Turbulent fluxes of scalars

A velocity sensor placed in a turbulent flow in the ocean will record a vector time series of motion $\underset{\sim}{u}(t)$ which can be regarded as a combination of a mean flow vector $\underset{\sim}{U} = (U, V, W)$ and a fluctuating turbulent component $\underset{\sim}{u}' = (u', v', w')$. This separation of the mean and the turbulent flow components, a procedure first introduced by Osborne Reynolds, is illustrated for motion in the x direction in Fig. 4.9. Here, the instantaneous u component consists of the mean U, averaged over some selected time scale, plus the turbulent component u' which, by definition, has zero mean when averaged over the same period, i.e.

$$u = U + u'; \quad \overline{u'} = 0. \tag{4.29}$$

Analogous relations for v and w complete the separation of turbulent and mean components of the flow vector $\underset{\sim}{u}$. It is also possible to regard the variation of a scalar property, e.g. salinity s, as consisting of a mean S and fluctuating component s' with zero mean:

$$s = S + s'; \quad \overline{s'} = 0. \tag{4.30}$$

The mean flow components U, V and W are the motions described in the basic dynamical balances in the previous chapter. The turbulent components are something extra and, as we shall see, are responsible for the enhanced diffusion brought about by turbulence and contribute to the dynamics by generating frictional stresses.

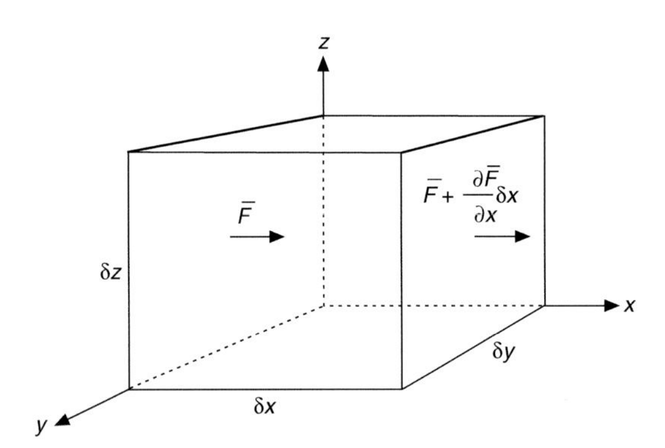

Figure 4.10 Fluxes of a scalar property due to flow in the x direction through a small control volume.

Consider first how the turbulent fluctuations contribute to the flux of a scalar property s through the small control volume of Fig. 4.10. The instantaneous flux (rate of transport per unit area) of the scalar in the x direction across the left face of the cuboid is just the product of velocity and scalar concentration:

$$F = (U + u')(S + s') \tag{4.31}$$

which, since $\overline{u'} = \overline{s'} = 0$, has an average value of:

$$\bar{F} = US + \overline{u's'}. \tag{4.32}$$

The first term US is the transport due to advection in the mean flow, and $\overline{u's'}$ is an additional turbulent flux due the covariance[2] of the fluctuations of u and s. Consideration of the scalar transport in the y and z directions leads to analogous advective and turbulent flux components VS, WS and $\overline{v's'}, \overline{w's'}$.

4.3.3 The advection–diffusion equation

Combining all the above fluxes, we can write a conservation statement for the scalar s in the control volume by comparing transports across opposing faces of the cuboid. In Fig. 4.10 you can see that the net input due to flow in the x direction is the change in F between the left and right faces multiplied by the area of the face $\delta y \delta z$, which is just:

$$\left(\bar{F} - \left(\bar{F} + \frac{\partial \bar{F}}{\partial x}\delta x\right)\right)\delta y \delta z = -\frac{\partial \bar{F}}{\partial x}\delta x \delta y \delta z = -\frac{\partial}{\partial x}\left(US + \overline{u's'}\right)\delta x \delta y \delta z. \tag{4.33}$$

Adding the equivalent contributions from flow through the other four faces of the cuboid, we can write the conservation statement for s as:

$$-\left(\frac{\partial}{\partial x}\left(US + \overline{u's'}\right) + \frac{\partial}{\partial y}\left(VS + \overline{v's'}\right) + \frac{\partial}{\partial z}\left(WS + \overline{w's'}\right)\right) + \Sigma = \frac{\partial S}{\partial t} \tag{4.34}$$

where $\partial s/\partial t$ is the rate of change of the mean concentration S and we have introduced Σ to represent any non-conservative processes by which the scalar may be produced or decay.

[2] The covariance of two quantities is the time average of their product. Unrelated quantities exhibit zero covariance.

Using the fact that the mean flow obeys the continuity of volume (Equation 3.1), we can write this statement as:

$$\frac{\partial S}{\partial t} = \overbrace{-U\frac{\partial S}{\partial x} - V\frac{\partial S}{\partial y} - W\frac{\partial S}{\partial z}}^{\text{advection}} \overbrace{-\frac{\partial(\overline{u's'})}{\partial x} - \frac{\partial(\overline{v's'})}{\partial y} - \frac{\partial(\overline{w's'})}{\partial z}}^{\text{diffusion}} + \Sigma$$
$$= \underset{\sim}{U}.\nabla S - \nabla.\underset{\sim}{G} + \Sigma \tag{4.35}$$

where $\underset{\sim}{G} = \left(\overline{u's'}, \overline{v's'}, \overline{w's'}\right)$ is the diffusive flux vector due to turbulence. Equation (4.35) is known as the advection–diffusion equation and describes the transport and dispersal of any scalar property. Its general solution requires knowledge of both the mean flow field $\underset{\sim}{U}$ and the diffusive fluxes $\underset{\sim}{G}$ throughout the domain of interest. In most cases, such detailed information is not available and we are frequently forced to simplify the problem by further assumptions. We shall look at such idealized solutions in Section 4.3.5, but first we need to examine the role of turbulent fluctuations in the dynamics of fluid flow.

4.3.4 Diffusion of momentum

In the same way that scalars are transported by turbulence, so is momentum. The components of momentum are given by $\rho u = \rho(U + u')$, $\rho v = \rho(V + v')$, $\rho w = \rho(W + w')$. If we replace the scalar s by, for example, the y component of momentum in Equation (4.32), we see that there will be a net flux of y momentum in the x direction of:

$$\bar{F} = \rho UV + \rho\overline{u'v'}. \tag{4.36}$$

The first term is the transfer of momentum in the mean flow and is accounted for in the dynamics of motion without turbulence, but the second is an additional momentum flux arising from the covariance of the u' and v' turbulent fluctuations.

This transfer of momentum constitutes an internal stress (a force per unit area). It is in this way that the frictional stresses τ, which we included in the equations of motion (Section 3.2.3), arise. Altogether there are nine possible stress components arising from the transfer of three components of momentum in three directions:[3]

$$\rho\begin{pmatrix} \overline{u'u'} & \overline{u'v'} & \overline{u'w'} \\ \overline{v'u'} & \overline{v'v'} & \overline{v'w'} \\ \overline{w'u'} & \overline{w'v'} & \overline{w'w'} \end{pmatrix}$$

These components are conveniently summarized in the Reynolds stress tensor:

$$\tau_{ij} = \rho\overline{u'_i u'_j} \tag{4.37}$$

[3] Think of the different stresses you can exert on a table top. Slide your hand backward and forward for 1 component, side-to-side for the 2nd component, and push downward for the 3rd component. Do the same on the wall facing you for another 3 components, and the wall next to you for the final 3 components.

where suffices 1, 2, 3 correspond to x, y, z components. (e.g. $u'_1 = u'$; $u'_2 = v'$; $u'_3 = w'$) Here the first index of τ, i, indicates the direction which is normal to the plane in which the stress operates, while the second, j, indicates the direction of the stress. So, for example, $\tau_{zx} = \rho\overline{w'u'} = \rho\overline{u'_3 u'_1}$ is the component of stress acting in the x-y plane along the x direction. Since $\tau_{ij} = \tau_{ji}$, only six of the nine components are independent. The three stress components with $i = j$ (e.g. $\tau_{xx} = \rho\overline{u'_1 u'_1}$) represent normal stresses (i.e. components of pressure) while those with $i \neq j$ are tangential stresses.

These Reynolds stress components can be shown to arise more formally by inserting the velocity decomposition (Equation 4.29) into the equations of motion and averaging. When this is done in a rigorous way (see for example Kundu, 1990) with the inclusion of the forces associated with molecular viscosity, the turbulent stresses appear in parallel with the viscous stresses. In the ocean, however, the Reynolds stresses are almost always much greater than the viscous stresses, so that the latter can frequently be omitted from the equations.

As we saw in the Ekman theory of Section 3.5, to solve problems involving frictional stresses, we need to know how the Reynolds stresses relate to the properties of the mean flow. Generally, however, we have rather little knowledge of these stresses and the way in which they vary with the properties of the mean flow. To get around this problem, it is commonly assumed that a stress component can be related to the velocity gradient by analogy with the relation for viscous stresses in Equation (4.28). For example, we define an *eddy viscosity* N_z which relates the horizontal stresses to the corresponding vertical velocity gradients according to:

$$\tau_{zx} = -\rho N_z \frac{\partial U}{\partial z}; \quad \tau_{zy} = -\rho N_z \frac{\partial V}{\partial z}. \tag{4.38}$$

So we are making an assumption, without any sound theoretical basis, that turbulent stresses act in a similar fashion to molecular stresses, but with a generally higher value for the appropriate viscosity because large turbulent eddies are mediating the transfers of momentum rather than small molecular collisions. Although mathematically convenient and widely used, this assumption reflects the immense difficulties in observing turbulent fluxes and it still leaves us with the question of how to determine the appropriate numerical value of N_z for each flow situation.

4.3.5 Fickian diffusion

There is, of course, the same problem in specifying the scalar fluxes due to turbulence. We noted above that molecular diffusion is described by Fick's law in which the property flux is proportional to the gradient of concentration (Equation 4.27). A proportionality similar to that underlying Equation (4.38) is again often assumed for the turbulent scalar flux components so that we write, for example, the x component of the turbulent diffusive flux as:

$$\overline{u's'} = -K_x \frac{\partial S}{\partial x} \tag{4.39}$$

where K_x is the *eddy diffusivity*, a property analogous to the molecular diffusivity k_m. However, unlike k_m, the eddy diffusivity is not a property of the fluid but of the flow and any further advance requires some assumptions about the value of K_x. The simplest assumption is that K_x is constant (K) and has the same value for all three flux components. If this is so, we have so-called Fickian diffusion for which the advection–diffusion equation is:

$$\frac{\partial S}{\partial t} = -\underset{\sim}{U}.\nabla S + K\nabla^2 S + \Sigma. \tag{4.40}$$

Further simplification is possible if the flow is 2D or even 1D. A valuable reference solution is that for diffusion of a conservative scalar property in one dimension (x) only and without advection ($U = 0$). Consider the instantaneous release of a quantity of dye of mass M_s into a canal at $x = 0$ at time $t = 0$. The canal is narrow and shallow so we can assume that the concentration of dye $S(x, t)$ is uniform across each section of the canal. In this case the advection–diffusion equation is just

$$\frac{\partial S}{\partial t} = K\frac{\partial^2 S}{\partial x^2}. \tag{4.41}$$

The solution of this equation is the well-known Gaussian function (Fischer *et al.*, 1979):

$$S(x, t) = \frac{M_s}{\sqrt{4\pi Kt}} e^{-x^2/4Kt}. \tag{4.42}$$

The behaviour of this solution is illustrated in Fig. 4.11. As the patch expands with time, the peak concentration (at $x = 0$) decreases as $t^{-1/2}$. The scale of the patch is set by the variance, σ^2 (standard deviation σ) of the dye from $x = 0$ which is given by:

$$\sigma^2 = \frac{\int_{-\infty}^{\infty} x^2 S(x, t)dx}{\int_{-\infty}^{\infty} S(x, t)dx} = 2Kt \tag{4.43}$$

so that the size of the patch increases as $t^{1/2}$. Approximately 90% of the dye lies within a distance 2σ of the origin so the patch size is $\sim 4\sigma$.

This basic solution may be used to build solutions to more complex problems in two or three dimensions (see Fischer, List, *et al.*, 1979). For example, multiple inputs of dye at different locations or at different times may be represented by simple addition of the Gaussian patches from the individual inputs. After a long enough time, the solution for such multiple inputs takes the form of a single Gaussian function. (To see this and examples of multiple patches and the solution with advection, see the supporting software at the book website.)

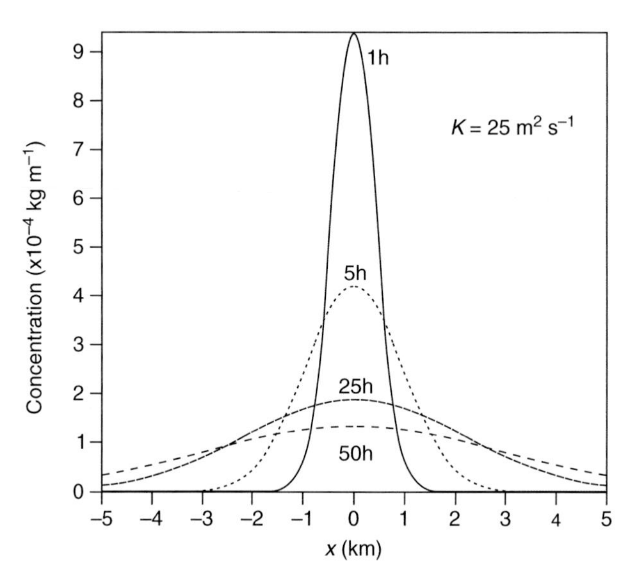

Figure 4.11 Distribution of 1 kg of dye released at $x = 0$ into a canal extending in the x direction. There is no advection ($U = 0$) and the diffusion is Fickian with an eddy diffusivity of $K = 25\ \text{m}^2\ \text{s}^{-1}$. The plot shows the concentration of dye (kg m^{-1}) at times in hours after the release.

Suppose that we make a series of dye release experiments corresponding to the above solutions of the diffusion equation. Because of the random nature of turbulence, each individual release will result in a dye distribution different from the others and not conforming exactly to the concentration function derived. What is the relation between these individual realisations of the experiment and the solution of the differential equations? The answer is that the concentration functions represent the average of a large ensemble of identical dye experiments and can be thought of as the probability distribution of dye. So they give us a rather smooth picture of diffusion which does not include the random nature of individual dye releases.

An alternative method, which better catches the random aspect of turbulence, relies on a random walk approach. Instead of deriving solutions to the diffusion equation, we concentrate on the motion of individual 'particles' of dye. To represent the chaotic movement of particles in turbulent flow, the particles are subjected to a large number of small random movements.

As an example, consider the 1D dispersion of a large number of particles released at the origin ($x = 0$) along the x axis corresponding to the canal diffusion scenario. At each time step Δt, the particles move a distance Δx in the positive or negative x direction according to the toss of an unbiased coin. After n time steps, the particles will spread out over a range $-n\Delta x$ to $+n\Delta x$ with the most likely position being at the origin. The particles will be distributed along the x axis with a probability given by the binomial distribution with a mean square deviation after time $t = n\Delta t$ given by

$$\sigma^2 = n(\Delta x)^2 = \frac{(\Delta x)^2}{\Delta t} t = 2Kt. \tag{4.44}$$

So the particles disperse from the origin with σ increasing as $t^{1/2}$ and a diffusivity $K = (\Delta x)^2/2\Delta t$. For a large enough number of steps n, the binomial distribution

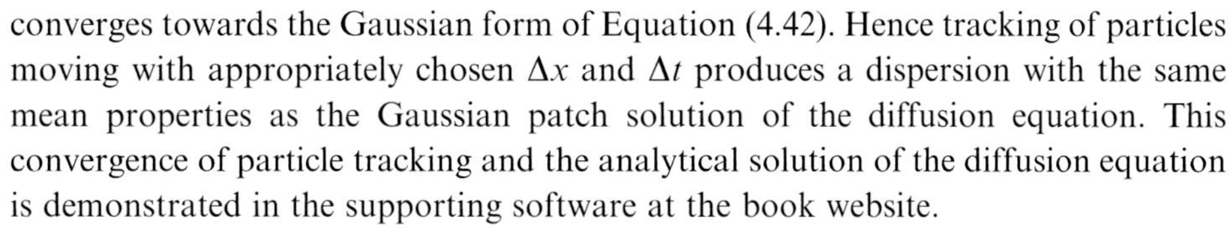
converges towards the Gaussian form of Equation (4.42). Hence tracking of particles moving with appropriately chosen Δx and Δt produces a dispersion with the same mean properties as the Gaussian patch solution of the diffusion equation. This convergence of particle tracking and the analytical solution of the diffusion equation is demonstrated in the supporting software at the book website.

Notice that individual particle tracking experiments differ from each other in a way that is more representative of the random reality of turbulent diffusion than the diffusion equation approach. Particle tracking has the further advantage of allowing advection by the mean flow to be readily combined with diffusion. In particular, the random walk approach has important applications in the modelling of the movements and growth of phytoplankton, as we shall see in later chapters.

4.3.6 When is diffusion in the ocean Fickian?

The solutions of the Fickian diffusion equation and the corresponding random walk methods discussed in the last section provide valuable reference models for the study of mixing in the shelf seas, but their rigorous application requires that the mixing is Fickian, i.e. it has a constant K and we know the appropriate value of K. A wide range of studies of ocean mixing, however, indicates that the diffusivity is generally not constant but tends to increase with the scale of the dispersing patch. In a classic work, Okubo (Okubo, 1971) compiled data from a large number of horizontal dye diffusion experiments at different scales in conditions where the diffusion was effectively two-dimensional because horizontal movement was unrestricted by boundaries and vertical diffusion was confined to the surface layer by stratification. Figure 4.12 shows Okubo's log-log plot of horizontal dispersion K_H against the scale of the diffusing patch L which indicates an increase in the diffusivity with $K_H \propto L^{1.15}$. This result is close to the scale dependence of $L^{4/3}$ proposed on empirical grounds by Richardson (Richardson and Stommel, 1948) and consistent with dimensional analysis for isotropic turbulence (see later in Section 4.4.3). The scale dependence of diffusion changes the time course of dispersion considerably. Okubo found that the variance of patch size increases as t^m with $m \sim 2.3$ in contrast to the Fickian case, where $m = 1$.

In shelf seas, there is an important exception to this picture of non-Fickian behaviour. The principal form of horizontal dispersion in tidally dominated situations is essentially Fickian and arises from the interaction of vertical shear of the horizontal velocity with vertical mixing. As we saw in Chapter 3, the flow driven by the tides moves particles around paths which are elliptical. In strong tidal flows, particle excursions may extend tens of kilometres from their mean position. Such movements decrease with depth because of frictional drag at the seabed, so there is generally a strong vertical shear even when the water column is homogeneous in density. In stratified conditions, the vertical shear may be considerably enhanced. Vertical shear in the tidal currents interacts with vertical turbulent diffusion to bring about horizontal dispersion. The essential mechanism was originally described for the case of steady flow in pipes by Taylor (Taylor, 1953; Taylor, 1954) and applied to

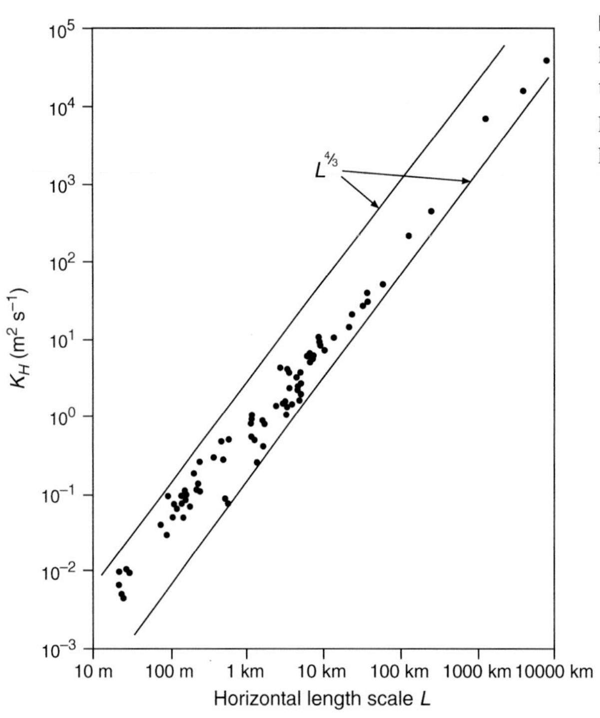

Figure 4.12 Variation of horizontal eddy diffusivity with the scale of the diffusing dye patch (from (Okubo, 1971), with permission from Elsevier).

oscillating currents in the shelf seas by Bowden (Bowden, 1965). The mathematical theory is rather intricate and will not be rehearsed here, but the interested reader can find an accessible account of the theory in (Fischer *et al.*, 1979). If you do not want to pursue the detail of the theory (and even if you do), you can get a visual picture of how shear diffusion works from numerical simulations in the associated software (www.cambridge.org/shelfseas).

The important result of the theory is that the dispersion by the shear-diffusion mechanism of a passive tracer of concentration s can be represented by a horizontal flux in the Fickian form:

$$\underset{\sim}{G} = \left(-K_x \frac{\partial s}{\partial x}, -K_y \frac{\partial s}{\partial y}\right). \tag{4.45}$$

If K_z is the vertical eddy diffusivity, the time to mix the water column of depth h is $T_m \approx h^2/K_z$. Providing $T_m \ll T_2$ the semi-diurnal tidal period, the components of the diffusivity K_x and K_y are determined by

$$\boxed{K_x = c_s u_T^2 T_m; \quad K_y = c_s v_T^2 T_m} \tag{4.46}$$

where u_T and v_T are the amplitudes of the surface to bottom tidal velocity differences in the x and y directions and c_s is a numerical factor depending on the form of the velocity profile. Dispersion increases with T_m and is thus inversely proportional to the vertical diffusivity K_z, a result which, at first, may seem counterintuitive. The inverse dependence on the vertical diffusivity results from the fact that, for low vertical mixing, the horizontal shear in the current is able to generate greater horizontal stretching of the tracer, whereas increased vertical mixing counters horizontal stretching of the tracer.

The general problem of determining the diffusivity in an oscillatory flow with period T_2 for an arbitrary mixing time T_m is more difficult, but can be solved (see (Fischer *et al.*, 1979)). The results show that maximum dispersion occurs for a mixing time of $T_m = T_2$. With this optimal matching of the vertical mixing time to the tidal period, the shear diffusion components are

$$K_x \simeq 0.0067 u_T^2 T_2; \quad K_y \simeq 0.0067 v_T^2 T_2 \tag{4.47}$$

which sets an upper bound for K due to shear dispersion.

Since u_T and v_T are proportional to the tidal stream amplitudes, Equation (4.46) implies an anisotropic dispersion which is strongest in the direction of the major axis of the *tidal ellipse*. Tidal shear dispersion of this kind has been widely invoked in studies of scalar transport in shelf seas. For example, in a simulation of the spreading of the ^{137}Cs on the European shelf from a source in the Irish Sea, and its subsequent transport around the north of Scotland and into the North Sea, Prandle (Prandle, 1984) used a combination of tidally controlled dispersion and advection by the mean flow including the component due to the rectification of the tide. It achieved a convincing representation of the observed large-scale distribution of the Caesium tracer shown earlier in Fig. 3.17.

4.4 The energetics of turbulence

As we noted above, turbulence is necessarily dissipative, i.e. it consumes energy. In the turbulent motions of a homogeneous fluid (ρ = constant), energy is transferred to smaller and smaller scales until it is eventually converted to heat at the smallest scale by molecular friction. In order to maintain turbulence, it is necessary to supply energy at a sufficient rate to satisfy this frictional demand.

In a stratified fluid there is an additional energy demand because turbulence acts to bring about vertical mixing, which increases the potential energy of the fluid. If there is insufficient energy being supplied to the turbulence to meet this demand, the turbulent motion will diminish and the flow will revert to the laminar form. This competition between the energy supply and the turbulence-suppressing influence of stratification, which is formulated in the following section, leads to an important criterion for when flow in the ocean is, and is not, turbulent.

4.4.1 Buoyancy versus shear production of turbulence: the Richardson number

In a turbulent stratified fluid, water particles with varying density differences will be moving up and down across each horizontal plane with turbulent vertical velocity $w' = Dz_p/Dt$, where z_p is the height of a particle above a reference level, as shown

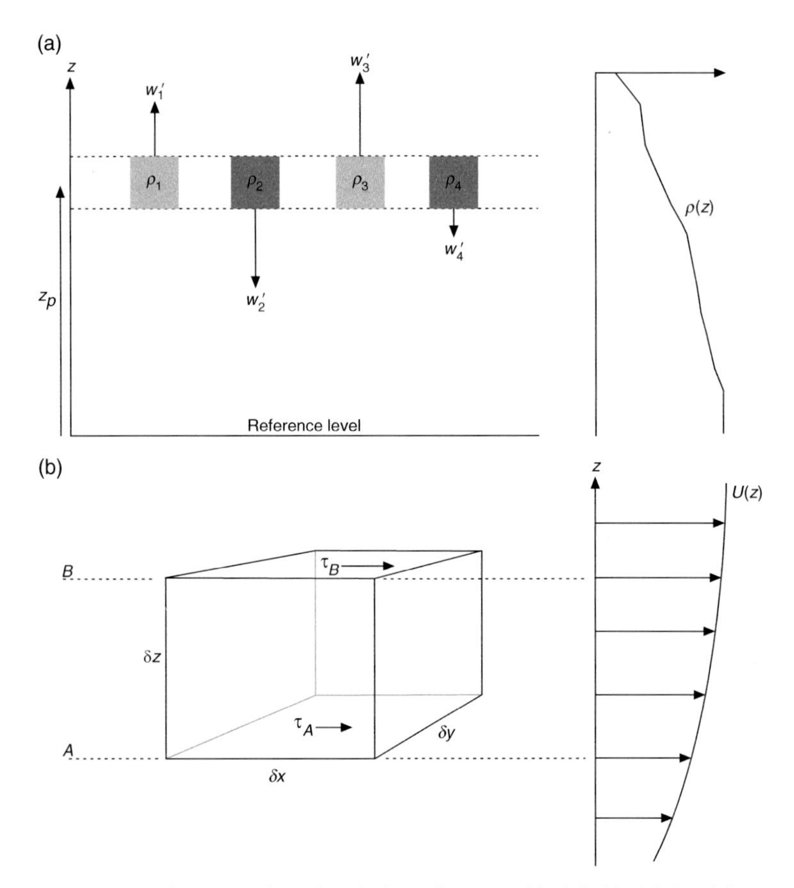

Figure 4.13 The energetics of turbulence in a stratified fluid. (a) Particles moving vertically with turbulent velocity w' and turbulent density deviation ρ' in a stratified fluid. The average product $g\overline{\rho' w'}$ is the rate of increase of the potential energy in vertical mixing. (b) In order to maintain turbulence, the energy consumed in mixing must be supplied by the shear flow. To determine the energy available to drive turbulence we first compare the rate of working by the shear stress τ on the cuboid surfaces in planes A and B and then subtract the rate of working in the mean flow.

in Fig 4.13a. A particle of volume ΔV_p will be changing its potential energy (PE) at a rate given by:

$$\frac{d(PE)}{dt} = \Delta V_p(\bar{\rho} + \rho')gw' \tag{4.48}$$

where $\rho' = \rho - \bar{\rho}$ is the density difference from the mean density $\bar{\rho}$ in the horizontal plane and we are assuming that the mean vertical velocity $W = 0$. The average rate of PE change per unit volume for all particles crossing a plane is then:

$$\frac{1}{\Delta V_p}\overline{\frac{d(PE)}{dt}} = g\overline{\rho' w'} = -gK_z\frac{\partial\bar{\rho}}{\partial z} = K_z\rho_0 N^2 \tag{4.49}$$

where, in the last step, we have written the turbulent flux of density in terms of the eddy diffusivity for density as in Equation (4.39). If turbulence is to continue, this energy demand must be met by a power source which in many cases is to be found in the mean flow.

Let us assume that the mean flow consists of a steady, vertically sheared flow in the x direction with a mean velocity profile $U(z)$. Consider the frictional stresses τ_{zx} acting on the upper and lower faces of a small cuboid in the fluid shown in Fig. 4.13b. The rate of working on the cuboid by the stress on the lower face is given by the force $\tau_{zx}\delta x\delta y$ times the velocity U, i.e. $\tau_{zx}U\delta x\delta y$. This differs from the rate of working at the upper face by:

$$P_1 = -\frac{\partial(\tau U)}{\partial z}\delta z\delta y\delta z. \tag{4.50}$$

where we are omitting the subscripts from T.

The quantity P_l is the net rate of input of energy to the cuboid. Not all of this power is, however, available to produce turbulence since some of the work done contributes to the energetics of the mean flow, i.e. the frictional forces may increase or decrease the kinetic energy of the mean flow $U(z)$. As we saw in (3.2.3), there is a net force on a cuboid due to the frictional stress of $-(\partial\tau/\partial z)\delta x\delta y\delta z$ which works on the mean flow at a rate of:

$$P_2 = -\frac{\partial \tau}{\partial z}U\delta x\delta y\delta z. \tag{4.51}$$

The difference between the rates of working P_1 and P_2 is:

$$\begin{aligned} P_1 - P_2 &= \left(-\frac{\partial(\tau U)}{\partial z} + \frac{\partial \tau}{\partial z}U\right)\delta z\delta y\delta z = -\tau\frac{\partial U}{\partial z}\delta z\delta y\delta z \\ &= -\tau\frac{\partial U}{\partial z} \text{ per unit volume.} \end{aligned} \tag{4.52}$$

This is the power available to fuel turbulence. Relating the stress to the velocity shear by the eddy viscosity as in Equation (4.38), the available power per unit volume can then be written as:

$$P_a = -\tau\frac{\partial U}{\partial z} = \rho N_z\left(\frac{\partial U}{\partial z}\right)^2. \tag{4.53}$$

In order to maintain turbulence, the power supply P_a must exceed the demand, i.e.:

$$\boxed{\rho N_z\left(\frac{\partial U}{\partial z}\right)^2 \geq -gK_z\frac{\partial \bar{\rho}}{\partial z}} \tag{4.54}$$

which may be written in the form:

$$Rf = \frac{K_z}{N_z} \frac{-\dfrac{g}{\rho}\dfrac{\partial \bar{\rho}}{\partial z}}{\left(\dfrac{\partial U}{\partial z}\right)^2} = \frac{K_z}{N_z} Ri \leq 1. \tag{4.55}$$

Rf, termed the *flux Richardson number*, is the ratio of power demand to power supply which must be less than unity if turbulence is to be sustained. Ri is the *gradient Richardson number* defined as:

$$Ri = \frac{-\frac{g}{\rho}\frac{\partial \bar{\rho}}{\partial z}}{\left(\frac{\partial U}{\partial z}\right)^2} = \frac{N^2}{S_v^2} \tag{4.56}$$

where $N^2 = -\frac{g}{\rho}\frac{\partial \rho}{\partial z}$ is the square of buoyancy frequency N (see also Section 3.4.2) and $S_v = \partial u \ / \ \partial z$ is the vertical shear. The two forms of the Richardson number are related by the ratio $Pr = K_z/N_z$, called the *turbulent Prandtl number*. Pr is usually close to unity so that the condition $Ri < 1$ is often used as a criterion for the maintenance of turbulence.

We should remember, however, that the Richardson number criterion sets a minimum criterion for the maintenance of turbulence since the above argument has concentrated on the power required for mixing and does not allow for the dissipation of energy into heat. It should also be noted that our energy analysis does not mean that turbulence will always develop whenever the gradient Richardson number is less than one. Starting from an initially laminar flow, the motion will only become turbulent if a more severe Ri criterion is met. The critical number here is $Ri < 0.25$, a result which emerged from the stability analysis of shear flows by Miles and Howard (Howard, 1961; Miles, 1961) who showed that for linear gradients of density and velocity between two homogeneous layers, a sinusoidal perturbation of the interface would decay if Ri remained above one quarter. Only if Ri was less than the critical value of 0.25 for sufficient time would the perturbation grow into an instability and generate turbulence. $Ri < 0.25$ is therefore usually taken as the appropriate, necessary condition for the development of turbulence.

Our energy analysis tells us only about the power requirements for fuelling turbulence and does not tell us anything about the mechanisms involved in transfer of momentum and energy into the turbulent velocity field. One way in which turbulence develops as Ri falls below the critical value is through the generation of more or less regular trains of billows. A laboratory example (Thorpe, 1971) is shown in Fig. 4.14. Initially, the laboratory tank is horizontal and the density structure consists of two homogeneous layers; the upper layer of freshwater overlies a denser brine solution which is dyed; the two layers are separated by a thin transition zone. After tilting the tank, the buoyant upper layer accelerates to the right while the lower layer accelerates to the left, generating an increasing velocity shear between the layers and reducing Ri until the flow becomes unstable and billows develop, in this example with a 'wavelength' of 0.12 m. The development of turbulence is illustrated in the

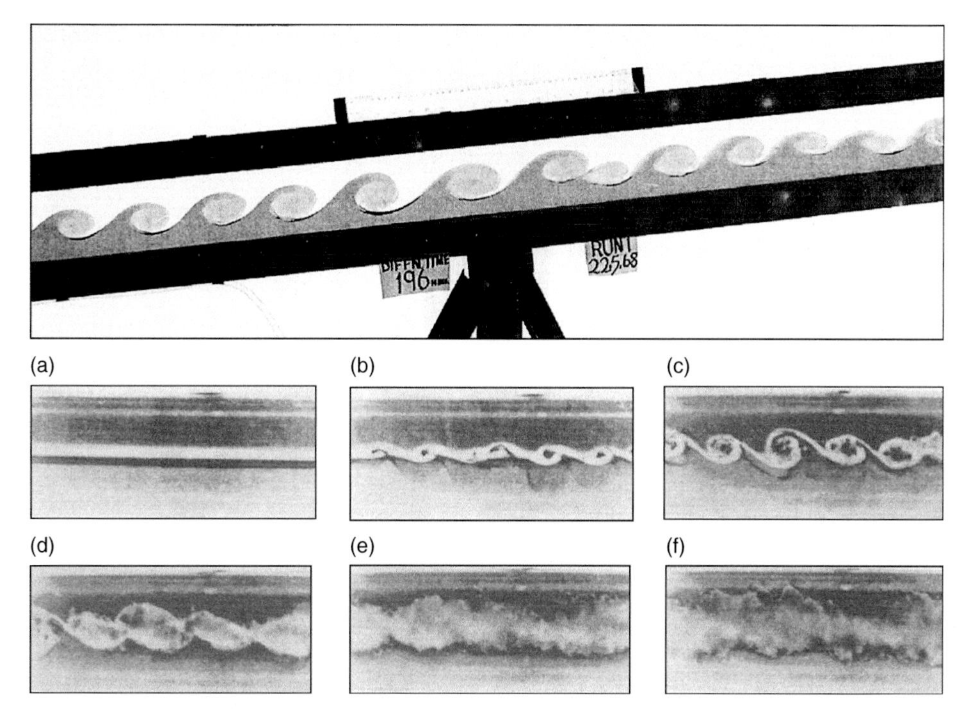

Figure 4.14 The development of turbulence through the formation of billows. Top panel shows billows developing in an accelerating stratified flow in a tilted tank. The lower panels (a–f) show stages in the transition to turbulence in a stratified fluid in a tilting tank experiment. From (Thorpe, 1971), with permission from Cambridge University Press.

lower panels, Fig. 4.14a–f. The tank has been returned to the horizontal after the flow has been accelerated. The flow is still laminar but unstable as $Ri \sim 0.06$. Fig. 4.14a–c show the evolving instability and the emergence of billows with initiation of turbulence in the centre of the billows. Fig. 4.14d–f show the amalgamation of turbulent patches to form a turbulent layer. Turbulence will continue only for a limited period as, with the tank horizontal, no further energy is supplied to mean flow. Such experiments have confirmed the validity of the Ri criterion for the onset of turbulence and have also shown how the scale of the billows depends on the initial thickness of the interface. Billows are frequently observed in cloud formations in the atmosphere and have been directly observed in the thermocline of the Mediterranean Sea (Woods, 1968). It has been postulated, although not yet confirmed, that billow turbulence is the source of the majority of turbulence observed in the ocean.

We now have two very different conditions for the occurrence of turbulence encapsulated in the Reynolds and Richardson numbers. For the ocean, if we take the water depth as the relevant lengthscale, the Reynolds number is generally very large ($\gg 10^6$) so we might conclude that turbulence will be ubiquitous except in thin regions very close to the boundaries where the length is restricted. In a homogeneous (i.e. completely unstratified) ocean this would be the case and the intensity of the turbulence would be limited only by the energy supply which is required to supply the dissipation of energy to heat by viscosity. In the real ocean, however, stratification is the norm and so the relevant criterion for whether turbulence will develop is generally set by the energetics considerations embodied in the Richardson number.

4.4.2 The turbulent kinetic energy equation and dissipation

In the previous section we focused on the competition between the production of turbulence in a shear flow and its conversion to potential energy in mixing. We now proceed to take a more general view of the energetics which combines all the important processes influencing the budget of turbulent energy. We define the *turbulent kinetic energy* (TKE) as:

$$\boldsymbol{E_T} = \frac{1}{2}q^2 = \frac{1}{2}(u'^2 + v'^2 + w'^2). \tag{4.57}$$

$\boldsymbol{E_T}$ is a scalar property which is subject to change by advection and diffusion as well as being produced and dissipated in the fluid motion. If we assume that conditions are horizontally uniform and that there is no vertical advection ($W = 0$), then the evolution of $\boldsymbol{E_T}$ will be described by:

$$\underset{}{\frac{\partial \boldsymbol{E_T}}{\partial t}} = \underset{\text{diffusion}}{-\frac{\partial(\overline{w'\boldsymbol{E'_T}})}{\partial z}} \underset{\text{production (P)}}{-\frac{1}{\rho_0}\left(\tau_x\frac{\partial U}{\partial z} + \tau_y\frac{\partial V}{\partial z}\right)} \underset{\text{mixing (B)}}{-\frac{g\rho' w'}{\rho_0}} \underset{\text{dissipation}}{-\varepsilon} \tag{4.58}$$

where $\boldsymbol{E'_T} = \boldsymbol{E_T} - \bar{\boldsymbol{E}}_{\boldsymbol{T}}$ and ε is the rate of energy conversion to heat. The production term is a generalised form of Equation (4.52) for flow in x and y directions. The mixing or "buoyancy production" term represents work done against, or by, buoyancy forces as expressed in Equation (4.49). For a stable stratification it represents a loss of TKE to potential energy. For an unstable situation, as in convection, this term is a positive input of TKE fuelled by the loss of potential energy.

The diffusion term is the divergence of the vertical diffusive energy flux which can be expressed as $-K_q\partial\bar{\boldsymbol{E}}_{\boldsymbol{T}}/\partial z$ where K_q is the eddy diffusivity for TKE. Writing the production and mixing terms in similar form (Equations 4.53, 4.49), we can express the TKE equation as:

$$\frac{\partial \boldsymbol{E_T}}{\partial t} = \frac{\partial}{\partial z}\left(K_q\frac{\partial \bar{\boldsymbol{E}}_{\boldsymbol{T}}}{\partial z}\right) + N_z\left(\left(\frac{\partial U}{\partial z}\right)^2 + \left(\frac{\partial U}{\partial z}\right)^2\right) + \frac{K_z g}{\rho_0}\frac{\partial\bar{\rho}}{\partial z} - \varepsilon. \tag{4.59}$$

In many situations, the diffusion of TKE is small relative to the leading terms. If we neglect it and assume a steady state, we have:

$$N_z\left(\left(\frac{\partial U}{\partial z}\right)^2 + \left(\frac{\partial U}{\partial z}\right)^2\right) + \frac{K_z g}{\rho_0}\frac{\partial\bar{\rho}}{\partial z} - \varepsilon = 0 \tag{4.60}$$

which is a balance referred to as *local equilibrium* in which turbulent kinetic energy is produced and consumed in the same location by dissipation or mixing. We can write this in shorthand form as:

$$\boldsymbol{P} + \boldsymbol{B} = \varepsilon \tag{4.61}$$

where P and B represent shear production and buoyancy terms, respectively. Using the fact that the ratio B/P = *Rf* (from Equation 4.55), we have, on re-arranging:

$$\boldsymbol{B} = \frac{Rf}{Rf + 1}\varepsilon. \tag{4.62}$$

Since $B = -g\overline{\rho' w'}/\rho_0 = K_z N^2$, we can write:

$$K_z = \frac{\boldsymbol{B}}{N^2} = \frac{Rf}{(1 + Rf)}\left(\frac{\varepsilon}{N^2}\right) = \Gamma\left(\frac{\varepsilon}{N^2}\right) \tag{4.63}$$

where $\Gamma = Rf/(1 + Rf)$. This important relation, derived by Osborn (Osborn, 1980), facilitates the estimation of vertical diffusivity from measurements of dissipation ε and the buoyancy frequency N^2. The relevant flux Richardson number is not known accurately but is widely assumed to be ~0.25, making $\Gamma \simeq 0.2$. We shall see in Chapter 7 how the diffusivity K_z measured in this way provides the basis for estimating the fluxes of nutrients which drive primary production in the seasonal thermocline.

4.4.3 Scales of turbulence: the Kolmogorov microscales

We noted in Section 4.1 that turbulence involves the transfer of energy from the large scale motion through a series of successively smaller eddies to the scales at which energy is converted to heat by molecular friction. In this section we will develop this idea further by considering the spectrum of turbulence and show how a general form for part of the spectrum can be derived by dimensional reasoning, following the radical hypothesis of Andrey Kolmogorov (Kolmogorov, 1941).

A spectrum is a way of representing the distribution of energy across different spatial or time scales. For example, a record of the u component of current $u(t)$ over a period of time T can be converted to a frequency spectrum $\Psi(\omega)$ by the standard methods of power spectral analysis (Emery and Thomson, 1997) to show how variations at different frequencies $\omega = 2\pi/\Delta t$, where Δt is a time scale, contribute to the variance of u. Similarly, if we have a record of u as a function of the spatial coordinate $u(x)$, power spectrum analysis will give us a spatial spectrum $\Psi(k)$ where $k = 2\pi/\lambda$ is the wave number corresponding to spatial scale λ.

In practice it is rather difficult to measure u as a function of the spatial coordinates (you need a fast moving submarine to get u as a function of x only!). A useful alternative is to transform from a frequency spectrum determined at a fixed point into a wave-number spectrum by using the *Taylor hypothesis*, in which the velocity fluctuations are assumed to be swept past a fixed sensor as a frozen field, i.e. a spatial pattern not evolving in time. This assumption is valid when the mean flow is large compared with the r.m.s. (root mean square) velocity of the fluctuations. The spatial and time scales (λ and Δt) are then simply related through advection in the mean flow at speed U, i.e.

$$\Delta t = \lambda / U; \quad \omega = \frac{2\pi}{\Delta t} = \frac{2\pi U}{\lambda} = kU. \tag{4.64}$$

Using this relation, we can convert $\Psi(\omega)$ to the wavenumber spectrum $\Psi(k)$.

Many different types of turbulence velocity spectra can be defined in terms of different combinations of velocity components depending on wavenumber or frequency variables. For our purpose it will be sufficient to use the scalar energy spectrum $\boldsymbol{E_S}(k)$ which expresses the distribution of TKE at different scalar wave numbers k. In other words, within a wavenumber interval dk, centred on wave number k, there is an amount of energy per unit mass $\boldsymbol{E_S}(k)dk$. Summing over all k gives the total TKE:

$$\int_0^\infty \boldsymbol{E_S}(k)dk = \frac{1}{2}(u'^2 + v'^2 + w'^2). \tag{4.65}$$

$\boldsymbol{E_S}(k)$, if we can determine it, tells us how energy is distributed over different spatial scales regardless of direction.

The Equilibrium Spectrum

In an important advance in turbulence theory, Kolmogorov argued that, in a steady state, $\boldsymbol{E_S}(k)$ should have a general equilibrium form for part of the wavenumber range. He did this on the basis of a bold hypothesis about how energy is transferred through the spectrum, from the largest eddies (wave number $\sim k_0$) which contain most of the energy, to small scales. He postulated that the energy transfer occurs by a large number of short steps from one spatial scale to an adjacent smaller scale, as in the schematic picture of Fig. 4.15.

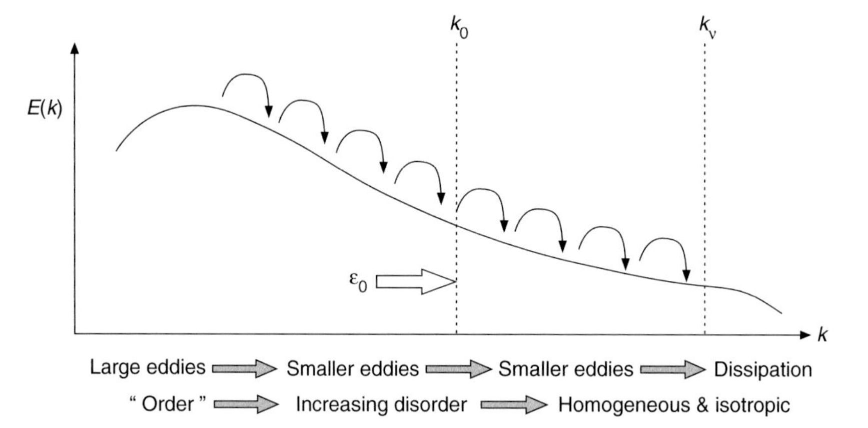

Figure 4.15 Schematic of energy transfer in the spectrum of turbulence. In the Kolmogorov cascade, the energy is transferred from the large scales of the mean flow via eddies of decreasing size to the smallest scales where it is dissipated as heat.

Because of the multiple steps involved in this *energy cascade* through the spectrum, the organised nature of the large-scale flow is lost as the energy moves to higher wave numbers ($k \gg k_0$) and the turbulence becomes independent of the large-scale flow which is driving it. Between k_0 and the microscale wave number k_ν, $\boldsymbol{E_S}(k)$ depends only on the net energy transfer through the spectrum ε_0 and the wave number k. Above k_ν, energy is dissipated to heat by viscous stresses. This type of turbulence will be homogeneous (uniform through the fluid) and isotropic (velocity components the same in all directions). Energy transfer through this region of the spectrum will continue until the eddies become so small that the wave number approaches the scale at which the viscous forces cause energy conversion to heat ($k \to k_\nu$).

In the sub-range of the spectrum, where $k_0 \ll k \ll k_\nu$, $\boldsymbol{E_S}(k)$ should be determined only by ε_0, the rate at which energy is being transferred through the spectrum, and the wave number k. The dimensions of these quantities are:

$$[\boldsymbol{E_S}(k)] = L^3T^{-2}; \quad [\varepsilon] = L^2T^{-3}; \quad [k] = L^{-1}$$

so that, on dimensional grounds, the spectrum must be of the form

$$\boldsymbol{E_S}(k) = A\varepsilon^{2/3}k^{-5/3} \tag{4.66}$$

where the proportionality $A \simeq 1.5$ is the Kolmogorov constant. This form of the spectrum, for what is termed the *inertial sub-range*, can be extended to include the dissipative region where the kinematic viscosity ν also becomes important. We might expect this to happen at spatial and velocity scales η_ν and V_ν set only by ε and the viscosity, so that, on dimensional grounds, we should have for the length and velocity scales:

$$\eta_\nu = \left(\frac{\nu^3}{\varepsilon}\right)^{1/4}; \quad V_\nu = (\nu\varepsilon)^{1/4}. \tag{4.67}$$

These are the *Kolmogorov microscales* which characterise the smallest eddies in a turbulent velocity field. Notice that while the velocity scale V_ν increases with the rate of dissipation, the microscale η_ν decreases, so the smallest eddies get smaller as the turbulence becomes more energetic. If we combine the microscales to form a Reynolds number (see Section 4.3.1) we find that $Re = V_\nu\,\eta_\nu/\nu = 1$ which means that, as the eddy scale becomes smaller than η_ν, the viscous forces start to exceed the inertial (mass $\times$ acceleration) forces with the consequent onset of dissipation.

At smaller scales (higher k), energy is extracted from the spectrum by this dissipation so that $\boldsymbol{E_S}(k)$ falls increasingly below the $k^{-5/3}$ form (Equation 4.66). A more general spectral form covering wave numbers $k > k_0$ allows for $\boldsymbol{E_S}(k)$ depending on the three quantities ε, k and ν may be written in the form:

$$\boldsymbol{E_S}(k) = F(k\eta_\nu)\varepsilon^{2/3}k^{-5/3} \tag{4.68}$$

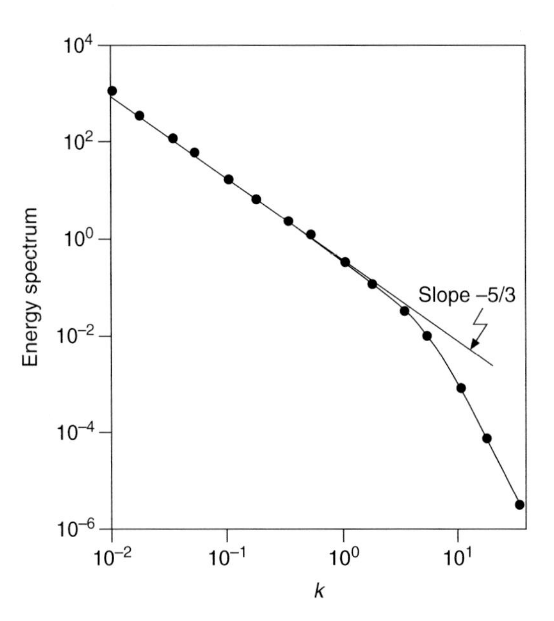

Figure 4.16 Energy spectrum measured off Vancouver Island by Grant *et al.* (1962). The straight line has a slope of −5/3. Reproduced courtesy Cambridge University Press.

where $F(k\eta_v)$ is a function of the non-dimensional variable $k\eta_v = k\left(\nu^3/\varepsilon\right)^{1/4}$. This is the universal equilibrium form of the spectrum predicted by Kolmogorov. It was first shown to apply to turbulence in the ocean by a famous series of measurements by Grant and co-workers in the early 1960s (Grant *et al.*, 1962) who used hot film anemometers to measure turbulent fluctuations of velocity in a tidally energetic channel off the west coast of Canada. The measurements were made from an anchored research vessel so the data constituted time series of velocity fluctuations from which the frequency spectrum can be determined. In order to transform from a frequency spectrum determined at a fixed point into a wavenumber spectrum, the Taylor hypothesis was used.

The wavenumber spectrum resulting from these measurements is shown in Fig. 4.16, along with a fit to $k^{-5/3}$ extending over two decades of k and a fall-off in the spectrum for wave numbers above $k\eta_v \sim 1$. Many other spectra observed in the atmosphere and ocean have been found to fit a $k^{-5/3}$ form over a wide range of scales, indicating the presence of an inertial subrange and seeming to confirm Kolmogorov's hypothesis about the nature of the cascade. Some doubts, however, have been raised about the accuracy of the fit to $k^{-5/3}$ (e.g. Long, 2003) and about the finding that, in the presence of weak stratification, the motion at high wave numbers is not fully isotropic as it should be in the Kolmogorov model. Nevertheless, the Kolmogorov theory stands as a key reference model in investigations of turbulence, and spectral fitting to the $k^{-5/3}$ form is widely used in the estimation of dissipation.

In the shelf seas, where dissipation in the tidal flow may exceed 1 W m^{-2} in 100 metres depth, $\varepsilon \approx 10^{-5}$ W kg^{-1} so, with the kinematic viscosity $\nu = 10^{-6}$ m^2 s^{-1} the microscale η_v may be as small as ≈ 0.6 mm. Even at much lower dissipation rates typical of low energy stratified conditions, where $\varepsilon \approx 10^{-8}$ W kg^{-1}, the microscale is only $\sim$ 3 mm. The inertial subrange of turbulence extends from these millimetre scales up to a maximum scale which is limited either by stratification or the proximity of the boundaries. In stratified conditions, buoyancy forces oppose the vertical extension of eddies beyond a point where their turbulent kinetic energy is all converted to potential energy in the overturning motion. On dimensional grounds this limiting scale should be set by the *Ozmidov length* (Ozmidov, 1965):

$$L_o = \left(\frac{\varepsilon}{N^3}\right)^{1/2} \tag{4.69}$$

where N is the buoyancy frequency. For strongly stratified conditions ($N^2 \sim 10^{-3}\ \mathrm{s}^{-2}$) where turbulence is relatively weak ($\varepsilon \sim 10^{-6}\ \mathrm{W\ kg}^{-1}$) L_o is only a fraction of a metre. Where stratification is very weak, L_o may increase to tens of metres and exceed the water depth. A closely related scale to the Ozmidov length is the *Thorpe scale* L_T which can be determined from observations by re-ordering particles in the density profile to remove instabilities in the water column and calculating the r.m.s. displacement of particles (Thorpe, 2005). Since it has been shown (Dillon, 1982) that $L_o \sim 0.8L_T$, profiles of L_T and N^2 derived from high resolution observations of density can provide estimates of the dissipation rate using Equation (4.69).

4.4.4 Turbulence closure

We have noted already that, in order to make progress in solving the equations of motion and diffusion, it is necessary to represent the turbulent stresses and the fluxes of scalars in terms of the mean properties of the flow. This requirement for turbulence closure can be met by a wide variety of assumptions and parameterisations. Frequently the approach is based on the notion of an eddy viscosity or eddy diffusivity which relates the stresses and fluxes to the relevant gradients of velocity or scalar concentration (see earlier in Sections 4.3.4 and 4.3.5). The simplest closure is then to assume that the eddy parameters are constants which may be adjusted to match observations. Such simple closure has obvious advantages in simplifying the maths if we are seeking analytical solutions.

We have seen, however, that turbulence is strongly influenced by buoyancy forces in a way which is encapsulated in the Richardson number, *Ri*. As a first step in putting closure on a sounder footing, it seems desirable to make the eddy parameters N_z and K_z depend on *Ri*. Parameterisations of this kind (e.g. Pacanowski and Philander, 1981) are generally too difficult to implement in analytical studies but have been widely used in numerical models. A more rigorous general approach to closure, which has emerged in recent years (see Burchard, 2002), involves expressing the eddy parameters in terms of the product of the r.m.s turbulent velocity q and a characteristic length scale L of the turbulence:

$$N_z = S_M qL; \quad K_z = S_H qL \tag{4.70}$$

where S_M and S_H are stability functions which allow for the effects of stratification on the eddy parameters. In numerical models, the eddy parameters are recalculated from these relations at each time step using the instantaneous mean flow, density and turbulent kinetic energy (TKE) profiles. The new N_z and K_z are then inserted in the equations of motion, the transport equation and the TKE equation to step forward in time and so on. We shall see more of how such models work in Chapter 7, when we use a turbulence closure model to study the evolution of dissipation and mixing in the shelf sea water column. Such a turbulence closure scheme also lies at the heart of the physics-primary production model which is utilised later in this book and is available at (www.cambridge.org/shelfseas).

Summary

This chapter described two very different classes of motion which play important roles in the shelf seas and are of particular importance in relation to the mixing of heat, salt, nutrients and other properties through the water column. We first dealt with surface wave motions which dissipate large amounts of energy and promote mixing in the near surface layers. The orbital motions of water particles in surface waves, the energy content of such waves and its propagation at the group velocity were discussed in some detail as a preliminary to understanding the analogous but more complex motions involved in internal waves propagating in a two-layer system. Internal waves travel much more slowly than surface waves but transmit significant amounts of energy whose dissipation may make an important contribution to internal mixing in low energy regions. The velocity field of internal waves in a two-layer system involves patterns of convergence and divergence at the surface which may be manifest through changes in surface roughness.

Turbulent motions are random, dispersive and lead to energy dissipation. The fluxes of scalar properties which arise from turbulent fluctuations in the flow, contribute to the transport of properties which is summarised in the advection–diffusion equation. Turbulent fluxes of momentum are similarly important in the dynamics, where they appear as the internal frictional forces. Fickian diffusion of scalar quantities provides an idealisation involving an eddy diffusivity which is uniform and independent of scale. Most dispersion in the ocean involves diffusivities which increase with scale, but an important exception in shelf seas is the process of tidal shear diffusion.

In a stratified fluid, buoyancy forces act to suppress turbulence by extracting energy from the flow through vertical mixing. Turbulence can then only be sustained if the mean flow supplies sufficient power to outcompete the buoyancy forces, a fact expressed in the Richardson number criterion. This is a fundamentally important result that will be relevant to understanding constraints on the biogeochemistry of the water column: pycnoclines are good at inhibiting vertical mixing. A more general description of the mechanisms transporting, generating and eroding turbulent kinetic energy (TKE) is given in the TKE equation. Turbulence closure schemes are needed to determine the turbulent diffusion and eddy viscosity parameters. The lengthscales of turbulent eddies are discussed in terms of Kolmogorov theory of energy transfer from large eddies to progressively smaller ones through the multiple steps of a cascade. The small scale eddy limit is reached when the energy is dissipated to heat at the microscale (~ mm). An upper limit for the scale of isotropic eddies in a stratified water column (the Ozmidov length) is set by the competition between the energy dissipation rate and the buoyancy forces.

FURTHER READING

The Dynamics of the Upper Ocean, by Owen M. Phillips, Cambridge University Press, 1966. [Chapters 4 and 5]

An Introduction to Ocean Turbulence, by Steve A. Thorpe, Cambridge University Press, 2007.

Marine Turbulence: Theories, Observations and Models, edited by H. Z. Baumert, *et al.*, Cambridge University Press, 2005.

Problems

4.1. Using expressions from Equation (4.9) for the particle velocities u and w of surface waves in deep water ($\lambda \ll h$), show that the kinetic energy of the waves per unit area is given by

$$\boldsymbol{T_w} = \frac{\rho g A_0^2}{4}$$

where A is the wave amplitude. Find an equivalent formula for the average potential energy per unit area V_w and hence determine the total energy density of such waves. If the wave amplitude $A_0 = 1.5$ metres, estimate the rate of energy transport by the waves given that the wave period $T_p = 6$ s.

4.2. Consider a train of internal waves propagating horizontally in a two-layer system like that depicted in Fig. 4.1. Using the velocity potential functions for the two layers (Equation 4.17), show that maximum horizontal velocity difference between top and bottom layers at the interface is given by:

$$\Delta u = \omega A_0(\coth kh_1 + \coth kh_2).$$

Determine Δu for waves of amplitude $A_0 = 5$ metres, wavelength 1.2 km and period 20 minutes if $h_1 = 25$ metres and $h_2 = 65$ metres.

4.3. 2D Fickian diffusion in x and y of a conservative substance with no advection ($U = V = 0$) is described by a reduced version of the advection–diffusion Equation (4.40), namely:

$$\frac{\partial s}{\partial t} = K\left(\frac{\partial^2 s}{\partial x^2} + \frac{\partial^2 s}{\partial y^2}\right)$$

where $s(x, y, t)$ is the concentration of the substance. Show that, if we can express s as a product of two functions $s = s_1(x, t)s_2(y, t)$, the diffusion equation can be separated into two equations, one in x only, the other in y only, which have the same form as the 1D equation. Hence, show that the radial spreading of a mass M_s of dye dumped at $x = y = 0$ is described by:

$$s(r, t) = \frac{M_s}{4\pi Kt} e^{r^2/4Kt}$$

where $r = \sqrt{x^2 + y^2}$ is distance from the origin and t is time from moment of dye release.

4.4. A dump of 25 kg of a conservative dye is spreading, by diffusion only, at a point in the shelf seas where the depth is 35 metres and the water column is well mixed. Using the result of the previous problem, estimate the radius of the patch and the maximum volume concentration 5 hours after the release if the effective diffusivity is $K = 12\ \mathrm{m^2\ s^{-1}}$.
At what time will the local maximum concentration occur 2 km from the release point, and what will be its value at this time? (Take the patch radius to be that of a circle which contains 95% of the dye.)

4.5. Two layers of uniform density in the ocean are moving relative to each other at a velocity which increases from zero to a steady speed of 25 cm $\mathrm{s^{-1}}$. If the interface between the layers has a thickness of 12.5 cm and the layers differ in density by 1.5 kg $\mathrm{m^{-3}}$, determine the relative velocity at which the initially laminar flow will start to become unstable.
Describe the subsequent development of the motion.

5 Life in the shelf seas

Much of the physics in the preceding chapters can be traced back to the fundamentals of fluid flow encapsulated in the equations of motion and continuity, along with the eddy description of turbulence. By contrast, describing the basics of life in the sea presents us with the difficulty of trying to distil a broad set of concepts from a system which is inherently very complex. Our experience of working at sea alongside biologists has been stimulating and fruitful, but there is always a tension: physicists can get exasperated at the complexity of the systems that biologists like to describe, while the biologists roll their eyes at the physicists' insistence on boiling problems down to as simple a level as possible. In this chapter we will take more of a physicist's view of biology in the ocean, focusing mainly on those aspects of the biology that are relevant to understanding how organisms' access to resources and growth are controlled by the structure and motion of the fluid environment.

Broadly, we are aiming to understand how organic compounds are produced in the ocean, and their fate. The schematic illustration of Fig. 5.1 provides us with a framework for the chapter; you could also have a look at the final schematic in Fig. 5.19 if you would like some idea of the details that we will be adding to this framework. We will begin by describing the fundamental biochemistry that lies at the heart of the growth of the *autotrophs*, the single-celled, photosynthesising phytoplankton which produce the organic matter and so power both the rest of life in the ocean and the cycling of carbon. In contrast, the *heterotrophs* consume organic material, either recycling it back to inorganic matter or passing it further up the food chain by being food for larger heterotrophs. Heterotrophs are much more varied in form and in their methods used for finding and consuming their prey, so instead of trying to detail this variety we will identify their broad roles in the ecosystem and some of the constraints that life in a turbulent fluid imposes on them. The biological processes that we will describe are in general common to the open ocean and to the shelf seas. We will use shelf sea examples to illustrate the processes, and identify where the important contrasts are between shelf and open ocean biology.

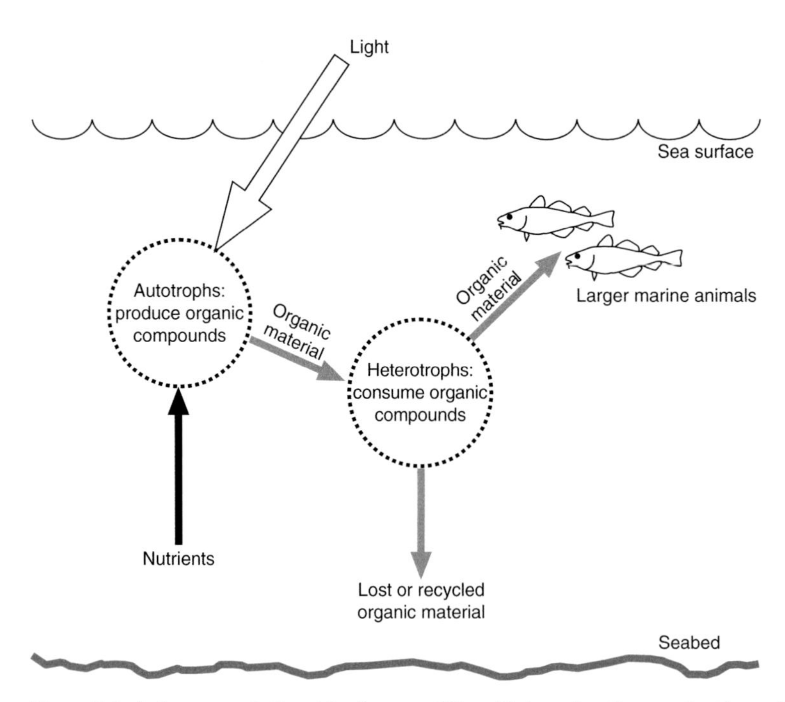

Figure 5.1 A framework for this chapter. We will describe the production of organic compounds and how those compounds can either reach larger marine animals or be exported/recycled.

5.1 Primary production in the sea: photosynthesis and nutrients

The photo-autotrophs, or phytoplankton, fix *dissolved inorganic carbon* (DIC) into the organic components that the rest of the marine food chain requires. If sufficient DIC is fixed in the surface water, reducing the concentration below the saturation value,[1] then CO_2 is transferred across the sea surface from the atmosphere to the sea. Thus, by fixing carbon and affecting the air-sea transfer of CO_2, the phytoplankton also play a fundamental role in the cycling of carbon within the Earth's climate system.

Growth of the phytoplankton, referred to as 'primary production', uses sunlight as the energy source, and also requires various nutrients. Let us first look at the common requirements of light and nutrients needed by all photo-autotrophs, and consider some of the methods we use to measure autotrophic growth at sea. At the end of this section we will also describe the main phytoplankton groups, some of their important

[1] DIC in seawater is made up of dissolved 'aqueous' CO_2, carbonate and bicarbonate. Dissolved CO_2 only accounts for $<1\%$ of DIC. The chemistry of what happens to CO_2 as it is absorbed by the seawater is beyond the scope of this book (see Chapter 11 of Williams and Follows 2011), but there are two key points to bear in mind. First, the amount of DIC that can be held by seawater is strongly temperature dependent, with warmer water able to hold less DIC. Second, the flux of CO_2 across the sea surface depends on the gradient between the atmospheric CO_2 and the aqueous CO_2, rather than with the total DIC, and so the time scale for DIC to reach equilibrium with the overlying atmosphere is much longer than, say, the time it takes dissolved oxygen to equilibrate.

characteristics, and provide a broad explanation for why the species of phytoplankton found in phytoplankton communities on the shelf and in the open ocean might differ.

5.1.1 Photosynthesis: light, pigments and carbon fixation

We learned in Chapter 2 that the tides produced by the Moon and the Sun dissipate about 3.7 TW of energy in the ocean, a rate that leads to a measurable slowing of the Earth's daily rotation. The photosynthesis of all plants on the Earth uses energy at a rate of about 100 TW (Nealson and Conrad, 1999), which is roughly 0.05% of the total energy supply from the sun (see Section 2.2.1). About half of the photosynthesis occurs in the oceans, and so, using the estimate of the contribution of shelf sea primary production to the global total from Chapter 1, about 8 TW of solar energy is utilised by the photo-autotrophs in the shelf seas. This corresponds to about half of humanity's total energy demand, so primary production is a remarkable utiliser of renewable energy.

Photosynthetic organisms use sunlight to convert carbon dioxide (CO_2), or 'fix' carbon, into organic compounds, particularly sugars (carbohydrates). Converting CO_2 into carbohydrate is a reducing reaction, which requires a source of energy and electrons. The energy is sunlight, while in aquatic photosynthesis the electrons are taken from water (H_2O). A by-product of this reaction is oxygen (O_2) which is released back into the seawater. Dissolved oxygen concentrations above the 'saturated' concentration that seawater can hold are an indication of active primary production. This excess oxygen will, once exposed to the atmosphere, transfer to the air; in this way marine primary production supplies us with about half of the oxygen that we breathe.

Fixing carbon takes place in two stages, shown schematically in Fig. 5.2. In the first, 'light-dependent' stage light is intercepted and used within one of many photosynthetic reaction centres in the cell. Each reaction centre has a light-gathering 'antennae' which uses pigments to absorb light energy. The energy is then used in two photosystems (referred to as photosystems I and II) to make the molecules ATP (adenosine triphosphate) and NADPH (reduced nicotinamide adenine dinucleotide phosphate). Next comes the 'light-independent' stage (also known as the 'dark reactions') of the Calvin-Benson cycle, where the ATP and NADPH are used to reduce CO_2, and ultimately to produce carbohydrates, amino acids and lipids. A summary of the chemistry of the photosynthetic reactions is:

$$2H_2O + CO_2 + 8 \text{ photons} \rightarrow CH_2O + H_2O + O_2. \quad (5.1)$$

CH_2O on the right of Equation (5.1) is the building block for carbohydrates; 6 of them yield glucose, while 12 of them form sucrose.

The details of photosynthesis can be found in Falkowski and Raven, 2007, and Kirk, 2010. For our purposes there are four main points to bear in mind.

(i) Not all photons are useful to photosynthesis. Light between wavelengths of about 400 and 700 nm, referred to as *photosynthetically available radiation* (PAR), is required by the phytoplankton. These wavelengths are in the green and blue parts of the visible spectrum (e.g. Section 2.2.1), which penetrate the furthest through

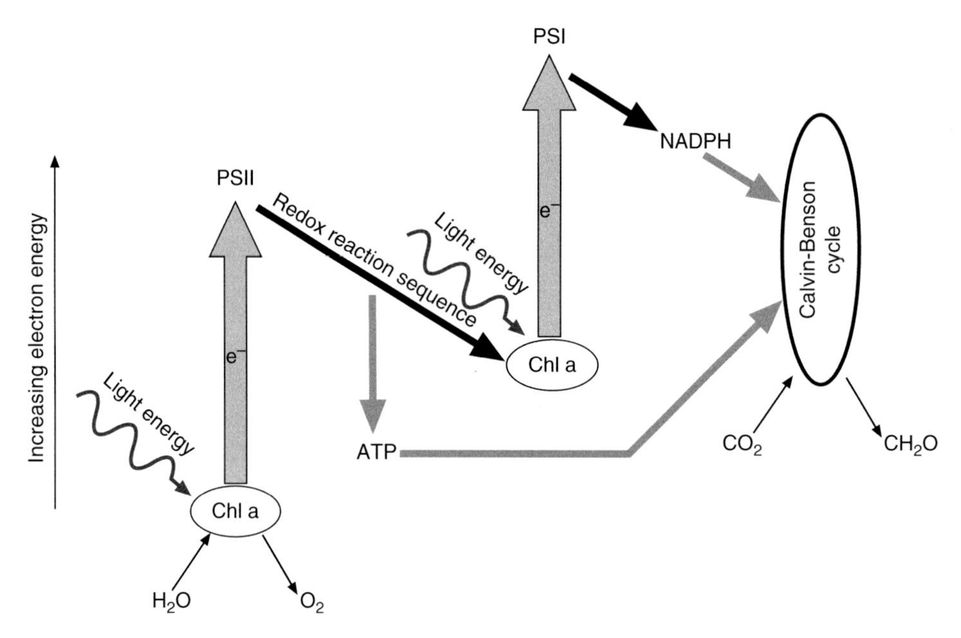

Figure 5.2 Schematic of the sequence of energy transfer through photosynthetic reaction centre. The two photosystems, PSII and PSI, both require light energy to excite an electron in the chlorophyll molecule, which is replaced by an electron from a water molecule, so releasing oxygen. The electron energy is used in a series of redox reactions that produce the molecules ATP and NADPH, required to fix carbon from CO_2 and produce CH_2O.

seawater. As we saw in Chapter 2, light attenuation is exponential through the water column, so we describe the vertical change in I_{PAR}, the downward flux of PAR, as:

$$\frac{\partial I_{PAR}}{\partial z} = K_{PAR}\, z \tag{5.2}$$

where K_{PAR} is an attenuation coefficient for PAR in seawater. At this stage we often take into account the effect on light attenuation of 'shading' due to the phytoplankton themselves by setting K_{PAR} to be the sum of the attenuation of PAR due to the seawater plus an extra attenuation proportional to the amount of phytoplankton chlorophyll in the water. For now, however, we assume K_{PAR} is constant through the water column, so that integrating Equation (5.2) yields:

$$\boxed{I_{PAR} = I_0 e^{K_{PAR} z}} \tag{5.3}$$

with I_0 the incident PAR on the sea surface. Typically we see $K_{PAR} \sim 0.1\ \mathrm{m}^{-1}$ in the surface waters of a summer stratified shelf sea so that PAR decreases to 10% and then to 1% of its surface value at depths of 23 and 46 metres. These changes with depth imply strong vertical gradients in PAR which will be important when we come to deal with photosynthesis in turbulent environments. There is an important contrast between the light attenuation in open ocean and that in the shelf seas. In the open ocean, variability in light attenuation is dominated by the contribution from the phytoplankton. In shelf seas we also need to include the effects of riverine sources of organic material as well as the re-suspension of sediments from the seabed. Particularly in the vertically mixed shallower waters on the shelf, these additional components can lead to K_{PAR} exceeding of 0.3 m^{-1}.

(ii) We tend to focus on chlorophyll *a* for the good reason that it is the most important pigment in photosynthesis. However, other pigments are also involved. Different phytoplankton groups utilise different *accessory pigments* as a part of their light-gathering antennae. These pigments work alongside chlorophyll *a*, extending the range of wavelengths which can be used by the photosystem. Measurements of accessory pigments provide a useful means of detecting key phytoplankton groups in the ocean (e.g. Barlow, *et al.*, 1993; Lohrenz *et al.*, 2003; Qian *et al.*, 2003)). The different pigments have response peaks at slightly different wavelengths of light, which allows scope for competition between phytoplankton groups. For instance, as the light spectrum changes with depth through the water column, contrasts in pigment composition can result in vertical separation of phytoplankton groups; you will see an example of this later in Chapter 7.

(iii) The light-dependent stage of photosynthesis involves photons exciting electrons within chlorophyll. The capacity for photosynthesis can become saturated under high light conditions if light energy is intercepted by the chlorophyll at a rate greater than the rate at which electron energy can be transferred and used within the photosystems. The continuing supply of light will keep exciting electrons within the chlorophyll and has the potential to damage the photosystem, so the chlorophyll has two mechanisms for shedding the excess light energy. A small amount (~1%) of absorbed energy can be re-emitted as light at a longer wavelength. This is the basis for the use of chlorophyll fluorescence as an indicator of phytoplankton biomass. Most of the excess energy that needs to be shed by the chlorophyll is transformed to heat in a process called *non-photochemical quenching*. High light levels can lead to a shift from fluorescence to non-photochemical quenching, which is important to remember when using fluorometers to assess phytoplankton biomass. For instance, at the sea surface, phytoplankton chlorophyll could fluoresce more during the night than during daylight hours, so that a time series of fluorescence could by misinterpreted as showing a diurnal signal in phytoplankton concentration.

5.1.2 Cell respiration: net and gross primary production

Some of the energy fixed chemically during photosynthesis is used in *cellular respiration*, maintaining the internal structure and operating capability of the cell. In the photo-autotrophs respiration is aerobic, meaning that it consumes oxygen. Respiration uses up some of the carbohydrates, lipids and amino acids generated by photosynthesis. We can describe carbon fixation as either *gross* or *net primary production*. Gross primary production is the total amount of carbon fixed by the phytoplankton, while net production is the gross production minus the amount of carbon consumed in cell respiration. Thus net primary production reflects the energy available for phytoplankton growth.

5.1.3 Techniques for measuring primary production and respiration

There are two commonly used methods for direct measurements of rates of carbon fixation, both based on incubating samples of seawater. After describing these two methods and identifying some of their limitations. we shall look briefly at some emerging new techniques.

Dissolved oxygen, primary production and respiration

We mentioned earlier that photosynthesis releases O_2 into the water. This provides the basis for a technique to measure rates of primary production by tracking changes in dissolved oxygen, first suggested in 1927 by Gaarder and Gran (Gaarder and Gran, 1927). Three bottled samples of water are required. The first is immediately analysed for its dissolved oxygen content. The other two are incubated, one illuminated by some representative intensity of radiation and the other kept in the dark. After some time the dissolved oxygen concentration in the 'light' and 'dark' bottles is measured. The increase in oxygen in the light bottle over the initial concentration will be due to both photosynthesis (adding oxygen to the water) and respiration (removing oxygen), and so provides a measure of the *net community production* in the bottle. The dark bottle should show a reduction in oxygen as without light only respiration should have been taking place, hence it provides a measure of *net community respiration*. An estimate of *gross community production* is formed by adding the amount of oxygen consumed in the dark bottle to that gained in the light bottle. Note here that we have added the word 'community' to all of these rate measurements; we will discuss this in more detail below.

The results can be expressed in terms of oxygen production per unit time, or they can be converted to carbon fixation rate by assuming a ratio of oxygen production to carbon fixation called the *photosynthetic quotient* with a value between 1.0 and 1.36 (Williams and Robertson, 1991). The basic oxygen method is simple and inexpensive, and widely used (e.g. Robinson and Williams, 1999). A development of the method uses $H_2{}^{18}O$, i.e. water with a stable isotope of oxygen (Bender *et al.*, 1987). The isotope ^{18}O amounts to only ~0.2% of naturally occurring oxygen, which is mainly made up of ^{16}O. Photosynthesis will release the ^{18}O into the water, while respiration consumes the ambient ^{16}O. This isotopic method thus separates the effects of autotrophs and heterotrophs, and provides a measure of gross photosynthetic production.

^{14}C: the standard measurement of primary production

Currently, the standard method for measuring rates of carbon fixation is to use radioactive carbon-14 (^{14}C), first suggested by Steeman-Nielsen in 1951 (Steeman-Nielsen, 1951). A bottle of seawater is 'spiked' with ^{14}C (in the form of sodium bicarbonate), and then incubated at the required intensity of light. After a few hours the phytoplankton in the sample are filtered and the filter analysed in a scintillation counter. The radioactivity of the filter indicates the amount of carbon fixed by the phytoplankton, so it provides a measure of photosynthetic production. The advantages of this method over the oxygen method described above are that the measure of carbon fixation is direct, i.e. not requiring an estimate of the photosynthetic quotient, and it is more sensitive. The

Figure 5.3 The interior of the radioisotope laboratory aboard RRS *Discovery*, Celtic Sea, June 2010. The laboratory, part of the UK National Marine Facilities, is installed within a standard shipping container so that it can be installed on different research vessels as required. (Photo by J. Sharples.)

disadvantage is practical; taking even the tiniest amounts of radioactive material to sea requires conforming with a host of regulations, and dedicated, expensive laboratory space is usually needed (an example is shown in Fig. 5.3). A recent alternative is to use the stable isotope ^{13}C instead of ^{14}C (Hama *et al.*, 1983), which avoids the regulatory constraints, but in this method the analysis of the filters requires a mass spectrometer and a larger amount of carbon biomass per sample.

Phytoplankton or community?

When seawater is incubated we observe the combined effects of all of the components of the plankton community trapped in the bottle; i.e. the photo-autotrophs (producing oxygen when photosynthesising, and also consuming oxygen in respiration) and the heterotrophs (largely consuming oxygen as they oxidise organic carbon). The simple oxygen incubation technique measures the impacts of oxygen production and consumption by the whole community (autotrophs and heterotrophs). Using the ^{18}O or ^{14}C technique allows measurement of production by the autotrophs, as only the photosynthesisers will be fixing ^{14}C or releasing the ^{18}O. Whether we are measuring gross or net photosynthetic production depends on how long the incubations are carried out. When using ^{14}C, the incubation time could be anywhere from 1 hour to 1 day. Short ^{14}C incubation times are usually viewed as providing something close to a gross photosynthetic production rate, while longer times (particularly where a day-night cycle is included) will measure net photosynthetic production.

At first glance none of the above techniques appears to allow us to separate the autotrophic and heterotrophic contributions to the community respiration. Often this is not necessary. For instance, a key question in marine biogeochemistry which can be answered without separating the respiration contributions is whether a whole community is net autotrophic or heterotrophic; the former will result in a net drawdown of CO_2 from the atmosphere to the ocean while the latter will lead to a net release of inorganic carbon into the seawater and potentially back to the atmosphere. If, however, we want to parameterise phytoplankton growth in a numerical model, we will require a value for the autotrophic respiration rate. Careful consideration of the information yielded by dawn-to-dusk and day-night ^{14}C incubations has suggested how autotrophic respiration can be separated from that of the whole community (Marra and Barber, 2004).

Incubations in a turbulent layer

Incubations are typically done over a fixed time interval at a fixed irradiance. In the real ocean, turbulence will mix phytoplankton vertically, presenting them with a rapidly changing light environment. Imagine a water sample collected at a depth of 5 metres in a surface layer of thickness 20 metres. Clearly if we were to incubate the sample at a fixed irradiance as measured at 5 metres, then the phytoplankton in our sample would experience a very different light regime compared to when they were being mixed continuously between the sea surface and the base of the surface mixed layer. This can lead to an incorrect estimate of primary production in the incubated sample by as much as 40%, though it is possible to compensate for the effect (Barkmann and Woods, 1996).

Other techniques for measuring production

In Chapter 1 we showed a global map of net primary production based on a satellite technique (Behrenfeld and Falkowski, 1997). As we noted, an important drawback of satellite techniques is that they only view conditions at the sea surface, and so miss any activity deeper than a few metres (about an optical depth, $1/K_{PAR}$). We will see in Chapter 7 that an important component of a shelf sea's annual primary production occurs in the thermocline, which is several optical depths below the sea surface, so satellite imagery is likely to under-estimate shelf sea productivity without applying some technique to account for sub-surface production (e.g. Longhurst *et al.*, 1995). Satellite techniques will also be undermined by the difficulty in accurately determining chlorophyll concentrations against the signals caused by dissolved organic substances and suspended sediments, a problem particularly associated with coastal seas. Coastal waters are categorised optically as 'case 2' waters, in contrast to the clear, oceanic 'case 1' water. Accurate chlorophyll detection using satellite sensors in case 2 waters still remains a key challenge in satellite oceanography (e.g. Blondeau-Patissier *et al.*, 2004).

A technique with particular promise is the use of *active fluorescence* measured with fast repetition rate (FRR) fluorometers. The fluorometers that we normally use to estimate chlorophyll biomass flash a strong light source into the water which saturates that phytoplanktons' photosystems, resulting in the maximum fluorescence. Fast repetition fluorometry interrogates the photosystem more subtly by gradually saturating it with a rapid series of light pulses. The electron transport rate within the photosystem can be calculated knowing the fluorescence characteristics, leading to a measure of photosynthesis in terms of oxygen evolution which can then yield a rate of carbon fixation (Kolber and Falkowski, 1993). The great attraction of this technique is that it has the potential to provide instantaneous, *in situ* measurements; the instrument can be fitted alongside a profiling CTD and could yield productivity data on scales compatible with our knowledge of the physical structure and short-term temporal variability of the water column. However, while comparisons with ^{14}C incubations do show good agreement (Kolber and Falkowski, 1993; Moore *et al.*, 2006), the technique is not yet ready to be employed as an alternative to ^{14}C. So far the real strength of the FRR fluorescence technique is in laboratory analyses of water samples collected by the CTD, allowing determination of the photo-physiological state of the phytoplankton community (Suggett *et al.*, 2009).

5.1.4 Measuring water column production and the photosynthesis-radiation curve

There are two basic aims of collecting measurements of primary production at different depths. The first is to be able to calculate a depth-integrated estimate of primary production (in units of $g\,C\,m^{-2}\,d^{-1}$), where the integration goes from the sea surface to the base of the *photic zone* (generally taken to be where PAR drops to 0.1% or 1% of its surface value). Second, by plotting how the primary production rate varies under different levels of light we can measure several parameters that provide us with information on the physiology of the phytoplankton, which are vital inputs to numerical models.

At sea, collecting the samples for experiments on primary production begins with the biogeochemists carrying out a CTD profile before dawn. In our work in the northwest European shelf seas, this means typically at 0300. The reason for this very early start is that the water samples need to be stored in bottles without being exposed to daylight; taking samples from deep in the surface layer and exposing them to daylight as they are taken from the CTD rosette bottles could damage the phytoplankton photosystems. Once the samples are collected there are two procedures commonly used. The first is mainly carried out in order to estimate net rates of water column photosynthesis, while the second allows the calculation of a curve describing the variation of photosynthesis with light. We will look at them both in some detail.

Simulated *in situ* productivity measurements

In the *simulated in situ* ^{14}C method, samples are taken from depths set by the light distribution, e.g. close to the surface, and at, say, 0.1%, 1%, 14%, 33%, 55% and 97% of surface PAR; the choice tends to be driven by the capacity for making the measurements and the filters available to mimic the light gradient. Determining the corresponding sampling depths requires a measure of K_{PAR}, which can be based on a light profile collected the previous evening. The bottles containing the water to be

Figure 5.4 Biochemists exercising their plumbing skills, setting up the apparatus for a simulated *in situ* incubation experiment on the aft deck of the RRS *Discovery*, June 2010. The boxes will contain the bottles of incubating samples, with surface seawater continuously pumped through them to keep the samples cool. Different light filters are fitted over the boxes to reduce the light received by the samples, mimicking the light variation with depth below the sea surface. (Photo by J. Sharples.)

incubated are spiked with ^{14}C and are then placed in large containers on a clear area of the ship's deck, exposed to daylight between dawn and dusk. Seawater from the sea surface is continuously pumped around the sample bottles to prevent them from over-heating. The light received by the samples is controlled using neutral-density and blue filters over the containers, attempting to simulate both the reduction of light and the change in the spectrum with depth. Each light 'treatment' will have several replicates to provide some statistical reliability to the production estimates. At dusk, some of the replicates at each light level are removed and analysed, yielding the net uptake of carbon during daylight hours. The others are left in the incubators overnight, subsequently analysed to yield net carbon uptake over 24 hours and, by comparing with the daylight uptake values, provide a measure of carbon loss during the night. An example of such an experimental set-up is shown in Fig. 5.4.

Short-term incubations and the photosynthesis-PAR curve

Understanding how photosynthesis of an autotroph population varies in response to light can be achieved using short-term (1–2 hour) incubations with ^{14}C. A large number of samples are collected from one depth, either set by the light level or perhaps targeting a particular layer of phytoplankton. The water samples are placed in small bottles along with the isotope spike. The bottles are then placed in a laboratory incubator called a 'photosynthetron' where they are incubated, each at

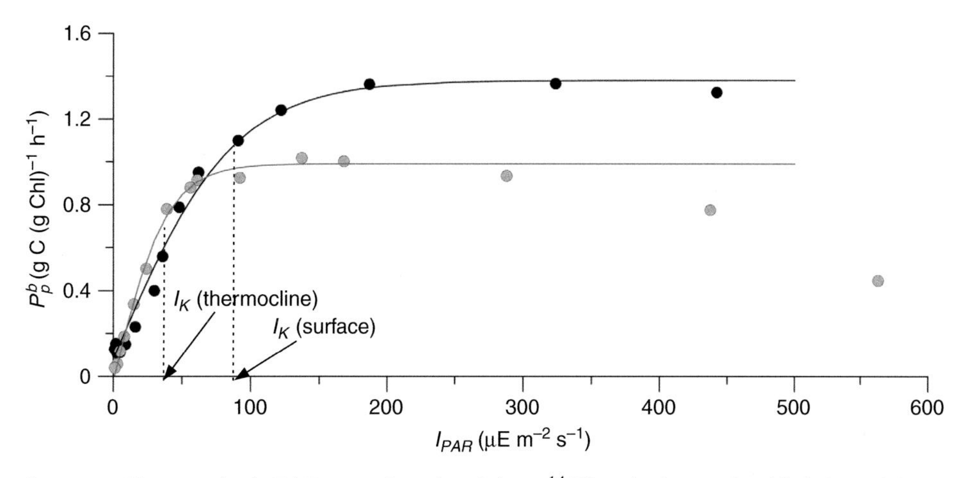

Figure 5.5 Photosynthesis-PAR curve based on 2-hour ^{14}C incubations at fixed light intensities. Samples were from the Celtic Sea surface layer (black) and the thermocline (grey). The curves are the best fits to Equation (5.4). Data courtesy of Anna Hickman, University of Essex UK.

a different light intensity, and then analysed for the amount of carbon taken up. An example of a photosynthesis-PAR curve generated using this technique is shown in Fig. 5.5. As the resource (light) increases, the photosynthesis rate gradually increases to eventually reach a maximum value. This type of response to a resource is common to many other ecosystem components. It is similar to a *Monod growth curve*, named after after the Nobel prize–winning French molecular biologist Jacques-Louis Monod. The curve shape is an example of a type II functional response in ecology, where the capacity for growth is eventually limited by the rate at which the organism is able to process the resource (in this case, the ability of the photosystem to utilise light energy and fix carbon). We can think of the curve as summarising the end result of the host of biochemical processes going on within the phytoplankton cells. It can be fitted to a number of mathematical descriptions of the curve shape (Jassby and Platt, 1976), with the most common being the hyperbolic tangent function. The data in Fig. 5.5 were fitted to:

$$P_p^b = P_{max}^b \tanh\left(\frac{\alpha_q I_{PAR}}{P_{max}^b}\right) + P_{dark}^b \quad \left[\mathrm{g\,C\,(g\,Chl)^{-1}\,h^{-1}}\right] \tag{5.4}$$

with P_p^b the phytoplankton specific photosynthesis rate (photosynthesis rate normalised by the phytoplankton chlorophyll biomass), P_{max}^b is the maximum specific photosynthesis rate and P_{dark}^b is the specific photosynthesis rate in the dark. The parameter α_q (units: g C (g Chl)$^{-1}$ m^2 s µE^{-1} h^{-1}) is the maximum light utilisation coefficient, and represents the slope of the initial part of the photosynthesis-PAR curve. The value of PAR where the photosynthesis roughly reaches its maximum value, called the saturating light, is given by:

$$I_K = \frac{P_{max}^b}{\alpha_q} \tag{5.5}$$

You can see in Fig. 5.5 that at high light intensities the rate of carbon fixation begins to decline away from Equation (5.4); this is particularly clear for the samples taken from the thermocline. This shows the effect of the high light beginning to damage the phytoplankton photosystem, and is called *photoinhibition*.

The deeper, thermocline samples in Fig. 5.5 show the greatest photoinhibition because they were acclimated to the low light at the thermocline; the peak noon light intensity at the thermocline (depth about 45 metres) was 165 $\mu E\ m^{-2}\ s^{-1}$, compared to a mean intensity within the surface mixed layer at noon of 487 $\mu E\ m^{-2}\ s^{-1}$. Thus, exposing the thermocline phytoplankton to high light in the incubation experiment would likely overwhelm their repair mechanisms. Photoacclimation is often seen in terms of changes in I_k, with deeper phytoplankton layers having lower values of I_k compared to populations that experience higher light. This ability to alter their physiology enables cells to find an optimum compromise between light absorption and photosynthesis, and photodamage. Applying Equation (5.5) to the data in Fig. 5.5 shows the surface layer phytoplankton to have $I_k = 87\,\mu E\,m^{-2}\,s^{-1}$ and the thermocline population to have $I_k = 39\ \mu E\ m^{-2}\ s^{-1}$.

Work in stratified shelf seas using active fluorescence has suggested that the acclimation to low light is achieved by increasing the number of reaction centres available to process light energy through photosynthesis (Moore *et al.*, 2006). In other words, the PAR photon flux is low, so the cells make sure that they can utilise the energy of as many of the photons that they intercept as possible, rather than wasting energy through non-photochemical quenching. This increase in the number of reaction centres in a cell will mean that there is an increase in the amount of chlorophyll per cell. Thus photoacclimation is an important consideration when attempting to interpret chlorophyll fluorescence in terms of phytoplankton biomass; calibration formulae calculated by comparing analysis of water samples with fluorometer data can be different at the sea surface compared to deeper in the thermocline, and differ between vertically mixed and stratified regions.

There is one final point to make concerning Fig. 5.5. You may have noticed something a little odd about the intercept of the photosynthesis curves at zero light intensity. We might have expected that P^b_{dark} would be negative; we would not expect photosynthesis in the absence of light. But P^b_{dark} is found to be positive, suggesting that carbon is being fixed by the plankton community in the dark. The reasons for this are not yet entirely clear, though uptake of carbon by heterotrophic bacteria and phytoplankton have been implicated (Li *et al.*, 1993). Estimating P^b_{dark} by fitting Equation (5.4) to the incubation data allows the maximum carbon fixation rate to be corrected for this non-photosynthetic carbon uptake.

5.1.5 Photosynthesis in a turbulent environment: triggering blooms

We will finish our look at the role of light in the growth of the autotrophic phytoplankton by considering how turbulence affects the light environment experienced by the phytoplankton cells, and how rapid phytoplankton growth (phytoplankton blooms) can be triggered. Let us ignore the effects of non-photosynthetic carbon

Box 5.1 A note on units in photosynthesis and primary productivity

Physicists describe energy flux in units of W m^{-2}, which correctly uses the scientific standard of the International System of Units. Biochemists often use an alternative unit for light called the 'Einstein', E, with a flux of light energy described in E m^{-2} s^{-1}. More typically the flux is multiplied by 10^6 and quoted in μE m^{-2} s^{-1}. Named after Albert Einstein, who explained the photoelectric effect in terms of light behaving as discrete 'quanta' or photons, an Einstein is one mole of photons (i.e. 6.022×10^{23} photons). In biochemistry it is one mole of photons within the PAR range of wavelengths. A physicist might initially find that an odd choice of unit, but there is good reason for using moles of photons as a unit. Reactions within the photosystem use photons to convert energy into numbers of atoms or molecules of different substances (e.g. O_2, CH_2O); notice in Equation (5.1) that 8 photons are required to carry out one complete photosynthesis sequence through to carbon fixation. Biochemists use the mole as a fundamental number to describe quantities of atoms or molecules, so the Einstein makes sense as a unit of packets of energy that correspond to numbers of atoms or molecules in reactions. To convert a flux of PAR from μE m^{-2} s^{-1} to W m^{-2}, multiply the flux by 0.2174.

Notice also that in the list of symbols at the beginning of the book we have not been specific about the time units used when measuring rates of primary production. Generally rates of photosynthesis are measured in per hour (h^{-1}) or per day (d^{-1}). Carbon fixation rates could also be quoted in per year (a^{-1}), as in Fig. 1.1b. The choice depends largely on the problem being addressed; just make sure you remember to check that units are consistent when working with, for instance, Equation (5.4).

uptake, and write Equation (5.4) just in terms of how phytoplankton carbon fixation can increase phytoplankton biomass:

$$P_p^b = P_{max}^b \tanh\left(\frac{\alpha_q I_{PAR}}{P_{max}^b}\right) - \left(r_p^b + g_p^b\right). \tag{5.6}$$

The parameter $r_p^b + g_p^b$ on the right of the equation represents the loss of biomass through phytoplankton cellular respiration and grazing by heterotrophs. For the growth rate, P_p^b, to be positive there needs to be sufficient photosynthesis being carried out to account for these losses, so that

$$\begin{aligned} &P_{max}^b \tanh\left(\frac{\alpha_q I_{PAR}}{P_{max}^b}\right) \geq r_p^b + g_p^b \\ &\Rightarrow I_{PAR} \geq \frac{r_p^b + g_p^b}{\alpha_q} \end{aligned} \tag{5.7}$$

where we have noted that at low light intensity $\tanh(\alpha_q I_{PAR}/P_{max}^b) \to (\alpha_q I_{PAR}/P_{max}^b)$. The light intensity given by $\left(r_p^b + g_p^b\right)/\alpha_q$ is the *compensation light intensity*. If we held

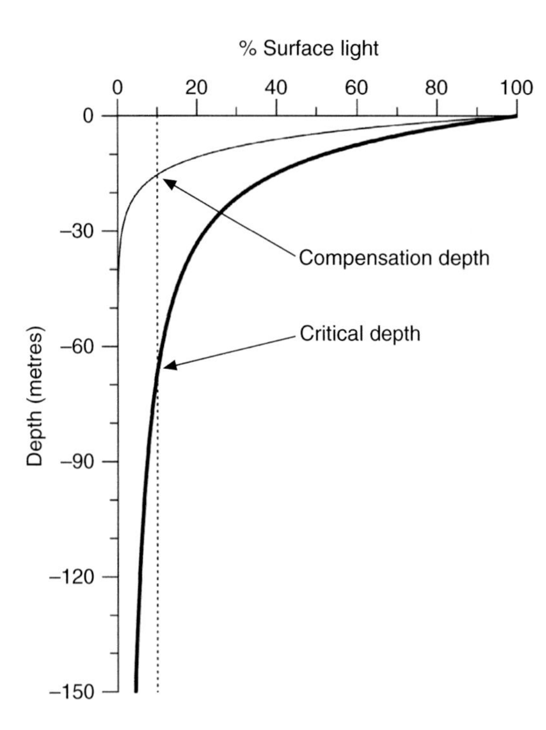

Figure 5.6 Comparing the compensation and critical depths. The thin curve is a light profile with k_{PAR}=0.15 m^{-1}. The bold line is layer-averaged light: at a point on the curve, depth=z, the light is averaged between z and the sea surface. If a cell requires light of 10% of the surface value in order to balance respiration and grazing losses, it would either have to be held at the compensation depth or alternatively mixed continuously between the critical depth and the sea surface.

a phytoplankton cell at the depth at which we found this light level, referred to as the *compensation depth*, photosynthesis would just balance the requirements of cellular respiration and heterotroph grazing.

Note that both r_p^b and g_p^b are not constant. Respiration rate will be temperature dependent, with cellular energy demands typically higher in warmer water. The grazing rate will have a seasonal signal, lower in winter and highest in summer (e.g. Lee *et al.*, 2002). Grazers will also respond to an increase in the biomass of their food supply, which is a topic we will return to later in this chapter.

For much of the time, especially in the near surface layers, the ocean is turbulent, so a phytoplankton cell is more likely to be travelling a random path in response to vertical turbulent mixing. Sometimes it will be near the sea surface, and receiving more than enough light to satisfy Equation (5.7), while at other times it could be much deeper and need to use stored energy to fuel its respiration. It is possible for this turbulent trajectory to take the cell below the compensation depth, but the respiration could later be exceeded by growth when nearer the surface and on average the cell may be able to grow. The depth at which the average light received is just able to balance respiration is the *critical depth*, a concept first introduced by Harald Sverdrup in 1953 (Sverdrup, 1953). In Fig. 5.6 you can compare these concepts of critical and compensation depths. Notice that they are substantially different: the critical depth is typically about a factor of 4 greater than the compensation depth.

Since Sverdrup's seminal paper, the concept of critical depth has been pivotal in our understanding of how rapid growth of phytoplankton (blooms) are triggered in the ocean. We will look at phytoplankton blooms in more detail in Chapter 6. For

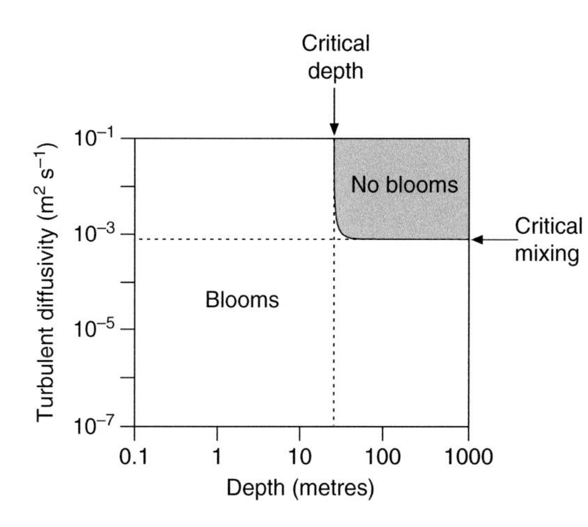

Figure 5.7 Results from a numerical model investigation of the depth or turbulent mixing requirements for bloom development. Blooms can develop in all of the white areas. After Huisman *et al.*, 1999, with permission from ASLO.

now, remember from Section 4.4 that vertical turbulent mixing will be suppressed in parts of the water column where there is a strong vertical density gradient. The development of a thermocline will limit the vertical transport of phytoplankton cells. The key point of Sverdrup's hypothesis is that if a thermocline develops above the phytoplankton critical depth, then the cells in the mixed layer above the thermocline will be able to grow and accumulate biomass.

Now think about phytoplankton cells in a mixed water column, but one in which the strength of the mixing is very weak. The water column may be deeper than the critical depth, but the weak turbulence could result in the residence time of the phytoplankton above the critical depth being long enough to allow growth, and so the phytoplankton population could accumulate biomass in the upper water column. This idea has been formulated as the *critical mixing* hypothesis: a bloom can develop in a mixed water column if the timescale for phytoplankton growth is less than the cells' residence time in the portion of the water column where net growth can be achieved (Huisman *et al.*, 1999). This concept is particularly pertinent to shelf seas as the proximity of the seabed provides a limit to the deep mixing of the phytoplankton cells. We can estimate this residence time as h^2_{crit}/K_z, with h_{crit} the critical depth. The roles of critical depth and critical mixing in bloom formation are summarised in Fig. 5.7 based on model experiments. At high levels of vertical turbulent mixing, a bloom will develop when the depth of water column, or mixed layer, decreases below the critical depth. Alternatively, in a deep water column a bloom can develop if the level of vertical mixing drops below some critical value. The common feature of both hypotheses is a mechanism that allows phytoplankton to receive sufficient light for enough time to allow them to grow.

5.1.6 Nutrient requirements and nutrient sources

Within the phytoplankton cell, nutrients are required in order to synthesise amino acids and nucleic acids (DNA and RNA) (see Falkowski and Raven, 2007). The principal requirements are for nitrogen and phosphorus, with silica also needed by

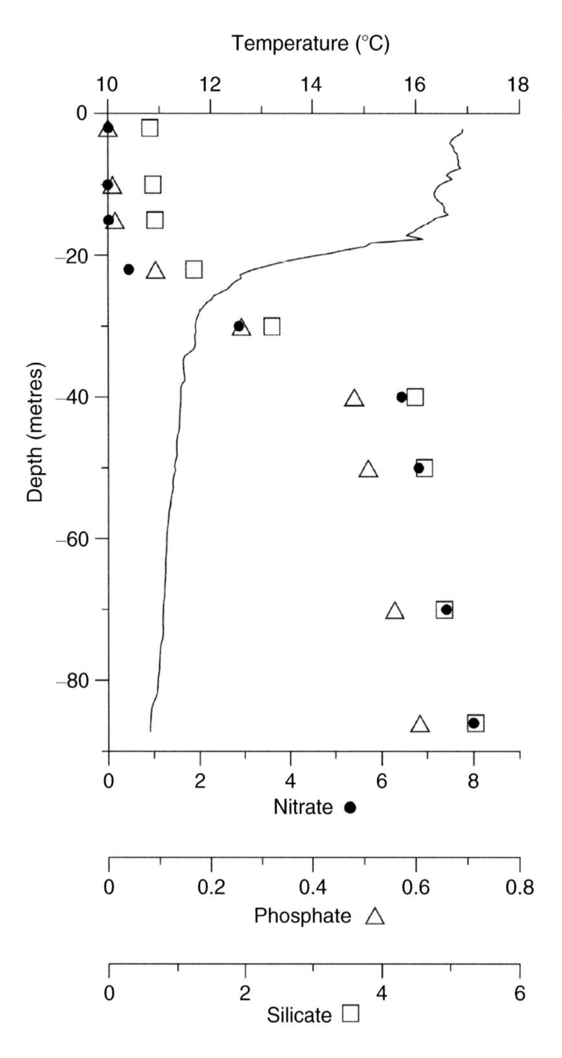

Figure 5.8 A CTD profile from the Celtic Sea, July 2003. The solid line is the temperature profile. Symbols are the nutrient concentrations, all mmol m^{-3}.

the *diatoms*. The dissolved inorganic nutrient compounds that provide these (e.g. nitrate, ammonium, phosphate and silicate) are referred to as *macronutrients*. In addition, a number of other nutrients are required in small quantities. The role of iron in controlling phytoplankton growth has received particular attention (e.g. (Martin and Fitzwater, 1988; Coale *et al.*, 1996; Boyd *et al.*, 2000)), but copper, nickel and zinc are also known to be important (Morel *et al.*, 1991). These are the *micronutrients*. Our focus is on the inorganic macronutrients, mainly because in coastal and shelf seas the combination of river inputs and suspension of material from the seabed mean that an adequate supply of micronutrients is almost always available.

The limiting nutrient in shelf seas

In Fig. 5.8 you can see a typical vertical profile of the macronutrients from our work in the Celtic Sea. The samples shown in Fig. 5.8 were collected in a thermally

Box 5.2 Units of nutrient concentration

Nutrient concentrations are normally quoted in terms of moles per unit volume of seawater; for instance, mmol m^{-3} (millimoles per cubic metre) which is the same as μmol l^{-1} (micromoles per litre, also written as μM and referred to as 'micromolar').

If concentrations are quoted in terms of mass per unit volume, we need to make it clear that we are dealing with either the mass of the nutrient molecule, or the mass of the nutrient atom within its molecule. For instance, the mass of 1 mol of nitrate (NO_3^-) is the mass of 1 mol of N plus 3× the mass of 1 mol of O (the 3× because we have 3 oxygen atoms in O_3):

$$\text{mass of 1 mol } NO_3^- = 14 + (3 \times 16) = 62 \text{ grams}$$

In this case we would quote the units as, for instance, μg nitrate l^{-1}. Alternatively we could identify the mass of the nitrogen component within the nitrate, in which case the amount in mol would be multiplied by 14 to get the mass in grams. The concentration would then be quoted in units of, for instance, μg $NO_3^- - N\ l^{-1}$.

The same approach can be used to convert molar concentrations of phosphate and silicate into mass concentrations of phosphate (or phosphorus) and silicate (or silica).

stratified water column, with the thermocline situated above the critical depth; this situation will be described in more detail in Chapter 6. The difference between the upper and lower layer macronutrient concentrations is due to the uptake of the nutrients by the phytoplankton in the surface layer. The demand for inorganic nitrogen, available in the form of nitrate, NO_3^-, has resulted in complete removal of nitrate from the surface layer. Phosphate (PO_4^{3-}) and silicate (SiO_4) are also diminished in the surface layer but they are still found in measureable concentrations, although phosphate is barely detectable at the sea surface. This pattern of nutrient distributions, with nitrate being the *limiting nutrient* to phytoplankton growth, is a common feature of shelf seas. Note also that, while the silicate still appears to be relatively high, diatom growth typically becomes limited for silicate concentrations of less than ~2 mmol m^{-3} (Egge and Aksnes, 1992).

Nutrient sources

There are four important boundaries in a shelf sea across which nutrients can be supplied: the coast, the shelf edge, the seabed sediments and the air-sea interface. The largest fluxes of nitrate and phosphate are thought to come from the deep ocean, supplying about 80–90% of the nitrate and 50–60% of the phosphate required by phytoplankton growth in the shelf seas (Liu *et al.*, 2010). In coastal regions, large amounts of nitrogen and phosphate are also supplied down rivers as a result of runoff from fertilised agricultural land and wastewater treatment. Averaged over an entire shelf sea, these riverine sources are generally not as important as the shelf edge. However, the localised nature of river sources means that the nutrient concentrations in the adjacent coastal waters can be very large, resulting in *eutrophication* and significant water quality problems arising from the phytoplankton growth that they fuel. We will discuss eutrophication in more detail in Chapter 9.

Nitrogen can also be supplied from the atmosphere, in the form of ammonium (NH_4^+) from the treatment of animal wastes and nitric oxide (NO_x) from the combustion of fossil fuels (Jickells, 2006). The impact of atmospheric ammonium inputs tends to occur close to the coast, while the reactions in the atmosphere required to transform nitric oxides to nitrate can result in long transport times before deposition to the ocean. Down-wind of heavily populated regions the atmospheric source of nitrogen can rival river inputs. However, unlike riverine nutrient sources, there is so far no evidence of atmospheric sources triggering excessive phytoplankton growth (Jickells, 2006).

Finally, bacterial action within the sediments of the seabed will remineralise organic material, such as dead phytoplankton, and supply the water column with nutrients. Measurements of nitrogen fluxes out of sediments suggest that nitrogen is released mainly in the form of ammonium in shallow waters, with the importance of nitrate increasing in deeper shelf waters (Rowe and Phoel, 1992; Hopkinson *et al.*, 2001). Nitrate fluxes out of the sediments of the Irish Sea have been measured between 0.3 and 0.8 mmol m^{-2} d^{-1} (Trimmer *et al.*, 2003). This sediment source of nitrogen to the overlying water can be important in regions with extensive shelves, where the weak residual flows (see Section 3.7) are likely to limit the direct influences of either fluxes across the shelf edge or nutrients supplied by rivers.

The discussion above has mainly concentrated on nitrate as the source of nitrogen to the phytoplankton. Ammonium is generally found in very small quantities, partially because of the oxidation process but also because it is energetically cheaper for phytoplankton to assimilate nitrogen from ammonium than from nitrate: when using nitrate, phytoplankton have to first convert it to ammonium, so if ammonium is available they will preferentially take it up. If it is not used quickly, for instance when organic material is being recycled in the sediments, ammonium is oxidised by bacteria to form nitrite (NO_2^-), and further oxidised to yield nitrate, a process known as *nitrification*. Another obvious source of nitrogen to seawater is nitrogen gas from the atmosphere, which is dissolved in seawater and forms by far the most abundant form of nitrogen in the marine environment. The chemical bonds holding the N_2 molecule together are very strong, making it inaccessible to most of the photo-autotrophs. However, there are specialised organisms in the ocean, the nitrogen-fixers or diazotrophs, which are able to fix atmospheric nitrogen to ammonium. The most important of the marine diazotrophs is the cyanobacterium *Trichodesmium* (Capone, Zehr, *et al.*, 1997), a species which occurs mainly in the near-surface waters of the deep ocean where nitrogen is severely limiting in the surface layer. Nitrogen fixation also occurs in coastal environments as seen, for example, in the western English Channel (Rees *et al.*, 2009). This source of atmospheric nitrogen to the ocean is vital. Up until the

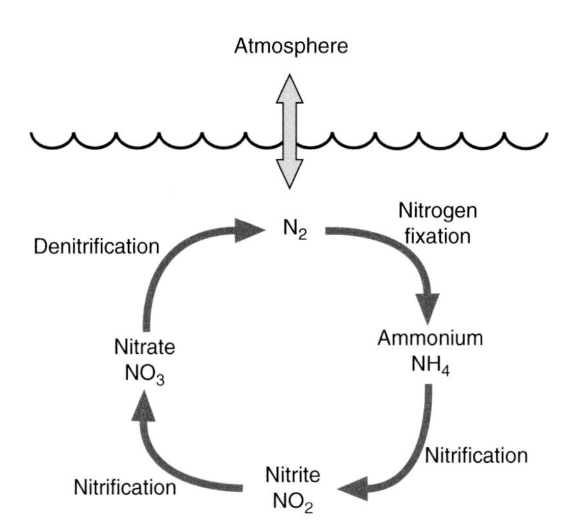

Figure 5.9 A schematic of the nitrogen cycle. Each of the curved arrows within the cycle represents a process carried out by bacteria.

invention of the Haber process,[2] nitrogen fixation by marine and terrestrial plants was the dominant source of fixed inorganic nitrogen to the ocean (Galloway *et al.*, 2004), and on very long timescales the nitrogen-fixing bacteria are thought to govern the global pool of inorganic nitrogen available to the photosynthetic autotrophs (Karl *et al.*, 2002).

The reduction of nitrate back to nitrogen gas, N_2, by bacteria is called *denitrification*. This sink for nitrate generally takes place in regions depleted in oxygen. It is of particular importance in estuaries, removing much of the nitrate before it reaches the shelf seas, and also within sediments and the water column on the shelf (Nixon *et al.*, 1996; Hydes *et al.*, 2004). You can think of the nitrogen cycle as beginning with atmospheric N_2 being fixed by the diazotrophs, with different species of bacteria then controlling the transfers of nitrogen from ammonium, through to nitrite and nitrate, and eventually denitrification returning N_2 to the atmosphere. This is summarised in the schematic illustration in Fig. 5.9. This bacteria-mediated cycle is independent of the rest of the marine primary production. The autotrophic phytoplankton simply tap into the nitrogen cycle, temporarily diverting some of the fixed nitrogen to aid their own growth.

One final point to note concerns the supply of silicate to the ocean. About two-thirds of the silicate that enters the ocean is from the weathering of rocks on land and subsequent transport down rivers into the coastal ocean, with another one-third from volcanism and hydrothermal vents (Demaster, 1981). The cycling of silicate within the shelf seas, and the physical transports towards and across the shelf edge, are thus important controls on the amount of silicate that reaches the open ocean.

[2] The Haber process, developed by the German chemist Fritz Haber in early 20th century, is a vitally important industrial method of fixing atmospheric nitrogen to ammonium in the production of agricultural fertilisers. Today global anthropogenic fixation of nitrogen by the Haber process, along with a smaller contribution from fossil fuel burning, fixes a similar amount of nitrogen as all marine and terrestrial plant nitrogen fixers. See J. N. Galloway *et al.*, (2004). Used to grow crops, this large amount of fixed nitrogen eventually finds its way into rivers and the coastal ocean, leading to eutrophication.

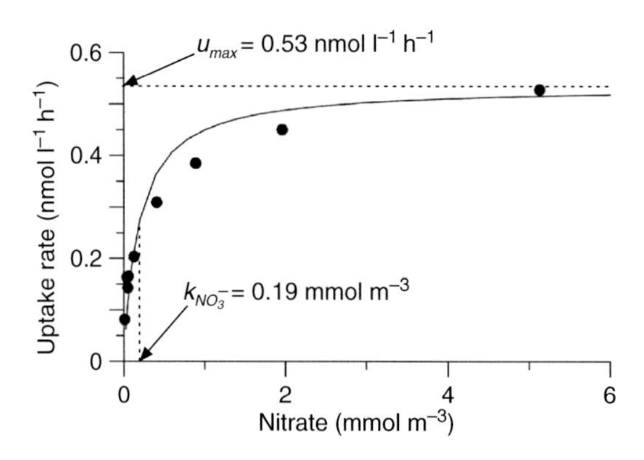

Figure 5.10 Nitrate uptake data for *in situ* incubations of phytoplankton samples at different nitrate concentrations. The fitted curve corresponds to Equation (5.8). Data provided by Mike Lucas (University of Cape Town, South Africa) and Mark Moore (University of Southampton, UK).

5.1.7 Nutrient uptake by phytoplankton cells

While a phytoplankton may have plenty of light for photosynthesis in the surface layer of the ocean, limiting access to nutrients could inhibit growth. In other words, there might be energy available to excite electrons in chlorophyll and begin the photosynthetic process, but without nutrients the lipids and amino acids required for growth cannot be generated.

Nutrient quota in phytoplankton cells

Nutrients are absorbed from the surrounding seawater, across the cell membrane and into the cell. Nutrients are stored within the cell, with different phytoplankton groups, different species and even different sizes of phytoplankton within the same species having varying capacities to store vital nutrients. One particular bacterium has been found to store nitrate in vacuoles at concentrations 3000 to 4000 times that in ambient seawater (McHatton *et al.*, 1996). Storing nitrate in this way has been suggested as a mechanism by which diatoms might compete for nitrate in environments with varying nitrate supply (Stolte and Riegman, 1995). This general ability to take up nutrients beyond the immediate demands of photosynthesis has led to the *cell quota* description of phytoplankton nutrient status and its impact on growth, first suggested by Michael Droop on the basis of extensive laboratory experiments (Droop, 1968). Such laboratory experiments are carried out on phytoplankton cultures of a single species, so that the quota of nutrient can be expressed in terms of an amount per cell or as mass of nutrient per mass of phytoplankton biomass.

Measuring nutrient uptake

Measurements of nutrient uptake on samples of phytoplankton collected at sea are made using similar incubation techniques to those employed in determining carbon fixation. Samples are 'spiked' with different quantities of filtered seawater with a known nutrient concentration, thus increasing the samples' nutrient concentrations by different amounts. At the end of the incubation period the reduction in nutrients in the samples is taken as a measure of the uptake rate by the phytoplankton. An example of an uptake curve generated by such an experiment is shown in Fig. 5.10. The data have been fitted to a curve describing *Michaelis-Menten uptake kinetics*:

$$u_{NO_3^-} = \frac{u_{\max} s_{NO_3^-}}{k_{NO_3^-} + s_{NO_3^-}} \tag{5.8}$$

with $u_{NO_3^-}$ the nitrate uptake rate, u_{max} the maximum nitrate uptake rate and $s_{NO_3^-}$ the nitrate concentration. The parameter $k_{NO_3^-}$ is the half-saturation concentration for nitrate uptake. The shape of the curve in Fig. 5.8 looks similar to the growth-PAR curves of Fig. 5.5, with u_{max} playing a similar role to P^b_{max}, and $k_{NO_3^-}$ analogous to I_k. As with photosynthesis Equation (5.8) is an empirical description of a complex set of biochemical processes associated with uptake (see Chapter 5 of Williams and Follows, 2011 for a detailed analysis of why the curve has this form).

The role of cell size in nutrient uptake

The uptake of nutrients through the cell wall from the ambient dissolved pool surrounding the cell is dependent on the size of the cell. Uptake results in a depleted zone of nutrients close to the cell wall, with molecular diffusion of the nutrient molecules transporting nutrients down the concentration gradient towards the cell. A simple analysis of this situation leads to the following dependence of the maximum cell volume-specific nutrient uptake rate, u^v_{max} in mmol m^{-3} s^{-1} (see Chapter 2 of Kiørboe, 2008 for a derivation):

$$u^v_{max} = \frac{3k_m s_\infty}{a_p^2} \tag{5.9}$$

where k_m is the molecular diffusivity, a_p is the phytoplankton cell radius and s_∞ is the concentration of the dissolved nutrient well away from the influence of the cell.

Equation (5.9) immediately tells us that small cells are better adapted than larger cells at acquiring nutrients, and explains the dominance of small phytoplankton throughout the open ocean *oligotrophic* gyres where nutrients are in very low concentrations. Uptake by larger cells is limited by the rate at which molecular diffusion can resupply the cell wall with nutrients. Indeed, with the inverse squared dependence on cell radius we might reasonably ask why large cells are ever able to compete successfully; we will return to this question when we discuss grazing and phytoplankton communities.

Phytoplankton growth: incorporating light and nutrient limitation

We now have two possible descriptions of phytoplankton growth rate. Growth in response to light is formulated as:

$$P^b_p = P^b_{max} \tanh\left(\frac{\alpha_q I_{PAR}}{P^b_{max}}\right) \tag{5.10}$$

Using the cell nutrient quota $Q^N = P_{Nut}/P_C$, with P_{Nut} the amount of nutrient in the cell and P_C the amount of cell carbon, growth rate in response to the internal nutrient supply, μ^N, can be written as:

$$\mu^N = \mu^N_{max}\left[\frac{1-(Q^N_{min}/Q^N)}{1-(Q^N_{min}/Q^N_{max})}\right] \tag{5.11}$$

with μ^N_{max} the maximum growth rate. The quotas Q^N_{min} and Q^N_{max} are the minimum (sometimes referred to as 'subsistence') and maximum nutrient quotas within the cell. Remember that cell respiration and grazing would have to be applied to both calculations of growth rate. Now imagine a phytoplankton cell in the nutrient environment of Fig. 5.8. In the bottom layer, the cell may have a high nutrient quota and a correspondingly high growth rate from Equation (5.11). However, there is not a lot of light in the bottom layer, so the growth rate predicted by Equation (5.10) would be very low. Conversely, a cell in the surface layer has access to plenty of light, but could be starved of sufficient nitrate, in which case Equation (5.11) would predict a lower growth rate. One method of incorporating the possibility of either light or nutrients being limiting to growth is to calculate growth rates based on Equations (5.10) and (5.11) and use whichever is the lower (e.g. Tett *et al.*, 1986). This is an application of 'Liebig's Law of the Minimum', where growth is controlled by the scarcest resource. An alternative approach is to use the cell nutrient status to moderate the maximum photosynthetic rate, i.e. in Equation (5.10) we replace P^b_{max} (e.g. Geider *et al.*, 1998):

$$P^b_{\max} \rightarrow P^b_{\max}\left[\frac{Q^N - Q^N_{min}}{Q^N_{max} - Q^N_{min}}\right]. \tag{5.12}$$

Redfield ratios

In 1934 the American oceanographer Alfred Redfield reported a remarkable consistency in the relative amounts of carbon, nitrogen and phosphorus in marine organic matter (Redfield, 1934). Observations show that, on average, the numbers of carbon, nitrogen and phosphorus atoms in organic material are found in the ratio C:N:P = 106:16:1. This basic piece of information about marine microbial life, called the *Redfield ratio*, has become fundamental to our understanding of ocean biogeochemistry. Observations can sometimes show departures from this Redfield ratio (Arrigo, 2005), often as a result of the phytoplankton optimising resource allocation during growth or because many phytoplankton are able to store nutrients internally. Also, notice in Fig. 5.8 that the N:P ratio in the bottom water is about 11, rather than the Redfield value of 16. This will be the result of denitification by bacteria. However, on average, the observed C:N:P ratio is frequently close to the Redfield value. Why the ratio should have the particular values that it does is not entirely understood, though it must be telling us something fundamental about the interactions within microbial communities and between the microbes and the physical and chemical marine environment.

5.1.8 Phytoplankton species

There are possibly about 5000 species of phytoplankton in the ocean (Sournia *et al.*, 1991) ranging from <0.5 μm to >100 μm in size. We will see below that their role in the ecosystem is largely determined by cell size. Small cells do not sink and are food for similarly small heterotrophs, while large cells can sink rapidly, exporting carbon to depth. The larger phytoplankton also provide a food source for larger

Figure 5.11 See colour plates version. Some examples of autotrophic phytoplankton. (a) The coccolithophorids *Emiliania huxleyi* and *Gephyrocapsa mullerae*; (b) An armoured dinoflagellate *Gonyaulax*; (c) A naked dinoflagellate *Gymnodinium*; (d) An armoured dinoflagellate *Protoperidinium*; (e) The warm water diatom *Planktonella sol*; (f) A diatom *Thalassiosira*. Images supplied by Alex Poulton (National Oceanography Centre, UK), Jeremy Young (University College London, UK) and Daria Hinz (University of Southampton, UK).

heterotrophs, including fish larvae and small fish, and so are the main link to higher *trophic levels* (i.e. progressively larger organisms) and ultimately commercial fisheries. We will briefly describe the characteristics of the main autotrophic groups, starting with the smallest. Figure 5.11 shows some examples of the phytoplankton.

Cyanobacteria

The smallest photosynthesisers in the ocean are the *cyanobacteria*, roughly 0.2–2 μm cell size. These are *prokaryotes*, meaning that they have no cell nucleus. Numerically, they are the most dominant group of phytoplankton and they carry out almost all of the primary production within the surface waters of the open ocean. Their dominance of the open ocean is a result of the advantage in being small in regions with very low nutrient concentrations (i.e. remember the $1/\text{radius}^2$ dependence of nutrient uptake in Equation 5.9). The cyanobacteria have existed for about 3.5 billion years and have played a vital role in the evolution of life on Earth. Photosynthesis by cyanobacteria led to our oxygen-rich atmosphere. The photosynthesising *eukaryotes*, cells with a nucleus including all the larger photo-autotrophs that we describe below and all terrestrial plants, evolved by incorporating cyanobacteria into their own cells (a process called 'endosymbiosis').

Flagellates and dinoflagellates

Flagellates and *dinoflagellates* are named because of having one (flagellates) or two (dinoflagellates) flagella, whip-like appendages that drive locomotion of the cell through the water. They are eukaryotes with cell sizes ranging roughly from 2 to a few $\times 10$ μm, with the flagellates generally being at the smaller end of the size range. It has been suggested that the ability to propel themselves through the water may promote nutrient uptake by steepening the nutrient gradient close to the cell wall, although this has been shown to have a negligible effect for typical dinoflagellate cell sizes (Kiørboe and Saiz, 1995; Karp-Boss *et al.*, 1996). Swimming of a phytoplankton cell also allows vertical migration along light and nutrient gradients. Such behaviour has been seen in response to cell nutrient requirements, both in laboratory and natural conditions (Eppley *et al.*, 1968; Cullen and Horrigan, 1981).

A typical range of swimming speeds achievable by phytoplankton is $v_c \approx 0.1$–$0.5\ \text{mm s}^{-1}$ (Ross and Sharples, 2007). Given the frequent occurrence of significant turbulence in shelf seas, we might ask if such motility could lead to a phytoplankton cell having any appreciable control over its position in the water column. In terms of swimming being a useful mechanism for positioning plankton vertically within the water column, we need to compare the strength of the swimming capacity with the turbulent mixing that could disrupt it. A basic assessment of whether or not plankton can swim successfully against turbulent mixing can be made by calculating the *Péclet number*:

$$Pe = \frac{\text{mixing time scale}}{\text{swimming time scale}} = \frac{L_s^2/K_z}{L_s/v_c} = \frac{v_c L_s}{K_z} \tag{5.13}$$

where L_s is a suitable length scale, such as the thickness of a mixed layer or the thickness of a nutrient gradient. If $Pe \ll 1$ then turbulence dominates over motility, while for $Pe \gg 1$ motility directed towards resources will be a viable strategy. If we set L_s as 10 metres and using the swimming speeds for a dinoflagellate noted above,

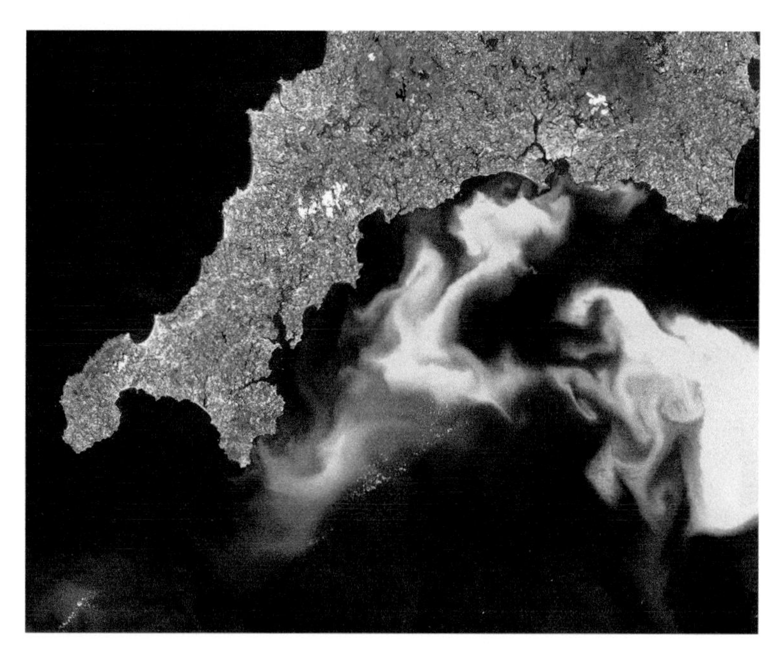

Figure 5.12 See colour plates version. A bloom of the coccolithophores *Emiliania huxleyi* in the western English Channel, June 1999. Image Courtesy NASA/JPL-Caltech.

then we might expect flagellates or dinoflagellates to be able to have some control on their position in a turbulent environment with $K_z < 10^{-3}\ m^2\ s^{-1}$. Tidal flows in shelf seas can drive turbulent diffusivities well in excess of this value in the bottom layer of the water column. However, in the thermocline we typically find $K_z \sim 10^{-5}\ m^2\ s^{-1}$, so swimming could be a useful trait. An observational assessment of different plankton swimming capabilities against the turbulent environment has demonstrated how the interaction between swimming speed and turbulence controls the contrasts in the vertical distributions of different plankton groups (Maar *et al.*, 2003). Clearly the vertical structure of turbulent mixing and the gradients in light and nutrient resources are important considerations. In Chapter 7 we will describe some examples where motility can provide access to light or nutrients in a shelf seawater column.

Coccolithophores

The coccolithophores range between about 10 and 100 μm in size, and are covered in small calcite disks (coccoliths) secreted out of the cell (see Fig. 5.11a). The function of these disks is not known, though blooms of coccolithophores have a substantial effect on the scattering of light (Balch *et al.*, 1991) and can raise surface water temperatures by ~1°C compared to adjacent non-bloom waters (Garciasoto *et al.*, 1995). The most abundant coccolithophore is *Emiliania huxleyi*, which is frequently seen to develop large blooms in stratified shelf waters during summer, as in the example in Fig. 5.12. The coccoliths are very resistant to decomposition, thus leading to the formation of limestones such as the White Cliffs of Dover, UK, and also providing the basis for a paleoceanographic proxy of past sea surface conditions (e.g. Colmenero-Hidalgo *et al.*, 2004). Coccolithophores do have a pair of flagella, similar to the dinoflagellates, so may be able to position themselves depending on resource needs in low turbulence environments.

Diatoms

The diatoms tend to be among the largest of the phytoplankton, with single cells between 10 and 100 μm and with some species forming large aggregations or colonies up to 1 mm or so in size. Diatoms frequently appear as the first and dominating species in phytoplankton blooms, just as conditions become favourable for phytoplankton growth, and are, therefore, regarded as the 'opportunists' within the phytoplankton. A defining characteristic of a diatom is that the cell is enclosed in a silica shell (a frustule), and as a result the diatoms are the only phytoplankton group that requires silicate as a nutrient. The reasons for the silica frustules are not entirely clear. The energy required to synthesise a frustule is less than that needed to make an organic cell wall by about 8% (Raven, 1983), while the silica may act as a chemical buffer aiding the absorption of carbon from seawater (Milligan and Morel, 2002). The mechanical strength of the silica walls, compared to an organic membrane, could also provide protection against grazing (Hamm *et al.*, 2003).

Diatoms do not have any ability to swim, but cell buoyancy has been linked to physiology. Actively growing diatoms are often found to be neutrally or positively buoyant, while physiological stress (e.g. nutrient deficiency) leads to cells sinking (Richardson and Cullen, 1995). This physiological switching of cell buoyancy is similar in outcome to the dinoflagellate swimming capability described earlier, with downward motion taking the cells to higher nutrient concentrations. As with the dinoflagellates, we might consider motion through the fluid as an aid to nutrient uptake by steepening the nutrient gradient across which molecular diffusivity acts to transport nutrient molecules to the cell wall. Analysis does show that for large diatoms (cells or colonies ~1000 μm) there could be an increase in nutrient supply by a factor of 10–100 (Kiørboe and Saiz, 1995). However, sinking to deeper water when in need of nutrients means that diatoms are dependent on vertical turbulent mixing to bring them back towards the sea surface. There are also observations of some species using positive buoyancy to return to the photic zone (Villareal, 1988; Villareal, 1992).

Contrasts in community structure

There is no simple rule to define contrasts between the relative abundances of phytoplankton groups in shelf and open ocean phytoplankton communities. Representatives from all the groups are found in the open ocean and on the shelf. Contrasts in the size-structure of phytoplankton communities are instead more correlated with contrasts in turbulence: larger-celled species tend to be found in more turbulent environments, as turbulence is necessary to transport sinking cells back towards the photic zone and nutrient fluxes are typically greater in more turbulent environments. Thus shelf phytoplankton communities, while including cyanobacteria, will have a larger proportion of dinoflagellates and diatoms compared to the subtropical oligotrophic gyres. Conversely, in the temperate subpolar gyres, turbulence is frequently injected into the surface ocean by extra-tropical storms and so phytoplankton community structure can be similar to that on the shelf.

5.2 The fate of organic matter: recycling, carbon export or food for heterotrophs

Broadly, there are three possible routes for organic material to take once it has been fixed by the phytoplankton. First, it could be recycled by the heterotrophic bacteria, releasing inorganic compounds back into the water column. Over some time scale dependent largely on where in the shelf seawater column the recycling takes place, e.g. the photic zone, the deep water or the sediments, the inorganic carbon and nitrogen will again become accessible to the autotrophs. Second, if the organic material can reach the waters below the winter thermocline of the open ocean, or be buried rapidly in the sediments of the open ocean or the seafloor of the continental slope, then it has been exported away from the phytoplankton and from contact with the atmosphere for some considerable time. Quantifying such long-term export of organic carbon is a key part of determining the role of the oceans in absorbing and removing carbon from the atmosphere. Third, the organic material could be consumed by larger heterotrophs, supporting a marine food chain that reaches up to large predatory fish, marine mammals and commercial fisheries.

We will first describe the export of carbon, and point to the role that the shelf seas are thought to play globally. Then we will look at how organic material can support the different components of the heterotrophic community.

5.2.1 Carbon export

When quantifying how much carbon is removed, or exported, from the surface ocean, we need to understand how the primary production that fixed the carbon was supported by nutrients. The bacterial recycling of organic material that generates ammonium will also be remineralising organic carbon. If this recycling is taking place in the photic zone, then both the ammonium and the recycled carbon can be utilised by the phytoplankton and the primary production will not require a source of carbon from elsewhere. Importantly this means that the growth will not require any flux of CO_2 from the atmosphere to the ocean. This primary production is referred to as *regenerated production*. If, however, there is a source of nitrate to the phytoplankton community, say from below the seasonal thermocline, then additional carbon may be fixed. Primary production using such a new supply of nitrogen is called *new production* and is important because it has the potential to result in a flux of CO_2 from the atmosphere to the ocean as the phytoplankton utilise the *dissolved inorganic carbon* (DIC) in the surface water. The ratio of new production to total production is the *f-ratio*, first defined by Richard Eppley and Bruce Peterson in an investigation of the flux of *particulate organic carbon* (POC) to the seafloor (Eppley and Peterson, 1979). If the phytoplankton using this source of nitrate are able to leave the surface layer to the deeper water or to the sediments (e.g. by sinking), then the POC has been exported. Assuming a steady state over the region and time in question, new production is also treated as *export production*, and the *f*-ratio equated with the *export ratio* (Laws *et al.*, 2000).

The role of cell size

The sinking speed of a phytoplankton cell can be estimated by looking at the balance between the frictional force between the cell and the water, the buoyancy of the cell arising from the contrast in density between the cell and the water and the gravitational force pulling the cell downward. The physicist George Stokes calculated the frictional force for the case of very small Reynolds numbers (see Chapter 4), which is applicable to phytoplankton cells in water. The resulting settling velocity of the cell is given by:

$$w_{cell} = \frac{2}{9}\frac{(\rho_p - \rho)}{\mu} g a_p^2 \tag{5.14}$$

with ρ_p the density of the cell surrounded by water of density ρ, $\mu = 1.1 \times 10^{-3}$ N m^{-2} s is the dynamic viscosity of seawater, and remembering that a_p is the radius of the phytoplankton cell. Note that there is a strong dependence on cell size; larger cells sink faster than small cells. In the phytoplankton community, this typically means that dinoflagellates, coccolithophores, and particularly diatoms can export carbon, while small flagellates and the cyanobacteria do so much less efficiently. A cell needs to be bigger than about 5 μm before it is viewed as playing a significant role in the vertical transport of carbon to depth (Legendre and Rivkin, 2002).

The Redfield ratio and achieving carbon export

We need to be careful in determining this potential for carbon export arising from new primary production. The water that is mixed upward with a quantity of nitrate will also contain DIC. If everything was in a strict Redfield balance then that source of DIC would be exactly what is required by the phytoplankton when they utilise the nitrate, and additional carbon from the atmosphere would not be required. Carbon export requires the Redfield balance to be contravened at some stage. For instance, source waters could have an excess of nitrate compared to the Redfield C:N. In the open ocean deep water can contain 'pre-formed' nutrients arising from before the water sank away from the surface, plus nutrients added later due to the recycling of sinking detritus (Redfield *et al.*, 1963). Alternatively, the C:N ratio of the particulate organic material that sinks to the seabed could have an excess in C due to different rates of recycling of the C and N components of the particles (Kiriakoulakis, Stutt, *et al.*, 2001; Schneider, Schlitzer, *et al.*, 2003), leading to a relative excess of C potentially buried and an excess of inorganic nitrogen left behind in solution after the material's journey downward through the water column. There is also an important contrast in the fate of sinking POC between the margins of the ocean and the open ocean. Slowly sinking down into the deep ocean, POC has plenty of time to be recycled back to DIC before it reaches the seabed. As a result, relatively little particulate carbon reaches the bed sediments, and the deep ocean is an enormous reservoir for DIC which is locked away from the atmosphere for the time scales of ocean circulation and overturning (decades to centuries). The proximity of the seabed of the shelf and continental slope to the photic zone makes the margins of the oceans important regions for longer-term particulate carbon export. Sinking POC has less time to be recycled, and so organic carbon has more chance of reaching the sediments than in the open ocean. In areas with high sedimentation rates (e.g. canyons, to which we will return in Chapter 10) or where bacterial processes are limited by low dissolved oxygen, the carbon has the potential to be buried in the seabed sediments and locked away from the atmosphere for millions of years.

Shelf seas and carbon export

The export of fixed POC is called the *biological pump*. We would expect the exported carbon to be the same as the amount of carbon taken from the atmosphere, so quantifying the biological pump is a vital part of understanding the fate of carbon within the climate system. Shelf seas play a disproportionately large role in global export production. The average global export ratio for carbon is estimated to be about 20% (Laws *et al.*, 2000). However, export ratios in shelf regions could be larger. For instance, the proximity of riverine sources of minerals that can either ballast POC (making it sink faster) or protect POC from bacterial degradation as it sinks, would raise export ratios (Dunne *et al.*, 2007). Also, temperature effects on autotrophic and heterotrophic growth rates suggest the mid to high latitude shelves can have export ratios of over 60% (Laws *et al.*, 2000). Taking a reasonable increase of the continental shelf export ratio of a factor of 3 above the oceanic mean, POC export from the ocean margins could account for 44% of the global total (Jahnke, 2010).

In the open ocean the concept of new production is tied to a supply of new nitrogen, i.e. nitrogen as nitrate transported upward to the photic zone from below the winter surface mixed layer. The nitrate is described as 'new' because its residence time in these deep waters is of the order of decades to centuries. In shelf seas this concept of new nitrate is subject to some ambiguity. Close to the shelf edge the nitrate may well be sourced from deeper in the ocean. However, particularly in regions of wide continental shelf, the weak residual flows in many shelf seas may mean that nitrate mixed upward into surface waters may have been remineralised from organic matter only a few months previously. It is then arguable whether the nitrogen is 'new' or 'recycled'. Generally we stick with the original spirit of the Eppley-Peterson definition and use the *f*-ratio to describe the proportion of primary production that uses nitrate.

The proximity of the deep water and seabed recycling communities to the photic zone is important in shelf seas. It means that recycled organic material is always within reach of the sea surface, particularly in temperate and high latitude shelf seas where winter convective mixing results in efficient connection between the seabed and the sea surface. This provides one of the important contrasts with the open ocean, and a reason for the relatively high production rates that we see on the shelf.

In broad shelf seas, away from the edge of the continental shelf, carbon export to the shelf sediments may only remove the carbon from access to the atmosphere for seasonal time scales. In temperate shelf seas particularly, the connection between the sediments and the atmosphere in winter allow DIC in the deep shelf water or released from the sediments to achieve equilibrium with the CO_2 in the atmosphere. Carbon arriving in the bottom layer of a shelf sea water column, or settling on the seabed, thus may be exported only for a few months, until the next winter period of convective mixing. Long-term export of carbon requires the carbon to reach the deeper waters of the open ocean below the winter mixed layer depth, or the sediments of the continental rise and slope. The physical processes that might lead to such export of shelf primary production are the subject of Chapter 10.

5.2.2 Food for the heterotrophs

Heterotrophic organisms need to consume organic material in order to survive and grow, and so they are ultimately reliant on the ability of the autotrophs to fix carbon. Heterotrophs encompass an enormous size range, with the consumption of organic matter being carried out by organisms ranging from bacteria to whales (and humans). It is difficult to root our discussion in some common metabolic process more specific than the basic need for organic fuel. However, the relative sizes of heterotroph grazers compared to their autotrophic food particle provides a basis for understanding the fate of the carbon fixed by the autotrophs. One useful ecological rule is that on average a planktonic predator tends to be about ten times larger than its prey (e.g. Sheldon *et al.*, 1977). This predator:prey size ratio does vary between heterotroph groups (Hansen *et al.*, 1994), but the generally consistent larger predator size has important implications for structuring the phytoplankton community that we will examine in more detail later. This size-dependence of who eats who also has implications for the size of phytoplankton cells that are more efficient at transferring organic matter upward to higher trophic levels. The 5 μm cell size that we earlier noted as important in determining the potential for carbon export via cell sinking is also viewed as the lower size limit for phytoplankton organic material to be transferred rapidly up to larger heterotrophs (Legendre and Rivkin, 2002). As we will see later in this chapter, the same size is viewed as setting the lower limit to particles likely to be grazed by fish larvae. The different size of heterotrophs will consume different sizes of the phytoplankton, and so will play different roles in the recycling, export and trophic transfer of organic material.

As with the phytoplankton, the heterotrophs cover a range of sizes, from bacteria (<1 μm) upwards. We will introduce heterotrophs from the bacteria up to the

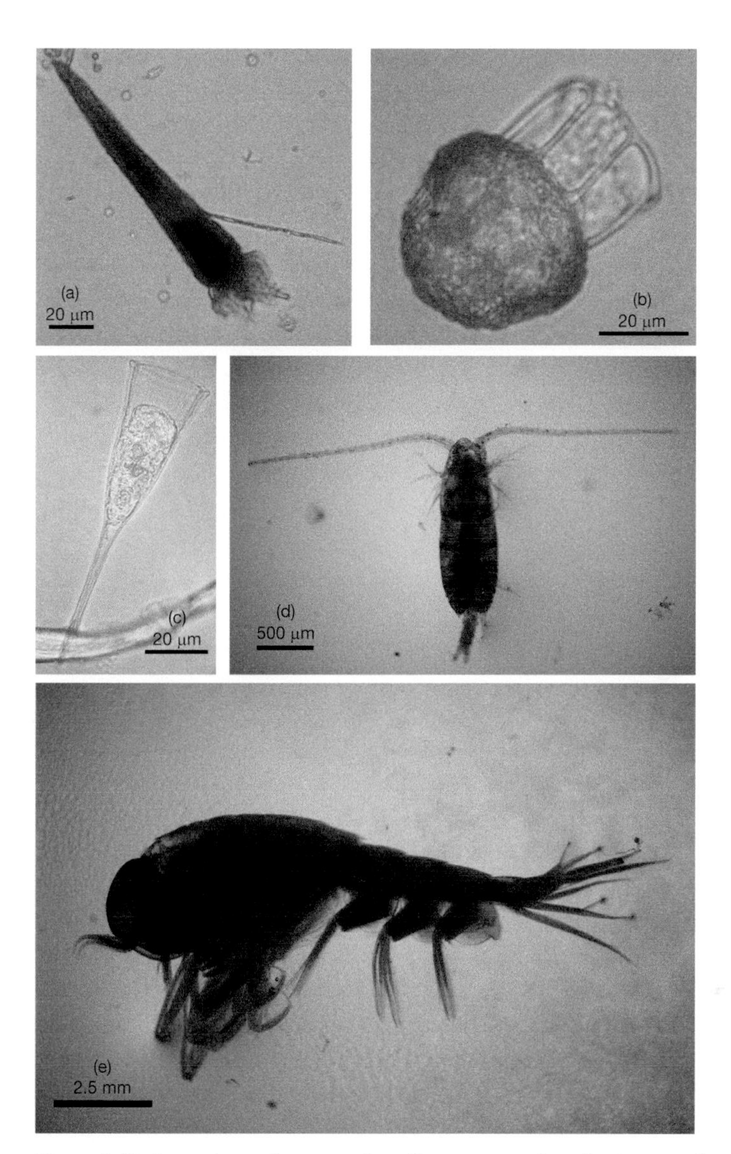

Figure 5.13 See colour plates version. Some examples of micro- and meso-zooplankton. (a) naked planktonic ciliate *Strombidium*; (b) planktonic tintinnid *Dictyocysta*; (c) planktonic tintinnid *Rhabdonellopsis*; (d) a *Calanus* copepod; (e) planktonic *Amphipod*. Images supplied by Alex Poulton (National Oceanography Centre, UK) and Sari Giering (University of Southampton, UK).

mesozooplankton (which include fish larvae), though as consumers of organic material heterotrophs obviously include the fish, marine mammals and ultimately the top predator of them all, we humans. Our focus on the smaller end of the heterotrophic size range is predicated on identifying the organisms that are most likely to be at the mercy of turbulence and flows in the ocean. Figure 5.13 illustrates a few of the many diverse many heterotrophic organisms.

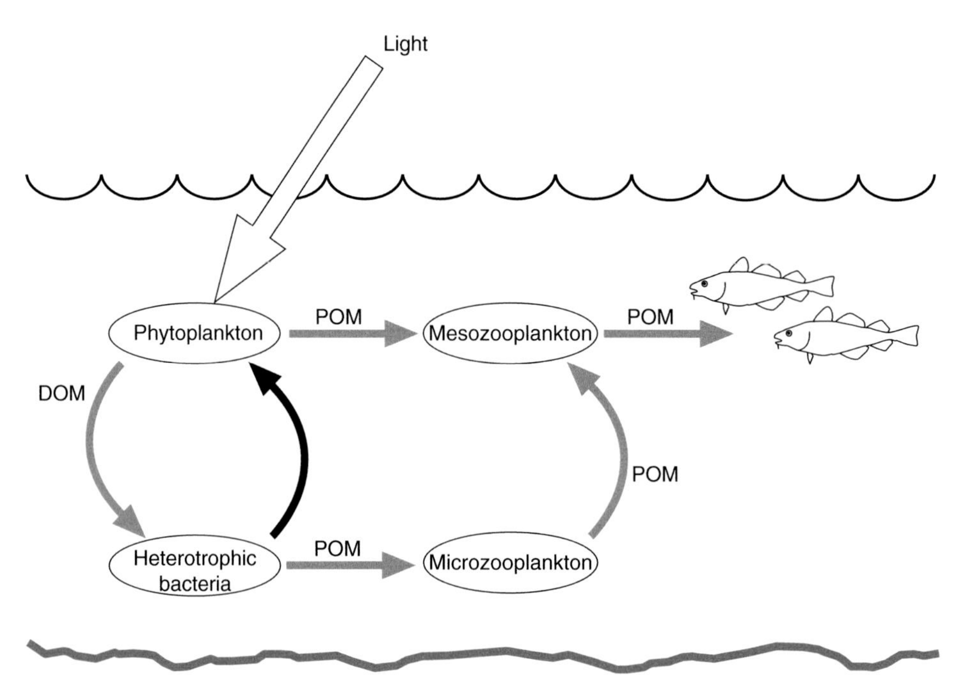

Figure 5.14 Schematic summary of the microbial loop, where heterotrophic bacteria transfer dissolved organic material (DOM) back to the phytoplankton and to the larger heterotrophs. Grey lines indicate pathways of organic material, the black arrow shows the return of remineralised carbon and nitrogen back to the phytoplankton.

Bacteria (size ~ 0.2–2 μm)

In the earlier discussion of nutrient sources and of regenerated production we already met the heterotrophic bacteria as important recyclers of organic material. Part of their recycling role involves mopping up components of organic material that would otherwise never be re-accessible to larger organisms. Much of the organic material generated by the phytoplankton ends up as *dissolved organic matter* (DOM), including *dissolved organic carbon* (DOC) and *dissolved organic nitrogen* (DON). DOM is released into the water by a range of processes, including the breakdown of cells (called cell lysis), leakage from the phytoplankton and inefficient feeding by larger heterotrophs. The distinction between 'dissolved' and particulate organic material (POM) in the water is operational, set by a filter pore size of 0.45 μm. Such tiny particles of organic material are not available to larger heterotrophs, so the bacteria that can consume and recycle DOM play an important role in returning DOM back to the rest of the ecosystem. This return pathway for DOM, referred to as the *microbial loop* (Azam *et al.*, 1983; Pomeroy *et al.*, 2007) and summarised in Fig. 5.14, either supplies organic material into the heterotrophic food chain when the bacteria are consumed by the microzooplankton, or accelerates the remineralisation of organic components which are then again available to the autotrophs. A significant amount of the DOC taken up by the bacteria is released as dissolved inorganic carbon back into the water column, and potentially back to the atmosphere to add to the pool of CO_2. Thus the bacteria help determine the air-sea flux of carbon. The oxidation of organic carbon by bacteria to form DIC removes dissolved oxygen from the surrounding seawater. In coastal regions influenced by riverine sources of nutrients, leading to eutrophication, the subsequent reduction in dissolved oxygen can have serious effects on marine life; we will show examples of this in Chapter 9.

The microbial loop described above is the most general mechanism for returning organic material back to a form suitable for uptake by the autotrophs. Some species of phytoplankton are able to take up dissolved organic nitrogen (DON) and phosphorus (DOP) directly. In regions where inorganic nitrogen has become limiting to autotrophic growth, the largest pool of fixed nitrogen will be the DON, and an adaptation to be able to utilise the DON directly is clearly advantageous. The significance of this nutrient route in shelf seas is not clear; there are some observations suggesting that DON in the form of urea and dissolved amino acids could be important (Tett *et al.*, 2003), particularly for the cyanobacteria (Heil *et al.*, 2007). The source of DON and DOP will dictate if its assimilation by phytoplankton leads to new or regenerated production; for instance, DON arising from the diazotrophs would fuel new production. While the role of dissolved nutrients in shelf seas is not yet clear, there is growing evidence that the shelf seas can be a source of dissolved nutrients that can be transported into the open ocean where inorganic nutrients are severely limiting to phytoplankton production (Torres-Valdes *et al.*, 2009).

Microzooplankton (size ~ 2–100 μm)

The microzooplankton include single-celled heterotrophic flagellates, dinoflagellates, tintinnids and ciliates (Fig. 5.13a–c), along with the early stages of some of the mesozooplankton (e.g. copepod larvae, called nauplii). These tiny grazers feed on bacteria (both the cyanobacteria and the heterotrophic bacteria) and on the autotrophic phytoplankton. The smaller microzooplankton are particularly dependent on the microbial loop as a source of organic material, and so act as a key step in the transfer of DOM back up towards larger heterotrophs. An important point to note about the microzooplankton is that their growth rates tend to be similar to those of their autotrophic prey. As a result we rarely see blooms of small phytoplankton because their predators are able to match their growth rates and so maintain a grazing pressure sufficient to damp down any increases in small-celled phytoplankton biomass. Generally in the ocean, the concentrations of autotrophic bacterial biomass appear to be limited to <0.5 mg Chl m^{-3} as a result of this responsive grazing impact (Chisholm, 1992; Armstrong, 1994).

One other group of grazers that needs to be mentioned are the *mixotrophs*. These organisms, typically bacteria or microplankton, are able to utilise light, inorganic nutrients and prey. They are able to shift the balance between autotrophic and heterotrophic modes of growth depending on resource availability. Observations typically show mixotrophs to form significant components of ecosystems where nutrients are low (Stoecker *et al.*, 1989; Lavrentyev *et al.*, 1998) and in some circumstances

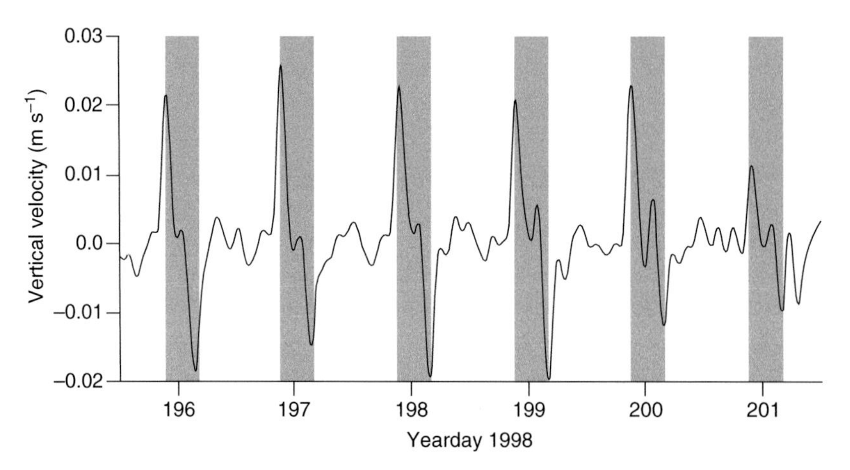

Figure 5.15 Vertical swimming velocities of zooplankton population measured using an upward-looking ADCP on the Malin shelf west of Scotland. The dashed boxes indicate the hours of darkness. From Rippeth and Simpson, 1998, with permission from ASLO.

could out-compete the plankton that rely solely on autotrophy or heterotrophy (Katechakis and Stibor, 2006).

Mesozooplankton (size ~ 100 μm – a few mm)

The mesozooplankton are multicellular animals that feed on the microzooplankton and on phytoplankton. Mesozooplankton can be either holoplanktonic, meaning that their entire life is spent in the plankton, or meroplanktonic, which means spending only the early part of their life as plankton. Holoplankton include the copepods (Fig. 5.13d), krill, ostracods (a small crustacean), chaetognaths (predatory worms), tunicates (or sea squirts, including the salps) and amphipods (Fig. 5.13e). The meroplankton are the larvae of larger marine organisms, particularly fish but also crustaceans and other benthic organisms. We could also include the young of the larger gelatinous plankton (e.g. jellyfish and larger tunicates) which often target the same food supply as the mesozooplankton.

Many of the mesozooplankton are capable of maintaining a sustained swimming speed that is able to compete successfully with turbulent velocities in shelf seas. For instance, a copepod might be able to sustain speed of 10 m hr^{-1}, which if used in Equation (5.13) suggests that the copepod should be able to maintain directed swimming in turbulence as strong as $K_z \sim 3 \times 10^{-2}$ m^2 s^{-1}; mixing at this rate is generated close to the seabed during strong tidal flows, but is rarely seen in the interior of the shelf seawater column. A common feature of mesozooplankton swimming behaviour is vertical migration with a daily periodicity. The most common migration is that of nightly excursions into the surface waters, with rapid downward migration to deeper waters as dawn approaches. An example is shown in Fig. 5.15 based on ADCP measurements of vertical velocity on the Malin shelf west of Scotland. This vertical migration is often viewed as an anti-predation strategy: the plankton are actively avoiding visual detection by predators in the well-lit surface waters during the day. Migration is also seen in the opposite sense: upwards to the surface during the day, and downward during the night. This seems like a dubious strategy for avoiding predators. However, zooplankton undergoing this 'reverse migration' have been seen to switch their migration behaviour to downward during daylight in response to an increase in predators (Frost and Bollens, 1992). Other reasons for a migration strategy have been proposed (e.g. see the summary in Lampert, 1989), including taking advantage of different metabolic rates in warm surface water and deep colder water. Thinking of the physical environment within which migrations take place, we might also suggest that the vertical shear between surface and deeper currents would lead to the zooplankton sampling a different body of surface water at each upward migration. As zooplankton horizontal distributions tend to be patchier than those of the phytoplankton (Abraham, 1998) this would increase the food supply to the zooplankton by giving them access to phytoplankton communities that have potentially received lower recent grazing.

An issue to which we will return to in Chapters 6 and 9 is that many large marine organisms begin their life in larval stages that cannot control their horizontal position in the ocean and so are at the mercy of ocean currents. However, the interaction between vertical position and shear in mean flows can be utilised to achieve net movement. Vertical movement of meroplankton can be *ontogenic*, meaning that their position within the water column depends on their age or stage. We will see some examples later where this ontogenic shift in position within depth-varying mean flows can be a vital component in the survival of populations.

We have already described the importance of new primary production in the carbon cycle. All mesozooplankton play a very important role in the biological pump of carbon from surface waters to the deep ocean and to shelf sediments (e.g. Wassmann, 1998). The mesozooplankton eat relatively small food particles, but they concentrate the waste products into faecal pellets that have much faster sinking speeds than the original prey items. Rapid sinking helps particles reach the seabed and be buried (sequestered) before microbial activity can recycle the organic material. Faecal pellets are also an important food source for heterotrophic bacteria, whose main foraging strategy involves waiting in the water column to be collected by a passing pellet.

The other role for the mesozooplankton, especially important in shelf seas, is the support of fish, including commercial fish stocks. As an example, consider the analysis of the diet of juvenile mackerel found at the shelf edge of the Celtic Sea (Conway *et al.*, 1999), shown in Fig. 5.16. The gut contents show a progression from feeding on phytoplankton for larvae of lengths < 5 mm, to ingestion of copepod eggs for larval lengths between 4 and 7 mm, and consistent ingestion of copepod nauplii for larval lengths > 4 mm. The availability of suitable phytoplankton for the early larval stages, followed by a population of reproducing copepods, is required to sustain the larvae towards recruitment to the adult stock. The semi-digested remains of the phytoplankton inside the larvae were not

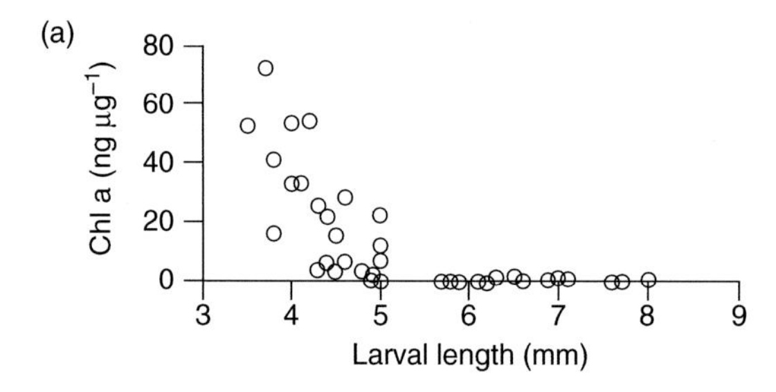

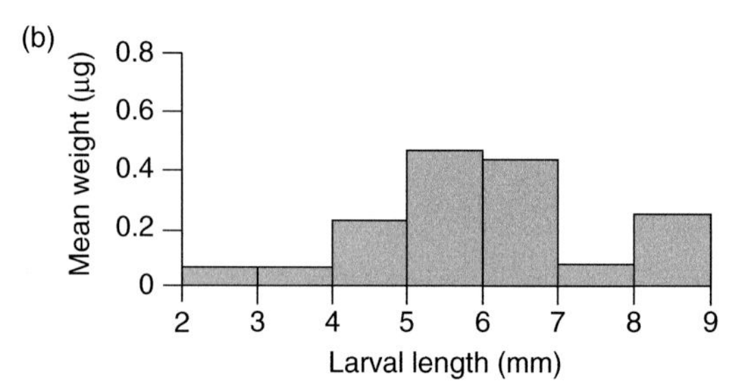

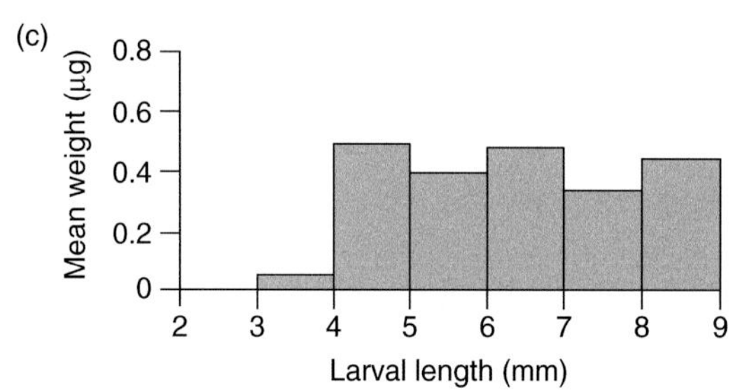

Figure 5.16 (a) Concentrations of chlorophyll in the guts of mackerel larvae, normalised by larval weight. (b) The proportion of gut contents made up by copepod eggs. (c) The proportion of gut contents made up by copepod nauplii. From Conway, *et al.*, 1999, with permission from *Vie et Milieu*.

identifiable. However, it is recognised that first-feeding fish larvae will need food particles > 5 µm (Cushing, 1995), so the larval food will be made up of eukaryotic phytoplankton. We will discuss the relation of these dietary requirements to oceanographic conditions in Chapter 10.

5.2.3 Finding prey in a viscous environment

A fascinating aspect of the heterotrophs is how planktonic predators find and catch their prey in the ocean. The problems posed by locomotion through water, the detection, tracking and grabbing hold of prey items has led to the evolution of a remarkable diversity of mechanisms and foraging strategies amongst the plankton (see Kiørboe 2008 for an illuminating analysis of planktonic life in a viscous environment).

As we are terrestrial organisms it is difficult for us to fully appreciate the challenges presented to tiny organisms living in the sea. The basic problem which plankton face is that their world is dominated by viscosity: almost all of the planktonic heterotrophs live at very low Reynolds numbers. Remember from Chapter 4 that we defined the Reynolds number as:

$$\frac{\text{flow speed} \times \text{length scale}}{\nu} \qquad (5.15)$$

with ν the kinematic viscosity of seawater (about $10^{-6}\ m^2\ s^{-1}$). A 10 μm cell swimming at 0.1 mm s^{-1} would have a Reynolds number of 10^{-3}. By contrast, a 10 cm fish swimming at O(1 m s^{-1}) has a Reynolds number of 100 000. The practical result of this is that a fish (or a human) swims through water by pushing water in the opposite direction to their motion (an example of conserving linear momentum). A plankton instead grabs hold of the water and pulls itself past it (and then has the tricky problem of how to release the water in order to prepare for the next 'stroke'). In the case of a flagellum the cell is not so much swimming as twisting itself through the viscous water, similar to when you drive a screw into a block of wood. Only at the upper end of the plankton size range are some organisms able, at least temporarily, to shift themselves out of the control of fluid viscosity. As an escape response to an approaching predator, a copepod (e.g. Fig. 5.13d) can put on a burst of speed of a few 10s of mm s^{-1}, briefly raising the Reynolds number to 100 or so (van Duren and Videler, 2003).

A good deal of attention has been focused on the role that turbulence might play in altering the encounter rates between predators and prey. Imagine a completely quiescent water column with non-motile predators and prey suspended within it; encounters between the predators and their prey would be rare. If we add some turbulent mixing into the water, then we would expect predators and prey to encounter each other much more often. So, as we gradually increase turbulence we might expect predator-prey encounter rates to increase. That is half of the story. During an encounter, a prey particle needs to remain in detection range of the predator long enough for the predator to respond and attempt to catch the particle. Also, once a prey item has been caught it needs to be hung on to and transferred to the predator's mouth. So, increasing turbulence might be expected to have a negative impact on feeding rate. Figure 5.17 summarises the net effect of these processes, based on a modelling study of cod larvae feeding on copepod nauplii (Mackenzie *et al.*, 1994). Figures 5.17a and 5.17b show the separate effects of turbulence on encounter rates and on catch success, with Fig. 5.17c showing the classic dome-shaped response when the two separate processes are combined. Notice the wind speeds in Fig. 5.17a, based on the wind required to generate the turbulent velocities at a depth of 20 metres in the surface mixed layer. Gale force winds have a speed of about 20 m s^{-1}, so the peak in successful predation occurs at typical wind speeds over the ocean. If we equate wind stress with a stress generated at the seabed by a current, then a 20 m s^{-1} wind is about the same as a 0.5 m s^{-1} current. So moderate tidal flows can lead to an increase in successful predation, but stronger tides will hinder feeding. Further work has shown that the effect of turbulence depends on the feeding strategy employed by the predator, and that for realistic values of turbulence the enhancement of predator-prey interactions only works for predators that are close to the Kolmogorov lengthscale (see Section 4.4.3) (Kiørboe and Saiz, 1995).

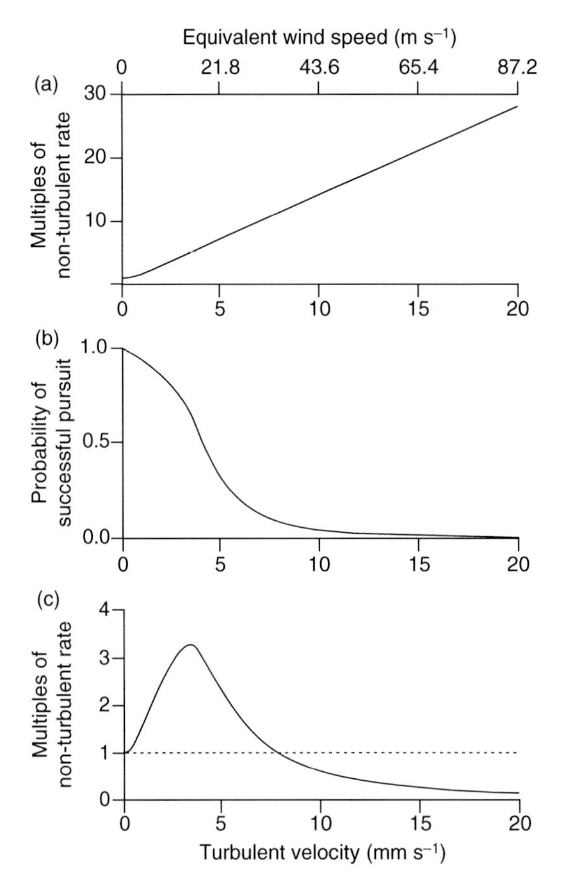

Figure 5.17 (a) The increase of encounter rates between larval cod and copepod nauplii as turbulence increases. (b) The reduction in the success of larvae catching nauplii as turbulence increases. (c) The net effect of turbulence on successful larva-nauplii encounters resulting from combining (a) and (b). From Mackenzie *et al.*, 1994, with permission from ASLO.

Box 5.3 Recycling fish

We tend to view the transfer of organic material through the heterotrophs as a progression of small organisms being eaten by larger organisms, which are in turn eaten by yet larger animals. An interesting observation during a research cruise in the Celtic Sea in summer 2008 illustrates a contrast in the flow of organic matter. PhD student Iñigo Martinez carried out several deployments of a near-bed camera to capture images of fish in the region. The camera was suspended above a weight baited with four mackerel, and took a picture every 1 minute during a deployment that typically lasted for 24 hours. Initial images from the first two deployments were obscured by suspended sediments, caused by the tidal flow, but when the water cleared the images showed that all that was left of the bait were four clean skeletons. Eventuality Iñigo managed to capture images of the culprits: within half an hour of deployment a swarm of isopods (the marine equivalent of woodlice) would find the bait and pick it completely clean within 2 hours. The image below shows the bait, along with a *nephrops* lobster (upper left). The reddish animal in the lower left is an isopod. If you look carefully at the mackerel bait you can also see dozens of isopods attached to the fish.

The isopods were obviously able to detect the bait, swim to it and hang on to it despite tidal flows of over 0.5 m s^{-1}. Isopods are also known to parasitise live fish. The observation reminds us that the fate of carbon that reaches higher in the marine food web is eventually to be consumed and recycled.

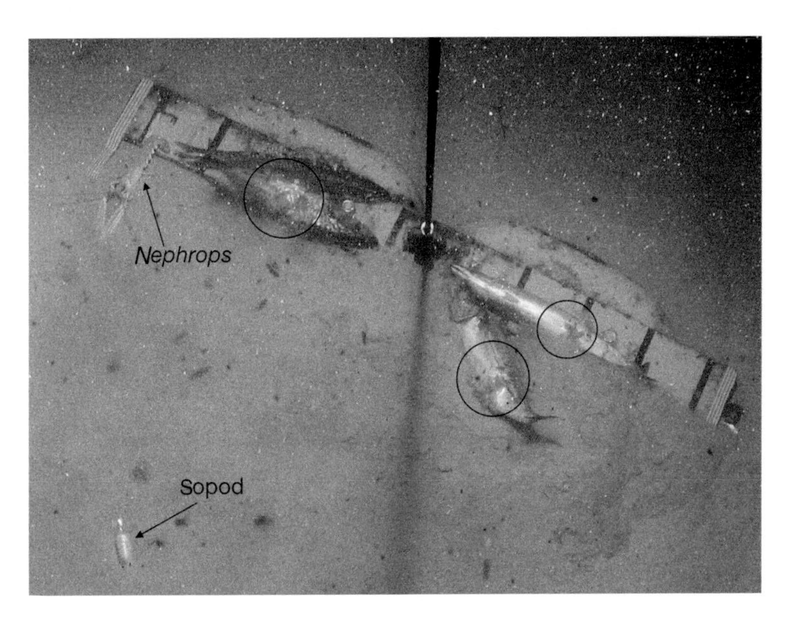

Figure B5.3. An image of the seabed of the Celtic Sea showing isopods feeding on mackerel bait. The circles indicate groups of feeding isopods. Image provided by Iñigo Martinez, Marine Scotland.

5.2.4 The role of grazing in the structure of phytoplankton communities

We can now return to a problem that initially appeared when we considered the role of phytoplankton cell size in the uptake of nutrients. Equation (5.9) shows that when competing for nutrients there is a tremendous advantage to being small in the ocean. So why do we see phytoplankton ecosystems that have a diverse spectrum of cell sizes? In particular, how can large cells such as diatoms compete for nutrients and grow in the ocean? We have seen that motility, either swimming or sinking, only appears to confer an advantage in nutrient uptake for very large cells. The best case would be a factor of about $\times 100$ increase in uptake for a cell or colony of diameter 1 mm, or a $\times 10$ increase for a 100 μm cell. However, such apparently large effects of movement cannot compensate for the $\times 10^4$ or $\times 10^5$ advantage that a 1 μm

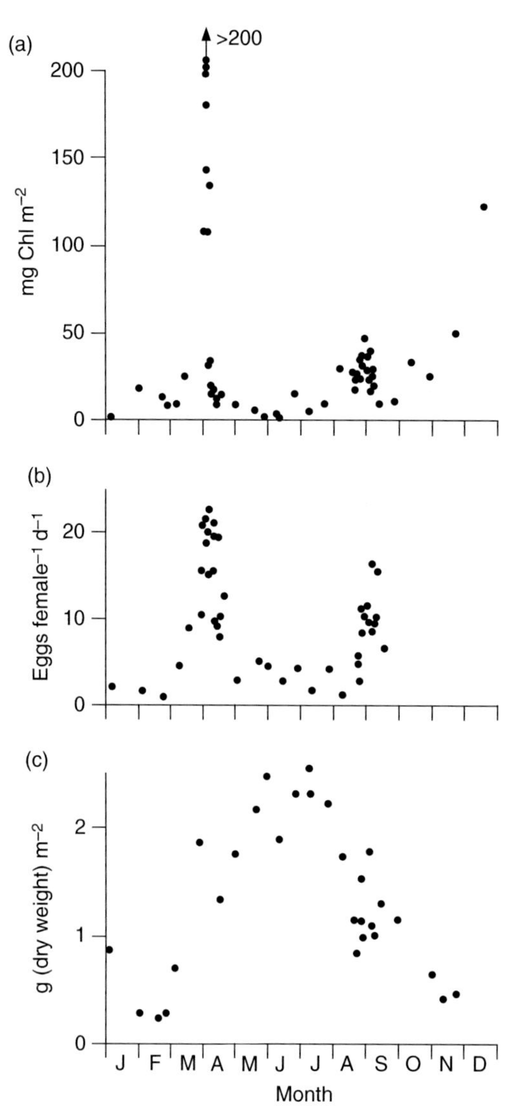

Figure 5.18 Observations from the coastal waters of the southern Kattegat, Denmark, over 1 year. (a) Phytoplankton chlorophyll biomass for cells >11μm. (b) Egg production rate for one species of copepod (*Acartia*). (c) Total adult copepod abundance. Adapted from (Kiørboe, 1993), with permission from Elsevier.

cyanobacteria has in comparison with these large cells because of the radius^{-2} dependence in the maximum uptake rate.

The answer to why diatoms can exist lies with the differences in the grazers of the small and large cells. Earlier we described how the microzooplankton that graze small phytoplankton cells have comparable growth rates to their prey, and so are able to provide a significant *top-down control* on concentrations of small phytoplankton. The production rates of the microzooplankton are often lower than those of the larger phytoplankton that they eat. For instance, there could be a significant delay between the start of conditions more favourable to microzooplankton production (e.g. the onset of a diatom bloom) and the generation of increased numbers of grazers, as the microzooplankton need to mate and produce eggs, and the eggs need to hatch. Depending on the species, the entire process from mating through to new adult copepods could take between a week and several months. This decoupling between the microzooplankton population and their large-celled phytoplankton food resource is nicely illustrated in Fig. 5.18. Notice that the first stage of the copepod life cycle, the production of eggs by the females, is correlated with bloom timing (Fig. 5.18a, b), but by the time the copepods become adults there is no correlation with the bloom (Fig. 5.18c). This explains why diatoms exist; the larger celled phytoplankton have larger but slower-growing predators, and these predators cannot keep up with a sudden spurt of growth in their prey.

Summary

The autotrophic fixation of carbon, driven by the energy provided by sunlight, lies at the heart of the marine food chain and determines the biologically mediated fluxes of carbon between the atmosphere and the long-term storage pools of carbon in slope sediments and the deep ocean. Physical processes, particularly vertical turbulent mixing, control the light and nutrients available to the autotrophs. In the shelf seas the high levels of turbulent mixing, combined with the proximity of the seabed sediments to the photic zone, help to drive the high rates of primary production compared to the open ocean.

The autotrophic phytoplankton range in size from cyanobacteria of O(1μm) size up to diatoms of O(100μm) size. There is a fundamental physical constraint on the potential nutrient uptake by phytoplankton, governed by cell size. Nutrient input to larger cells is severely limited by the rate at which molecular diffusion is able to replenish the zone around the cell wall. Taking advantage of the autotrophic supply of organic food is a grazing community with a size range from the heterotrophic bacteria O(1μm) in size, through to the microzooplankton (O(1–10 mm)) and extending up to fish, marine mammals and humans. In plankton food chains, we can say roughly that predators, or grazers, tend to be a factor of 10 larger in size compared to their prey (a notable exception to this rule-of-thumb is the grazing by bacteria on sinking faecal pellets). The heterotrophic bacteria drive a part of the food chain called the microbial loop, providing a recycling pathway that can return dissolved organic material back to the microzooplankton. Without this microbial loop, much of the organic material in the ocean would never make it into the higher trophic levels.

The existence of large phytoplankton cells is dependent on differences in the rate at which the grazers of different phytoplankton are able to respond to changes in the biomass of their food source. The grazers of the diatoms, mesozooplankton such as copepods and early-stage fish larvae, take time to respond to increases in diatom biomass which allows the diatoms to briefly bloom in response to an influx of

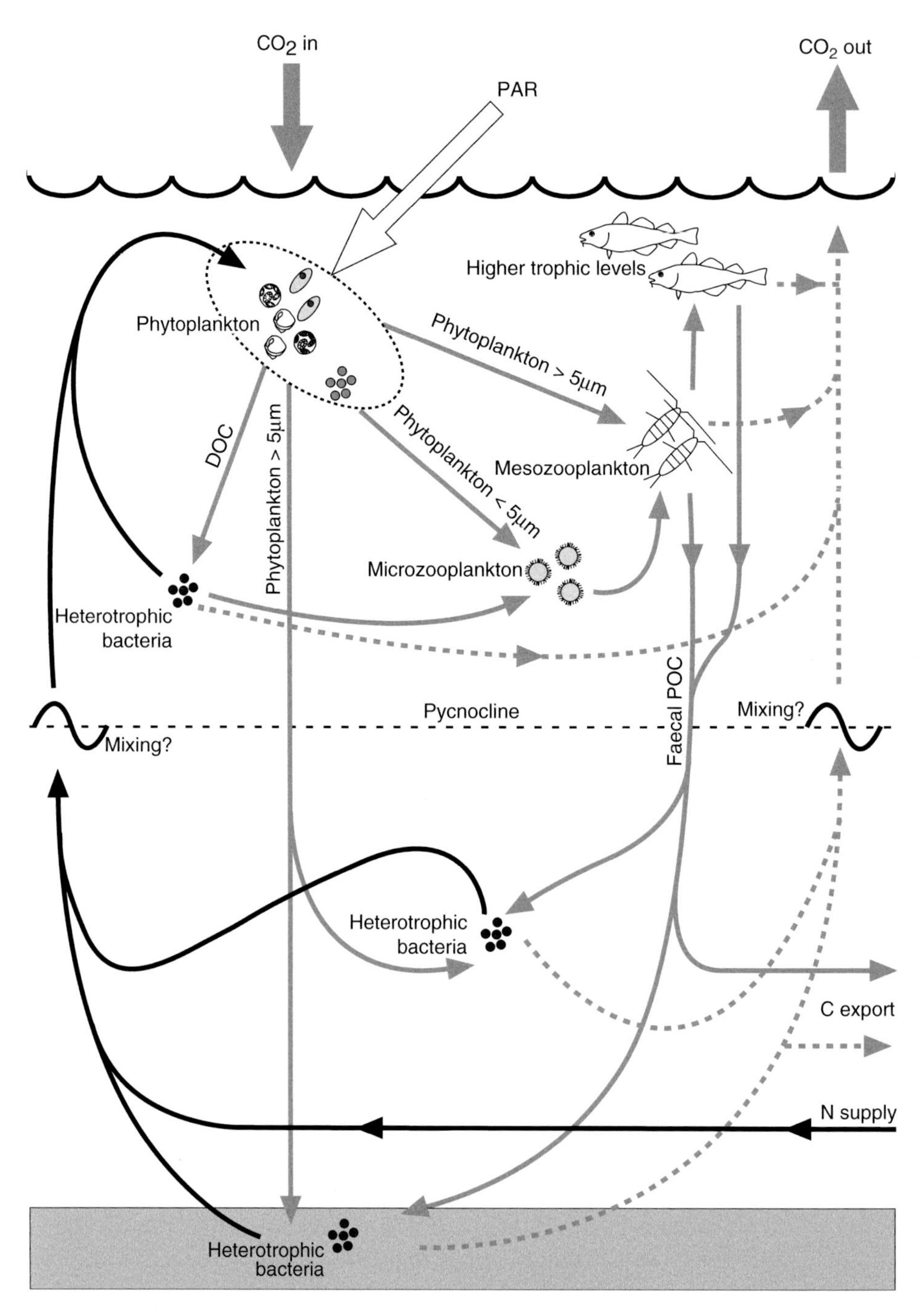

Figure 5.19 The main pathways of carbon and nitrogen in the water column. Black lines show routes for dissolved inorganic nitrogen, solid grey lines show the routes taken by organic material. The dashed grey lines represent dissolved inorganic carbon. The long-term export of carbon to the deep ocean or the continental slope and the source of new nitrate from across the shelf edge will determine the balance of the CO_2 flux across the sea surface. Note that mixing across the pycnocline controls the transfers of dissolved nitrogen and carbon back into surface waters.

nutrients. The large phytoplankton cells, and the faecal pellets released by the mesozooplankton that graze them, form particulate organic carbon that sinks out of the photic zone, so providing a vital export of carbon away from the air-sea boundary and contributing significantly to the flux of CO_2 from the atmosphere to the sea.

Big phytoplankton cells play two key roles in the ocean: exporting carbon and providing food (directly, or indirectly through the mesozooplankton) for higher trophic levels. Both the biogeochemists and the fisheries scientists use a key particle size of 5 μm. Particles above this size are needed for first-feeding fish larvae (Cushing, 1995), and can sink out of the surface layers to export carbon (Legendre and Rivkin, 2002). Particles smaller than 5 μm will take part in the microbial loop; ultimately this can result in their organic material being available to higher trophic levels or to sinking into deeper water, but the additional steps required mean that efficiencies of these transfers will be reduced. Figure 5.19 provides a final summary of the generation and fate of organic carbon in the shelf seas.

FURTHER READING

A Mechanistic Approach to Plankton Ecology, by Thomas Kiørboe. Princeton University Press, 2008.

Aquatic Photosynthesis, by Paul Falkowski and John Raven. Princeton University Press, 2007.

Ocean Dynamics and the Carbon Cycle, by Richard Williams and Michael Follows. Cambridge University Press, 2011.

Population Production and Regulation in the Sea: a Fisheries Perspective, by David Cushing. Cambridge University Press, 1995.

Influence of nutrient biogeochemistry on the ecology of North-West European shelf seas, by P. Tett, R. Sanders, and D. Hydes, in *Biogeochemistry of Marine Systems*, ed. K. D. Black and G. B. Shimmield. Blackwell, 2003, 293–363.

Problems

5.1. Produce a plot that shows how the depth of the photic zone changes with the PAR attenuation coefficient, over the range $0.1 < K_{PAR} < 0.4$ m^{-1}. Take the base of the photic zone to be at 1% of the surface PAR.

5.2. Data shown in Lee *et al.*, 2002 suggest that the grazing rate in the North Sea in winter is 0.08 d^{-1}, and in summer, 0.35 d^{-1}. If the phytoplankton respiration rate is 0.1 d^{-1} and the maximum light utilisation coefficient is 0.12 d^{-1} (W m^{-2})$^{-1}$, calculate the compensation depths in winter and summer. Take the winter surface PAR to be 75 W m^{-2} and the summer surface PAR to be 300 W m^{-2}, and assume $K_{PAR} = 0.15$ m^{-1}.

5.3. Phytoplankton with a growth rate of 0.5 d^{-1} are mixed through a homogeneous deep water column within which the critical depth is 40 metres. What is the maximum vertical eddy diffusivity that would still allow growth of phytoplankton?

5.4. Convert 10 mmol m^{-3} of nitrate into a mass of nitrogen per litre.

5.5. For the eddy diffusivity that you calculated in problem 5.3, estimate the swimming speed that a dinoflagellate would need to have in order to outcompete a neutrally buoyant phytoplankton. (Assume they both have the same growth rate.)

6 Seasonal stratification and the spring bloom

We have now provided the foundations in physics and biology that can be used to understand how shelf sea ecosystems are driven by the underlying physics. In this chapter we will begin our journey through the links between physics and biology in shelf seas by identifying the role of physics in partitioning the shelf seas into biogeochemically contrasting regimes. Recall from Chapter 5 that the vertical structure of the water column is critical in determining the availability of both light and nutrients to phytoplankton. In this chapter we first consider the fundamental physical controls that determine whether or not a region will thermally stratify in spring. We derive a condition for seasonal stratification to occur which indicates that, during the summer season, the shelf seas will be divided into stratified and mixed regimes. We then consider the spring stratification as the physical trigger for the spring bloom of phytoplankton. This highlights the spatial partitioning into mixed and stratifying regions as controlling whether or not regions experience a spring bloom.

6.1 Buoyancy inputs versus vertical mixing: the heating-stirring competition

The vertical structure of the water column is the result of an ongoing competition between the *buoyancy inputs*, due to surface heating and freshwater input on the one hand and stirring by the tides and wind stress on the other. Our focus here is on temperate shelf sea regions where the dominant buoyancy input is a seasonally varying surface heat flux. We will deal with freshwater as a source of buoyancy in Chapter 9. During the winter months, when heat is lost from the surface, the buoyancy term contributes to stirring by increasing surface density and making all or part of the water column *convectively unstable*. As a result, apart from limited areas influenced by large positive buoyancy inputs from estuaries, the shelf seas are vertically well mixed during the winter months. This vertically mixed regime continues until the onset of positive heating at, or close to, the vernal equinox (21 March/September in the northern/southern hemisphere) after which the increasing input of positive buoyancy tends to stabilize the water column. Whether or not the water column stratifies is then dependent on the relative strengths of the surface heating

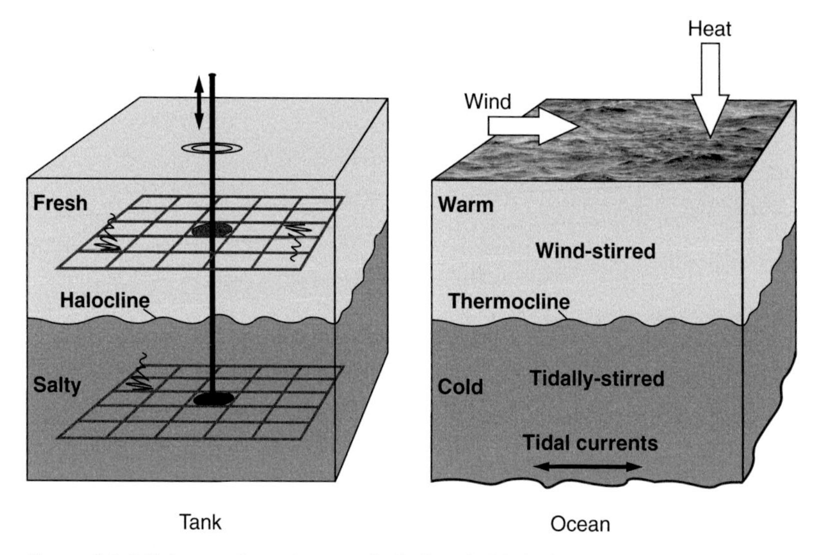

Figure 6.1 Mixing tank analogue of wind and tidal stirring in the ocean. The tank on the left is equipped with oscillating grids to simulate the shelf seawater column on the right which is stirred by wind and tides.

and the stirring due to frictional stresses imposed at the bottom boundary by the tidal flow and at the surface by wind stress.

6.1.1 Mixing and the development of mixed layers

The struggle for control of the water column can be readily visualized in terms of analogous laboratory experiments (Turner and Kraus, 1967; Thompson and Turner, 1975). A perspex tank can be equipped with stirring grids which oscillate vertically to produce turbulence and mixing in two layers, simulating the effects of wind and tide stirring in the ocean, as illustrated in Fig. 6.1. Either heat or salt can be used to set up the initial vertical density gradient. Heat is the more difficult as it leaks by conduction through the walls of the tank, unlike salt which is conserved in the tank.

Consider what would happen if we set up a stable density gradient with salinity increasing with depth in the tank, and then used only the upper grid to generate stirring. A layer of mixed water would develop near the surface with the salinity gradient continuing below it into the lower part of the tank, as shown in Fig. 6.2a. As stirring continues, the mixed layer deepens but at a decreasing rate as the base of the mixed layer moves farther away from the source of turbulent energy in the grid (i.e. more of the turbulence is dissipated within the mixed layer and is unable to work against the salinity gradient). This behaviour is analogous to the deepening of the surface mixed layer in the deep ocean, or in shelf seas with weak tides, as near-surface stratification is eroded by wind stirring.

The analogue for the shelf seas involves stirring by both grids, with the second, lower grid added in order to simulate stirring by turbulence from tidal friction at the bed. The experiment is again started from an initially stable condition in which the density increases uniformly with depth. Stirring by the two grids sets up top and

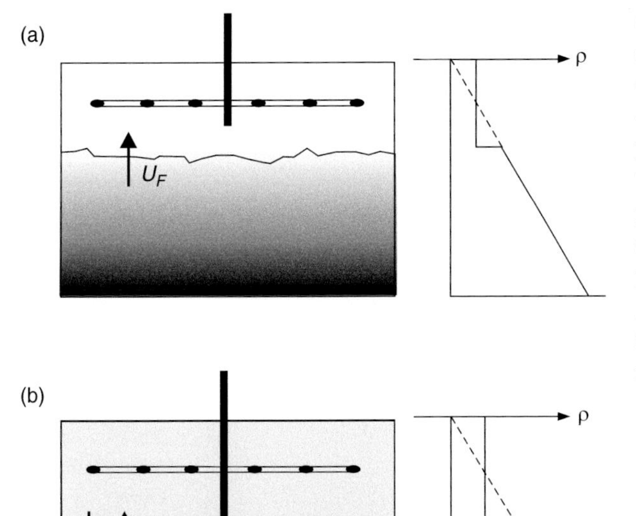

Figure 6.2 The action of (a) a surface grid only, and (b) surface and bottom grids, on an initial linear density profile in the experimental tank. The initial density profile is the dashed line; the solid line is the density profile some time after the grids have been oscillating. U_F is an entrainment velocity of fluid driven by the stirring.

bottom mixed layers separated by a sharp density interface (a pycnocline), as shown in Fig. 6.2b. The interface develops at a point where the turbulent intensities from the two grids are the same and upward and downward entrainment rates are equal; if the two grids are identical and driven at the same rate, this balance will occur halfway between them. If there is no further addition of buoyancy, mixing across this interface slowly erodes the density contrast and the associated stability until mixing is complete and the water column is vertically uniform. This is the first of two important conclusions to draw from this type of experiment: turbulent mixing by stress at a boundary will produce a homogeneous layer associated with that boundary, the thickness of which is dependent on the strength of the turbulence. Measurement of the entrainment velocity, the effective velocity at which fluid is being transferred across the pycnocline by the stirring, provides further insight into how turbulence acts on density gradients. Figure 6.3 shows the results from experiments carried out by J. S. Turner using grid-generated turbulence in a laboratory tank.

Notice in Fig. 6.3 that the ability of turbulence to transfer fluid across the density interface is significantly reduced as the stability and the Richardson number at the interface increase. This is a second important result from such experiments, which is fundamental to understanding how a water column responds to a buoyancy supply. If buoyancy is added to the surface layer at a sufficiently high rate, it may out-compete the reduction of density contrast by stirring and stratification may be sustained or even enhanced. Indeed, the laboratory results show that once stratification starts to increase, it acts to reduce the 'efficiency' of mixing and makes it easier for stratification to develop further. We met the theoretical basis for these laboratory results, encapsulated in the Richardson number, in Section 4.4.1. This positive

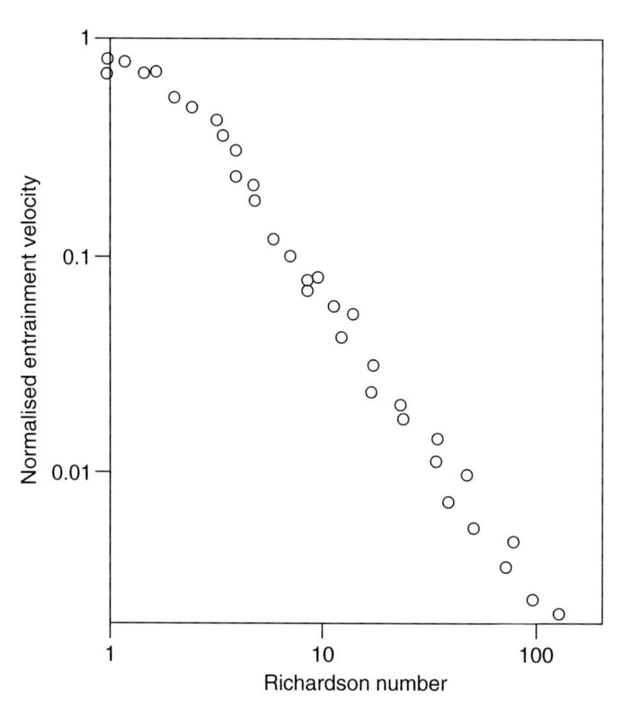

Figure 6.3 The entrainment velocity across the interface in a grid-stirred tank experiment, as a function of a measure of the Richardson number. The entrainment velocity has been normalized to a measure of the turbulent velocity generated by the oscillating grid. [Figure will use open circles only for the case of density controlled by salinity changes.] Adapted from Turner, 1973, with permission from Cambridge University Press.

feedback in the growth of stratification also occurs in the full-scale shelf sea system so that, once stratification is established, it tends to grow and is not easily destroyed by episodes of enhanced stirring until the onset of cooling and convective mixing in the autumn. We shall return to this interesting feedback aspect in Section 7.4, but we now turn to an analytical representation of the heating/stirring competition in its simplest form that we can then apply to the real ocean.

6.1.2 Criterion for water column stratification (the energetics of mixing by the tide alone)

We shall formulate the basic competition between the stratifying and stirring agencies in terms of a stratification parameter Φ (Simpson, 1981) defined as:

$$\Phi = \frac{1}{h}\int_{-h}^{0} (\hat{\rho} - \rho(z))gz\,dz \quad [\mathrm{J\,m^{-3}}] \tag{6.1}$$

with the water column mean density

$$\hat{\rho} = \int_{-h}^{0} \rho(z)dz \quad [\mathrm{kg\,m^{-3}}] \tag{6.2}$$

Here $\rho(z)$ is the density profile determined from temperature and salinity profiles. The parameter Φ, referred to as the *potential energy anomaly*, is a quantitative

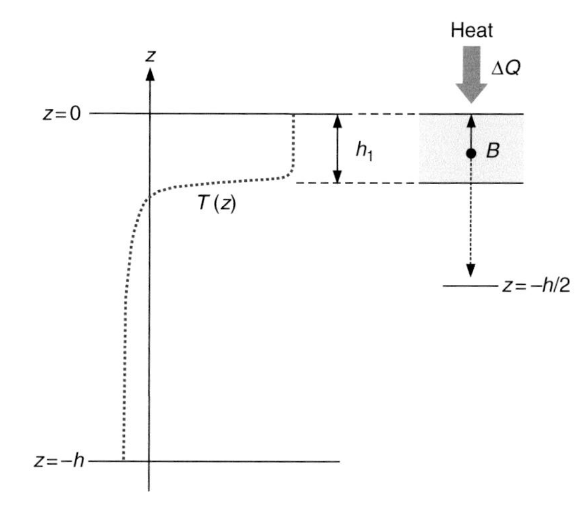

Figure 6.4 The input of buoyancy due to surface heating.

measure of stratification and represents the work required per unit volume to bring about complete mixing of the water column. We can see this by splitting the integral into its two components,

$$\begin{aligned}\Phi &= \frac{1}{h}\left\{\int_{-h}^{0} \hat{\rho} g z dz - \int_{-h}^{0} \rho g z dz\right\} \\ &= \frac{1}{h}\{\mathrm{PE(mixed)} - \mathrm{PE(strat.)}\} = \frac{\Delta \mathrm{PE}}{h},\end{aligned} \tag{6.3}$$

which are measures of the potential energy (PE) per unit area of the water column before and after mixing. Φ is zero for a fully mixed column and becomes increasingly positive as stratification increases.

Before proceeding further, we make two simplifying assumptions, namely (i) that buoyancy input is only via surface heat exchange and (ii) that stirring is due only to tidal flow. Both these restrictions will be removed later on, but for the moment they provide valuable simplifications of the problem.

We start the analysis by asking how the input of buoyancy by heating will change Φ. Consider an amount of heat per unit area ΔQ [J m^{-2}] entering the mixed water column as solar radiation at the start of the seasonal heating cycle (soon after the vernal equinox). As we saw in Section 2.2.1, this heat input will be absorbed close to the surface in a thin layer, shown in Section Fig. 6.4, which we represent as a mixed layer of thickness h_1. Within this layer the temperature will be increased by:

$$\Delta T = \frac{\Delta Q}{c_p \rho_0 h_1} \quad [^\circ\mathrm{C}] \tag{6.4}$$

where c_p is the specific heat capacity of seawater (4000 J kg^{-1} °C^{-1}). This rise in temperature will reduce the density by:

$$\Delta \rho = \alpha \rho_0 \Delta T = \alpha \rho_0 \frac{\Delta Q}{c_p h_1 \rho_0} = \frac{\alpha \Delta Q}{c_p h_1}, \tag{6.5}$$

where α is the volume expansion coefficient of seawater. We can think of this lower density layer as floating on the surface with a buoyancy force of:

$$b = g\Delta\rho = \frac{\alpha g \Delta Q}{c_p h_1} \quad [\mathrm{N\,m^{-3}}] \tag{6.6}$$

which implies a total force on the layer per unit area of:

$$B = bh_1 = \frac{\alpha g \Delta Q}{c_p} \quad [\mathrm{N\,m^{-2}}]. \tag{6.7}$$

Now suppose the water column is being mixed by mechanical stirring. How much work must be done against the buoyancy force in order to completely mix the stratification? As the density becomes uniform, the density deficit is distributed over the full depth h. You can think of this as being the same as moving the buoyant force down by a distance $\sim h/2$ on average, so the work that needs to be done is $Bh/2$ or

$$\Delta \mathrm{PE} = h\Delta\Phi = \frac{\alpha g \Delta Q h}{2c_p} \quad [\mathrm{J\,m^{-2}}]. \tag{6.8}$$

If the heat is supplied in a short time interval Δt at a rate Q_i then we have for the change of Φ:

$$h\Delta\Phi = \Delta \mathrm{PE} = \frac{\alpha g h}{2c_p} Q_i \Delta t \quad [\mathrm{J\,m^{-2}}] \tag{6.9}$$

or as $\Delta t \rightarrow 0$, we can write the time derivative of Φ as:

$$\left(\frac{\partial \Phi}{\partial t}\right)_{heat} = \frac{\alpha g Q_i}{2c_p} \quad [\mathrm{W\,m^{-3}}]. \tag{6.10}$$

This is the rate at which Φ increases for heating in the absence of stirring. The heat flux Q_i is the net flux of heat across the sea surface, as we described in Section 2.2.4. In order to maintain vertical homogeneity ($\Phi = 0$), the stirring must provide at least this amount of power.

Under our present assumptions, the power must come from the tidal flow, so the next step is to develop an appropriate representation of tidal energy input. We saw in Section 4.4.1 that for a shear flow $u(z)$ in the x direction, turbulent energy production occurs at a rate given by $-\tau\, \partial u/\partial z$ per unit volume. The total rate of production over the water column is therefore just:

$$P_T = -\int_{-h}^{0} \tau \frac{\partial u}{\partial z} dz \quad [\mathrm{W\,m^{-3}}] \tag{6.11}$$

which can be evaluated if the profiles $u(z)$ and $\tau(z)$ are known. Otherwise we can approximate the integral under the simplifying assumptions (i) that there is a uniform stress gradient so that $\partial u/\partial z$ = constant, (ii) $\tau = 0$ at the surface and $u = 0$ at the bottom. Under these conditions we have:

$$P_T = -\tau_b \hat{u} = k_b \rho_0 |\hat{u}|^3 \quad [\mathrm{W\,m^{-2}}] \tag{6.12}$$

where the bottom stress τ_b is related to the depth mean velocity $\hat{u}$ by the quadratic drag law $\tau_b = -k_b \rho_0 |\hat{u}| \hat{u}$ with a drag coefficient k_b ~0.0025 and $|\hat{u}|$ is the modulus (magnitude) of $\hat{u}$.[1] P_T is the power input to the turbulent motions which are responsible for stirring the water column and bringing about vertical mixing. Most of the turbulent energy is ultimately dissipated and hence transformed to heat within the mixed layer, but some of it is converted to potential energy in the mixing process. We shall hypothesise that a fixed fraction e of P_T is used in this way so that the power available to change the water column potential energy, and hence Φ, is just

$$\frac{\partial(\mathrm{PE})}{\partial t} = eP_T = ek_b \rho_0 |\hat{u}|^3 = h\left(\frac{\partial \Phi}{\partial t}\right)_{stir} \quad [\mathrm{W\,m^{-2}}]. \tag{6.13}$$

The parameter e can be thought of as an 'efficiency' of mixing. Combining Equations (6.10) and (6.13) we have for the net change in Φ due to heating and tidal stirring:

$$\frac{\partial \Phi}{\partial t} = \underset{\text{heating}}{\frac{\alpha g Q_i}{2c_p}} - \underset{\text{stirring.}}{\frac{ek_b \rho_0 |\hat{u}|^3}{h}} \quad [\mathrm{W\,m^{-3}}] \tag{6.14}$$

Equation (6.14) represents the essential competition between the promotion of stratification by surface heating and its erosion by tidal stirring. Note that the heating term in (6.14) can be either positive (net heat supply) or negative (net heat loss, such as in winter) as we discussed in Section 2.2; this is where the seasonality of the stratification enters. The stirring term is always negative; i.e. turbulence always acts to destroy stratification. If Q_i and $\hat{u}$ are known, Equation (6.14) can be integrated forward in time to predict the evolution of stratification. The tidal stirring term can be simplified if we assume that the tidal flow is dominated by the M_2 semi-diurnal constituent of the form $\hat{u} = \hat{u}_{M2} \sin \omega_{M2} t$, where $\hat{u}_{M2}$ is the depth-mean current amplitude of the M_2 constituent and ω_{M2} is the M_2 frequency. It can then be shown that, averaged over one or more whole tidal cycles:

$$\overline{|\hat{u}|^3} = \frac{4}{3\pi} \hat{u}_{M2}^3 \tag{6.15}$$

so that Equation (6.14) becomes:

[1] Alternatively, if we consider the motion *relative* to the water, the stress exerted by the seabed moves backwards at the near-bed velocity u_b so the work done on the water is $\tau_b u_b = k_b \rho_0 \hat{u}^2 |u_b|$ which gives an estimate comparable to that from Equation (6.12) but modified by a factor $u_b/\hat{u}$. Other profiles of u and τ all lead to the u^3 dependence but with different numerical factors.

$$\frac{\partial \Phi}{\partial t} = \frac{\alpha g Q_i}{2c_p} - \frac{4ek_b\rho_0\hat{u}^3_{M2}}{3\pi h}. \tag{6.16}$$

Stratification will only develop if $\partial\Phi/\partial t > 0$, i.e. if

$$\frac{\alpha g Q_i}{2c_p} > \frac{4ek_b\rho_0\hat{u}^3_{M2}}{3\pi h} \tag{6.17}$$

which we can re-arrange to give:

$$\boxed{\frac{3\pi\alpha g}{8c_p\rho_0 k_b e}\left(\frac{Q_i h}{\hat{u}^3_{M2}}\right) > 1.} \tag{6.18}$$

This is the condition for the development of seasonal stratification. If it is not satisfied, the water column will remain vertically mixed. When $\partial\Phi/\partial t = 0$ there is an exact balance between heating and stirring and the left side of Equation (6.18) is unity. This condition marks the position of *tidal mixing fronts* which are the transitional zones between regions which develop seasonal stratification and those which remain permanently mixed. Note that all the terms outside the bracket expression in Equation (6.18) are essentially constants (e by hypothesis) so that the control of stratification rests with the term $Q_i h/\hat{u}^3_{M2}$. For a particular region of the shelf seas, Q_i may be regarded as approximately spatially uniform, so that the positions of the tidal mixing fronts are determined simply by the parameter $h/\hat{u}^3_{M2}$ (Simpson and Hunter, 1974). Because of the large variations in depth and $\hat{u}^3_{M2}$, the range of $h/\hat{u}^3_{M2}$encountered in many shelf sea areas extends over four or more decades. In general, maps of this parameter are frequently presented in the form of $\log_{10}(h/u^3)$, with u being some measure of the strength of the tidal currents (see Box 6.1).

Box 6.1 Different options for *u* in *h*/*u*³

So far, we have utilized the depth mean M_2 amplitude, $\hat{u}_{M2}$, but other measures of the tidal current speed may be used. For instance, the surface amplitude of the current at spring tides, u_s, is convenient as it is given in tidal atlases. Typically the surface current is 15% greater than the depth mean, and for the northwest European shelf the ratio of amplitude at spring tides to the M_2 amplitude is close to 9/7 so that:

$$u_s = {}^9\!/_7 \times 1.15 \times \hat{u}_{M2} = 1.47\hat{u}_{M2}$$

which means that:

$$\log_{10}(h/u_s^3) = \log_{10}(h/\hat{u}^3_{M2}) - \log_{10}(1.47^3) = \log_{10}(h/\hat{u}^3_{M2}) - 0.50.$$

Another option is to use the average over the tidal cycle of the cubed depth mean current, i.e. $\overline{|\hat{u}^3|}$ which is a quantity readily available from numerical models. This option implies a further adjustment by $\log_{10}({}^4\!/_{3\pi})$ so that

$$\log_{10}\left({}^{h}\!/\overline{|\hat{u}^3|}\right) = \log_{10}(h/\hat{u}^3_{M2}) + 0.37 = \log_{10}(h/u_s^3) + 0.87.$$

The choice of u does not affect the shape of the contours of $\log_{10}(h/u^3)$ but does change their numerical values and hence the critical value of $\log_{10}(h/u^3)$ at which heating and stirring balance. A much-used map of the stratification parameter by Pingree and Griffiths (Pingree and Griffiths, 1978) for the European shelf uses $\log_{10}(h/k_b u^3)$ with h and u in cgs units and the drag coefficient $k_b = 0.0025$.

For the most widely used version of the h/u^3 parameter, we will use the shorthand form

$$SH = \log_{10}\left(h/\left|\overline{\hat{u}^3}\right|\right).$$

with h (m) and $\hat{u}$ (m s^{-1}).

6.1.3 Testing the stratification criterion using tidal mixing front positions

We shall consider the detailed physics of tidal mixing fronts and their importance for biological processes in Chapter 8, but here we can use the observed positions of fronts to test the theory of the previous section and, in the process, obtain an estimate of the efficiency of mixing e.

The argument of the last section indicates that the shelf seas should be divided into regions which exhibit seasonal stratification and others in which strong tidal stirring rapidly mixes buoyancy inputs and maintains a vertically uniform water column. This is shown schematically in Fig. 6.5a. Measurements of temperature and salinity in summer along sections between areas of strong and weak stirring reveal this transition from a single mixed layer to an approximately two-layer structure, as illustrated, for example, by an early set of observations from the western Irish Sea in Fig. 6.5b. This transition, which occurs rather rapidly, typically in a horizontal distance ~10km, is the tidal mixing front. For the example shown in Fig. 6.5b the front occurs at a value of $h/\hat{u}^3_{M2} \sim 220$ m^{-2}s^3. If we assume a typical summer heat supply to the sea surface of the western Irish Sea of $Q_i = 120$ Wm^{-2}, and taking $\alpha = 1.67\times10^{-4}\,^\circ$C^{-1}, $c_p = 4.0\times10^3$ J kg$^{-1}\,^\circ$C^{-1}, $\rho_0 = 1026$ kgm^{-3}, $k_b = 2.5\times10^{-3}$, we can estimate the tidal mixing efficiency, e, by re-arranging Equation (6.18):

$$e = \frac{3\pi\alpha g}{8\rho_0 c_p k_b}\left(\frac{Q_i h}{\hat{u}^3_{M2}}\right) \approx 0.0050. \quad (6.19)$$

Rather surprisingly, this means that only ~0.5% of the available stirring energy is utilised in mixing; most of the energy input goes into heat, much of it in the highly turbulent region near the bottom boundary. It is this concentration of turbulent energy in the near bed region that accounts for the low efficiency of the mixing process. In order to be effective in mixing, TKE needs to be injected close to the stable region of the water column, i.e. in the vicinity of the pycnocline which, in the typical summer scenario, is closer to the surface than the bottom as can be seen in Fig. 6.5b.

If our hypothesis of a constant efficiency of mixing is to be sustained, then we would expect that all other frontal transitions in adjacent areas should occur at the same critical

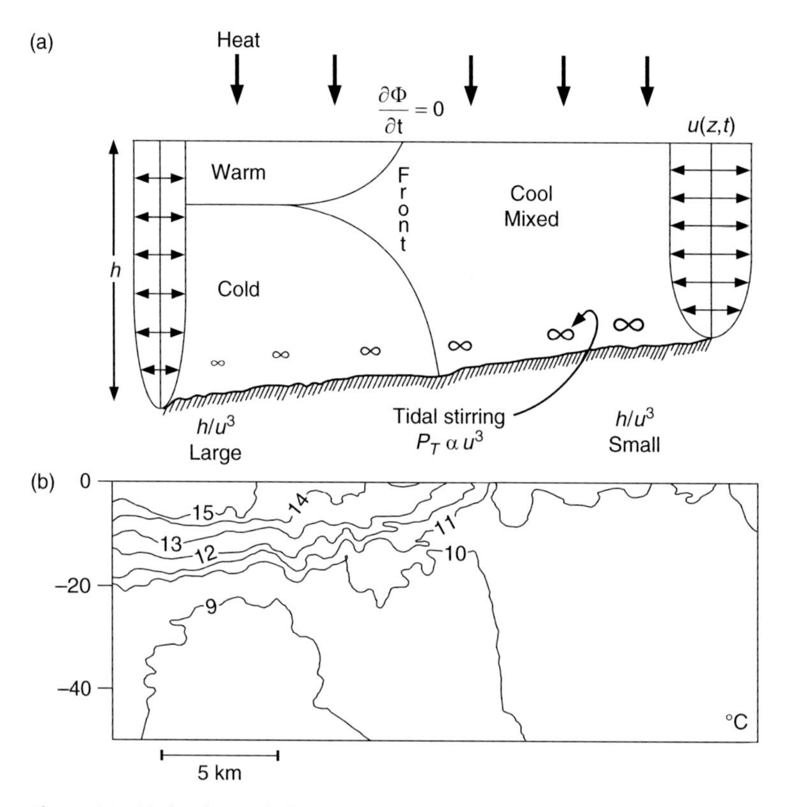

Figure 6.5 (a) A schematic illustration of the stratified and mixed regimes of a shelf sea, separated by a tidal mixing front, from Simpson and James, 1986, courtesy of the American Geophysical Union; (b) a section of temperature observed across the tidal mixing front in the western Irish Sea (Simpson, 1981), with permission from the Royal Society, London. The section in (b) was collected using a towed, undulating CTD resulting in a horizontal resolution of 500 metres.

value of $h/\hat{u}^3_{M2}$. This prediction can be tested by examining the stratification of the shelf seas using existing databases of temperature and salinity profiles to determine the stratification parameter Φ. An analysis of data for August from the shelf seas to the west of the UK, shown in Fig. 6.6, shows clearly the partitioning of the seas into stratified (Φ up to 250 J m^{-3}) and mixed ($\Phi \approx 0$) regimes separated by frontal boundaries. Comparison with the adjacent map of h/u_s^3 indicates that strong stratification is associated with values of $\log_{10}(h/u_s^3) > 3$, while low values (<1.5) correspond to complete vertical mixing. The transitions between the two regimes, i.e. the tidal mixing fronts, are seen to occur in all cases close to a value of $\log_{10}\left(h/u_s^3\right) \approx 1.9$. This consistency of the frontal positions, of which we shall hear more in Chapter 8, provides strong support for the idea that $e \approx$ constant, at least in the region covered by Fig. 6.6.

The mapping of the parameter h/u^3 has been undertaken for many regions of the shelf seas worldwide using u from numerical models. As in the case of the European shelf seas, a single value of h/u^3 appears to determine the positions of fronts in a given region (Garrett *et al.*, 1978; Lie, 1989; Glorioso and Flather, 1995; Kobayashi *et al.*, 2006) but there is some variation between regions in the critical value. This may be

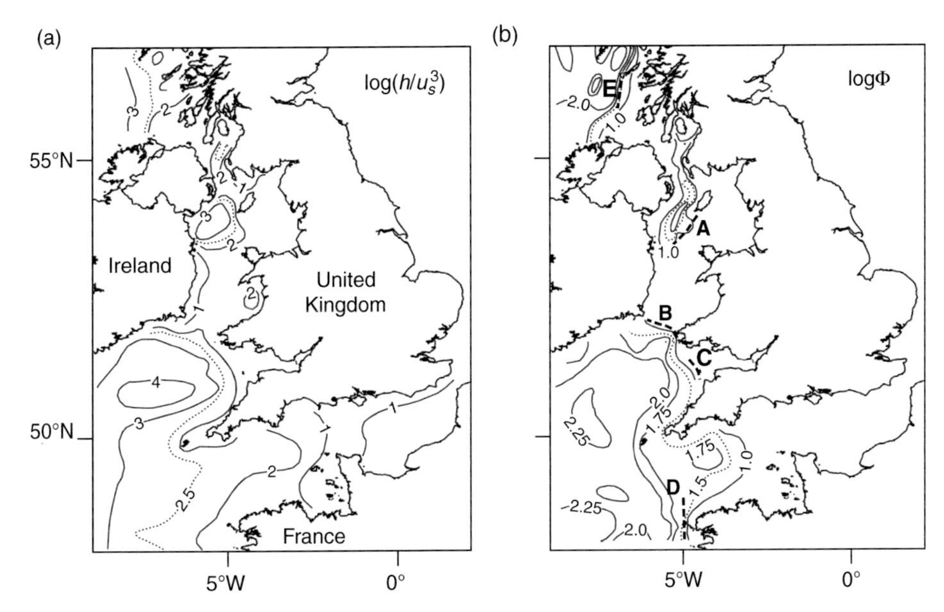

Figure 6.6 (a) A map of $\log_{10}(h/u_s^3)$, with u_s the amplitude of spring tide currents, west of the UK; (b) A corresponding map of the distribution of the potential energy anomaly as $\log_{10}\Phi$ in August. In (b) the dashed lines A–E mark the positions of the tidal mixing fronts. Based on Simpson and Pingree, 1978 with permission from Springer-Verlag.

due to differences in the heat input rate Q_i and also in the expansion coefficient α which varies significantly with temperature. Differences may also be attributable to changes in the relative importance of the other main source of boundary-driven turbulence, the winds. We shall address the impact of wind mixing in the following section.

6.1.4 Adding the effect of wind stress

In the development of Section 6.1.3, we included only mechanical stirring by the tides and neglected the influence of stirring by wind stress. The fact that the observed partitioning of the shelf seas and, in particular, the positions of the tidal mixing fronts, conforms to the simplified theory provides post facto justification for our assumption. Comparison of the stresses produced by the wind acting at the surface and tidal flow at the bed supports the idea that the tides will usually predominate in the stirring in many of the world's shelf seas. In the tidally energetic regions of the Irish Sea the stress in a tidal flow of 1 m s^{-1}, for example, is equivalent to that of a hurricane wind acting on the Earth's surface. However, in shelf areas with weaker tidal flows we might expect the wind stress to make a significant contribution especially in shallow areas (h~30 metres or less) where wind stirring alone may be sufficient to achieve complete mixing of the water column.

To include wind forcing in the argument, we need to insert a second stirring term in Equation (6.14). The rate of working by the wind stress τ_s (N m^{-2}) on the sea surface with a wind-induced surface current speed of u_s is just

$$P_W = \tau_s u_s = C_d \rho_a W^2 u_s \quad [\mathrm{W\,m^{-2}}] \tag{6.20}$$

where we have assumed that the stress is related to wind speed W by a quadratic drag law with a drag coefficient C_d ($\approx$0.0012, but see Section 2.4) and ρ_a ($\approx$1.3 kg m^{-3}) is the density of air. If we further assume that surface water velocity is a fixed fraction of the wind speed ($u_s = \gamma_s W$, with γ_s typically about 0.02), we then have for the stirring power input, per unit area, from the wind:

$$\begin{aligned} P_W &= C_d \rho_a W^2 \gamma_s W \\ &= C_d \gamma_s \rho_a W^3 \\ &= k_s \rho_a W^3 \end{aligned} \tag{6.21}$$

where $k_s = C_d \gamma_s$. Inserting P_w into Equation (6.14) gives:

$$\frac{\partial \Phi}{\partial t} = \frac{\alpha g Q_i}{2 c_p} - \frac{e k_b \rho_0 |\hat{u}|^3}{h} - \frac{e_s k_s \rho_a W^3}{h} \tag{6.22}$$

where e_s is an efficiency of wind mixing analogous to the tidal stirring efficiency e.

The condition for the frontal transition ($\partial \Phi / \partial t = 0$) then becomes:

$$\frac{\alpha g Q_i}{2 c_p} = \frac{e k_b \rho_0 |\hat{u}|^3}{h} + \frac{e_s k_s \rho_a W^3}{h}. \tag{6.23}$$

This relation gives a rather better fit than Equation (6.18) to observed frontal positions (Bowers and Simpson, 1987) although the good fit of tidal stirring alone does not leave a great deal of room for improvement. The influence of wind stirring is seen more convincingly when we use Equation (6.22) to predict the evolution of stratification over the shelf and compare with observations. Integrating forward in time from the vernal equinox, when Q_i becomes positive, up to time t_1, we have

$$\begin{aligned} \Phi_1 &= \frac{\alpha g}{2 c_p} \int_0^{t_1} Q_i dt - \frac{e k_b \rho_0}{h} \int_0^{t_1} |\hat{u}|^3 dt - \frac{e_s k_s \rho_a}{h} \int_0^{t_1} W^3 dt \\ &\cong \frac{\alpha g Q_1}{2 c_p} - \frac{4 e k_b \rho_0}{3\pi} \left(\frac{\hat{u}_{M2}^3}{h} \right) t_1 - e_s k_s \rho_a \left(\frac{\overline{W^3}}{h} \right) t_1 \end{aligned} \tag{6.24}$$

where Q_1 is the total heat input up to time t_1. The first term on the right of Equation (6.24) represents the stratification that would occur in the absence of stirring. The second term is the accumulated tidal stirring approximated, as before, by the cube of the M_2 tidal stream amplitude, while the third term is the equivalent wind stirring up to time t_1.

For a region where Q_1 may be assumed uniform, and amalgamating the constant quantities, we can write Equation (6.24) in the form:

$$\Phi_1 = c_0 - c_1 \left(\frac{\hat{u}_{M2}^3}{h} \right) - c_2 \left(\frac{\overline{W^3}}{h} \right). \tag{6.25}$$

Values of Φ determined from observations can be compared with the controlling variables $\left(\frac{\hat{u}_{M2}^3}{h}\right)$ and $\left(\frac{\overline{W^3}}{h}\right)$ by regression analysis to determine the constants and hence estimates of the efficiencies e and e_s. Results from a survey of summer stratification at 146 stations in the Irish and Celtic Seas (Simpson *et al.*, 1978) lead to efficiency estimates of $e \approx 0.004$ and $e_s \approx 0.023$. The higher efficiency of wind mixing may be explained in terms of the closer proximity of the wind stirring input to the main density gradient in the thermocline. The inclusion of the wind stirring term considerably increases the statistical significance of the fit relative to the regression on tidal stirring alone.

6.2 Seasonal cycles in mixed and stratifying regimes

In the previous section we developed a simple, energy-based argument that allows us to decide whether or not a region of shelf sea will be found to be thermally stratified or vertically mixed in summer. Our focus was on the summer spatial partitioning of the shelf sea environment. Now we will look at the temporal evolution of stratification over one year. If we neglect the influence of winds, we can revisit Equation (6.18) and generate the following condition that tells us what surface heat flux is required for a water column to thermally stratify:

$$\frac{3\pi\alpha g}{8c_p\rho_0 k_b e}\left(\frac{Q_i h}{\hat{u}_{M2}^3}\right) > 1$$
$$Q_i > \frac{8c_p\rho_0 k_b e\hat{u}_{M2}^3}{3\pi\alpha g h}. \tag{6.26}$$

At a particular place within our shelf sea, where we know the depth and the tidal currents, the collection of parameters on the right of Equation (6.26) is fixed and we have a condition for the rate of net surface heat flux required to trigger the onset of stratification. We described in Section 2.2 how the different components that determine Q_i lead to strong seasonality in heat flux, with negative flux (cooling) in winter switching to positive (warming) sometime in spring. Let us now look at how that seasonality in the heat flux generates annual cycles of water column structure.

6.2.1 The Two Mixed Layer (TML) model

So far we have represented stratification in terms of the variable Φ which summarises the state of water column stratification in a single number. Φ has the virtues of having a straightforward physical meaning and being easily incorporated into simple models of stratification like the one we developed in Section 6.1.2. You could take Equation (6.22) and, with sufficient knowledge of the seasonal heat flux, integrate it forward in time to track the evolution of the strength of stratification through one year (e.g. similar to the procedure in Equation 6.24).

On the other hand, Φ tells us only the magnitude of the stratification and nothing about the detail of water column structure. If we want to predict, or hindcast, the evolution of the density profile under the influence of heating and stirring, then we need to formulate a more explicit model. The simplest model of this kind (Simpson and Bowers, 1984) uses the same energy arguments embodied in Equation (6.22). It is based on a water column consisting of two mixed layers: a surface layer, heated and stirred by the atmosphere, and a bottom layer, stirred by the tides, with the two being separated by a high gradient region. Such a system is closely analogous to the laboratory tank model discussed in Section 6.1.1.

To simplify the problem we will assume that the vertical fluxes in the heating cycle tend to greatly exceed the heat inputs to the water column resulting from the horizontal transport (see Section 2.2). This allows us to model the one-dimensional vertical structure of the water column without worrying about horizontal gradients. We shall also make a number of other assumptions about the surface heat exchange processes to simplify the model formulation. From Equation (2.6) we represent the net heat flux into the water column, Q_i, as the difference between the solar radiation incident at the top of the atmosphere, Q_s (modified by the atmosphere's albedo, A) and the combined loss term Q_u:

$$Q_i = Q_s(1 - A) - Q_u. \tag{6.27}$$

The heat supplied to the sea surface by solar radiation, $Q_s(1 - A)$, is divided between a fraction (55%) which is assumed to be absorbed at the surface (i.e. in the uppermost layer) and the remainder (45%) which is absorbed exponentially in the underlying layers according to $e^{K_{av}z}$ where $K_{av} \sim 0.2\ \mathrm{m}^{-1}$ is an attenuation coefficient. Splitting the heat input in this way is a simple approach to including the wavelength-dependence of light attenuation, with the red end of the spectrum being absorbed far more rapidly than the blue end. All of the heat loss term is extracted from the immediate surface layer since the processes involved all draw heat from a very thin micro-layer at the surface. Q_u is a sensitive function of the difference between sea surface temperature T_s and the air temperature T_a and may be determined providing we also know the wind speed W, air pressure, cloud cover and the relative humidity (e.g. Gill, 1982; Sharples *et al.*, 2006). These driving meteorological parameters may conveniently be taken from sinusoidal fits to observational time series (Sharples, 2008). Notice that, since the heat flux is directly influenced by the surface temperature, there is a degree of feedback operating to control heat input.

As in the Φ model, we assume that the power available for mixing by wind and tidal stirring can be written as

$$e_s P_W = e_s k_s \rho_a W^3; \quad e P_T = e k_b \rho_0 |\hat{u}|^3 \tag{6.28}$$

where $\hat{u}$ is the depth-mean tidal velocity. P_w and P_T are assumed to be concentrated in the top and bottom layers respectively.

The operation of the numerical scheme for the two-layer model (see TML model at website) involves the application of successive heating and stirring inputs to modify

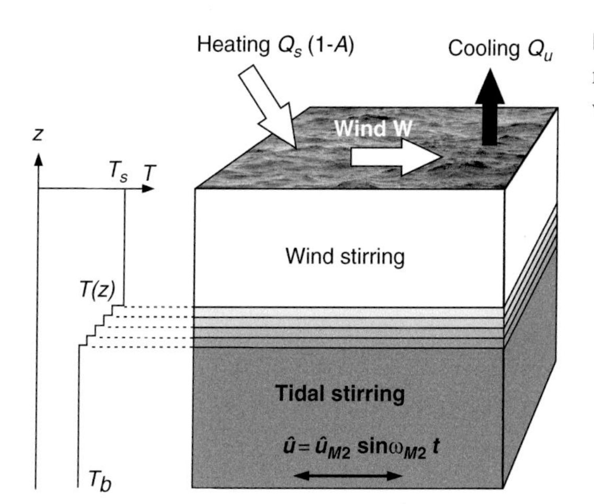

Figure 6.7 The main forcings of the model and the discrete resolution of the vertical temperature profile.

the temperature profile which is stored as discrete values for sub-layers of thickness Δz (typically 1 metre), shown in Fig. 6.7. The operating cycle starts with the input of solar heat. The heat loss from the sea surface Q_u is next extracted from the surface sublayer of the model vertical grid. The upper sublayers are then mixed by the available power P_w to produce a new surface mixed layer with uniform temperature which extends downwards as far as the available power allows; i.e. the increase in the potential energy of the water column matches the energy supplied by the mixing. The algorithm for the new layer depth includes any energy available from convective instability introduced by surface cooling. After surface mixing, a similar procedure is applied at the bottom to create a new bottom mixed layer, which utilises all the available tidal energy P_T arising from the oscillatory tidal currents, thus completing the heating-stirring cycle. At the next time step, the heat flux is modified according to the new surface temperature and the cycle is repeated.

Starting from a vertically uniform column at the spring equinox when Q_i becomes positive, the model steps forward in time to simulate the seasonal cycle of water column temperatures. An example for a mid-latitude site with rather weak tidal stirring ($SH = 3\ \mathrm{m}^{-2}\ \mathrm{s}^{3}$) is shown in Fig. 6.8a. Notice how the surface and bottom temperatures diverge soon after the onset of heating in the spring, which is when the condition for stratification in Equation (6.26) is met. The temperature difference and stratification continue to increase during the summer until they reach a maximum in September. However, the increasing surface temperature leads to large heat losses (Sections 2.2.2 and 2.2.3), and the net heat flux Q_i is significantly curtailed. After the temperature maximum, there is a rather rapid decrease as surface cooling commences and starts to induce unstable conditions which drive convective overturning in the water column. The mixing process is augmented by enhanced wind stirring but, in contrast to a widespread misconception, it is convection driven by cooling and not wind stirring which dominates the autumnal overturn and return to a vertically homogeneous state.

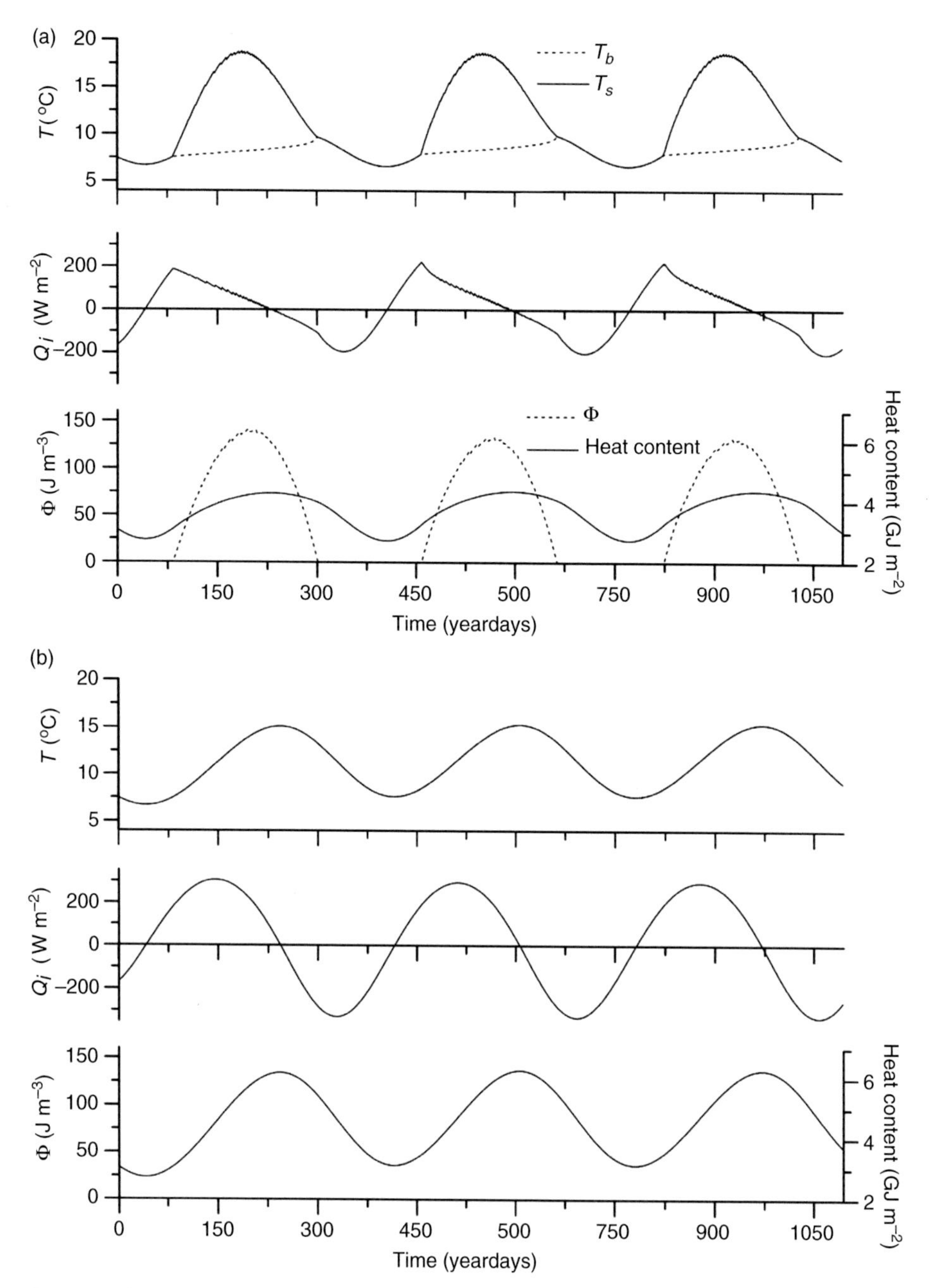

Figure 6.8 Annual cycles simulated by the heating-stirring model for (a) $SH = 3$ and (b) $SH = 1.7$ Middle panels show the net heating rate Q_i (W m^{-2})

By contrast, consider the case when tidal stirring is strong (SH= 1.7 m^{-2} s^3), as shown in Fig. 6.8b. Surface and bottom temperatures are equal throughout the year ($\Phi = 0$ all year) and vary sinusoidally in response to the meteorological forcing. The lower surface temperature, compared to Fig. 6.8a, leads to the heat loss term Q_u having far less effect on the net heat input.

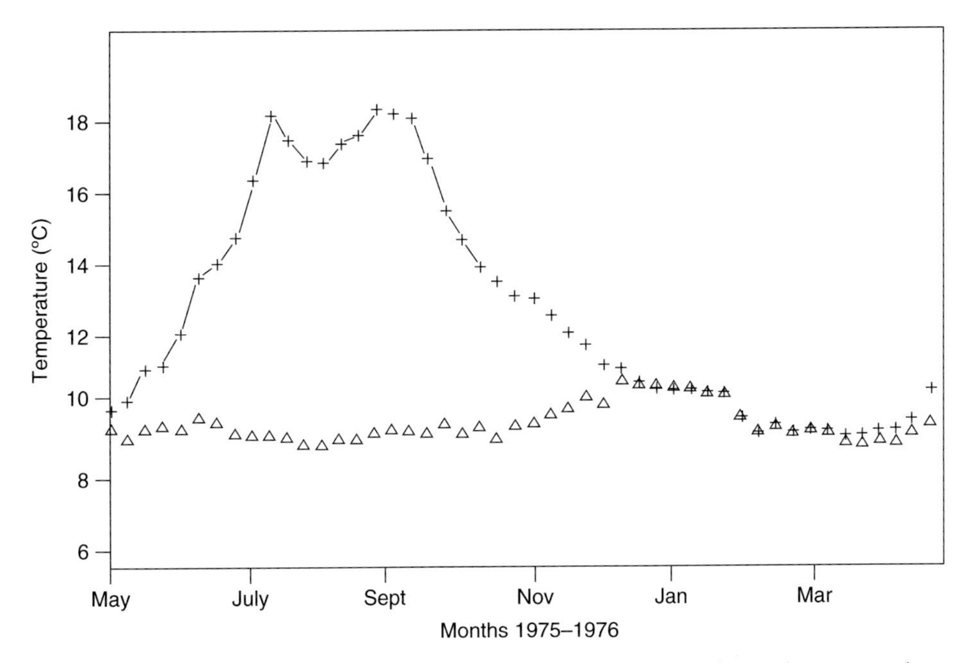

Figure 6.9 Annual cycle of surface (+) and bottom (Δ) temperature observed at a strongly stratified site on the Nymphe Bank, Celtic Sea (lat. 50° 40' N, long. 7° 30' W). This is an early data set, collected by UK Admiralty vessels which, for secret naval reasons, visited the Nymphe Bank several times per week for a whole year and recorded the temperature profile. Release of the data for civilian use required weekly averaging to avoid giving away the position of Her Majesty's ships. Adapted from Simpson and Bowers, 1984, with permission from the European Geosciences Union.

An interesting feature of such simulations in the stratified regions is that the bottom temperature, which remains low for almost the whole summer, exhibits a short sharp rise to a maximum at the overturn. This 'cusp' in the bottom temperature cycle is confirmed by observations, an early example of which is shown in Fig. 6.9. It means that the temperature cycle of the benthic environment contrasts sharply with that of the waters above the thermocline.

In spite of its simplicity, the TML model does a surprisingly good job at providing a realistic account of the seasonal cycle in the different mixing regimes of the shelf seas. In the stratified regions, it predicts surface and bottom temperatures and mixed layer depths with reasonable accuracy (Elliott and Clarke, 1987). To see how a sequence of runs of the TML model can be used to simulate the structure of a Tidal Mixing Front, try running the TMFI module from the book website. The two-layer model also illustrates how seasonal heat storage in the water column is influenced by vertical mixing. Compare the heat storage time series in Fig. 6.8a, b. In strongly mixed regions (Fig. 6.8b) heat is stored uniformly throughout the water column, which limits the rise in surface temperature. In a stratifying system (Fig. 6.8a), the incoming heat is concentrated in the upper layer, leading to higher surface temperatures and greater heat loss back to the atmosphere, which reduces the total amount of heat absorbed by the water column. As a result, the mixed water column achieves much greater heat storage than strongly stratified regions. The contrast by a factor of ~2 in the amplitude of heat storage between mixed and stratified regimes predicted by the two-layer model is close to that observed (Simpson and Bowers, 1984; Simpson and Bowers, 1990). On the down side, the ability of the two-layer model to represent the evolution of vertical structure is ultimately limited by the strong assumptions on which it depends. The model does not give us a detailed picture of the vertical structure, nor does it indicate where and when mixing is occurring in the water column. It also neglects the fact that the efficiencies of mixing are not constant but depend on the degree of stratification, as we saw with the laboratory results of Turner (Turner, 1973) (Fig. 6.3). This latter point will be revisited in Chapter 8 when we discuss the adjustment of the tidal mixing fronts as the tides change over the spring-neap cycle.

Physics Summary Box

- The net heat input to the sea surface is the difference between the incident solar radiation and the loss of heat back to the atmosphere (by infra-red radiation, conduction and evaporation).
- As with light, heat flux declines exponentially with depth through the water column. In a typical shelf sea, 90% of the net heat crossing into the sea surface will be absorbed within the upper 5 metres of the water column. In the absence of any mixing, the temperature profile of the water would also be exponential; i.e. a stratified water column with surface water becoming much warmer than deeper water.
- Whether or not a shelf sea water column will become thermally stratified is dependent on the balance between the rate at which heat input can generate surface water warmer than the deep water, and the rate at which mixing processes can re-distribute the heat throughout the water column.
- In areas where mixing is dominated by turbulence generated by friction between tidal currents and the seabed, the balance between heating and mixing results in h/u^3 being a useful parameter to map the horizontal distribution of stratified and mixed waters (h is the water depth and u is some measure of the strength of the tidal currents). High h/u^3 (deep water and/or weak tides) is associated with regions that stratify during summer, low values of h/u^3 (shallow water and/or strong tides) are found in regions that remain vertically mixed all year.
- The competition between heating and stirring divides the shelf seas into mixed and stratified regimes. The heating and stirring effects balance at a critical value of h/u^3, leading to the formation of the shelf sea fronts (see Chapter 8).
- In parts of the world where tides are generally weak, the effects of wind-driven mixing should be included in assessing the likelihood of seasonal stratification in shelf seas.

6.3 Primary production in seasonally stratifying shelf seas

The tides impose an almost fixed pattern of stirring on the shelf that is dictated by the dynamical response of the shelf to forcing at the shelf edge imposed by the oceanic tidal wave. We have just seen how this pattern is reflected in the physical partitioning of the shelf into regions that experience seasonal stratification and those that remain vertically mixed throughout the year. We will now look at how the primary producers respond to the seasonal cycle of thermal stratification, addressing the phenomenon of the spring bloom and the pivotal role of the thermocline in inhibiting vertical turbulent mixing.

6.3.1 The spring bloom

One of the most important events in the annual cycle of primary production of temperate shelf seas is the sudden flush of phytoplankton growth in spring known as the *spring bloom*. The rapid growth of near-surface phytoplankton provides the first substantial food supply of the year to much of the rest of the marine food chain. The development of the bloom is dependent on the changes in the vertical profile of turbulent mixing in the water column in response to the development of stratification. These changes in the physical environment control the availability of resources (light and nutrients) to the phytoplankton. This important link between the physical environment and the growth of the phytoplankton was first recognised by G. A. Riley in the early 1940s (Riley, 1942), and led to Sverdrup's formulation of the concept of the critical depth (Sverdrup, 1953) that we described in Section 5.1.5.

As we have seen already, during winter the water column of most temperate shelves is kept vertically homogeneous by convective overturning, as the surface ocean loses heat to the atmosphere. This convective overturning is a whole-depth turbulent motion that maintains vertical homogeneity of water constituents, including phytoplankton cells and inorganic nutrients. The vertical movement of the phytoplankton cells between sea surface and seabed combined with weak winter sunlight leads to the phytoplankton not receiving enough light to allow them to grow significantly. The phytoplankton will still try to grow, but the photosynthesis that they can achieve while close to the sea surface cannot compete against the losses due to respiration and mortality in the darker bottom water and the impacts of grazing. In other words, the critical depth for positive net phytoplankton growth lies somewhere within the interior of the water column, and the turbulent mixing prevents the phytoplankton from staying above that critical depth.

As the sun's elevation increases through late winter and spring, the water column eventually begins to receive more heat than it loses back to the atmosphere. Initially the tidal (and wind) mixing may prevent a warm surface layer from becoming established; notice in Fig. 6.8a how the water column begins to warm when the heat flux switches to net warming, but that there is a delay before stratification begins. In many areas the condition described in Equation (6.18) eventually is met.

The resulting appearance of an initially weak thermocline, separating the warming surface layer from the colder, deeper water, is the key to the spring phytoplankton bloom. Thermoclines inhibit the vertical turbulent transfer of water and its constituents between the surface and bottom layers (see Turner's experiment described in Section 6.1.1).[2] As stratification develops, the phytoplankton are split into two separate communities. The 'luckier' cells are in the surface water as the thermocline develops, and so they become trapped in the new surface mixed layer. As long as the thermocline develops above the phytoplankton critical depth, then the surface layer phytoplankton will now receive enough light to achieve net photosynthesis exceeding the losses due to respiration and grazing. Growth rates are initially not reduced by a lack of dissolved inorganic nutrients, because, at the onset of stratification, nutrient concentrations are uniformly high throughout the water column. The less fortunate phytoplankton cells are trapped within the bottom mixed layer; the amount of light they receive drops dramatically, and unless they are able to re-access the surface layer, respiration and grazing losses will gradually reduce their numbers.

The rapid surface growth of phytoplankton is, however, short-lived. The thermocline isolates the surface layer, severely restricting the re-supply of nutrients from the deep water. The surface layer phytoplankton use up the finite store of nutrients which was trapped with them above the thermocline, and then their growth becomes nutrient-limited and the surface population decreases. The rate at which the cell concentration in the surface layer decreases is partially determined by the mortality of the nutrient-starved cells, and also by the increasing impacts of the grazers as they take advantage of this flush of food in the bloom.

This spring pattern of rapid phytoplankton growth, a sharp biomass peak, followed by a decrease towards a low-biomass surface layer in summer is a well-recognised feature of temperate and high-latitude oceans. The introduction of *in situ* measurements of chlorophyll fluorescence as a proxy for phytoplankton biomass in the 1970s provided a method of recording the development and decay of blooms at high temporal resolution. An example of a spring bloom in the North Sea, observed by instruments on a mooring, is shown in Fig. 6.10. Notice that the bloom peaks when surface-bottom temperature differences are still only about 0.5 °C. The daily signal in the chlorophyll fluorescence during and after the bloom is a result of non-photochemical quenching (see Section 5.1.1). Regional pictures of the formation and decay of spring blooms are available from satellite imagery (e.g. Thomas *et al.*, 2003).

Due to the sudden high light, and the abundant nutrients, carbon fixation rates during shelf sea spring blooms can be high. However, the transient nature of the spring bloom, combined with the difficulty in predicting it, means that there are few reliable *in situ* observations of carbon fixation rates during a bloom.[3] Based on

[2] It is remarkable how apparently weak the stratification can be and still have an appreciable effect on vertical turbulent mixing. As a rule of thumb a cross-thermocline temperature difference of only 0.5 °C (a density contrast of about 0.06 kg m^{-3}) is sufficient to severely inhibit mixing and trigger a biological response.

[3] The difficulties of observing such a transient phenomenon were underlined for one of us while modelling the inter-annual variability of primary production off the west coast of New Zealand. The local biological oceanographers claimed to have never seen a spring bloom despite several 'spring' cruises to

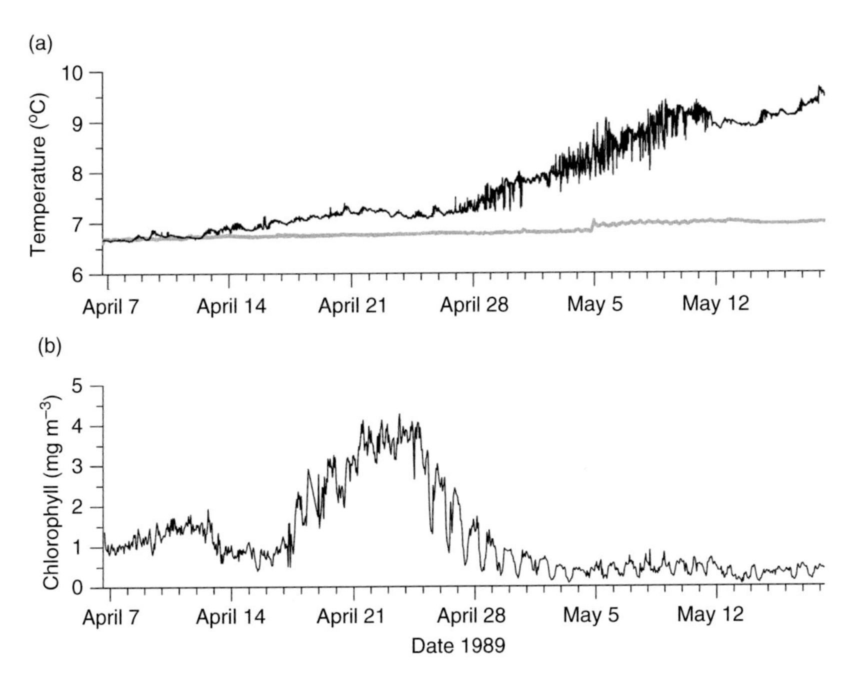

Figure 6.10 Time series illustrating (a) the surface (black line) and near-bed (grey line) temperature and (b) the surface chlorophyll over the spring bloom in the central North Sea. Stratification begins to be established from about April 13, with a surface-bed temperature difference of 0.5 °C occurring about April 21. The bloom begins on about April 16–17 when the surface-bed temperature difference is only 0.2–0.3 °C. Data were recorded during the United Kingdom's North Sea Project (see http://www.bodc.ac.uk/projects/uk/north_sea/) by instruments moored in a depth of 85 metres in the central North Sea.

^{14}C incubation experiments, a carbon fixation rate of 1.4 g C m^{-2} d^{-1} has been reported for the shelf edge of the Celtic Sea (Rees *et al.*, 1999), and a rate of 0.5 g C m^{-2} h^{-1} (6 g C m^{-2} d^{-1} if we assume 12 hours daylight and ignore nighttime losses) measured in the western English Channel (Pingree *et al.*, 1976). This spring bloom primary production is new production with a high *f*-ratio (see Section 5.2.1). A simple stoichiometric approach based on the Redfield ratios (Section 5.1.6) can allow us to estimate maximum potential growth rates during a spring bloom (see Problem 6.6 at the end of this chapter).

6.3.2 Phytoplankton species during the spring bloom

Spring blooms tend to be dominated by large cells (Pingree *et al.*, 1976; Joint *et al.*, 1986), mainly diatoms (Flagg *et al.*, 1994; Mills *et al.*, 1994; Townsend and

the region, and they felt sure that spring blooms predicted by the model were artefacts of the modelling process. The subsequent comparison between the model results and 7 years of cruise observations demonstrated that it was very likely that every spring cruise to the region had simply missed the short-lived bloom period (Hadfield and Sharples, 1996).

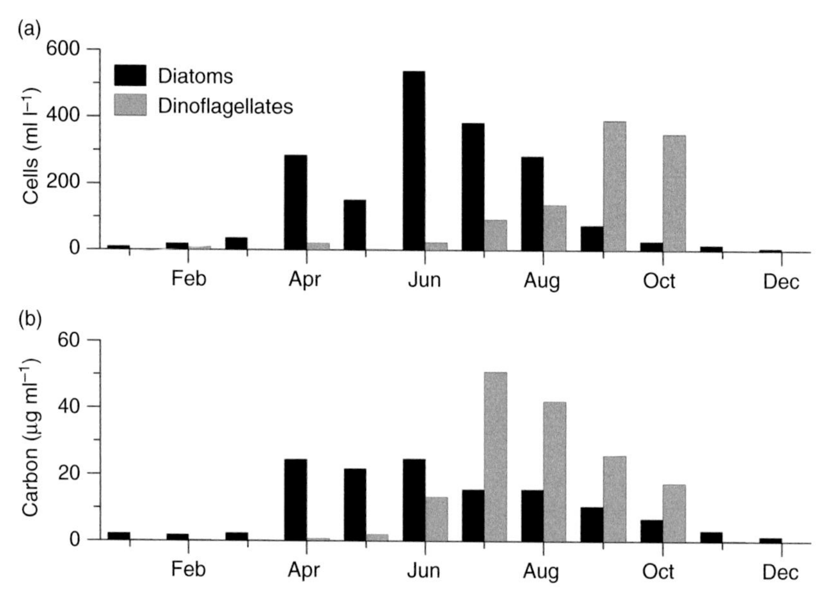

Figure 6.11 Time series of diatom and dinoflagellate cell numbers (a) and their carbon content (b) from the western English Channel. (Station L4 in the Western Channel Observatory maintained by Plymouth Marine Laboratory, see: http://www.westernchannelobservatory.org.uk/.) Data provided by Claire Widdicombe, Plymouth Marine Laboratory, UK.

Thomas, 2002). Observations from a long-term seasonally stratified station in the western English Channel by the UK's Plymouth Marine Laboratory (see http://www.westernchannelobservatory.org.uk/) are shown in Fig. 6.11 and illustrate how initial diatom dominance in spring shifts to dinoflagellates in the summer. There are several reasons thought to explain diatom dominance: (i) Diatoms are often viewed as a group of phytoplankton particularly suited to turbulent environments, and so may be well-positioned in the pre-bloom water column, poised to take advantage of the developing stratification; (ii) Diatoms are generally thought to have higher growth rates, so that in an environment where nutrients are not limiting (such as the initial stages of spring stratification) they might be expected to dominate; (iii) Diatoms are armoured with silicate, which will reduce the grazing impact on them (e.g. by reducing the range of grazers capable of eating them) and thus lower their mortality; even a small reduction in the effects of grazing will provide significant advantage as light increases in the pre-bloom phase and as stratification develops. However, remember from Chapter 5 (Section 5.2.3) that we found that smaller phytoplankton cells will always have an advantage in the competition for nutrients. The best explanation for sudden blooms of diatoms in high nutrient environments is that they are able, at least initially, to out-grow the mesozooplankton that graze on them (Kiørboe, 1993). Most model simulations of the spring bloom predict initial diatom dominance (e.g. Kelly-Gerreyn *et al.*, 2004), primarily because of one or more of the above advantages enjoyed by the diatoms. It is worth noting that in winter it is possible for a bright, clear, calm day to result in enough light to allow net phytoplankton growth. Such a calm period can potentially generate thin, short-lived near-surface stratification sufficient to drive a brief surface bloom of these opportunist phytoplankton. Even without stratification, the weak winter sunshine may at least be able to reduce the vigorous convective mixing in the surface layer, leading to the possibility of turbulence in the photic zone dropping below the critical mixing threshold (Huisman *et al.*, 1999) (see Section 5.1.5). Indeed, we could conjecture that such short-lived events of primary production in winter may be important in establishing the phytoplankton community that eventually responds to the spring stratification. However, at the moment the transient, unpredictable nature of the winter meteorology and the difficulty in observing winter time series in temperate regions means that our understanding of phytoplankton dynamics prior to the spring bloom is limited.

There are important consequences of spring blooms being dominated by diatoms. As a source of food, the relatively large size of diatom cells might seem to make them particularly useful to the rest of the ecosystem. Their silicate armouring tends to limit grazing by microzooplankton, while the mesozooplankton (e.g. copepods) have the strength and appendages suitable for cracking open the cells (Smetacek, 2001). Thus direct grazing of diatoms is carried out mainly by organisms that will themselves feed the higher trophic levels of the ecosystem, rather than by organisms that contribute more to the re-cycling of organic material. Indeed, the diatoms are generally considered to provide the basis for food chains that support important commercial fisheries (Irigoien, Harris, *et al.*, 2002). At the same time, however, diatoms tend to be negatively buoyant, and so the cells in the surface layer may not be available to the grazers for too long. This downward flux of diatom cells, combined with that of the rapidly sinking fecal pellets produced by the copepods, could mean that much of the spring bloom biomass may be exported to the deeper water and the seabed and so play an important role in supplying the benthos with food (Maar *et al.*, 2002) and exporting carbon. Modelling experiments have suggested that the proportion of the spring production that reaches the seabed can be increased if the pelagic grazing is reduced by colder water temperatures (Townsend *et al.*, 1994).

6.3.3 Variability in the timing of the spring bloom

With such strong tidal control we might expect the timing of the bloom to be almost fixed each year. However, year to year variations in weather conditions leading to differences in the air-sea exchange of heat and in wind stirring provide considerable potential for inter-annual variability in the timing of the spring bloom. Because of the bloom's importance as the first supply of significant concentrations of new organic material for the rest of the marine ecosystem, many organisms have evolved to time their breeding to take advantage of it. For instance, fish larvae may be dependent on bloom timing either directly through planktivorous fish larvae feeding on phytoplankton, or indirectly with larvae feeding on zooplankton that are themselves responding to the bloom (Platt *et al.*, 2003). Higher trophic levels, dependent on fish larvae for their food, can also be affected by the timing of the spring bloom; for instance seabird breeding success is thought to be partially driven by bloom timing (Frederiksen *et al.*, 2006) and the survival of young shrimp in the northwest Atlantic has been linked to the timing and duration of the spring bloom (Ouellet *et al.*, 2011). This apparent dependence of some organisms on bloom timing forms the basis of the well-known *match-mismatch* hypothesis proposed by the fisheries scientist David Cushing (Cushing, 1990). We do not expect fish larvae or seabirds to be capable of predicting the onset of stratification; instead it is more likely that they will have adapted to some long-term mean bloom timing.

We have already noted that Equation (6.26) provides a way of determining when the condition for the onset of thermal stratification is just met. Inter-annual variability in this energy balance reflects variations in Q_i (a more cloudy spring would delay the onset of stratification) and W (a windier spring would delay the onset of stratification). More subtly, Q_i can vary in response to changes in air temperature, with warmer air helping to increase heat flux into the ocean. Similarly Q_i is influenced by sea surface temperature; a cooler sea (controlled by heat exchanges in the months prior to spring) will lead to an increase in heat supplied to the sea surface. Year to year variability in meteorological conditions is the principal source of inter-annual fluctuations in bloom timing in the open ocean (Waniek, 2003), and plays a significant role in bloom timing in shelf seas. In the Gulf of Maine, variations in cloudiness and in the intrusions of nearby slope water (the latter probably a response to non-local meteorology and the North Atlantic Oscillation) are known to drive changes in bloom timing (Townsend *et al.*, 1994; Thomas *et al.*, 2003). In the northern North Sea the onset of spring stratification has been found to vary over a total range of one month, with much of the variability attributable to inter-annual changes in wind stress and air temperature during the spring months (Sharples *et al.*, 2006). Such strong control by atmospheric conditions has important, far-reaching implications in the context of our changing climate.

There is an additional, important, source of variability in the timing of shelf sea stratification and the spring bloom. Most shelf seas experience significant changes in tidal current amplitude in the form of the 14.8 day spring-neap cycle, due to the interaction of the M_2 and S_2 tidal constituents (see Section 2.5.1). Inter-annual shifts in the phase of the spring-neap cycle can influence the timing of spring stratification, for instance by delaying the onset of stratification if the stronger mixing of a spring tide coincides with the long-term mean date of stratification. A double, or prolonged, spring bloom is possible if the spring tidal currents are able to halt the early progress of the bloom as shown in the example of Fig. 6.12. The initial spring bloom, triggered at the neap tide of April 24, is interrupted and redistributed throughout the whole water column by strong spring tide mixing around May 3. The establishment of spring stratification is more likely to occur during the transition from spring to neap tides (i.e. as the tidal mixing is reducing) than as tides increase from neaps to springs (Sharples *et al.*, 2006).

Note that you can investigate the influences of different tidal and meteorological conditions on the timing of the spring bloom using the model associated with this book.

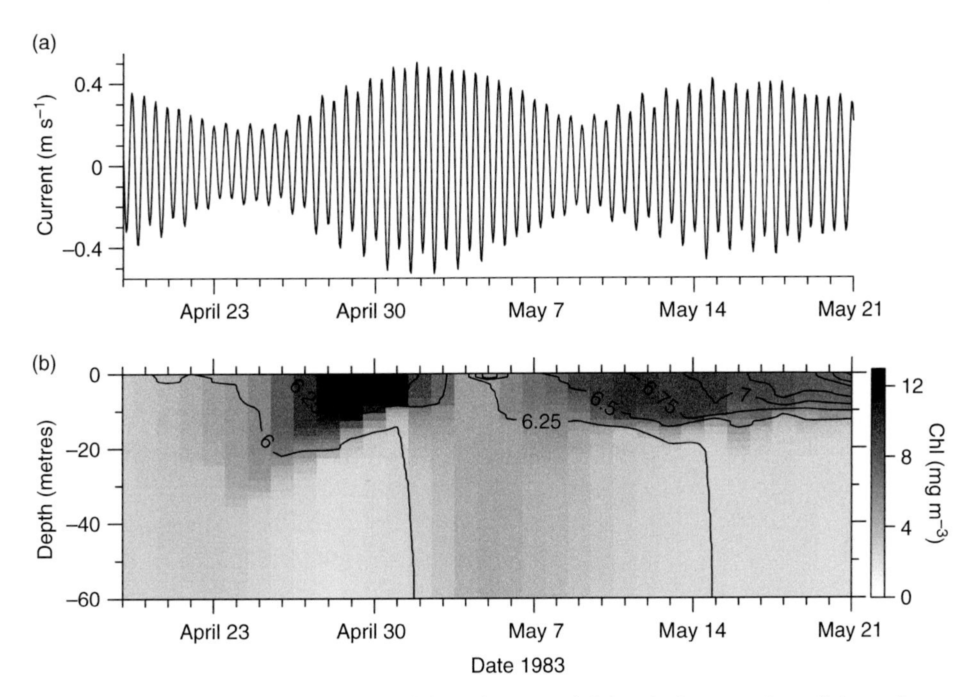

Figure 6.12 See colour plates version of (b). The potential for the interruption of the spring bloom caused by the spring-neap tidal cycle, based on model results for the northeastern North Sea in 1983. (a) Tidal current velocity, with neap tides on April 24 and spring tides on May 3; (b) Depth-time series of temperature (line contours) and chlorophyll (colours). Adapted from (Sharples *et al.*, 2006), with permission from Elsevier.

6.3.4 Surface layer phytoplankton after the spring bloom

What happens to the phytoplankton production and population in the surface layer after the spring bloom has utilised all of the available nutrients? A typical summer vertical profile of nutrients in a stratified shelf sea shows nitrate to be a limiting nutrient in surface waters (e.g. Fig. 5.8), so how does this influence (e.g. Fig. 5.8) primary production? The source of nitrogen for growth switches from nitrate to ammonium, with the ammonium supplied by the action of the microheterotrophs (ciliates and flagellates) and bacteria recycling organic matter. During this post-bloom period, there is an equilibrium in which the uptake rate of ammonium balances the rate at which ammonium is produced (Le Corre *et al.*, 1993). The system becomes almost entirely dependent on a supply of locally regenerated nitrogen. In our own work off the southwest of the United Kingdom, we often find the summer surface layer to have the clear blue-ness of the open ocean, and many of the attributes of the surface waters in the an oligotrophic subtropical gyres, including low attenuation of PAR (typically 0.08–0.09 m^{-1}) and low *f*-ratios. The phytoplankton population in this nutrient-starved surface layer shifts from the large cells of the spring bloom to much smaller cells (e.g. picoeukaryotes and the cyanobacterium *Synechococcus*) as size becomes an important factor in the competition for the limited supply of nitrogen.

An obvious way of triggering further new phytoplankton growth would be to rapidly but temporarily re-mix the water column with a strong burst of wind, re-supplying nutrients to the surface water. Subsequent surface heating after the wind event would then re-stabilise the water column and trigger a bloom analogous to the spring bloom. With well-established stratification in midsummer, is this feasible? A simple calculation based on the wind impulse required to homogenise a two-layer shelf sea water column (see Problem 6.4) suggests that very strong and/or prolonged winds are required. In temperate shelf seas in summer, we would normally expect such meteorological events to be very rare. The most likely regions to experience such summer winds are those in the tropical and subtropical hurricane zones, where there are examples of hurricanes re-supplying nutrients to the surface water and fuelling post-hurricane phytoplankton blooms (Fogel *et al.*, 1999; Sharples *et al.*, 2001a; Shi and Wang, 2007). The example of a wind mixing event based on mooring data from the northeast shelf of New Zealand shown in Fig. 6.13 illustrates how the passage of a tropical cyclone completely mixed the shelf water column. A nitrate analyser attached to the mooring near the sea surface recorded the mixing of the nitrate profile, and the replenishment of nitrate in the surface waters (Fig. 6.13c). The utilization of the surface layer nitrate took about 2 days as stratification became re-established after the cyclone. Note the semi-diurnal signal in the nitrate in Fig. 6.13c. This is caused by an internal tide driving the thermocline (and nitracline) up and down, a topic which will be addressed in Chapter 10.

6.4 Primary production in mixed water

So far we have concentrated on surface phytoplankton growth in regions that undergo seasonal stratification, where stratification plays the pivotal role in controlling primary production by modifying the profile of vertical turbulent mixing. What happens in the areas of the shelf seas that remain well mixed all year? The main problem for the phytoplankton here is that they are being continuously mixed from the sea surface to the seabed. If the depth of the water column is greater than the critical depth, then we would expect there to be no net phytoplankton growth.

6.4.1 Surface phytoplankton blooms in the absence of stratification

There have been instances where surface blooms appear to have begun prior to the onset of any stratification. An example from Massachusetts Bay is shown in Fig. 6.14 where a surface concentration of phytoplankton appeared to develop in contradiction to Sverdrup's critical depth hypothesis. How can biomass become concentrated in surface waters without the aid of stratification?

This apparent paradox can be resolved by noting the concept of *critical mixing* (Huisman *et al.*, 1999) that we discussed in Section 5.1.5. Remember that if we have estimates of the vertical diffusivity within the surface water of a mixed water column, K_z, and some knowledge of the critical depth of the phytoplankton population, Z_{cr}, we can estimate a residence time for cells above the critical depth as $T_{res} \approx Z_{cr}^2/K_z$. If this

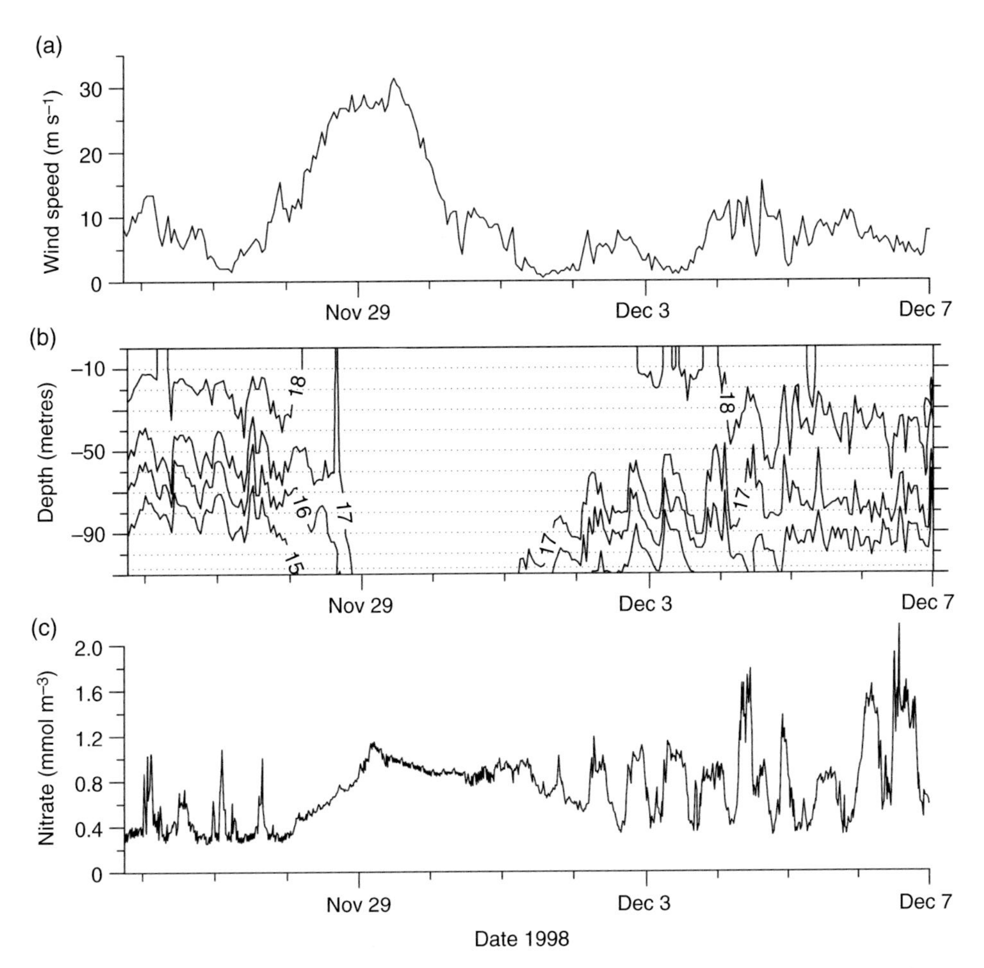

Figure 6.13 Data from a mooring off the northeast shelf of New Zealand. (a) Wind speed; (b) The vertical temperature structure on the shelf, with horizontal dotted lines marking the positions of the temperature loggers on the mooring; (c) Near-surface nitrate concentration recorded by a nitrate analyser on the mooring. The analyser was situated 25 metres below the sea surface. [Adapted from Sharples *et al.*, 2001a, courtesy of the American Geophysical Union.]

residence time is greater than the cell doubling time scale, there is the potential for biomass accumulation in the surface waters. The magnitude of the critical mixing could be raised if cells were buoyant and able to offset some level of turbulence. The important message here is that the growth of phytoplankton is a response to the turbulent environment. A vertical profile of temperature or density is an often-used proxy for providing some insight into the distribution of turbulence, but it is not completely reliable.

6.4.2 Phytoplankton growth in turbulent water

Strong tides and turbulent mixing produce high concentrations of suspended particulate material, thus increasing the attenuation of light and severely limiting the

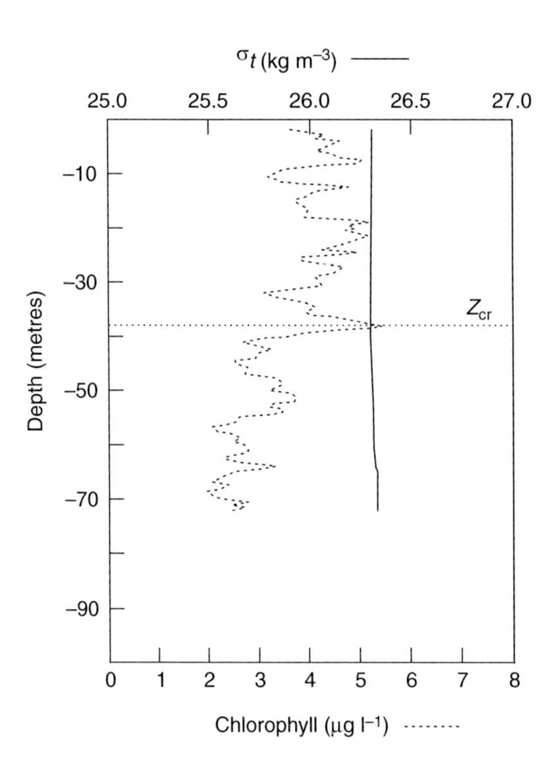

Figure 6.14 Vertical profiles of density and chlorophyll from Massachusetts Bay, showing increased surface chlorophyll despite the lack of any clear stratification above the critical depth, Z_{cr}. From Townsend *et al.*, 1992, courtesy Nature Publishing Group.

vertical extent of the water column within which photosynthesis is possible. In the Irish Sea, for example, attenuation of PAR in mixed waters can be 0.3–0.4 m^{-1} (Bowers *et al.*, 1998). Phytoplankton growth is, however, still often possible in summer if the high surface irradiance results in a critical depth that reaches the seabed. You can investigate this using the physics-primary production model associated with this book. The mixed waters of, for instance, Georges Bank and the Irish Sea often have chlorophyll concentrations of 1–2 mg m^{-3}. Note that this chlorophyll concentration is seen throughout the depth (40–70 m on Georges bank, 60–80 m in the Irish Sea), so the integrated biomass through the whole water column is high (around 100 mg m^{-2}).

The phytoplankton population in mixed environments is invariably observed to be dominated by diatoms. This is probably because there is usually an abundant supply of nutrients available from re-mineralisation in bottom sediments so that large, fast-growing cells with an ability to counter some of the grazing pressures will out-compete small cells that are better suited to low nutrient environments. One interesting feature of mixed-water phytoplankton is that they tend to have low Chl:Carbon ratios (~ 0.02 mg Chl (mg C)$^{-1}$) (Holligan *et al.*, 1984). This perhaps seems paradoxical. Surely, given the low average light received by a cell as it is mixed from the surface to the seabed and back again, the photosystem would optimize to the low light? However, the rapid mixing means that the cells are re-introduced to the surface of the water column frequently and unavoidably, experiencing low average light but with intermittent very high light. Figure 6.15 illustrates the light histories of cells

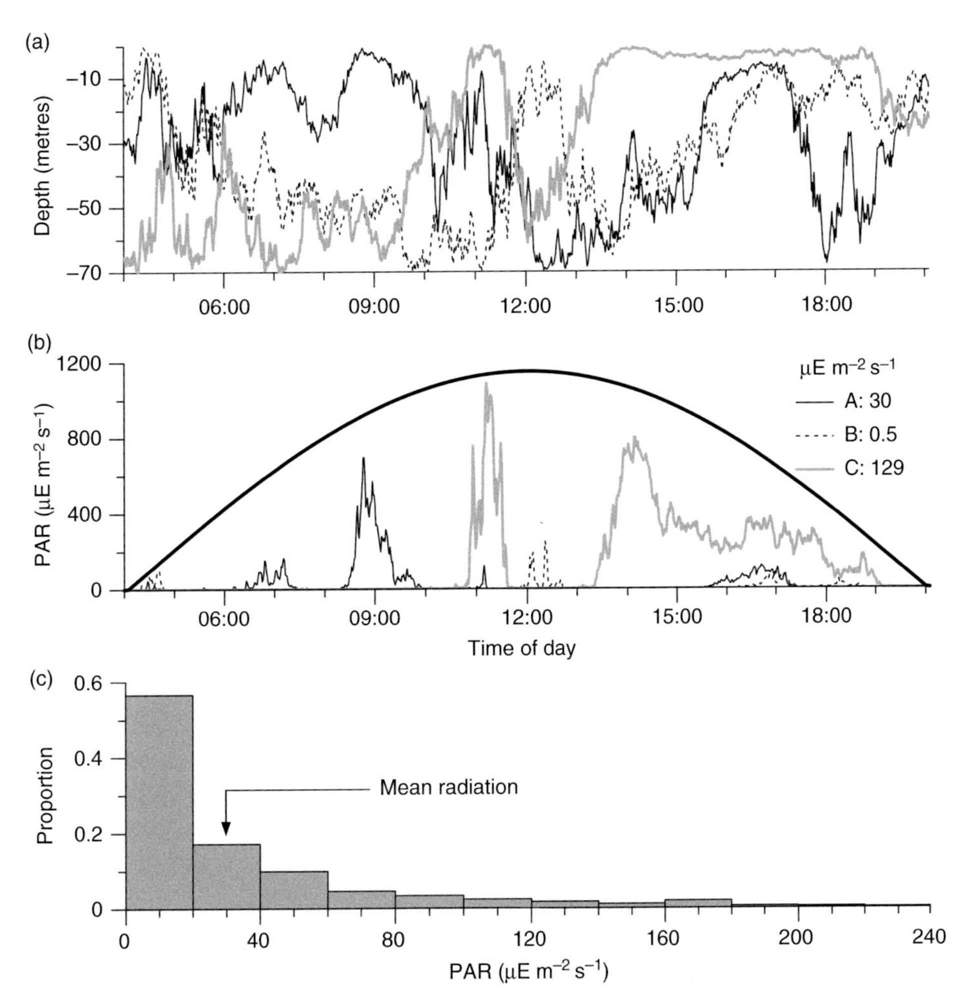

Figure 6.15 The light experienced by different phytoplankton cells as they are mixed through a homogeneous water column (depth = 70 m, tidal current amplitude = 0.8 m s^{-1}, K_{PAR} = 0.3 m^{-1}). (a) Trajectories of 3 phytoplankton cells during daylight; (b) The light experienced by the cells as they follow the trajectories in (a). The bold line is the surface incident PAR and the legend notes the average irradiances experienced by the cells during daylight hours (mean = 729 μE m^{-2} s^{-1} over daylight hours). (c) Distribution of the average light received by 1000 phytoplankton cells during daylight hours. The predicted depth- and time-averaged irradiance is marked by the arrow. The data are based on a Lagrangian model of neutrally buoyant cell trajectories (Ross and Sharples, 2004), courtesy of Oliver Ross (Institut de Ciències del Mar de Barcelona). The depth- and tidally averaged vertical eddy diffusivity was 0.08 m^2 s^{-1}.

being mixed through a turbulent water column, using a Lagrangian model to track the cell trajectories. The average light received by the population of cells is what we would expect if we calculated mean light over the exponential light profile. However, individual cells within the population experience dramatically different light histories. Cell A in Fig. 6.15 spent quite a lot of time near the sea surface, but mostly during times of low surface irradiance so that it received low light over the whole day. Cell B spent much more time away from the surface; around noon it did manage to get to about 10 metres from the sea surface, but the strong light attenuation still limited the total light it received. Cell C was mixed to the top of the water column briefly close to noon, and as a result received a blast of very high irradiance only 30 minutes after it had been in total darkness at the bottom of the water column. It is possible that phytoplankton may bias their acclimation status to cope with the extremes in light history such as those experienced by cell C in Fig. 6.15 (Moore *et al.*, 2006).

Summary

A competition between heating and stirring controls the occurrence of seasonal stratification of the water column in the shelf seas. When tidal stirring is dominant, the competition can be conveniently represented in an energetics model which leads to the parameter $Q_i h/u^3$ where h = depth (m) and u (ms^{-1}) is a measure of the tidal stream velocity. For a region where the net heating rate $Q_i \approx$ constant, control rests simply with h/u^3. For large h/u^3 (deep water or weak tidal currents) the water column will become stratified, while for low values (shallow water or strong tides) it will remain vertically mixed throughout the season. Hence, during the summer months, the shelf seas are divided into a two-layer stratified regime and a continually mixed regime separated by tidal mixing fronts. Wind stirring can also contribute to mixing the water column and its inclusion in the energetics model improves the fit to observations of stratification.

The two regimes provide contrasting environments for primary production. In regions of high h/u^3 the development of stratification is a key event in the growth of phytoplankton. The new thermocline confines the phytoplankton to the surface layer where they enjoy high levels of light. Moreover, at the onset of stratification, nutrients are initially abundant in the surface layer. With neither light nor nutrients limiting their capacity for growth, the phytoplankton grow rapidly in a spring bloom which is dominated by large cells, mainly diatoms. Production has a high *f*-ratio, utilising the supply of winter nitrate trapped in the surface layer. This bloom continues for a period during which the nutrients in the surface layers are rapidly depleted; as well as inhibiting removal of phytoplankton from the surface layer, the thermocline blocks replenishment of the depleted nutrients in the surface layer by the still high nutrient pool in the deeper water. Thereafter, there is much reduced growth in the nutrient-poor surface layers, in which diatoms are replaced by dinoflagellates and smaller cells as the surface layer production becomes dominated by the recycling of organic material; primary production here has low *f*-ratios. The onset of stratification and with it the timing of the spring bloom varies considerably from year to year because of variations in surface heat exchange, wind stirring and phase of the spring-neap cycle in the tides. In the lower layers of the stratified water column, nutrients remain high throughout the season but growth is inhibited by the lack of light.

In the continuously mixed (low h/u^3) regimes, phytoplankton are transported through the full depth of the water and so experience a highly variable and erratic light regime. Here growth is still possible during the high surface irradiances of summer, but the limited light reduces growth rates even though nutrients are abundant. In addition to continuous vertical motions, the average light supply may be further restricted by high levels of re-suspended sediments maintained by the strong vertical mixing which characterises these areas.

FURTHER READING

Circulation and Fronts in Continental Shelf Seas, ed. *John C. Swallow, et al.*, The Royal Society of London, 1981.

The Dynamics of Marine Ecosystems: Biological-Physical Interactions in the Oceans, by Kenneth H. Mann and John R. N. Lazier, Blackwell Publishing, 2006.

Problems

6.1. Consider an initially depth-uniform water column of depth h which receives an input of heat ΔQ per unit area which is absorbed in a thin near-surface layer of thickness $h_1 \ll h$. Calculate the potential energy of (i) this stratified water column and (ii) its value if the water column becomes fully mixed. Hence show that the energy required to bring about mixing is

$$\Delta(P.E.) = \frac{\alpha g h \Delta Q}{2c_p}$$

where α is the volume expansion coefficient and c_p is the specific heat of seawater.

6.2. Observations at the summer solstice (June 21), at a location where the depth h =95 m and the tidal stream amplitude is 0.5 m s^{-1}, indicate that the potential energy anomaly $\Phi = 55$ Jm^{-3}. If the net seasonal heat input to the sea surface is given by $Q_i = 140 \sin(2\pi t/365)$ Wm^{-2}, where t is time measured from the vernal equinox (March 21), determine the total heat input up to the time of the solstice and hence the value of Φ which would occur in the absence of mixing. Assuming that all mixing is due to tidal stirring, estimate the efficiency of mixing.

6.3. For a region of the shelf seas, the peak summer heat input is 105 Wm^{-2} and average cubed wind speed W^3= 520 m^3 s^{-3}. Estimate the maximum depth to which complete mixing can be maintained if the tidal stream amplitude is (a) 0.6 m s^{-1} and (b) 0.1 m s^{-1} using the combined Φ model for tidal and wind stirring.

(You may use the numerical values of constants and estimates of mixing efficiency given in Section 6.3 and 6.4, i.e. α=1.67 × 10^{-4} °C^{-1}, c_p=4.0 × 10^3 J kg^{-1} °C^{-1},

ρ=1026 kg m^{-3}, k=2.5 × 10^{-3}, ρ_a =1.3 kg m^{-3}, C_d = 0.0012, γ=0.02, e = 0.004 and e_s =0.023).

6.4. A water column of 80 metres depth is thermally stratified. The density profile can be approximated as a 20-metre surface layer with a density of 1025.0 kg m^{-3}, above a 60-metre bottom layer of density 1026.0 kg m^{-3}.

(a) Calculate the amount of mixing energy required to vertically mix this water column.

(b) Assume the energy is supplied by winds. For the following wind strengths, estimate the amount of time needed to achieve a vertically mixed water column:

(i) A gale (wind speed = 20 m s^{-1}).
(ii) A severe storm (wind speed = 30 m s^{-1}).
(iii) A hurricane (wind speed = 40 m s^{-1}).
Comment on your results.

6.5. If we assume that a phytoplankton population approximately maintains a Redfield C:N of 6.6 during a spring bloom, write down an equation that allows you to estimate the total carbon fixed (g C m^{-2}) during a spring bloom in a surface layer of thickness h_s (metres) and with an initial pre-bloom nitrate concentration of S_0 (mmol m^{-3}).

Estimate the total carbon fixed rate during the spring bloom with S_0 =7 mmol m^{-3} and h_s =20 m.

6.6. Using an estimate of the timing of a spring bloom from Fig. 6.10, calculate the rate of carbon fixation by the bloom in the previous problem (g C m^{-2} d^{-1}).

7 Interior mixing and phytoplankton survival in stratified environments

In the last chapter we developed an understanding of the basic seasonal competition between heating and stirring, in which the mixing was driven by frictional stresses at the water column boundaries. In this chapter we shall describe the generally far weaker mixing which occurs across density interfaces within the interior of the water column. We will illustrate the physics involved using more detailed models of the interaction between buoyancy input and vertical mixing processes in the seasonally stratified regime. We will show where the models fail in their descriptions of mixing and how correcting these failings is vital if we are to understand and model the survival and growth of phytoplankton in stratified waters.

Pycnoclines often separate biochemically distinct regimes in the water column: high light and low nutrients near the sea surface, low light and high nutrients near the seabed. The inherent stability of a pycnocline can provide a niche for phytoplankton that contains both sufficient light and nutrients for survival. We will describe the links between physical and biological processes that lead to the survival of phytoplankton; you will see that understanding the processes that drive turbulence within and across pycnoclines lies at the heart of the growth and distribution of the primary producers.

7.1 A 1D model of vertical mixing with turbulence closure (the TC model)

In order to overcome the limitations of the simple mixed-layer model of Section 6.2.1, we need a fuller representation of processes in the water column which permits more complex vertical structures and allows the mixing rate to depend on density and velocity structure in the flow. Such a detailed model of the vertical structure is required to give us a more complete picture of where and when vertical mixing is taking place, information which will be vital when we come to understand and use the model of the seasonal cycle of primary production.

To establish a more fundamental physical model (Simpson, Crawford, *et al.*, 1996) we start with the horizontal momentum Equations (3.9) which, in linearised form without external forces, may be written as:

$$\frac{\partial u}{\partial t} = fv - \frac{1}{\rho}\left(\frac{\partial p}{\partial x} + \frac{\partial \tau_x}{\partial z}\right); \quad \frac{\partial v}{\partial t} = -fu - \frac{1}{\rho}\left(\frac{\partial p}{\partial y} + \frac{\partial \tau_y}{\partial z}\right);. \tag{7.1}$$

The motion is forced by the tides through an oscillating pressure gradient which, in the absence of horizontal density gradients, may be written in terms of the surface slopes as:

$$\frac{\partial p}{\partial x} = \rho g \frac{\partial \eta}{\partial x}; \quad \frac{\partial p}{\partial y} = \rho g \frac{\partial \eta}{\partial y}. \tag{7.2}$$

The slopes are conveniently found as a sum of N harmonic constituents (see Section 2.5.1) of the surface slope induced by the tide; e.g. for the x component:

$$\frac{\partial \eta}{\partial x} = \sum_{n=1}^{N} A_n \cos(\omega_n t + \alpha_n - g_n) \tag{7.3}$$

where A_n and g_n are the amplitude and phase lag of the surface slope constituent n with frequency ω_n. The phase α_n is the lag of constituent n in the TGF.

Wind forcing enters at the surface boundary via the stress terms (τ_{Wx}, τ_{Wy}) which are related to the velocity shear by the eddy viscosity N_z as in Equation (4.37). At the surface ($z = 0$), the stress is set equal to the applied wind stress so that:

$$\tau_x = -\rho N_z \frac{\partial u}{\partial z} = -\tau_{Wx}; \quad \tau_y = -\rho N_z \frac{\partial v}{\partial z} = -\tau_{Wy}. \tag{7.4}$$

At the bottom boundary ($z = -h$), the stresses must match the bottom stress in the quadratic drag law

$$-\rho N_z \frac{\partial u}{\partial z} = -k_b \rho U u_b; \quad -\rho N_z \frac{\partial v}{\partial z} = -k_b \rho U v_b, \tag{7.5}$$

where $U = \sqrt{(u_b^2 + v_b^2)}$ is the modulus of the bottom velocity (u_b, v_b).

Density will change in response to changes in T and S which are governed by the advection-diffusion Equation (4.39). In the absence of horizontal gradients and assuming the mean vertical motion is zero, this reduces to:

$$\frac{\partial(T,S)}{\partial t} = \frac{\partial}{\partial z}\left(K_z \frac{\partial(T,S)}{\partial z}\right) \tag{7.6}$$

where K_z is eddy diffusivity which we take to be the same for heat and salt.

Surface heat exchange consists of solar radiation input of $Q_s(1-\text{A})$ and heat losses Q_u as in the TML model. At the surface boundary ($z = 0$), the diffusive heat flux must equal the net heat input to the top layer, i.e. $-c_p \rho_0 K_z \frac{\partial T}{\partial z} = 0.55\, Q_s(1 - \text{A}) - Q_u$. Also, if we assume that precipitation and evaporation are in balance, then there is no exchange of freshwater so that $\frac{\partial S}{\partial z} = 0$. Similarly at the bottom ($z = -h$) the fluxes of heat and salt are negligible, so we can set $\frac{\partial T}{\partial z} = \frac{\partial S}{\partial z} = 0$.

The Equations (7.1) and (7.6) may be numerically integrated forward in time for specified surface slope, wind stress and surface heat exchange providing we know the eddy viscosity N_z and diffusivity K_z. These parameters, which depend on the amount of turbulence in the flow, may be obtained from a *turbulence closure scheme* (see Box 7.1) which relates them to the product of the intensity of the turbulence q (m s^{-1}) and a length scale L (metres):

$$N_z = S_M q L; \qquad K_z = S_H q L. \tag{7.7}$$

We will refer to our new model as the 'turbulence closure' or TC model. The turbulent intensity q is the root-mean-square speed of the turbulent motions, with the turbulent kinetic energy $E_T = \frac{1}{2}q^2$. The evolution of turbulent kinetic energy (TKE) is determined by the TKE equation which we established in Chapter 4 (see Section 4.4.2 for details of the individual terms):

$$\frac{\partial E_T}{\partial t} = -\frac{\partial(\overline{w' E'_T})}{\partial z} - \frac{1}{\rho_0}\left(\tau_x \frac{\partial U}{\partial x} + \tau_y \frac{\partial V}{\partial x}\right) - \frac{g\overline{\rho' w'}}{\rho_0} - \varepsilon. \tag{7.8}$$

The stability functions S_M and S_H represent the inhibiting effect of stratification on mixing, i.e. simulating the response of mixing to stability seen in Turner's tank experiment (Section 6.1.1). Both S_M and S_H depend on the gradient Richardson number (Equation 4.56) which, in turn, is derived from the velocity and density structure. For our TC model, we shall adopt a relatively simple formulation of the

Box 7.1 Turbulence closure schemes

In order to set up a system of equations which unambiguously describes the evolution of the structure and flow of the water column, we need to add to the basic equations (i.e. dynamics, continuity, advection-diffusion) a prescription for relating the eddy diffusivity K_z and eddy viscosity N_z to other parameters of the system and so 'close' the problem. Closure schemes range from the specification of constant values for K_z and N_z, through their prescription as simple functions of the gradient Richardson number *Ri*, to more elaborate formulations based on one or two equations for turbulent quantities. Most of the latter involve an equation for the evolution of the turbulent kinetic energy q^2 (like our Equation 7.8) and a second equation for another turbulence parameter. Two apparently different and widely used approaches have evolved; the Mellor-Yamada (MY2.5) scheme which employs an equation for q^2L and the $k - \varepsilon$ approach which incorporates an equation for evolution of the dissipation ε. The two approaches, which have generated a large and sometimes arcane literature, are distinguished by different notations and apparently differing assumptions. However, it has been shown that, in physical terms, the two are essentially the same and in practice lead to closely similar results (Burchard *et al.*, 1998). For the reader who wishes to pursue the topic of turbulence closure further, we would recommend the overview by Burchard, 2002. A full description of the equations used in our TC model here can be found in (Sharples and Tett, 1994; Simpson *et al.*, 1996).

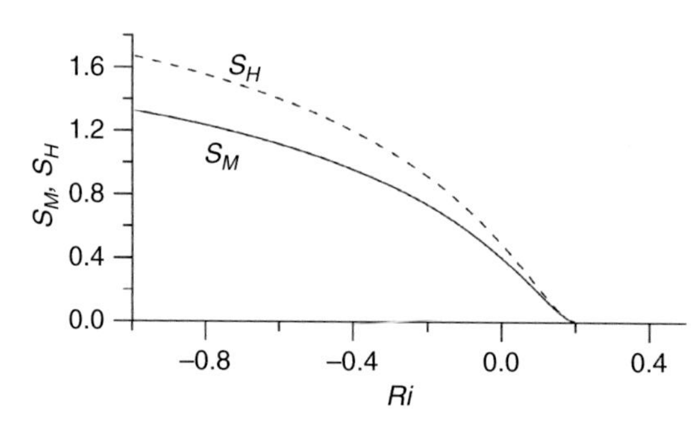

Figure 7.1 Variation of the stability functions in response to the Richardson number, Ri. Remember that positive Ri occurs for stable stratification, negative Ri implies convective instability.

stability functions (Galperin, Kantha, *et al.*, 1988) which is based on the assumption of local equilibrium (Equation 4.60) and leads to the dependence on Richardson number illustrated in Fig. 7.1. The sharp drop in S_M and S_H as Ri increases to a positive critical value is characteristic of most proposed closure schemes (e.g. Kantha and Clayson, 1994; Burchard *et al.*, 1998; Canuto *et al.*, 2001).

The length scale L can be thought of as some measure representing the size of the turbulent eddies. We shall use a simple algebraic representation in which L is controlled only by distance from the boundaries:

$$L = \kappa(z + h)\left(\frac{-z}{h}\right)^{1/2}. \tag{7.9}$$

Here κ is von Karman's constant ($\kappa \approx 0.41$); an important constraint on the behaviour of the turbulent lengthscale is that it should be scaled by κ near the water column boundaries. With this particular formulation, the eddy scale increases with distance from the top and bottom boundaries, reaching a maximum of $L = 0.38\kappa h$ at a height above the seabed of $2h/3$.

The set of equations (7.1) to (7.9) constitutes our TC model describing the physical processes acting in the water column at one location in the horizontal. It represents a much more complete account of the processes than was possible in the mixing model in Chapter 6, and it incorporates the various feedback mechanisms that are inherent in the interaction of flow and density structure. These feedbacks and the interdependence of the variables involved are evident in Fig. 7.2 which illustrates the logical flow used in the integration of the equations within the numerical model. At each time step the momentum, diffusion and TKE equations are stepped forward to generate new values for u, v, ρ and q which are used to determine the vertical profile of buoyancy frequency $N^2 = -\frac{g}{\rho}\frac{\partial \rho}{\partial z}$ and the velocity shear, which then determine S_M and S_H. These latter parameters are then combined with q and the length scale L to determine vertical profiles of N_z and K_z in readiness for the next time step.

Wind stress, surface heat exchange parameters and the initial density structure are obtained from observations. The tidal forcing can also be obtained from observations of bottom pressure gradient or extracted from a tidal model. A convenient alternative is to drive the model with surface slopes estimated from knowledge of the near surface tidal current ellipse (Sharples, 1999). The operation of a TC model forced in this way over three tidal cycles is illustrated in Fig. 7.3a for a seasonally stratified location in the

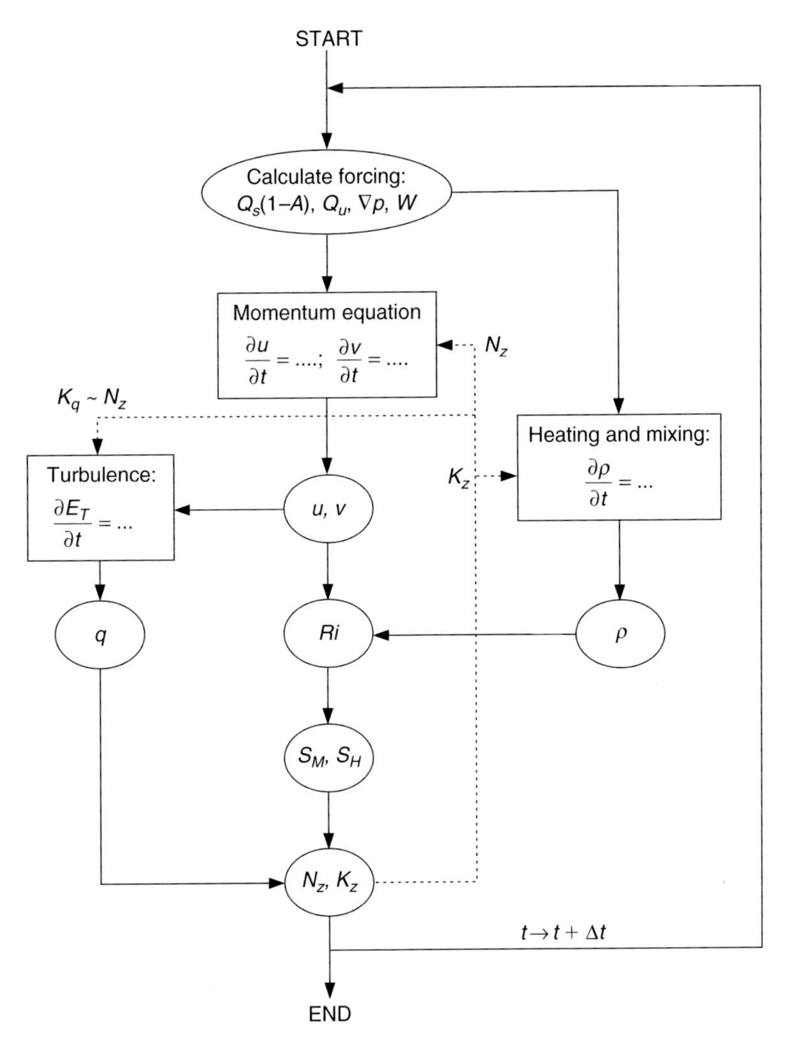

Figure 7.2 Simplified schematic of the links between the vertical velocity and density structure required within the TC model during one time step to generate the eddy diffusivities and viscosities.

western Irish Sea. Comparison with observations in Fig. 7.3b shows that the model generally does rather well in simulating the time evolution and vertical structure of the tidal flow, which was dominant in this case as winds were light throughout the observational period. The motion is largely barotropic with a frictional boundary layer near the seabed which is well reproduced by the model. To learn more about the TC model and run it on short and seasonal time scales, go to the book website.

7.2 Comparison with observations of turbulence

A more important question in the present context is how well the model performs in representing turbulent mixing. Accurate simulations of the turbulence are vital if the

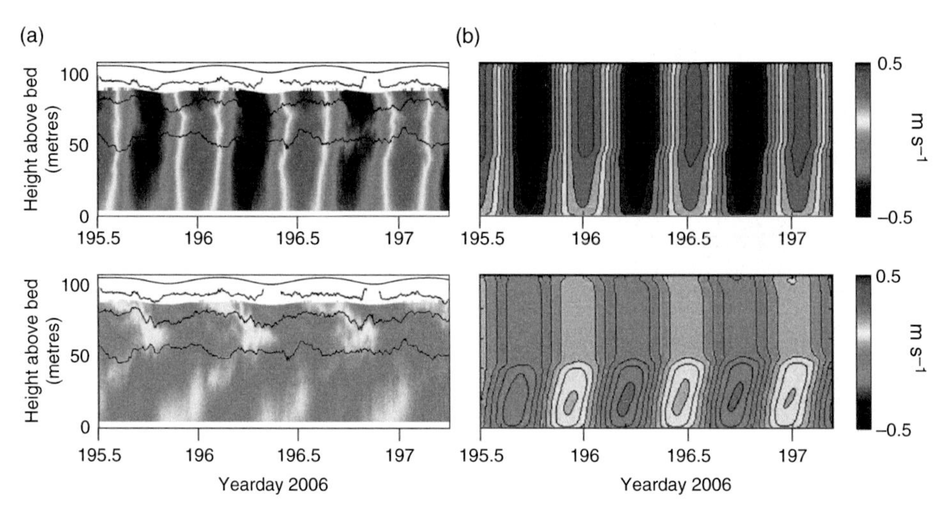

Figure 7.3 See colour plates version. Velocities along major (top) and minor (bottom) axes of the tidal current ellipse at station SWIS (53°52.00' N, 05°27.00' W) in the stratified western Irish Sea during July 2006: (a) measured using ADCP (from Simpson *et al.*, 2009, with permission of Elsevier), and (b) derived from the turbulence closure model driven by tidal and wind stress forcing. Grey scales are the same for all panels. Black lines in (a) are isotherms at 10, 12, and 14 °C, indicating the stratification of the water column.

model is to simulate the evolution of the density structure, the fluxes of nutrients through the water column which control primary production and the interaction between turbulent mixing and phytoplankton swimming.

7.2.1 Observing and modelling turbulent dissipation

We can assess the performance of the model in relation to mixing by comparing the model with observations of turbulence in the water column in seasonally stratified and continuously mixed regimes. The rate of production of turbulence is still difficult to measure in all but the most tidally energetic regimes. However, the rate of turbulent energy dissipation can be determined with free-fall shear profilers, such as that shown in Fig. 7.4. As the profiler falls at a steady speed of $w_p \sim 0.7$ m s^{-1}, it senses the relative horizontal velocity u' on scales down to < 1 cm via an airfoil shear probe (Osborn and Siddon, 1975; Osborn and Crawford, 1980) mounted in the nose. The gradient of the horizontal velocity is obtained by differentiating the velocity with respect to time to give a shear component $\partial u'/\partial z = (\partial u'/\partial t)/w_p$. The dissipation rate ε is then found from the relation for isotropic turbulence (Batchelor, 1960):

$$\varepsilon = 7.5\nu\overline{\left(\frac{\partial u'}{\partial z}\right)^2}. \tag{7.10}$$

This relation assumes that the turbulence is isotropic at the high wave numbers (small scales) which dominate the shear spectrum. The overbar denotes an average over a

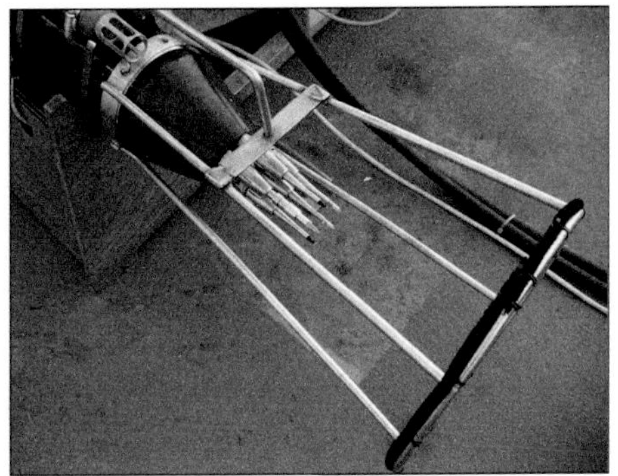

Figure 7.4 A free-fall turbulence microstructure profiler. Left: the profiler and its winch. The brush at the top of the profiler provides drag, maintaining the profiler vertical in the water column as it falls at 0.6–0.7 m s^{-1}. The turbulence sensors are inside the guard ring. The light Kevlar cable provides power and retrieves data as the profiler falls towards the seabed. Right: close up of the sensors and the guard ring which protects them when the probe hits the seabed. As well as the shear sensors, two fast response thermistors are used to record temperature fine structure. (Photos by J. Sharples.)

section of the water column, usually ~1 metre, the minimum needed to give an adequate sample of the turbulence, which is generally highly variable. Further improvement in the statistical reliability of estimates is achieved by averaging a rapid sequence of repeat profiles. At the end of each profile, signaled by impact with the bed, the profiler is hauled back to the surface by a light Kevlar tether which contains pairs of electrical conductors carrying the high data rate signals to a computer onboard the research vessel. As well as high resolution velocity, the profiler is also equipped with sensors for temperature and conductivity so that parallel profiles of temperature, salinity, density and hence the buoyancy frequency are available. The rate of dissipation ε (W m^{-3}) together with N^2 may be used to estimate the mixing rate in the form of the eddy diffusivity via the Osborn relation (Equation 4.63):

$$K_z = \Gamma\left(\frac{\varepsilon}{N^2}\right) \tag{7.11}$$

where $\Gamma = R_f/(1 + R_f) \approx 0.2$. Note that ε here is in units of W kg^{-1} ($\equiv m^2\ s^{-3}$). We will also quote dissipation values in units of W m^{-3}, which are widely used in the literature. Numerical values in the latter units are larger by a factor of ~1000, i.e. the density of seawater in kg m^{-3}. Dissipation can be measured by the profiler over most of the water column, except for a region very close to the surface (within about 8 metres) which is excluded because of possible interference from the ship's wake and the time it takes for the profiler to reach its terminal fall speed. Measurements continue to within ~0.15 metres of the seabed when the guard ring hits the bottom.

The results of a time series of observations of ε at the SWIS location are shown in Fig. 7.5a in the form of a contoured plot of $\log_{10} \varepsilon$ (Wm^{-3}) versus time and depth. This log presentation is preferable because of the wide range of dissipation rates involved. Have a look at Box 7.2 to gain some understanding of the rates of turbulent energy

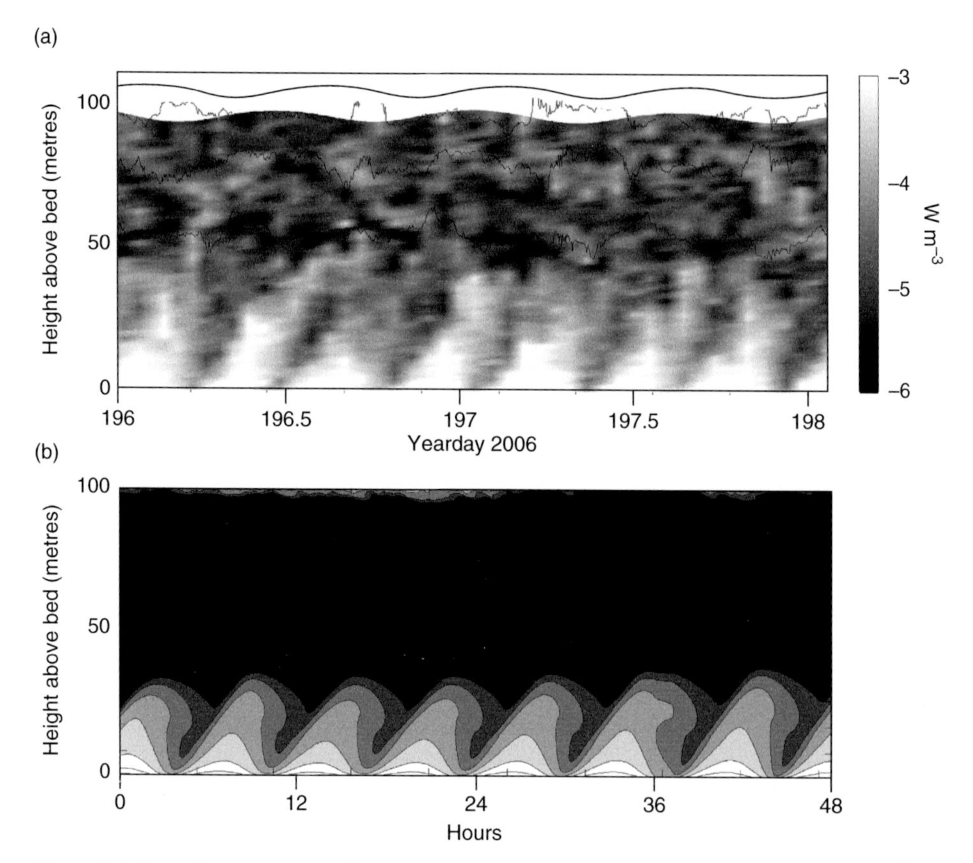

Figure 7.5 See colour plates version. 48-hour time series of dissipation as $\log_{10} \varepsilon$ (W m^{-3}) at station SWIS (53°52.00′ N, 05°27.00′ W) in a seasonally stratified region of the western Irish Sea. (a) Contour plot based on an average of 6 profiles per hour with the dissipation profiler, with isotherms (black lines) at 10, 12 and 14 °C (from Simpson *et al.*, 2009, with permission of Elsevier); (b) Dissipation simulated by a run of the TC model forced by the tide and local winds. Shade scales are the same for the two panels.

dissipation that we commonly observe in shelf seas. In Fig. 7.5a there is a region below 40 mab (metres above the bed) where ε exhibits a strong quarter diurnal (M_4) variation with peaks during the ebb and flood flow. Very close to the bottom boundary, peak dissipation exceeds 10^{-3} W m^{-3}, while in midwater it falls close to the instrument noise level of ~10^{-6} W m^{-3}. In the bottom boundary layer it can be seen that the time of maximum dissipation is increasingly delayed with height above the bed. This delay, which is ~4 hours at the top of the tidal boundary layer, is not primarily a result of upward diffusion of turbulence from the near-bed region but results from a systematic delay in the production of turbulence arising from the velocity shear having a phase lag that increases with height (Lamb, 1932; Simpson *et al.*, 2000).

The TC model simulation of the observed conditions is shown in Fig. 7.5b. The model reproduces some of the main features of the observations, with a similar boundary layer structure and a pronounced phase delay in ε with height. Because the winds were light, the model indicates only slight dissipation in the surface boundary which in this case was too thin to be observed by the profiler. The rates of dissipation in the bottom boundary layer are reasonably well simulated, as is the fall-off to low levels in the interior region of the water column. There is, however, a conspicuous failure of the model to reproduce the small but important episodes of dissipation in the interior which are apparent in the observations when ε rises, for short periods, by two orders of magnitude to $\sim 10^{-3.5}$ W m^{-3}. This intermittent dissipation is responsible for mixing properties into/out of pycnoclines. It is crucial in relation to the supply of nutrients to the upper part of the water column to drive primary production. It is also important when assessing the potential for directed swimming that some phytoplankton are capable of, and for the export of phytoplankton from a pycnocline into deeper water. In the present form of the TC model, this vital contribution of midwater mixing is missing since modelled ε falls several orders of magnitude below the observations. This interior turbulence and its associated diapycnal fluxes are missing from the model because the turbulence closure scheme is relying solely on energy inputs due to boundary stresses at the seabed and the sea surface. Other physical mechanisms can generate velocity shear, and hence mixing, away from the boundaries; these mechanisms need to be either fully resolved and simulated by the model, or at the very least parameterised within the model in a way that leads to realistic mixing.

Box 7.2 How turbulent are the shelf seas?

Rates of turbulent energy dissipation vary widely in shelf seas. In regions of strong flows (> 2 m s^{-1}) in shallow water, dissipation near the bottom boundary may exceed 1 W m^{-3}. In less energetic regions where the water column stratifies, as in Fig. 7.5a, there is a large contrast in dissipation of 2–3 orders of magnitude between the tranquil pycnocline region and the boundary layers near the surface and bottom. You can get an idea of the strengths of turbulent mixing in terms of something more familiar by making a comparison with, say, the mixing generated by a typical kitchen mixer. For instance, mean dissipation in the pycnocline may be as low as 10^{-5} W m^{-3}. This is roughly equivalent to two 100 W hand-held kitchen mixers operating in a square kilometer of pycnocline 20 metres thick (see Problem 7.1).

Before we continue, let us briefly convince ourselves that this 'missing mixing' that appears in our observations but not in our model of the interior of the water column is real. Estimates of the diffusivity in the water column for the SWIS time series of Fig. 7.5a using the Osborn relation (Equation 7.11) and the observed values of ε and N^2 are shown in Fig. 7.6. The average mixing rate corresponds to $K_z \sim 3 \times 10^{-5}$ m^2 s^{-1}. We might ask if this mixing rate through the pycnocline is sufficient to account for the observed rate of temperature rise in the deep water of the stratified

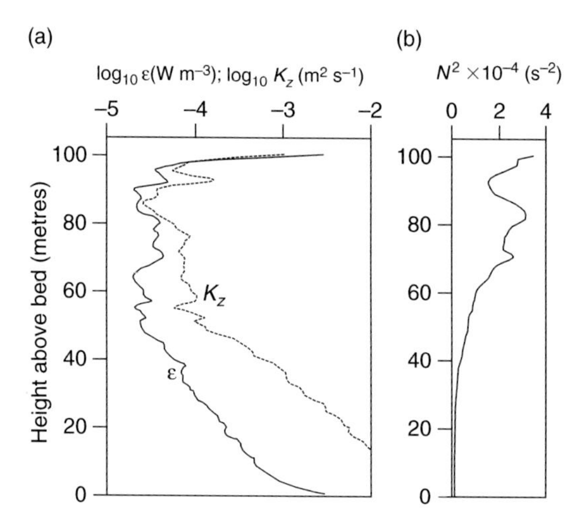

Figure 7.6 Average values of (a) dissipation ε (Wm^{-3}) and the vertical eddy diffusivity K_z ($\mathrm{m}^2\,\mathrm{s}^{-1}$), and (b) the squared buoyancy frequency N^2 (s^{-2}) used to calculate K_z ($\mathrm{m}^2\,\mathrm{s}^{-1}$) using the Osborn relation Equation (7.11).

water column during the midsummer period. If heat input to the bottom mixed layer is only by downward diffusion, then we can write the heat balance as:

$$h_b \frac{\partial T_b}{\partial t} = K_z \frac{\partial T}{\partial z} = \frac{\Gamma \varepsilon}{N^2} \frac{\partial T}{\partial z} \tag{7.12}$$

where $\frac{\partial T}{\partial z}$ is the temperature gradient in the pycnocline, h_b is the thickness of the bottom layer, and T_b is the bottom layer temperature. If the density depends only on temperature, then $N^2 = -(g/\rho_0)\partial\rho/\partial z = g\alpha\partial T/\partial z$ where α is the thermal expansion coefficient. Substituting for N^2, we have for the rate of temperature rise:

$$\frac{\partial T_b}{\partial t} = \frac{\Gamma \varepsilon}{g \alpha h_b} \tag{7.13}$$

which is, perhaps surprisingly, independent of the temperature gradient.

For $\varepsilon = 10^{-7}\ \mathrm{W\,kg}^{-1}$; $h_b = 50$ m; $\Gamma = 0.2$; $\alpha = 1.6 \times 10^{-4}\ {}^\circ\mathrm{C}^{-1}$ we find that: $\frac{\partial T_b}{\partial t} = 2.5 \times 10^{-7}\,{}^\circ\mathrm{Cs}^{-1} = 0.65\,{}^\circ\mathrm{C}$ per month. This accounts for most of the observed temperature rise of $\sim$1°C per month, so we may reasonably conclude that the observed dissipation in the pycnocline is a realistic measure of the vertical diffusion which is driving fluxes of heat and nutrients and other properties through the pycnocline.

7.2.2 Physical mechanisms responsible for mixing at pycnoclines

What is missing in the physics of the TC model which causes it to fail in the pycnocline region? There has to be one or more sources of energy for driving the weak internal mixing observed. There are three main contenders for supplying this mixing, namely:

(i) Internal wave motions

Large displacements of the isotherms are commonly observed to occur in the pycnocline as, for example, in Fig. 7.5a where there are variations of isotherm depths of ~10 metres (Green *et al.*, 2010). These movements of the pycnocline indicate the presence of internal wave motions which frequently involve large velocity shear across the pycnocline, with upper and lower layers moving out of phase (see Section 4.2). Such shear can contribute to reducing the Richardson number and hence to producing internal mixing. Internal waves can have a wide range of frequencies extending from the inertial frequency f to the buoyancy frequency N. Much of the energy input to internal waves comes from the barotropic tidal movement of a stratified water column over uneven bottom topography. The most obvious example of this mechanism in shelf seas is the particularly large internal waves of the semi-diurnal period (the *internal tide*) generated over the steep topography of the shelf edge (more on this will follow in Chapter 10). Waves generated in this way travel on to the shelf and become distorted through non-linear processes to form shorter period internal waves, termed solibores. These modified waves transmit significant amounts of energy on to the shelf and can contribute to mixing in the shelf sea pycnocline close to the shelf edge. Internal waves of tidal period may also be produced away from the shelf edge by flow over banks and depressions in the seabed (Moum and Nash, 2000).

(ii) Inertial motions

As we saw in Section 3.4, an impulse of wind at the sea surface generates a 'slab' motion of the surface layer over the thermocline, leading to inertial oscillations in surface currents. Where the water column is stratified, the pycnocline is a region of reduced frictional stress (low N_z) so the damping of inertial oscillations in the surface layer is weak and they may persist for many days. Moreover, in shelf seas where there are land boundaries influencing the flow, the wind stress can set up an opposing pressure gradient which induces an out-of-phase oscillation in the layer below the pycnocline (Rippeth *et al.*, 2002).

(iii) Spring-neap contrasts in tidal mixing

Turbulent dissipation arising from friction between tidal currents and the seabed can often be detected high within the water column (Sharples *et al.*, 2001b) (Fig. 7.5a), potentially impacting on the base of the thermocline. The intensity of this dissipation is modulated by the spring-neap cycle in the tidal currents, resulting in a fortnightly cycle in the mixing at the base of the thermocline (Sharples, 2008). As tidal currents increase towards spring tides, so does the strength of the current shear and mixing at the base of the thermocline and the lower few metres of the thermocline are eroded. As the tidal flow and current shear decreases between spring tides and neap tides, the base of the thermocline again deepens. The position of the base of the thermocline oscillates due to a fortnightly shift in the balance between the turbulent mixing generated by the tides in the bottom layer and by winds in the surface layer.

Processes (i) and (ii) above are mechanisms internal to the water column, while (iii) is a boundary-driven mechanism. The easiest process to simulate with a numerical model is the spring-neap cycle in the erosion/re-formation of the thermocline base. Our TC model can simulate the modulation of the tidal shear, and as long as the vertical resolution of the model is sufficient, the shifting position of the thermocline depth can be modelled.[1] Inertial oscillations can, at times, dominate observations of current shear (Palmer, Rippeth, *et al.*, 2008). Models are able to represent this, though achieving the correct mixing at the thermocline depends on having sufficient vertical resolution and on the temporal resolution of the driving winds. Simulating mixing by internal waves is more problematic. At present, large-scale models of shelf seas do not have the horizontal resolution to allow the proper representation of the internal waves, which can have components with wavelengths of 1 km or less which may invalidate the hydrostatic assumption. Only models covering limited domains and time scales are able to cope with the computational effort required for such small-scale processes (Lamb, 1994). Moreover, the sensitivity of internal wave mixing to changes in the slope of the seabed means that high-quality bathymetric information is required.

As an interim solution to the problem, models often 'parameterise' the impacts of unresolved processes. The challenge then is to represent the mixing impact of a process, based preferably on some broadly applicable relationships arising from observations of the process. For the case of interior mixing driven by internal waves or by meteorological variability, the simplest parameterisation is to limit the lowest allowable eddy diffusivity and viscosity (Sharples and Tett, 1994); this is the approach taken in our coupled physics-biogeochemistry model (www.cambridge.org/shelfseas).

Physics summary box

- In order to obtain a more complete picture of vertical exchange than that provided by the TML model of Chapter 6, we need to represent turbulent processes more explicitly in our numerical models.
- Such models require a 'turbulence closure' scheme which relates the eddy viscosity and diffusivity to mean properties.
- Measurements of turbulent energy dissipation by shear profilers provide a direct test of the ability of the models to simulate turbulent mixing.
- Sharp density gradients (pycnoclines) can severely limit the ability of turbulence to mix water properties into and across them, an effect which can be quantified by relating diffusivities to the gradient Richardson number using a turbulence closure scheme.
- Knowledge of the eddy diffusivity is fundamental to quantifying turbulent fluxes of, for instance, heat, nutrients and carbon. Measurements of dissipation rates and water column stability can be combined to estimate eddy diffusivities (e.g. Equation 7.11).
- Typical values of eddy diffusivity within a pycnocline can be 10^{-4}–10^{-6} m^2 s^{-1}. A wind-mixed surface layer might have values of 10^{-2}–10^{-4} m^2 s^{-1}, while the

[1] This is a difficult process to observe in the real ocean; the spring-neap signal needs to be separated from internal mixing mechanisms, and from the tidal advection of a horizontally variable thermocline structure. Numerical modelling (Sharples, 2008) suggests that the base of the thermocline only changes by 3–5 metres, implying that a model requires a vertical resolution of 1 metre or less in order to simulate this process.

bottom layer of a tidally energetic shelf region could have diffusivities of 10^{-1}–10^{-2} m^2 s^{-1}.

- Comparison of model predictions with observations of turbulent dissipation generally shows that the models do well at reproducing boundary-driven turbulence; e.g. turbulence generated by friction between water flow and the seabed and between the wind and the sea surface. However, away from the boundaries the models do far less well, potentially underestimating turbulent dissipation by several orders of magnitude.
- Model failure in the water column interior arises because of physical processes that the models do not resolve. Spring-neap changes in tidal mixing at the base of the thermocline, wind-driven motion of the surface layer at the inertial frequency and short-wavelength (1 – a few km) internal waves are all important potential sources of interior mixing that either need to be resolved or adequately parameterised.

More complex methods involve limitations imposed on the eddy diffusivity and viscosity or on the turbulent dissipation as a function of the gradient Richardson number (Kantha and Clayson, 1994; Canuto *et al.*, 2001).

7.3 Phytoplankton growth, distribution and survival in pycnoclines

In regions which stratify, the spring bloom (Chapter 6) often receives particular attention as the most conspicuous event of the primary production cycle. After the spring bloom, surface layer chlorophyll is usually seen to be very low throughout the summer and dominated by small cells in an efficient recycling system as the effect of the thermocline in inhibiting mixing prevents the surface from being replenished with nutrients. The dominant small cells are not efficient in supplying organic material to the rest of the marine food web, and the primary production of large cells during the spring bloom is not sufficient to fuel the known demands of the pelagic and benthic food webs during the rest of the year. We have also seen that, once stratification has become established, it is very hard to re-mix the water column with winds. In these circumstances, we might reasonably ask where and how is new primary production (i.e. production that utilises nutrients from below the thermocline) supported after the spring bloom.

Understanding phytoplankton survival within pycnoclines requires knowledge of the often-intermittent interior mixing, which we described earlier in this chapter, and how phytoplankton respond to it. We will first describe how turbulent mixing across the base of a pycnocline can both supply a phytoplankton layer with nutrients and entrain phytoplankton into the bottom waters. We will then demonstrate that primary production within the SCM is a significant contributor to the total annual production of a shelf sea. Some phytoplankton attempt to control their own resource needs by swimming vertically; we will describe how the success of this strategy depends on the strength of the turbulence in the bottom water and within the pycnocline. Finally, we will illustrate how changes in the light spectrum through a pycnocline can lead to layering of individual species within a single chlorophyll layer, again showing how the weakly turbulent environment provided by pycnoclines plays a vital role.

7.3.1 The subsurface chlorophyll maximum (SCM)

Below the sea surface, phytoplankton are often found to be located in well-defined layers where they are actively growing. In regions where mixing is dominated by the barotropic tide, these phytoplankton are generally found in a thin layer within the base of the seasonal thermocline, such as in the example from the Celtic Sea in Fig. 7.7a. In areas with weak barotropic tides, where internal mixing plays a more dominant role in subsurface structure, broader temperature and subsurface chlorophyll distributions are seen. Two examples are shown in Fig. 7.7b, c, from the northeast shelf of New Zealand and from the edge of Georges Bank, northeast United States. In regions of weak barotropic and internal mixing, highly concentrated chlorophyll peaks can be found, often dominated by a single species of motile phytoplankton, as seen off Monterey shown in Fig. 7.7d. A layer of phytoplankton situated within a region of vertical density gradient below the surface is often referred to as a *subsurface chlorophyll maximum* (SCM).

These layers can be very extensive, covering entire stratified regions as seen in a 500 km transect with a towed CTD through the Celtic Sea shown in Fig. 7.8. Several candidate mechanisms have been proposed as the cause of such layers of phytoplankton. Cells may sink to form a layer at a location within a pycnocline where they are neutrally buoyant, or alternatively they may sink until a supply of nutrients within the *nitracline* (the region of rapid changes in nitrate within the pycnocline) allows them to grow and increase their buoyancy (Steele and Yentsch, 1960). The ability of some species (particularly dinoflagellates) to swim in response to resource needs may allow them to actively form the layer (Margalef, 1978). Alternatively, growth of a layer of phytoplankton could be driven by a weak flux of nutrients into the thermocline from the bottom layer (Pingree *et al.*, 1977), as long as there is sufficient light to drive photosynthesis. It has also been argued that the SCM may not represent a true biomass maxima, but instead may be a result of changes in pigment concentration per cell (Kiefer and Kremer, 1981). Indeed, for SCMs in the open ocean it has been found that the entire signal of the chlorophyll peak can be attributed to higher pigment concentrations, with cells in similar numbers to the surface waters (Veldhuis and Kraay, 2004). In shelf seas, however, we generally find that both increased pigment per cell and increased cell numbers are evident.

7.3.2 Nutrient supply and primary production within the SCM

Let us focus first on a two-layer stratified system in a region with moderate or strong barotropic tides, ignoring the possibility of significant cell swimming so that phytoplankton growth requirements are dominated by the turbulent environment. For growth of the phytoplankton within a midwater layer, nutrients must be mixed into

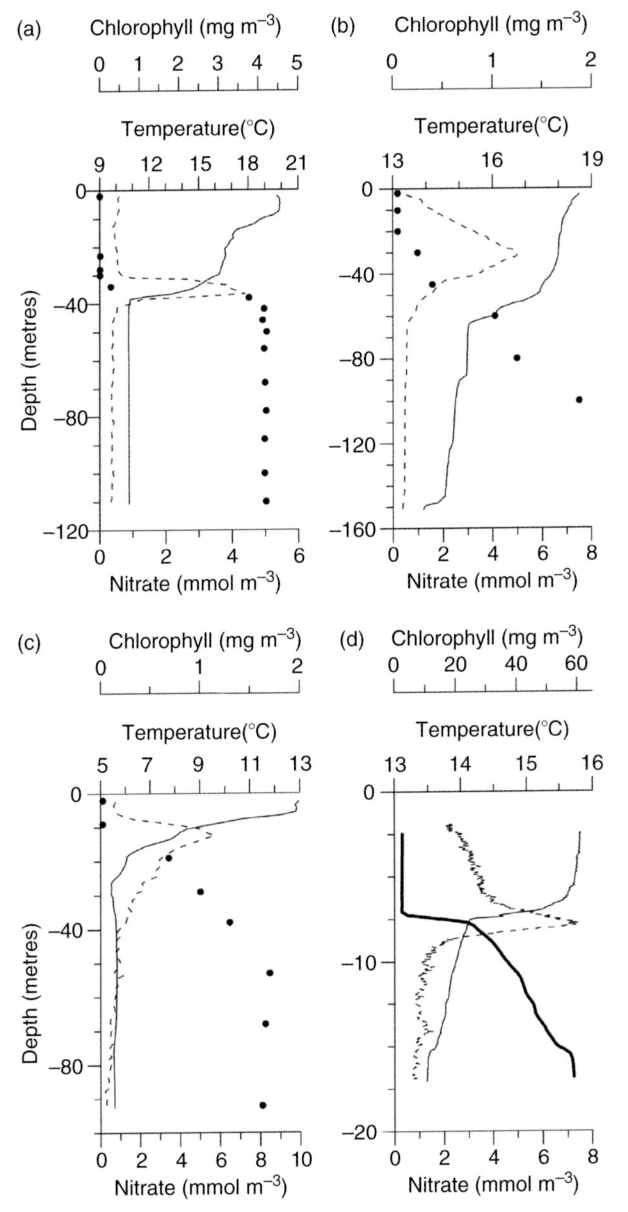

Figure 7.7 Examples of vertical profiles of temperature (solid line), chlorophyll (dashed line), and nitrate (dots in (a)–(c), bold line in (d). (a) The western English Channel (Sharples *et al.*, 2001b). (b) The northeast shelf of New Zealand (Sharples *et al.*, 2001a). (c) Northeast flank of Georges Bank. (Data courtesy of David Townsend, University of Maine, and the US GLOBEC Program.) (d) Monterey Bay, California. Adapted from Steinbuck *et al.*, 2009, data courtesy of Jonah Steinbuck, Stanford University, and co-workers with permission from ASLO.

the base of the layer (Sharples and Tett, 1994) and there must be sufficient light to drive photosynthesis. Moreover, as mixing drives nutrients into the layer, it will also be removing phytoplankton from the layer; growth within the layer needs to be able to offset this export to deeper water. A numerical model of phytoplankton growth within a framework similar to the TC model described earlier in Section 7.2.1 does show the development of a post-spring bloom SCM made up of neutrally buoyant phytoplankton. However, the SCM only develops if there is some 'background' mixing at the thermocline (Sharples and Tett, 1994), potentially attributable to one

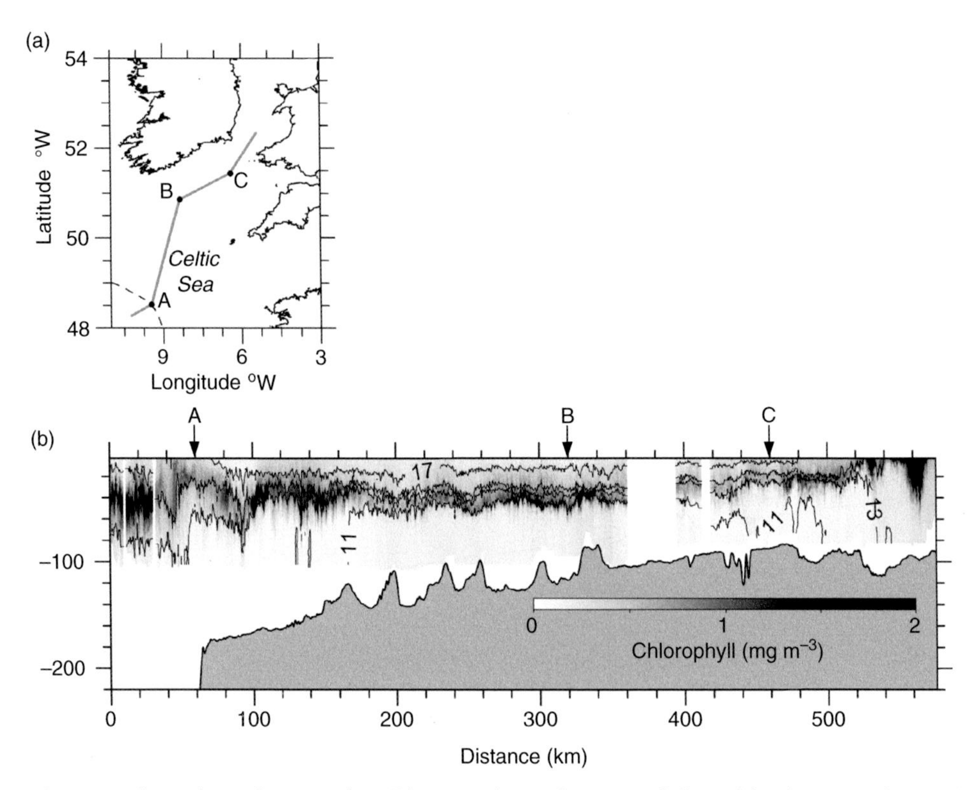

Figure 7.8 See colour plates version. The map shows the course followed by the research vessel RRS *James Clark Ross* in August 2003, towing the Seasoar undulating vehicle (see Fig. 1.5). The contoured image shows temperature (lines, every 2 °C) and chlorophyll concentration (shaded) along the transect. The dark grey region is the seabed, showing the dropoff at the shelf edge (60 km) and gradually shallowing towards the entrance to the Irish Sea (550 km). Data were collected between the surface and a depth of 100 metres. Vertical white bands indicate regions where there was a communication failure with the instrument. The arrows labeled along the top axis show the positions of the course changes marked on the map.

or more of the three interior mixing processes described in Section 7.2.2. Can we find evidence to indicate whether, or not, each of the three key thermocline mixing processes has a significant impact on the biochemistry of the thermocline?

Looking for such evidence is challenging; we need vertically well resolved time series of the SCM. The subsurface nature of these layers rules out satellite remote sensing as a convenient means to construct such time series. Cost and problems with biofouling usually mean that moored instrumentation does not have the vertical resolution to identify the evolution of the structure of subsurface chlorophyll layers. However, we do have some useful observational evidence, and can gain further information/insight by using numerical models.

Consider first the biochemical response to topographically generated internal waves. The clearest examples occur at the shelf edge (see Chapter 10), but there is some evidence of effects over the slopes of large banks on the shelf, as shown in Fig. 7.9. Stratified regions of the NW European shelf receive a typical daily-mean

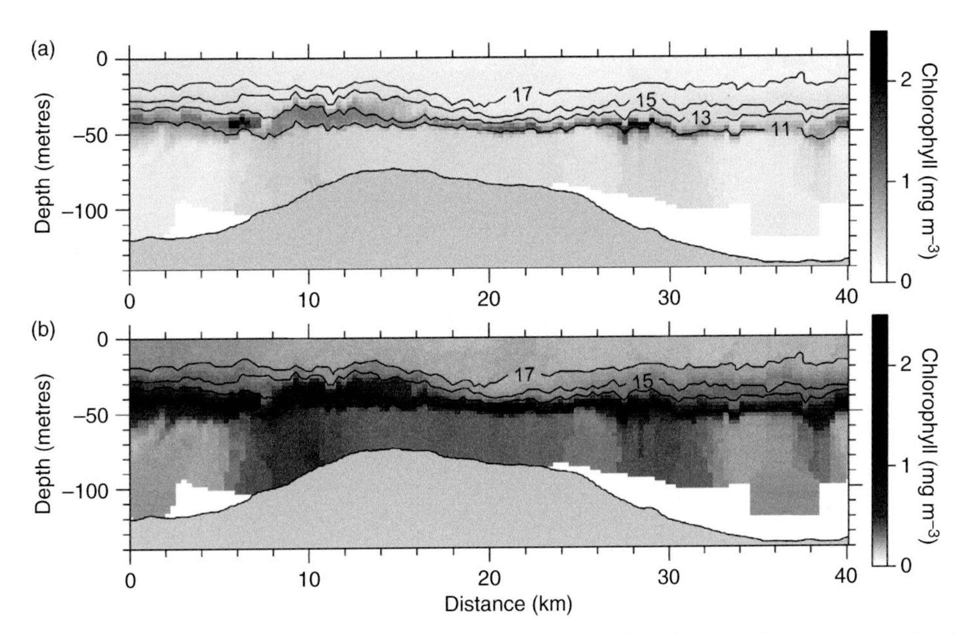

Figure 7.9 See colour plates version. A transect of chlorophyll (colours) and temperature (lines) over Jones Bank in the Celtic Sea in July 2005, carried out with the Seasoar vehicle (see Fig. 1.5). The data in (a) and (b) are the same, but in (b) the chlorophyll colour scale has been skewed to highlight the lower chlorophyll concentrations.

vertical flux of nitrate into the base of the thermocline of 2 mmol m^{-2} d^{-1}, while over the bank slopes we have seen the flux reach over 20 mmol m^{-2} d^{-1}. Note in Fig. 7.9a that there are two patches of elevated chlorophyll concentration over the slopes of the bank. The shift in the chlorophyll scale used in Fig. 7.9b highlights the increase in bottom layer chlorophyll underneath these patches which contrasts with the very low chlorophyll concentrations in bottom water away from the bank. The mixing process is thought to be the generation and breaking of internal waves as the stratified tidal flow moves on- and off-bank, similar to the mixing seen over similar seabed features elsewhere (Moum and Nash, 2000).

Observational evidence of a subsurface biochemical response to inertial mixing (mechanism (ii) in Section 7.2.2) is more difficult to capture, largely because of the intermittent nature of the meteorological forcing and the challenge of identifying a response against the slow mean advection of horizontal patchiness. The need to incorporate a reasonable amount of natural variability in wind stress has been identified in numerical models as driving possibly-important contrasts in annual primary production (Radach and Moll, 1993), but observational evidence of phytoplankton growth in response to wind variability is currently lacking. We could imagine that a pulse of wind might supply nutrients into the SCM and increase growth. On the other hand, we might also expect the pulse of wind to deepen the base of the thermocline and so disrupt and reduce the light received by the phytoplankton. This is a good example of the kind of problem amenable to carefully designed numerical model experiments.

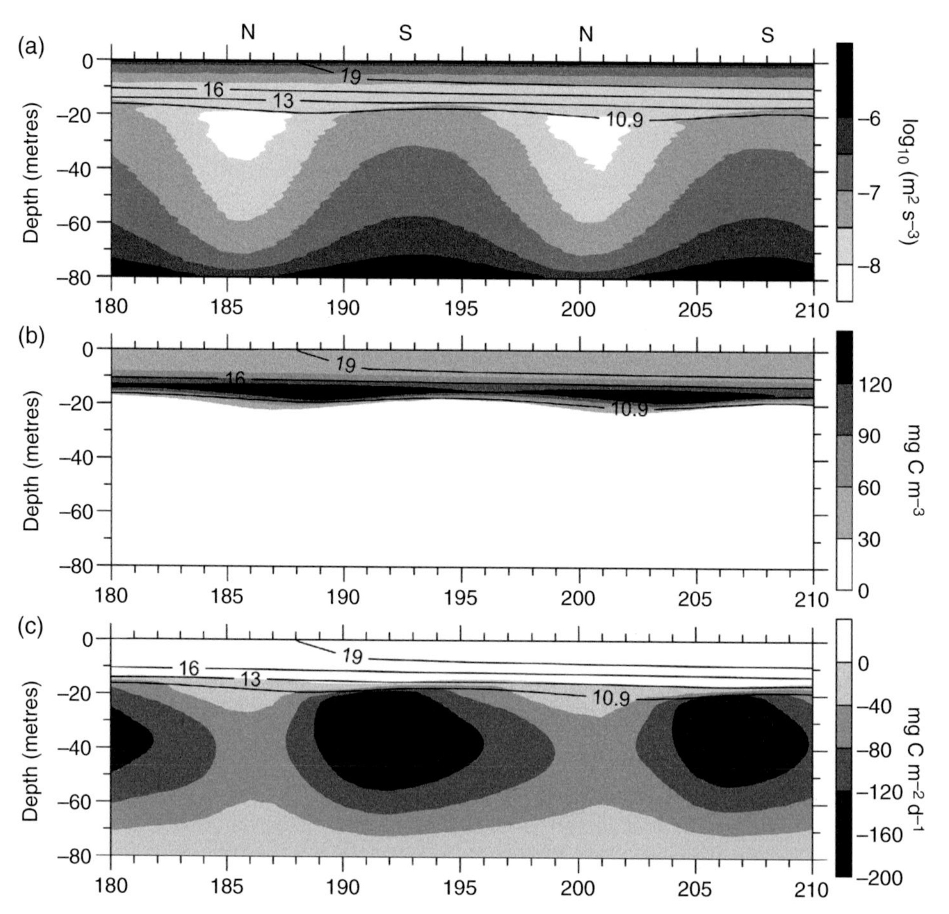

Figure 7.10 Model simulation of the effects of spring-neap modulated mixing on the base of the thermocline and the SCM, based on (Sharples, 2008). (a) Daily averaged vertical distribution of the turbulent diffusivity, K_z; (b) phytoplankton carbon concentration; (c) daily averaged vertical turbulent flux of phytoplankton carbon. In (a)–(c) the line contours are isotherms. Along the upper axis in (a) N show neap tides and S spring tides.

Spring-neap changes in bottom layer turbulent mixing (mechanism (iii) in Section 7.2.2) leading to erosion of the base of the thermocline and exporting any biomass in the base of the thermocline into the bottom mixed layer, is amenable to our 1D numerical model (Sharples, 2008). The model simulation is shown in Fig. 7.10. As turbulent mixing increases towards spring tides (Fig. 7.10a), the base of the thermocline is eroded and phytoplankton carbon is mixed downward into the bottom layer (Fig. 7.10c). As the tidal currents decrease towards the next neaps, solar heating and wind stirring push the stratification back downward and incorporate bottom-layer nutrients into the base of the thermocline. These nutrients are then available for primary production within the base of the thermocline; in the model results (Fig. 7.10b) this produces fortnightly pulsing of thermocline phytoplankton biomass. Carbon export from the thermocline to the bottom mixed layer as a result of increasing turbulent mixing towards spring tides has been observed (Sharples *et al.*, 2001b), and a modelling study with a more complex microbial system has suggested possible spring-neap changes in the community structure within the thermocline (Allen *et al.*, 2004). Phytoplankton growth rates at the depth of the SCM, typically $0.25\ d^{-1} - 0.5\ d^{-1}$, mean that growth or community responses to a physical perturbation will require a few days to become significant. This is evident, for instance, in the 2–3 day lag in the biomass response relative to the timing of neap tides that you can see comparing Fig. 7.10a and 7.10b. The time scales of physical change driven by the spring-neap cycle tend, therefore, to favour phytoplankton growth.

It is worth noting at this stage that in many cases the vertical mixing of nutrients is not strong enough to exceed the nutrient uptake capacity of the phytoplankton within the SCM. Thus little or none of the nutrient flux reaches the sea surface (e.g. look at all of the nitrate profiles in Fig. 7.9), suggesting that the nutrient supply is an important limiter of growth in the SCM. Consequently, direct measurements of vertical turbulent fluxes of nitrate into the base of the thermocline have been used to infer the potential new phytoplankton carbon fixation rate within the SCM (Sharples *et al.*, 2001b).

How much primary production occurs in the seasonal thermocline? Much of our knowledge of annual primary production rates globally is now based on the use of satellite images of surface chlorophyll (e.g. Fig. 1.1b). Such a method will capture the major production event of the spring bloom, but clearly there is phytoplankton growth continuing within the thermocline after the bloom. Light at the thermocline is generally very weak, typically 1–10% of the surface PAR, so primary production rates tend to be low. Studies in the thermocline of the Western English Channel in summer have yielded rates of primary production over the whole stratified water column of $0.5–0.8\ g\ C\ m^{-2}\ d^{-1}$ (Moore *et al.*, 2003), which is far less than the value of about $6\ C\ m^{-2}\ d^{-1}$ reported for the spring bloom in the same area (Pingree *et al.*, 1976). The differences are due to the low light experienced by cells at the thermocline, though the full impact is offset by the cells' ability to acclimate to the low light environment. Values of I_k are significantly lower within the SCM, so that production rates are able to approach their maximum values (Moore *et al.*, 2006) (see Fig. 5.6).

While the production rates appear small compared to the spring bloom, we should remember that they are maintained for many weeks. The 1D numerical model can be used to provide a rough answer to the primary production question, or we can make use of estimates of nutrient supplies to the spring bloom and to the thermocline during summer (see Problem 7.4). Observations of surface and subsurface primary production rates over a whole year are limited. Estimates based on observations of primary production rates within the thermocline near the Dogger Bank in the North Sea have suggested that subsurface carbon fixation contributes about 37% of the annual primary production of the stratified North Sea (Weston *et al.*, 2005), and could even exceed the production achieved during the spring bloom (Richardson *et al.*, 2000). Moreover, the SCM is likely to play a vital role in providing organic fuel for the pelagic food web during summer (Richardson *et al.*, 2000). Our own work in the Celtic Sea thermocline arose from estimates of spring primary production

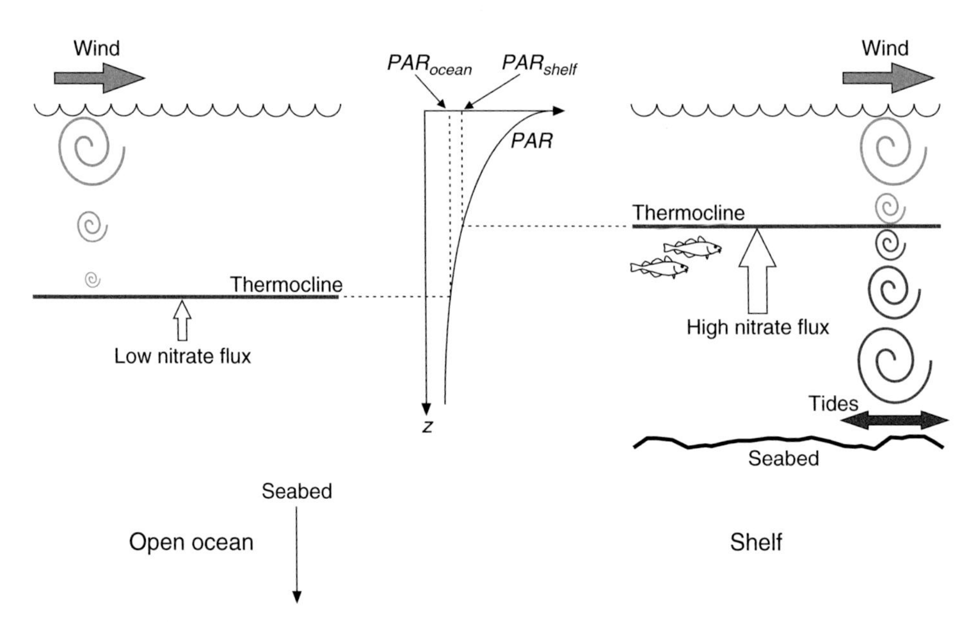

Figure 7.11 Schematic illustration of the role of tides on the shelf in altering the light and nutrient environment experienced at the seasonal thermocline. Tidal mixing increases both the light and the supply of nitrate in the shelf sea thermocline compared to the non-tidal environment of the open ocean.

suggesting that the bloom was insufficient to support the demands of the pelagic and benthic foods webs in the shelf sea. Also, remember from Chapter 6 that diatom-dominated blooms may rapidly sink out of the water column to the sediments, so the bloom is possibly only briefly accessible to pelagic organisms. The SCM provides pelagic organisms with an alternative strategy for acquiring food, with low phytoplankton concentrations available for several months in contrast to the concept of needing to match the timing of the spring bloom discussed in Section 6.3.3.

The role that the shelf sea SCM plays in augmenting annual primary production contrasts markedly with the SCM found in the open ocean. We noted earlier that the chlorophyll layer found in the thermocline of the major oceanic gyres is generally seen to be the result of changes in the cell chlorophyll content, not additional cell numbers. We can suggest two reasons for this arising from the different physics on the shelf, illustrated in Fig. 7.11. First, the depth of the seasonal thermocline in the open ocean is set by the amount of wind mixing acting to deepen the surface mixed layer. On the shelf, the thermocline depth is a balance between wind mixing trying to deepen it, and tidal mixing attempting to push it back upward. Thus the shelf thermocline tends to be shallower and so the SCM receives more light. Second, tidal mixing, or internal wave mixing originating from tidal flows over an uneven seabed, leads to relatively high fluxes of nitrate upward into the shelf sea SCM. Typically we see fluxes of 1–10 mmol m^{-2} d^{-1}. In the open ocean gyres the lack of such a strong source of turbulence at the thermocline leads to far lower nitrate fluxes of the order 0.1 mmol m^{-2} d^{-1} (Lewis *et al.*, 1986; Planas *et al.*, 1999).

7.3.3 Phytoplankton motility and phytoplankton thin layers

In the previous section we described phytoplankton growth within the SCM that involves the transfer of dissolved inorganic nutrients from the bottom layer into the thermocline, with the nutrients then assumed to be available for uptake by the phytoplankton. The phytoplankton were treated as non-motile organisms, with the thermocline providing a suitable niche for access to the required light and nutrient resources. Survival of phytoplankton within the SCM is then a balance between nutrient supply and growth on the one hand, and respiration, grazing and turbulent export on the other. What happens if the phytoplankton can swim in response to their resource requirements? For instance, dinoflagellates have been observed to swim downward when nutrient-deplete, gathering nutrients from deeper water before ascending back towards the light (Eppley *et al.*, 1968; Cullen and Horrigan, 1981).

Given the often strong tidal currents and associated turbulence in shelf seas, is the swimming capability of single-celled plants significant? For instance, typical swimming speeds of dinoflagellate cells range between $v_c = 0.1$ and 0.5 mm s^{-1} (Kamykowski and McCollum, 1986), which is two orders of magnitude lower than the typical turbulent velocities in the tidally energetic bottom layers of the NW European shelf. An assessment of the likely importance of motility in a turbulent environment can be made by calculating the cell Péclet number, as described in Chapter 5.

For tidal conditions similar to those of the NW European shelf, the strength of the turbulence in the bottom mixed layer overwhelms typical phytoplankton swimming capabilities (Ross and Sharples, 2008). However, swimming can have an impact close to and within the thermocline. Results from a random walk Lagrangian model of cell trajectories in a tidal, stratified water column, shown in Fig. 7.12, suggest that tidal turbulence periodically transports cells into the base of the thermocline, regardless of whether a cell can swim or not (Ross and Sharples, 2008). These cells will be nutrient-replete, having been able to take up bottom layer nutrients. Once in the base of the thermocline, swimming becomes advantageous as the weakly turbulent environment of the thermocline allows a motile cell to swim upward, into higher light and away from the risk of being entrained back down into the mixed layer by the next pulse of tidal turbulence. Notice in Fig. 7.12a how the motile particle is able to reach deeper into the thermocline, undertaking a number of migration cycles within the nitracline, compared to the non-motile particle (Fig. 7.12b).The results also show that the motile cells are able to migrate up and down within the base of the thermocline, swimming down to extract nutrients from the nitracline and then upward to photosynthesise. Experiments with this model suggested that a motile species can out-compete a non-motile species even if the motile cell has some relative growth rate disadvantage[2] (Ross and Sharples,

[2] It is interesting to note that a similar model experiment applied to the surface of the open ocean suggests that motility confers no advantage (Broekhuizen, 1999). This may reflect the role of the tidal turbulence, absent from the open ocean, in providing a rapid transport mechanism from deep water back up to the thermocline.

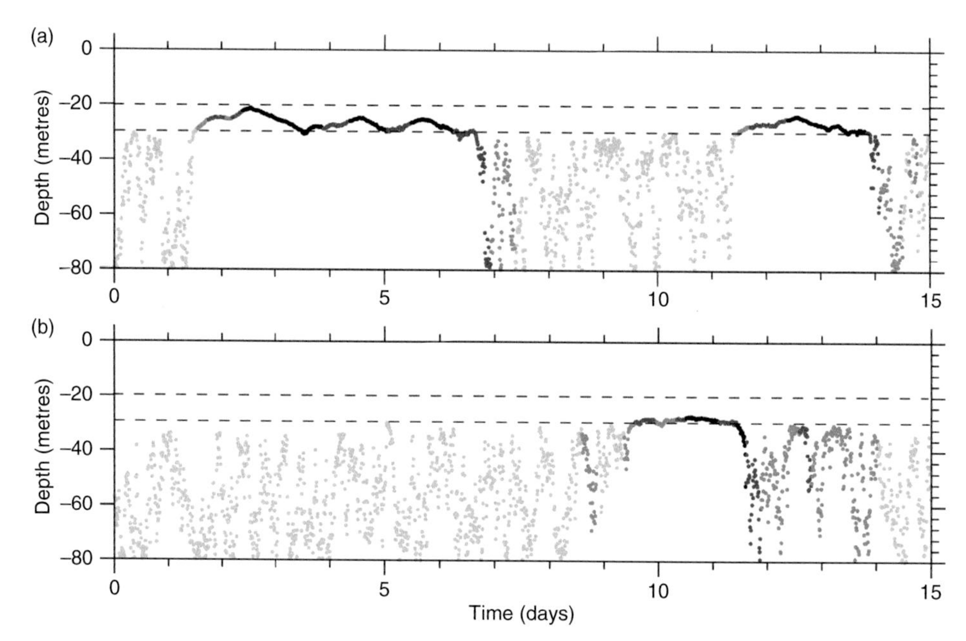

Figure 7.12 Time series of particle tracks in a stratified water column for (a) a motile particle that can swim at 0.1 mm s^{-1} in response to light and nutrient requirements, and (b) a neutrally-buoyant particle. The horizontal dashed lines mark the upper and lower bounds of the thermocline. The particles are shaded according to their nutrient status: lighter shading represents particles that are nutrient replete, darker shading indicates particles that need to take up nutrients. Based on the random walk turbulence model of (Ross and Sharples, 2008).

2007). So, while motility is not a requirement for the existence of the SCM of a tidally energetic region, it can play a role in the competition between phytoplankton species.

Motility can be more successful in regions with weak bottom-layer turbulence arising from small tides. Thin layers of phytoplankton, often dominated by one species of dinoflagellate, are observed in Monterey Bay off California (Ryan *et al.*, 2008). Formation of these layers is still not fully understood, but is thought to be controlled by one or more of: (i) velocity shear and straining across pycnoclines, (ii) cell motility and buoyancy, and (iii) internal waves (Cheriton *et al.*, 2009). Layers have also been seen to form, dissipate, and re-form within a diurnal cycle (Sullivan *et al.*, 2010). In this latter case (e.g. the layer of phytoplankton in Fig. 7.7d), and in contrast to the SCM in tidally energetic shelf seas, it was demonstrated that *in situ* growth cannot form the layer fast enough. Instead, the time scales of layer formation suggest directed cell swimming as the primary formation mechanism (Steinbuck *et al.*, 2009). These layers have attracted attention as possible sources of harmful blooms (McManus *et al.*, 2008).

7.3.4 Phytoplankton species in the SCM

While the spring surface bloom is dominated by diatoms, the phytoplankton population within the subsurface chlorophyll maximum is often found to be much more

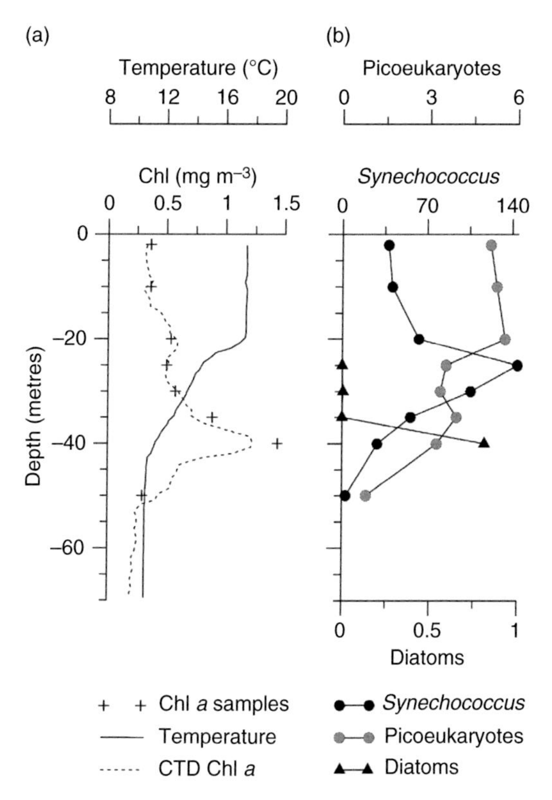

Figure 7.13 Observations from a site in the Celtic Sea show (a) a typical summer stratified water column, with a SCM towards the base of the thermocline. Analysis of water samples (b) illustrates how the SCM profile is made up of signals from different phytoplankton groups (expressed as cells $\times 10^3$ ml^{-1}). The distribution of the phytoplankton groups is consistent with their photosynthetic pigments and the changes in the light spectrum with depth. (Adapted from Hickman *et al.*, 2009, with permission from ASLO; data courtesy of Anna Hickman, University of Essex).

cosmopolitan in composition. The relatively quiescent, low nutrient regime tends to favour smaller cells (small dinoflagellates and flagellates, cyanobacteria). Diatoms are also sometimes seen, mostly at the base of the thermocline. We might conjecture that diatoms arise in response to short-term mixing events that temporarily alleviate the nutrient stress, though sufficiently long and well-resolved time series of the thermocline phytoplankton community that might allow testing of this hypothesis do not exist. The slight negative buoyancy of diatoms would lead to their being preferentially found in the lower thermocline. Also, Lagrangian modelling of diatom trajectories has shown that cells, replenished by nutrient uptake in the bottom layer, are transported by bottom layer tidal turbulence into the base of the thermocline, but with their lack of motility preventing them from reaching further upward they are at risk of being mixed back into the bottom layer (Ross and Sharples, 2008) (see Fig. 7.12b). Phytoplankton species have also been observed to be layered within the SCM in response to vertical changes in the spectral properties of light, as shown in the example from work in the Celtic Sea shown in Fig. 7.13 (Hickman *et al.*, 2009).

From an observational point of view we identify phytoplankton layers as 'chlorophyll' maxima primarily because of the ease of observing chlorophyll fluorescence. However, the ecology of these layers is far more complex than our simple observations of chlorophyll might at first lead us to believe. Turbulence, current shear, cell motility, cell size, and cell light requirements all play a role in the competition for resources, the balance between growth and loss rates, and the competition between different species in phytoplankton layers.

7.4 Zooplankton and larger animals at the SCM

We can re-visit the Péclet number and estimate the swimming speed needed to overcome the turbulence in a tidally mixed bottom layer, and see if we might expect larger motile organisms to be able to control their position vertically. Taking the bottom layer of the Celtic Sea to be 80 metres thick, and the turbulent diffusivity to be 10^{-1} m^2 s^{-1}, we can calculate a necessary sustained swimming speed to be >1 mm s^{-1}. According to a theoretical result of Okubo (Okubo, 1987), which relates swimming speed to the size of an organism, a speed of >1 mm s^{-1} sets a lower size limit for a migrating organism of about 0.2 mm. This is much larger than any motile phytoplankton, but it is about the lower limit of the size range associated with mesozooplankton. So, we can expect anything bigger than an ostracod (including all copepods) to be able to swim vertically through a tidally energetic bottom mixed layer.

The SCM is a site of relatively high concentrations of food for grazing zooplankton, particularly since it is where most of the new production occurs in summer and contains larger phytoplankton cells than the recycling microbial community in the surface layer. So the ability to swim vertically and find the SCM could be a useful trait. Do we see zooplankton migrating into the SCM? Certainly in our own work we have regularly observed signals in the backscatter from ADCPs showing the diurnal migration of scatterers, and often the apparent targeting of the SCM during the night. A detailed set of temperature, optical and acoustic observations from Monterey Bay, California, has identified night time layers of zooplankton associated with layers of phytoplankton (Holliday *et al.*, 2010). Fig. 7.14 shows an example, with an acoustic signal attributable to small copepods forming a night time layer and tracking the thermocline. Note in particular the scatterer layer rising from the seabed at dusk, and how it tracks the deepening thermocline just after midnight. The high vertical resolution (a few cm) of these observations showed that the zooplankton did not always simply track layers of high phytoplankton biomass; food quality, as well as quantity, played a role in determining where zooplankton layers formed.

The acoustic signals recorded in the time series of Fig. 7.14 were also analysed at very high spatial and time resolution, showing individual acoustic targets identified as small pelagic fish also following the thermocline. This supports an observation by the fisheries scientist Reuben Lasker, who identified the requirement of stable layers of dinoflagellates for successful first feeding of anchovy larvae off the California coast (Lasker, 1975). Correlation between high chlorophyll concentrations within the SCM and the distribution and foraging of larger predators, including seabirds and marine mammals, has also been seen, for instance in the North Sea (Scott *et al.*, 2010). It is clearly tempting to draw a simple connection between trophic levels, beginning with higher phytoplankton numbers and working through to the top

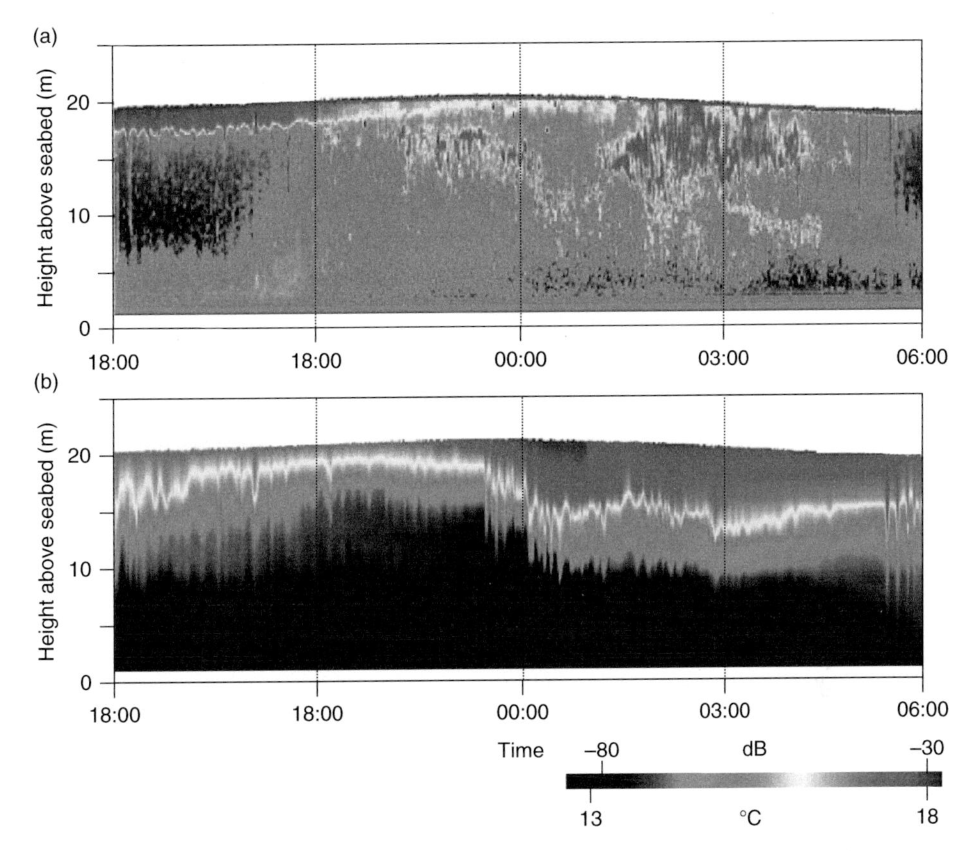

Figure 7.14 See colour plates version. Mooring observations of (a) the acoustic volume scattering strength (dB) at 265 kHz and (b) temperature (°C) in Monterey Bay. From (Holliday *et al.*, 2010), with permission from Elsevier.

predators. However, it is also possible that the phytoplankton and the top predators are responding in different ways to the same physical processes. For instance, foraging predators over a small seabed bank in the Gulf of Maine are not correlated with overall prey biomass, but instead are found at sites where internal waves act to concentrate patches of prey (Stevick *et al.*, 2008). Hence, predators are not necessarily responding to a sequence of trophic interactions that began with increased primary production, but are instead more immediately dependent on the mechanics of turbulence and prey availability. From an oceanographer's perspective, patchiness of the chlorophyll in the SCM is interesting because the phytoplankton are providing a relatively fast response indicator of interior mixing processes.

Summary

The seasonal cycle of thermal stratification in a shelf sea can be well reproduced by models based on the energetics arguments developed in Chapter 6. A more complete picture of the evolution of water column structure and flow is given by more advanced models incorporating a turbulence closure scheme. Such schemes provide a more explicit representation of turbulent processes, including the inhibiting effect of stratification on mixing. The models give a fuller account of the heating-stirring problem and furnish detailed information on the distribution of vertical mixing. In particular they provide profiles of the TKE dissipation rate and hence of the eddy diffusivity, a parameter which is needed for the estimation of vertical fluxes of nutrients and other properties. Turbulence closure schemes of varying degrees of complexity are now widely used in shelf-wide coupled physics-biogeochemistry-ecology models.

The period of 4–7 months during which many shelf sea regions are strongly stratified provides a stringent test of the capabilities of our models. Comparing the models with observations of, for instance, surface and bottom layer temperatures or profiles of currents generally shows good agreement. However, the more challenging test of comparison with observed turbulent dissipation rates indicates that the models fail to reproduce bursts of mixing activity in the interior of the stratified water column which is of fundamental importance to the biochemistry. The reason for the model-observation discrepancies is most probably that the models are failing to adequately represent one or more important sources of interior turbulence, such as internal waves, spring-neap changes in bottom layer turbulence and inertial oscillations.

Throughout the stratified period, subsurface layers of phytoplankton (subsurface chlorophyll maxima, or SCMs) associated with pycnoclines are a common feature of shelf seas. Turbulent mixing within and at the base of pycnoclines plays a fundamental role in the survival of the phytoplankton. Turbulence can supply bottom water nutrients to the pycnocline, where they support phytoplankton growth, or it can remove phytoplankton from the base of the pycnocline into the bottom water. Nitrate profiles frequently show a sharp nitracline within the phytoplankton layer, with negligible nitrate above the layer and in the surface waters, suggesting that the turbulent flux of nitrate into the base of a phytoplankton layer is a limiting factor to phytoplankton growth. SCMs are not visible from above the sea surface, i.e. they are not accessible to mapping by satellites. Estimates of carbon fixation rates within SCMs, either inferred from the supply of nitrate or from direct measurements using C^{14} incubations, suggest that they make an important contribution to annual primary production in shelf seas, potentially fixing the same amount of carbon as the spring phytoplankton bloom.

The low turbulent region within the pycnocline can be exploited by motile phytoplankton that are capable of using swimming to balance their resource requirements. In some weak tidal regions, turbulence is also low below the pycnocline, and swimming phytoplankton can aggregate to form concentrated layers far more rapidly than they could solely as a result of *in situ* growth. While observations of phytoplankton layers often concentrate on detecting chlorophyll, more detailed assessments of the community of phytoplankton show different phytoplankton groups exploiting different parts of a pycnocline.

Variations in turbulence, combined with the distribution of nutrients and modification of the light spectrum with depth, sets up an array of niches within which phytoplankton can achieve significant growth and where the community of phytoplankton species competes. Grazers and larger marine predators are often seen to be associated with patches of high phytoplankton biomass. Zooplankton and fish are able to position themselves within the water column to take advantage of phytoplankton layers. Larger predators may be correlated with elevated phytoplankton biomass, but rather than responding to a series of trophic links between their prey and the primary production they may instead be taking advantage of localized prey distributions driven by the interior turbulence.

FURTHER READING

Burchard, H. *Applied Turbulence Modelling in Marine Waters*, Springer-Verlag, Berlin-Heidelberg, 2002.

Cullen, J. J., The deep chlorophyll maximum – comparing vertical profiles of chlorophyll *a*. *Canadian Journal of Fisheries and Aquatic Sciences*, **39**(5), 791–803, 1982.

Sharples J., et al., Phytoplankton distribution and survival in the thermocline, *Limnology and Oceanography*, **46**(3), 486–496, 2001.

Problems

7.1. Estimate the total rate of kinetic energy dissipation per square kilometre of water in a layer 20 metres thick for (i) the stable pycnocline region where the mean dissipation $\varepsilon = 10^{-5}$ W m^{-3}, and (ii) in the bottom boundary layer where $\varepsilon = 3 \times 10^{-3}$ W m^{-3}. Express your results in terms of the number of hand-held 100W kitchen mixers required to produce an equivalent stirring effect.

7.2. Using the TML model of heating vs. stirring, investigate the dependence of the amplitude and phase of the annual cycle of heat storage for the following cases:

(i) Strong tidal stirring ($\log_{10} SH = 1.5$) in a range of water depths e.g. 10, 20, 40, 80, 160 metres.

(ii) Weak tidal stirring ($\log_{10} SH = 3.5$) in the same range of depths.

Plot amplitude and phase versus depth and explain the form of the curves and the marked differences between mixed and stratified conditions.

7.3. The Ozmidov length is a measure of the scale of the largest over-turning eddies in stratified turbulence. It is defined as:

$$L_O = \sqrt{\frac{\varepsilon}{N^3}}.$$

Using the data shown in Fig. 7.6, estimate L_O (i) in the pycnocline at 75 mab, and (ii) in the bottom boundary layer at 10 mab. Explain the significance of your result for plankton in the two situations.

7.4. A vertically mixed shelf seawater column has an initial spring concentration of 8 mmol N m^{-3}. In late spring it stratifies with a surface layer of about 30 metres thickness. When the water is stratified (May–September) there is a mean nitrate supply rate of 2 mmol N m^{-2} d^{-1} into the SCM.

Estimate the contributions that the spring bloom and the SCM make to annual carbon fixation. (Hint: remember the Redfield ratio.)

7.5. Consider the possible biological response to the mixing event described in Problem 6.4.

(i) Prior to the mixing event the water column has negligible surface layer nitrate and a uniform bottom layer nitrate concentration of 7 mmol m^{-3}. Following a wind event that vertically mixes the whole water column, estimate the potential total carbon that can be fixed by phytoplankton (expressed as g C m^{-2}).

(ii) Compare the additional carbon fixed as a result of the wind event to phytoplankton growth in the SCM through 5 months of seasonal stratification. [Assume a mean nitrate supply rate to the thermocline of 2 mmol m^{-2} d^{-1}.]

[Hint: remember the Redfield ratio.]

7.6. Taking a cell swimming speed of $v_c = 1 \times 10^{-4}$ m s^{-1}, a tidally averaged bottom layer diffusivity for the Celtic Sea of $K_z = 2 \times 10^{-2}$ m^2 s^{-1}, and L = distance travelled in 16 hours of daylight, calculate the Péclet number.

Do the same for the bottom water of Monterey Bay where the lower layer diffusivity is 10^{-5} m^2 s^{-1}.

What do the answers tell you about the likely role of cell swimming in the two environments?

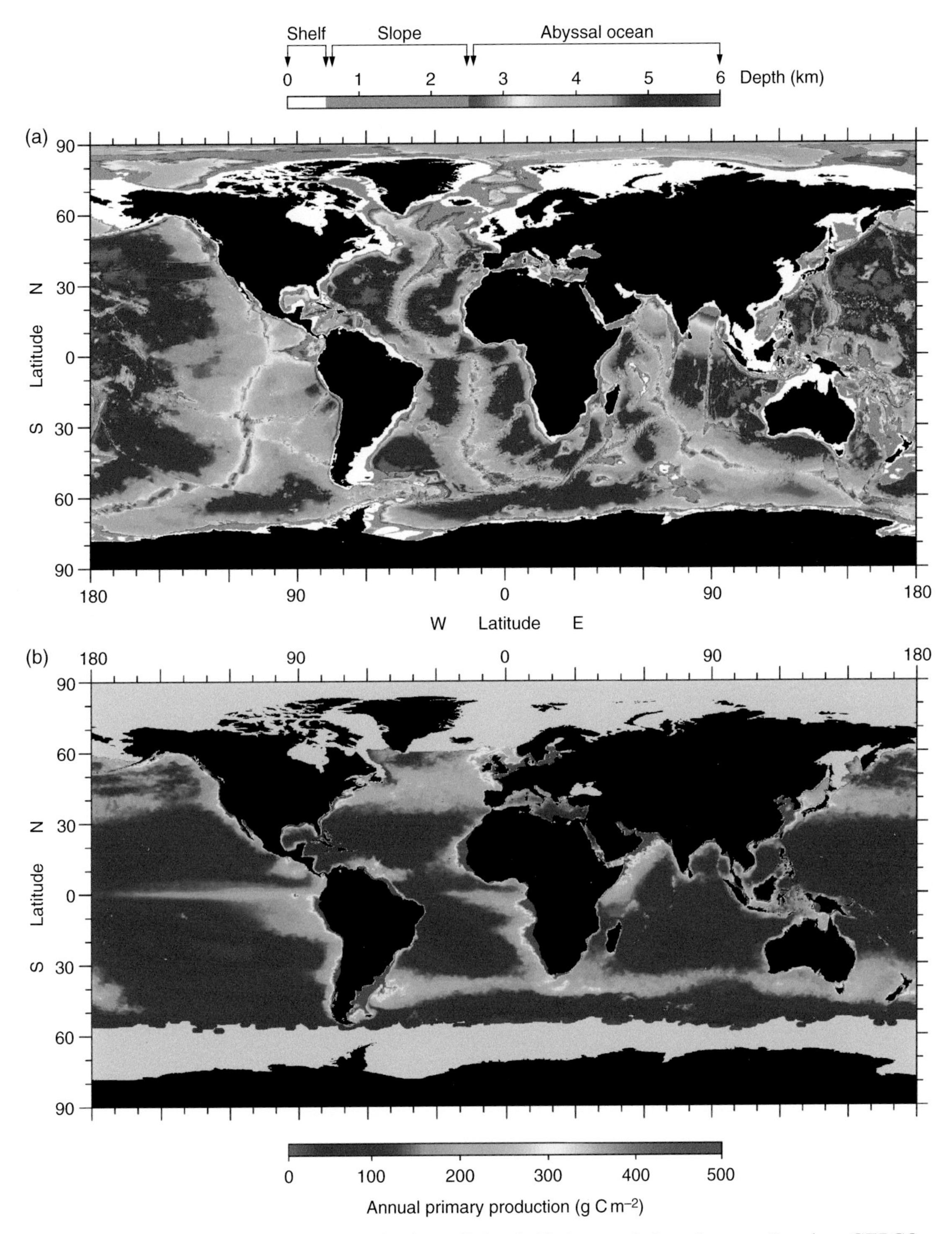

Figure 1.1 (a) Global ocean depths, split by shelf, slope and abyssal ocean. Based on GEBCO bathymetry; (b) global annual primary production, using imagery from the MODIS satellite (Behrenfeld and Falkowski, 1997); data courtesy of Mike Behrenfeld, Oregon State University, USA: http://www.science.oregonstate.edu/ocean.productivity/index.php.

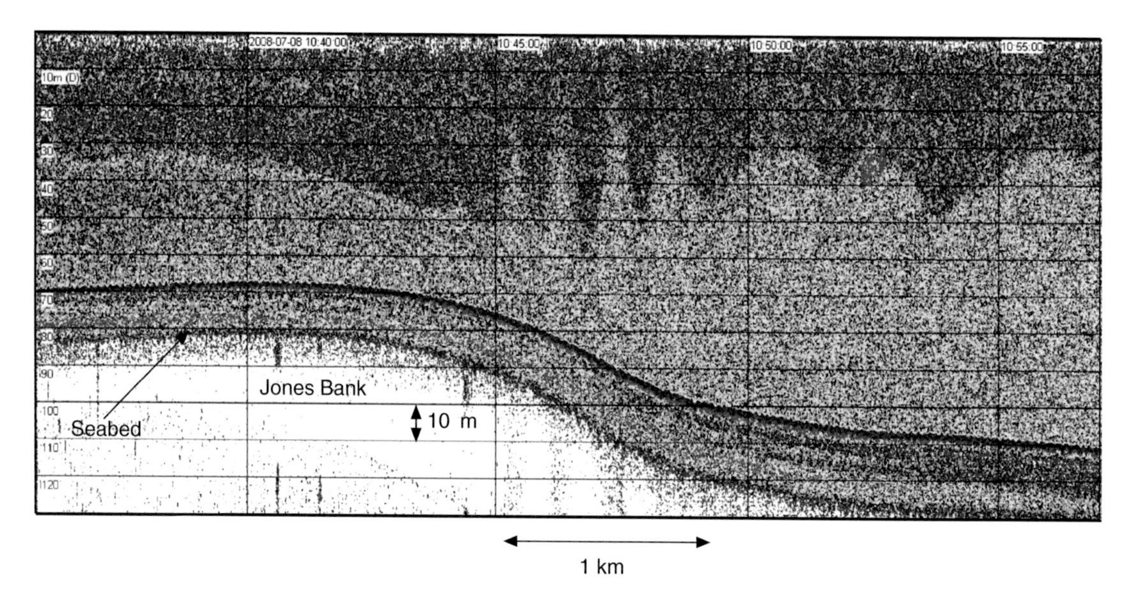

Figure 1.6 A series of short internal waves on the edge of Jones Bank, Celtic Sea, as seen in the data from the 200 kHz transducer of an EK60 echosounder. Image courtesy of Clare Embling, University of Aberdeen, UK.

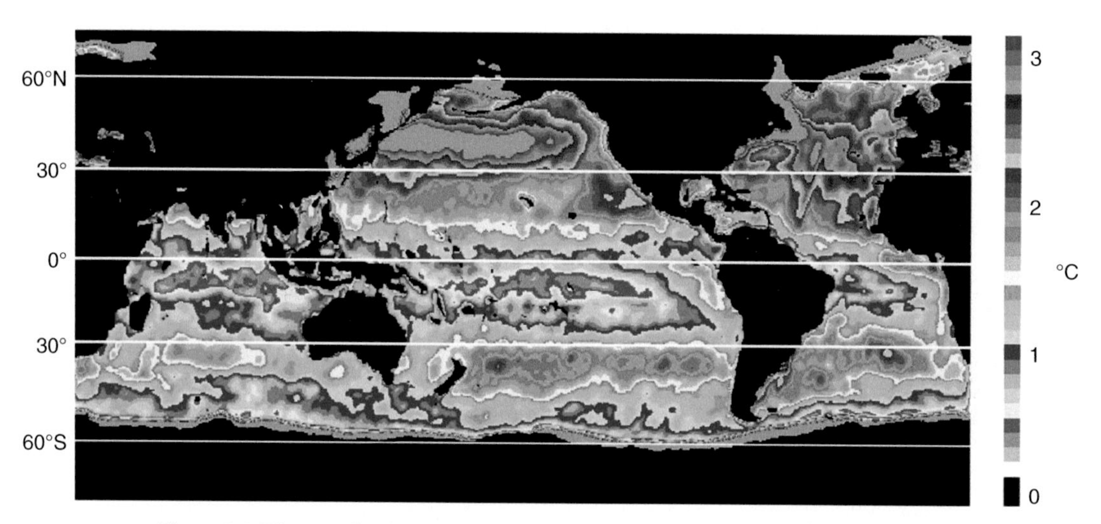

Figure 2.8 The amplitude of SST (°C) over the seasonal cycle derived from the scanning multichannel microwave radiometer (SMMR) on board the NASA Nimbus 7 satellite from 1978 to 1987. From (Yu and P. Gloersen, 2005), courtesy of Taylor and Francis.

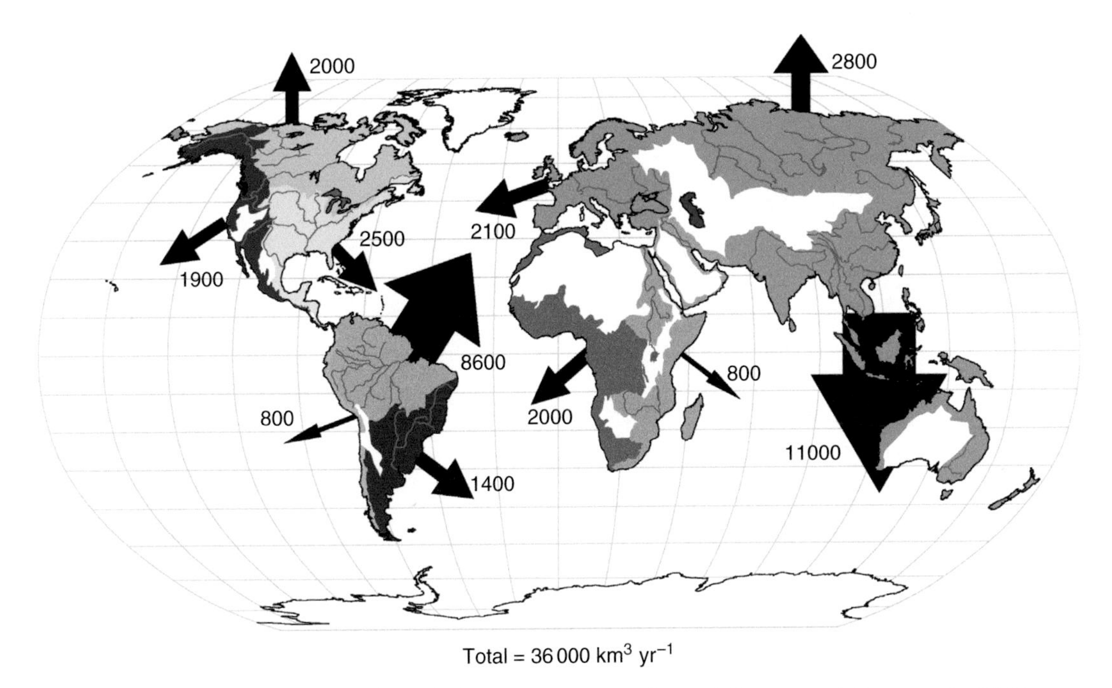

Figure 2.10 Annual river discharge to the ocean from (Milliman and Farnsworth, 2011), courtesy Cambridge University Press. Figures represent freshwater runoff from coloured land areas in cubic kilometres per year (1 $km^3\ yr^{-1} = 31.7\ m^3\ s^{-1}$).

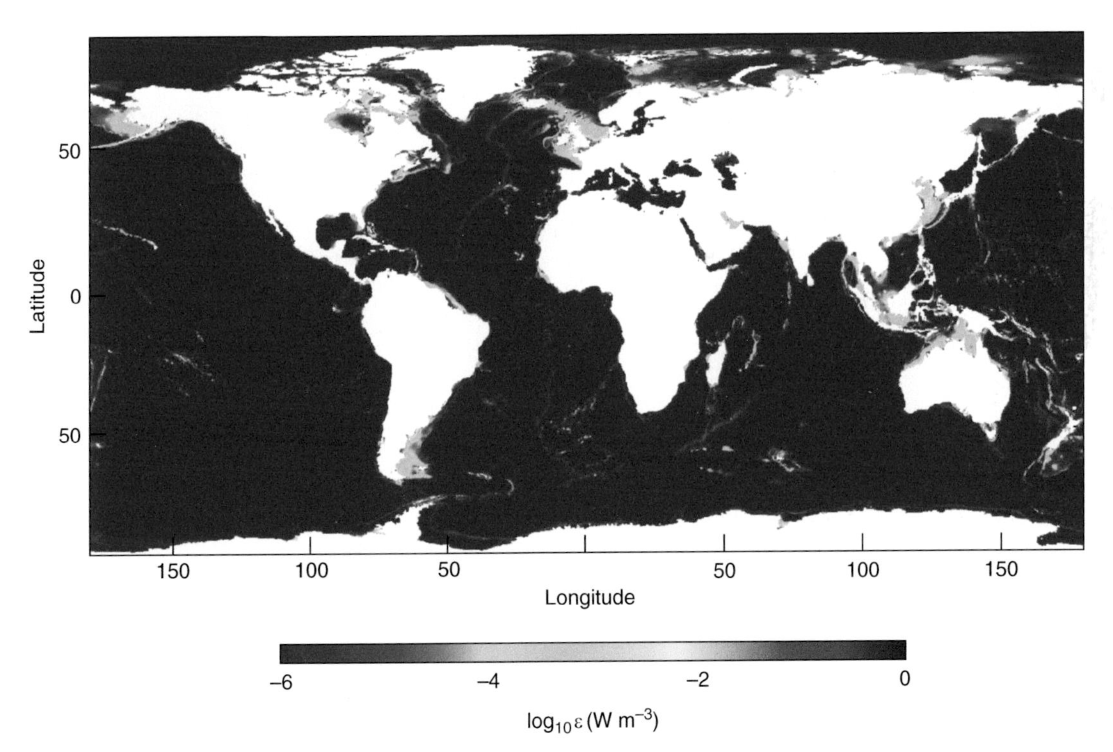

Figure 2.13 Distribution of tidal dissipation in Wm^{-3} derived from a numerical model of the global ocean. Note the large contrast (~6 decades) in the magnitude of dissipation between low dissipation in much of the deep ocean and high values exceeding ~1 Wm^{-3} occurring mainly in the shelf seas. Figure from (Green, 2010), with permission from Springer.

Figure 5.11 Some examples of autotrophic phytoplankton. (a) The coccolithophorids *Emiliania huxleyi* and *Gephyrocapsa mullerae*; (b) An armoured dinoflagellate *Gonyaulax*; (c) A naked dinoflagellate *Gymnodinium*; (d) An armoured dinoflagellate *Protoperidinium*; (e) The warm water diatom *Planktonella sol*; (f) A diatom *Thalassiosira*. Images supplied by Alex Poulton (National Oceanography Centre, UK), Jeremy Young (University College London, UK) and Daria Hinz (University of Southampton, UK).

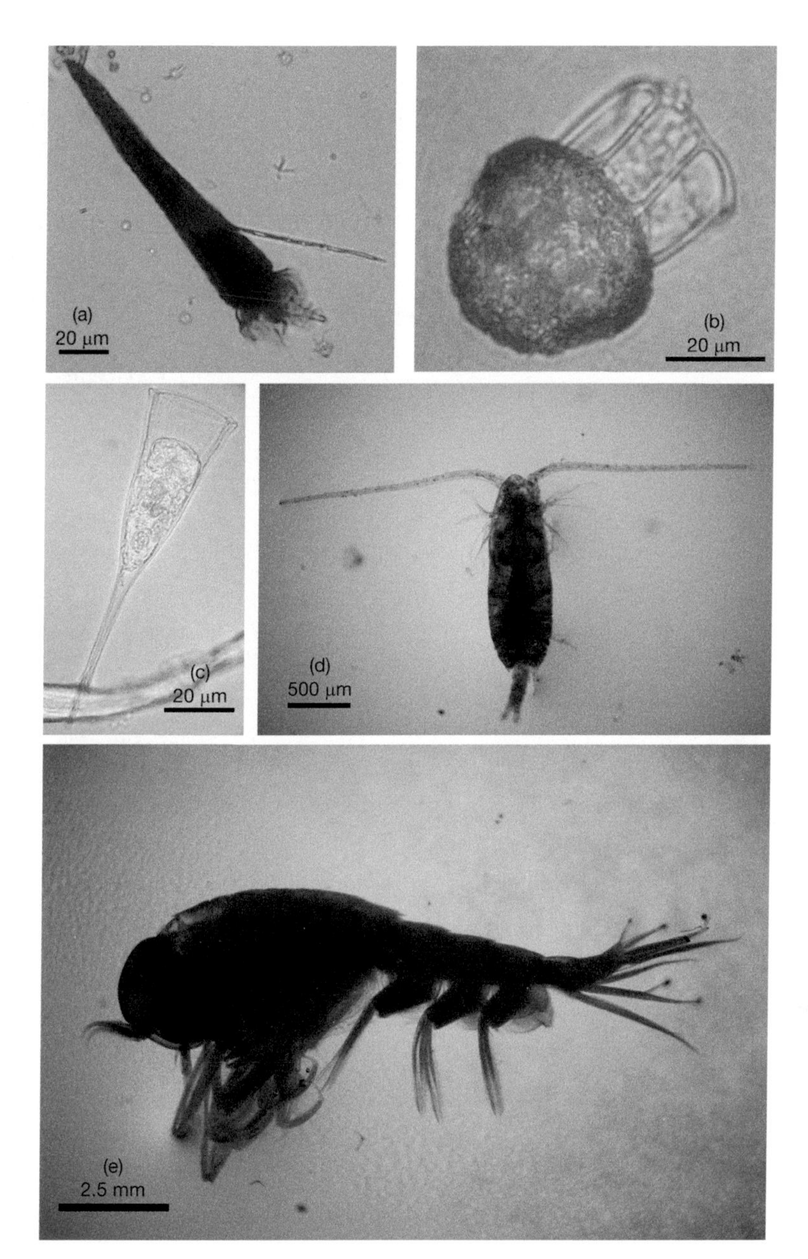

Figure 5.13 Some examples of micro- and meso-zooplankton. (a) naked planktonic ciliate *Strombidium*; (b) planktonic tintinnid *Dictyocysta*; (c) planktonic tintinnid *Rhabdonellopsis*; (d) a *Calanus* copepod; (e) planktonic *Amphipod*. Images supplied by Alex Poulton (National Oceanography Centre, UK) and Sari Giering (University of Southampton, UK).

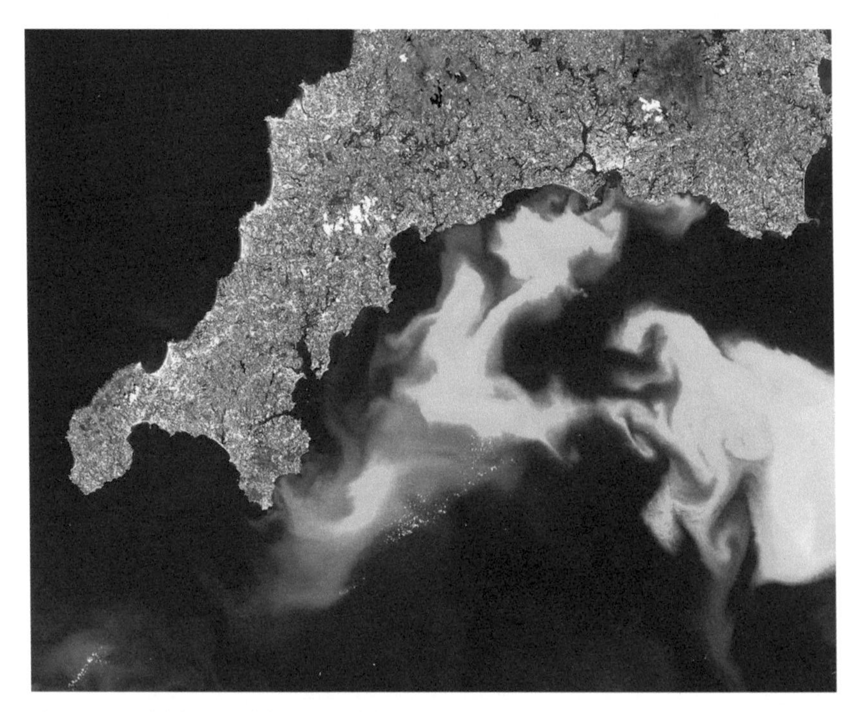

Figure 5.12 A bloom of the coccolithophores *Emiliania huxleyi* in the western English Channel, June 1999. Image courtesy NASA/JPL-Caltech.

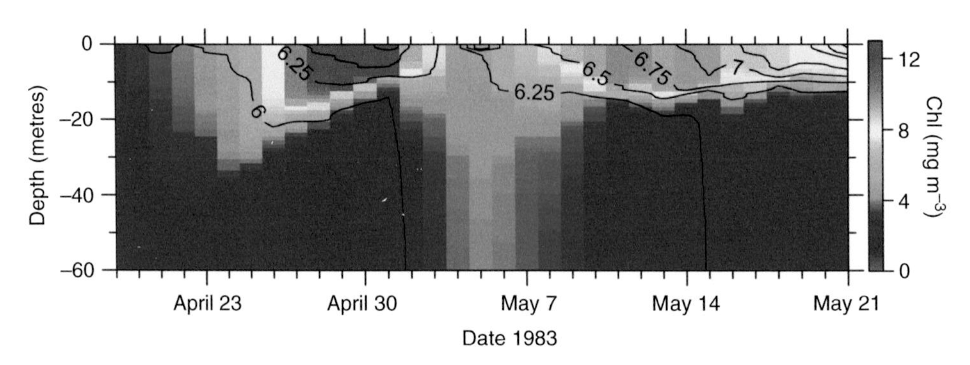

Figure 6.12 Depth-time series of temperature (line contours) and chlorophyll (colours). Adapted from (Sharples *et al.*, 2006) with permission from Elsevier.

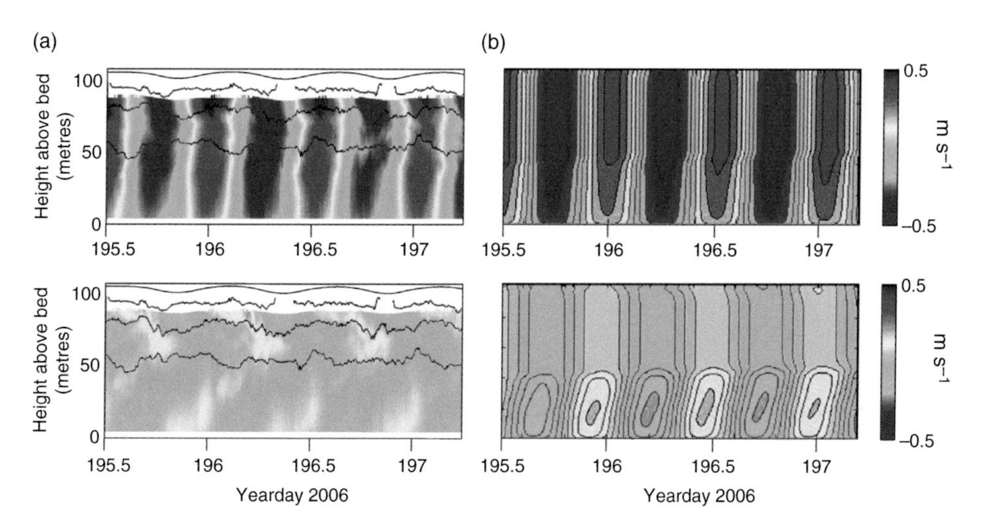

Figure 7.3 Velocities along major (top) and minor (bottom) axes of the tidal current ellipse at station SWIS (53°52.00′ N, 05°27.00′ W) in the stratified western Irish Sea during July 2006: (a) measured using ADCP (from (Simpson *et al.*, 2009) with permission of Elsevier), and (b) derived from the turbulence closure model driven by tidal and wind stress forcing. Colour scales are the same for all panels. Black lines in (a) are isotherms at 10, 12 and 14 °C, indicating the stratification of the water column.

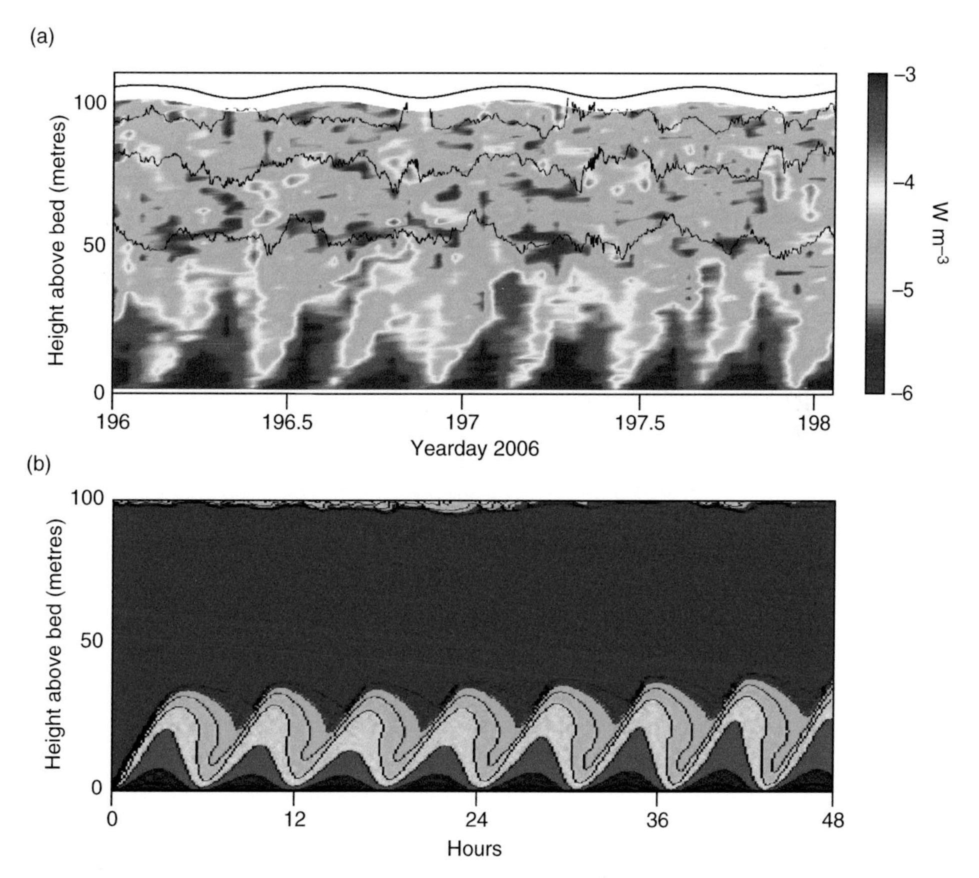

Figure 7.5 48-hour time series of dissipation as $\log_{10} \varepsilon$ (W m^{-3}) at station SWIS (53°52.00′ N, 05°27.00′ W) in a seasonally stratified region of the western Irish Sea. (a) Contour plot based on an average of 6 profiles per hour with the dissipation profiler, with isotherms (black lines) at 10, 12 and 14 °C (from Simpson *et al.*, 2009, with permission of Elsevier); (b) Dissipation simulated by a run of the TC model forced by the tide and local winds. Shade scales are the same for the two panels.

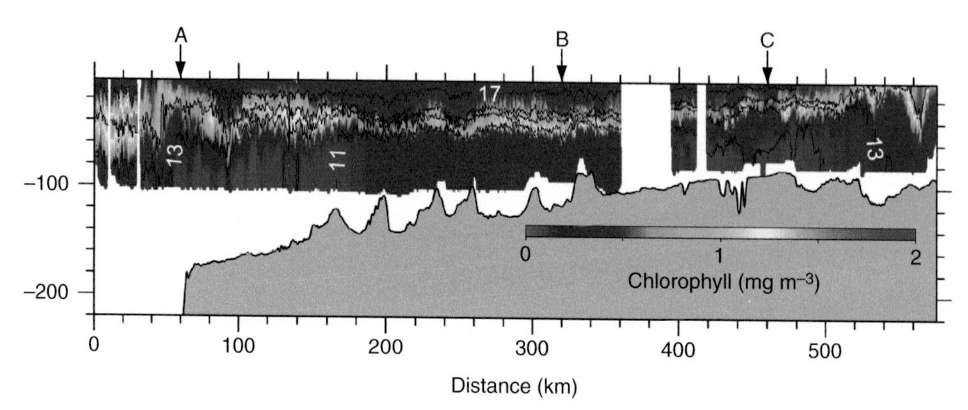

Figure 7.8 The map shows the course followed by the research vessel RRS *James Clark Ross* in August 2003, towing the Seasoar undulating vehicle (see Fig. 1.5). The contoured image shows temperature (lines, every 2 °C) and chlorophyll concentration (colours) along the transect. The grey region is the seabed, showing the dropoff at the shelf edge (60 km) and gradually shallowing towards the entrance to the Irish Sea (550 km). Data were collected between the surface and a depth of 100 metres. Vertical white bands indicate regions where there was a communication failure with the instrument. The arrows labeled along the top axis show the positions of the course changes marked on the map.

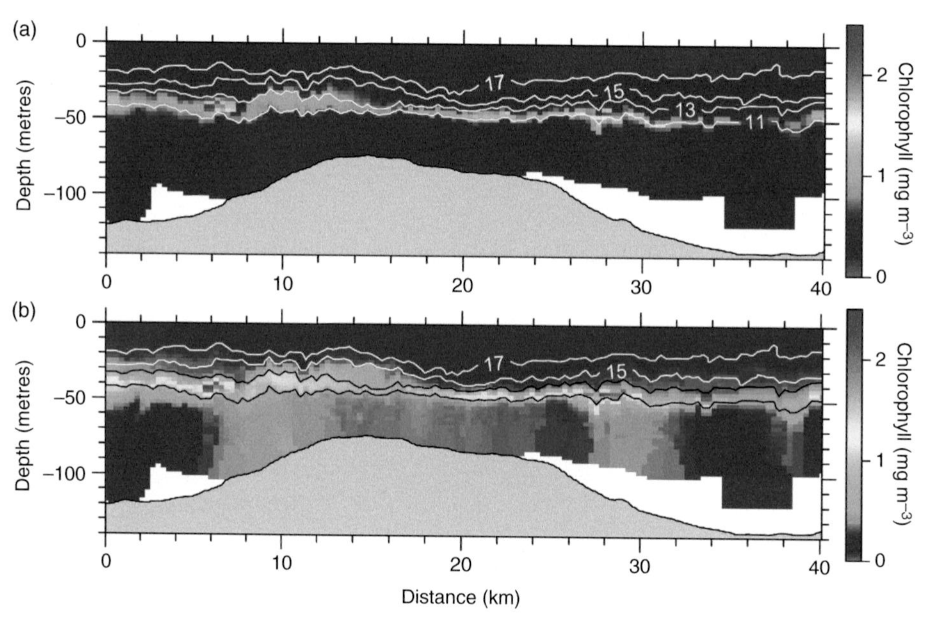

Figure 7.9 A transect of chlorophyll (colours) and temperature (lines) over Jones Bank in the Celtic Sea in July 2005, carried out with the Seasoar vehicle (see Fig. 1.5). The data in (a) and (b) are the same, but in (b) the chlorophyll colour scale has been skewed to highlight the lower chlorophyll concentrations.

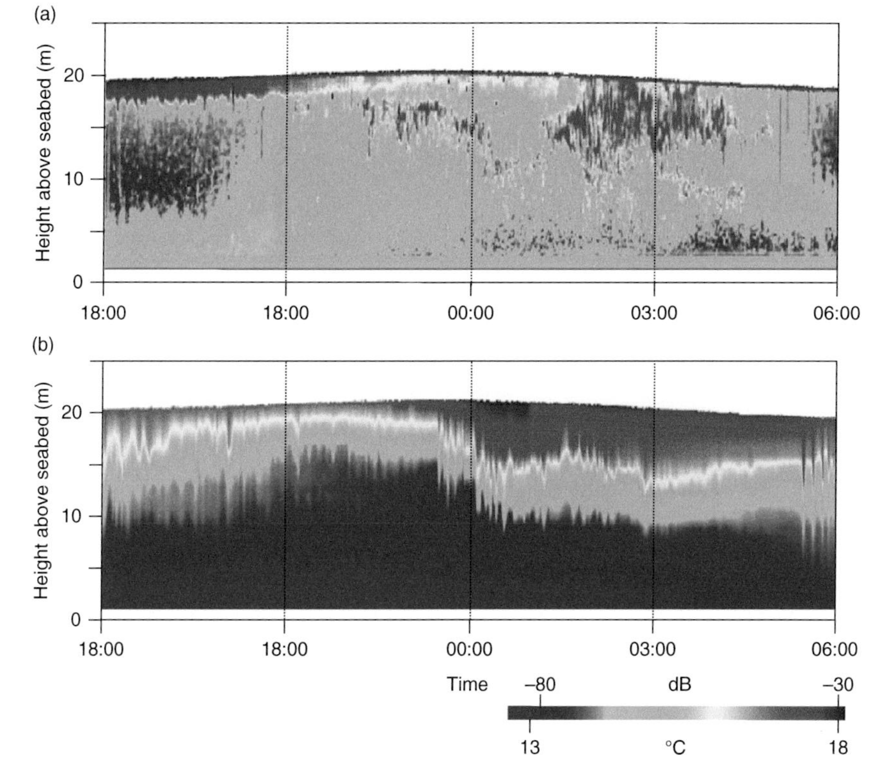

Figure 7.14 Mooring observations of (a) the acoustic volume scattering strength (dB) at 265 kHz and (b) temperature (°C) in Monterey Bay. From (Holliday *et al.*, 2010), with permission from Elsevier.

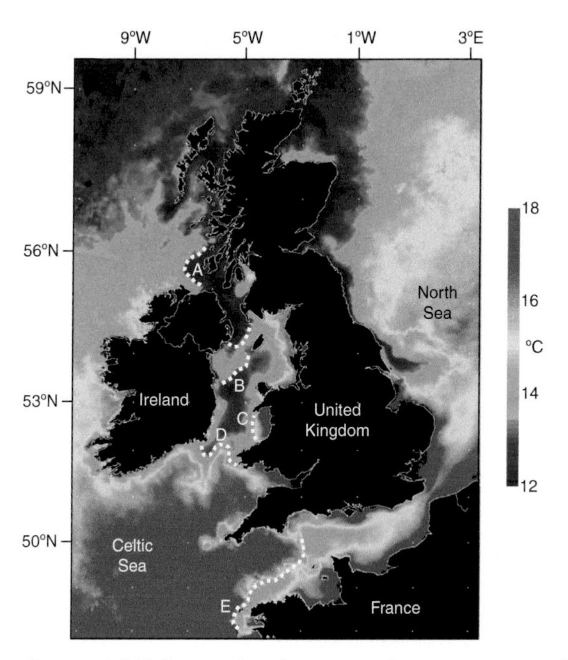

Figure 8.1 I-R image showing sea surface temperature. The image is a composite between July 9 and 15, 2006. The main fronts to the west of the United Kingdom are marked with white dashed lines. A: the Islay front, B: Western Irish Sea front, C: Cardigan Bay front, D: St George's Channel front, E: Ushant and Western English Channel front. Image courtesy of NEODAAS, Plymouth Marine Laboratory, UK.

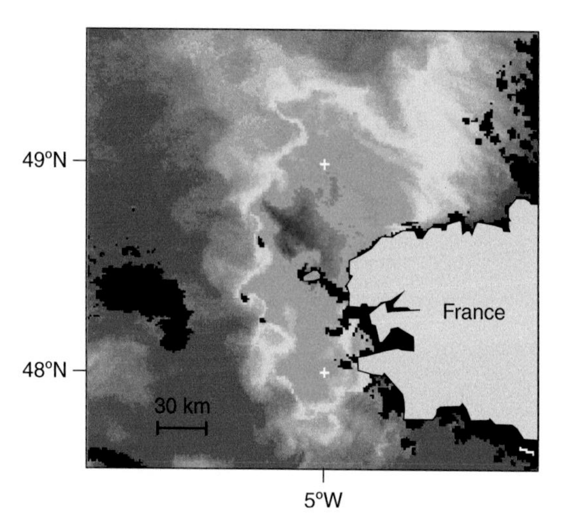

Figure 8.8 Infra-red image of the TM front to the west of the Brittany peninsula, NW France. Note the complex structure of the boundary between warm stratified (orange, red) and colder mixed (green, blue) waters. Convolutions of the front have length scales in the range 10–40 km, indicating the instability of geostrophic flow with the development of eddies. Image courtesy of NEODAAS, Plymouth Marine Laboratory, UK.

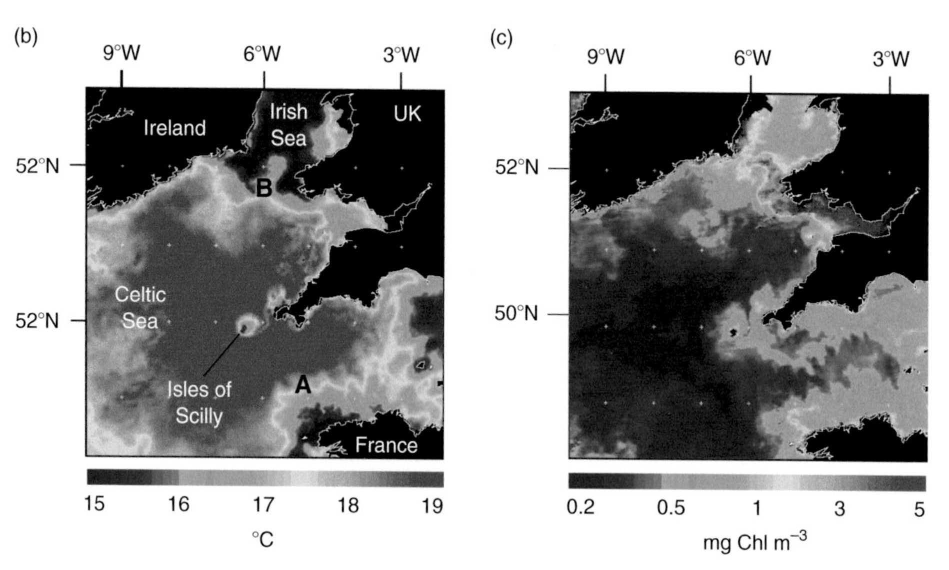

Figure 8.12 Satellite images of sea surface temperature and chlorophyll for the Celtic Sea. Images are composites over July 12–18, 2005, provided by NEODAAS Plymouth Marine Laboratory. Tidal mixing front A is the Ushant front across which the section in (a) was collected. Front B is the Celtic Sea front. Note also the 'island mixing front' and patch of mixed water surrounding the Isles of Scilly, along with an associated increase in sea surface chlorophyll.

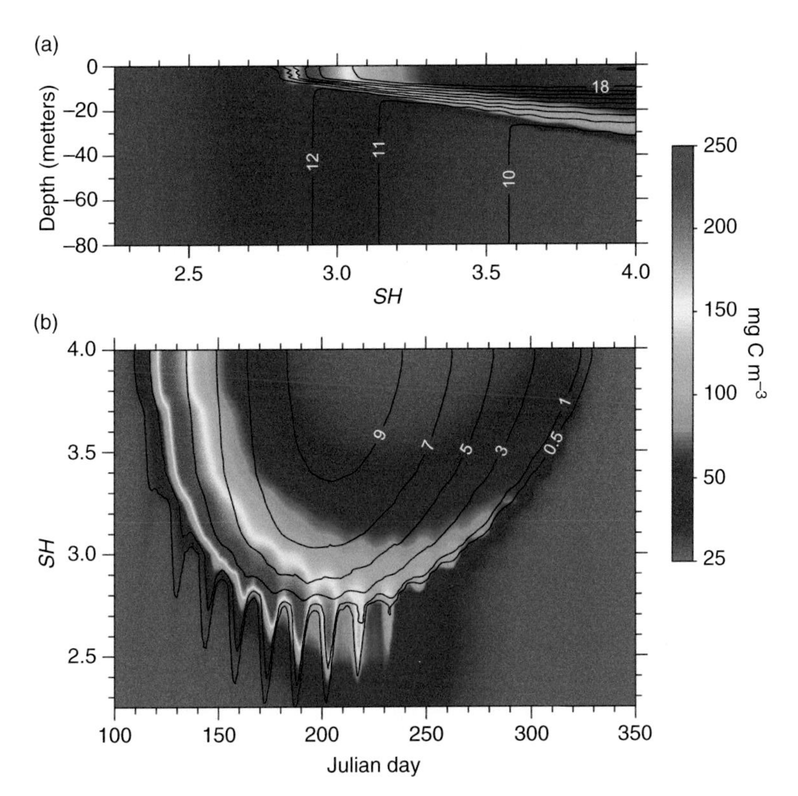

Figure 8.16 Model predictions of (a) cross-frontal temperature (°C, line contours) and phytoplankton carbon (shaded), and (b) variations in stratification (surface-bottom temperature difference, °C, line contours) and surface phytoplankton carbon (shaded) between mid April and late November for different *SH*. In (a) only, the M_2 tide was operating, so that phytoplankton within the front could only grow in response to changes in the vertical flux of nitrate at the front. In (b) both M_2 and S_2 tides operated to drive a spring-neap cycle in mixing. Adapted from Sharples, 2008, with permission from Oxford University Press.

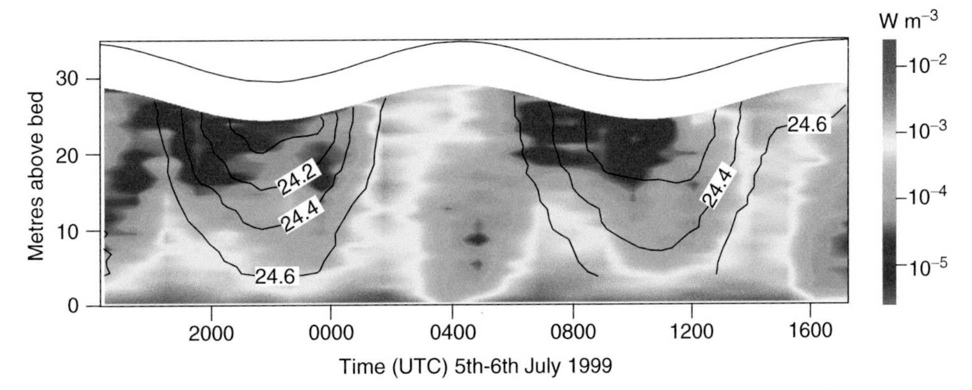

Figure 9.9 Two tidal cycles of σ_t (kg m^{-3}, lines) and turbulent dissipation (W m^{-3}, colours) measured from a vessel anchored in Liverpool Bay. The vessel position was the same as that of the mooring data in Fig. 9.7. Adapted from Rippeth *et al.*, 2001, with permission from the American Meteorological Society.

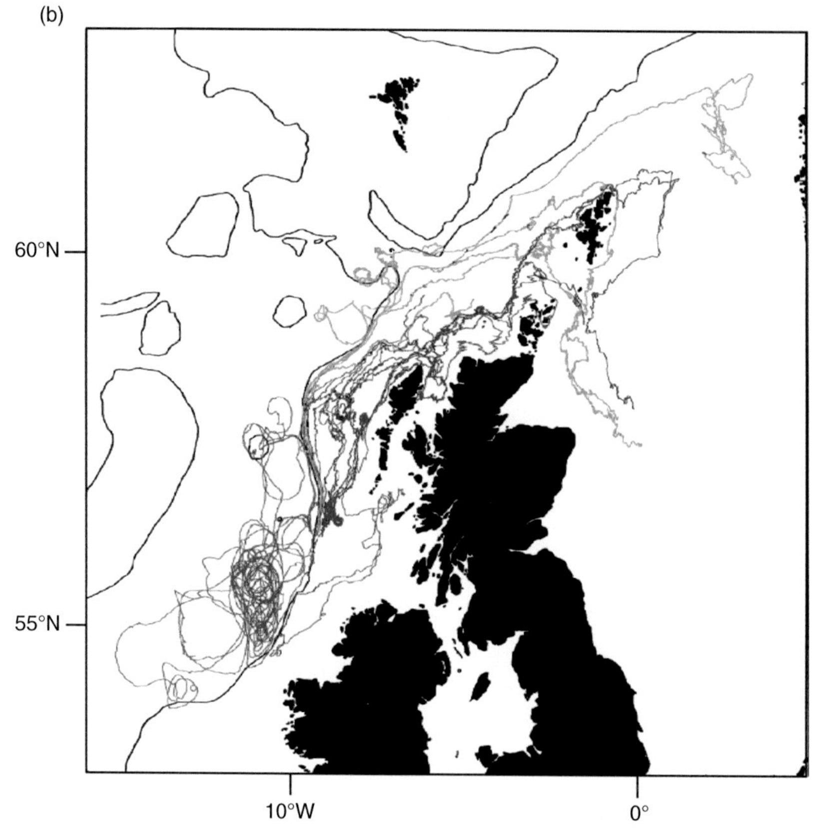

Figure 10.4 Tracks of satellite-tracked drifting buoys released during the SES programme. Surface floats were attached to high drag drogues to follow the currents at 50 m depth (Burrows *et al.*, 1999). Clusters of drogues were released at three locations: one on the shelf (blue), one over the slope (green) and a third in deep water at the bottom of the slope (red). Figures courtesy of Elsevier.

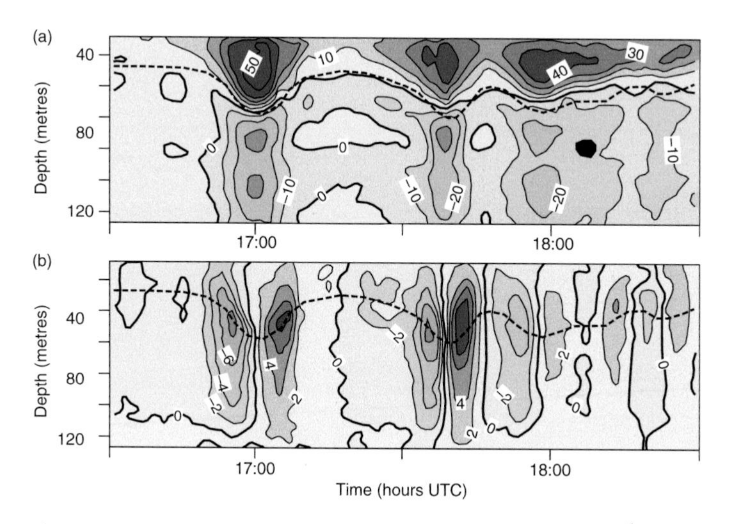

Figure 10.15 (a) Horizontal and (b) vertical velocities (cm s^{-1}) observed by a moored ADCP for a packet of waves travelling on to the Hebridean shelf. The dashed grey line indicates the vertical displacement of a particle initially at a depth of 25 m. Note the opposing horizontal flows in the top and bottom layers and the large up and down vertical velocities as each trough passes the mooring. Image courtesy of Mark Inall, Scottish Association for Marine Science.

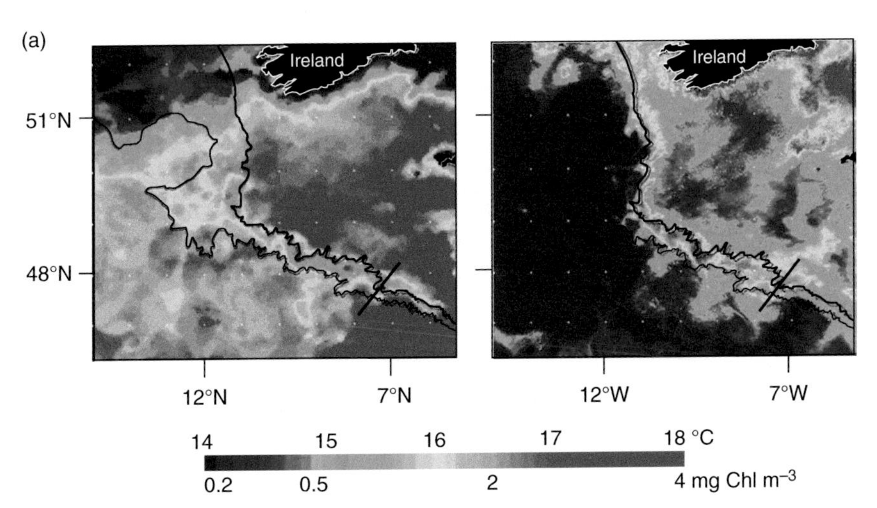

Figure 10.16 (a) Composite images of the Celtic Sea, June 13–19, 2004 courtesy of NEODAAS, Plymouth Marine Laboratory, UK. The left panel is sea surface temperature collected by AVHRR, the right panel is sea surface chlorophyll taken by MODIS. The bold contour line marks the 200 metre isobath at the shelf edge, the thin contour is the 2000 metre isobath.

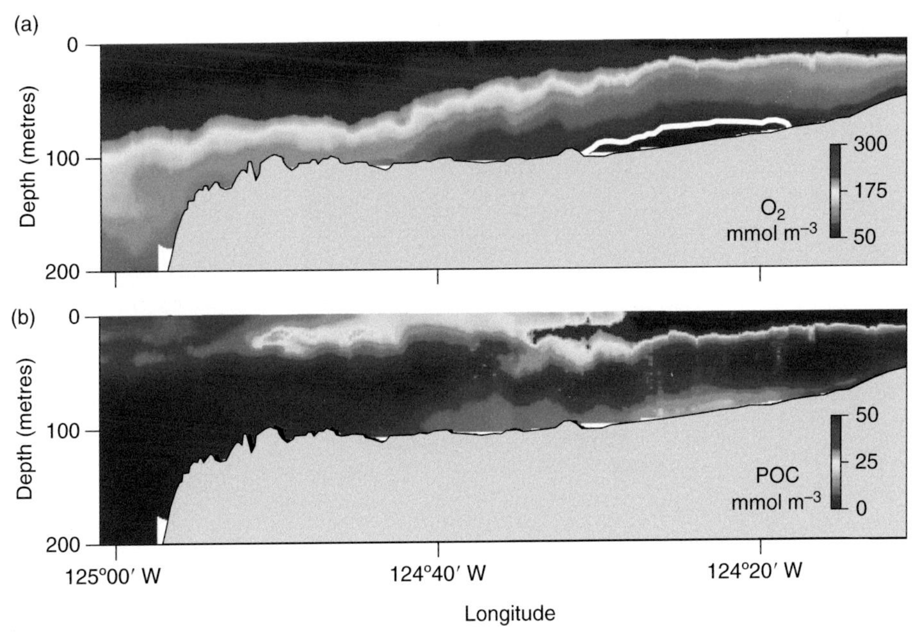

Figure 10.21 (a) Transect of dissolved oxygen concentration across the shelf and shelf edge off Oregon. The bold white contour line marks the extent of hypoxic bottom water; (b) Transect of particulate organic carbon (POC) across the Oregon shelf and shelf edge; (c) Annual cycle of the dissolved oxygen concentration (left) and % saturation (right) on the Washington shelf. (a) and (b) are high resolution observations collected in May 2001 during a period of upwelling-favourable winds. After Hales *et al.*, 2006, courtesy of the American Geophysical Union. (c) is adapted from Landry *et al.*, (1989), with permission from Elsevier.

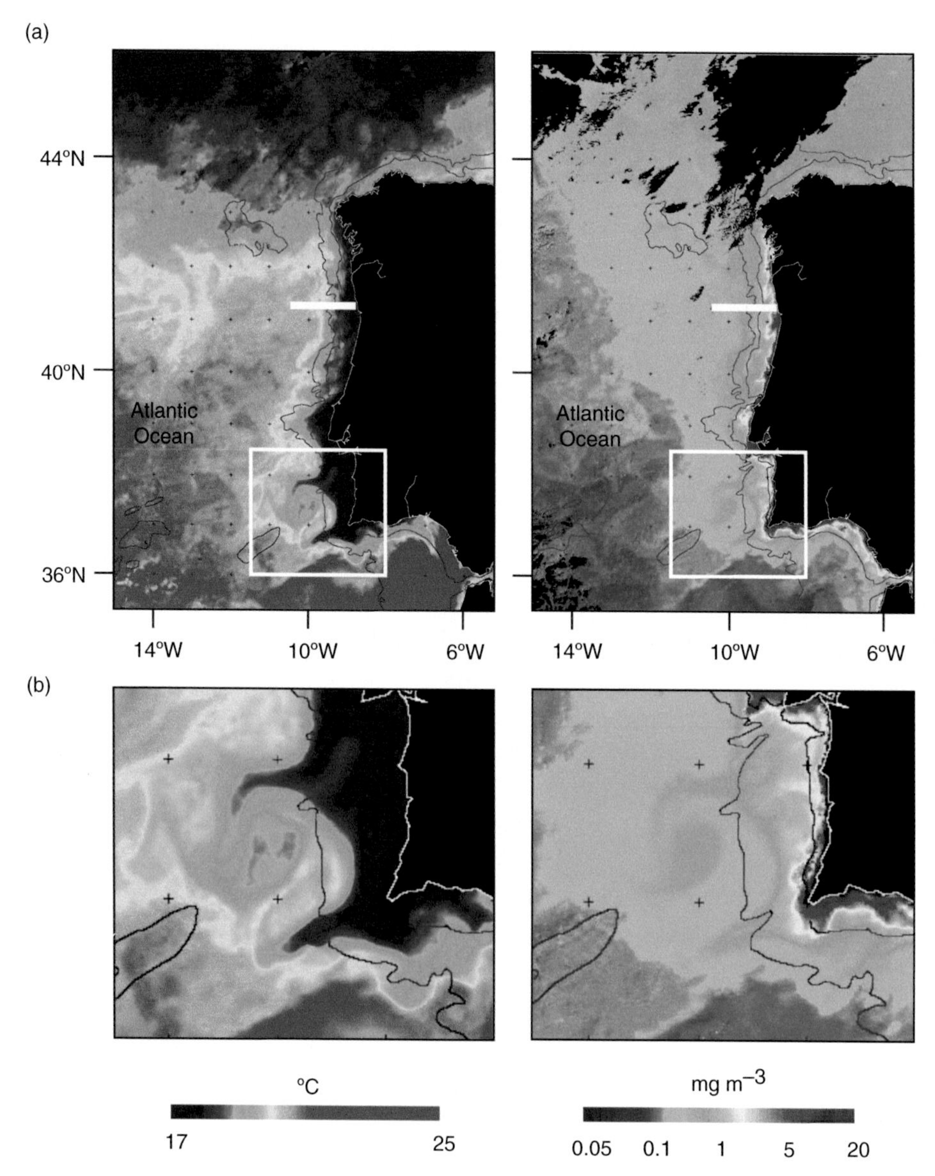

Figure 10.17 (a) Satellite images of sea surface temperature (left) and sea surface chlorophyll (right) off the Iberian Peninsula. Both images are composites taken over July 25–27, 2007; (b) Detail of the boxes marked in (a) showing two large filaments of upwelled water containing chlorophyll extending off-shelf. Images from NEODAAS, Plymouth Marine Laboratory, UK.

Figure 10.22 Photograph of a coccolithophore bloom in Jervis Bay, east Australia, following the upwelling of nutrients driven by a filament of the East Australia Current impinging against the shelf edge (Blackburn and Cresswell, 1993). Image provided by Susan Blackburn, CSIRO, photographer Ford Kristo, Australian National Parks. Reproduced with permission from CSIRO Publishing.

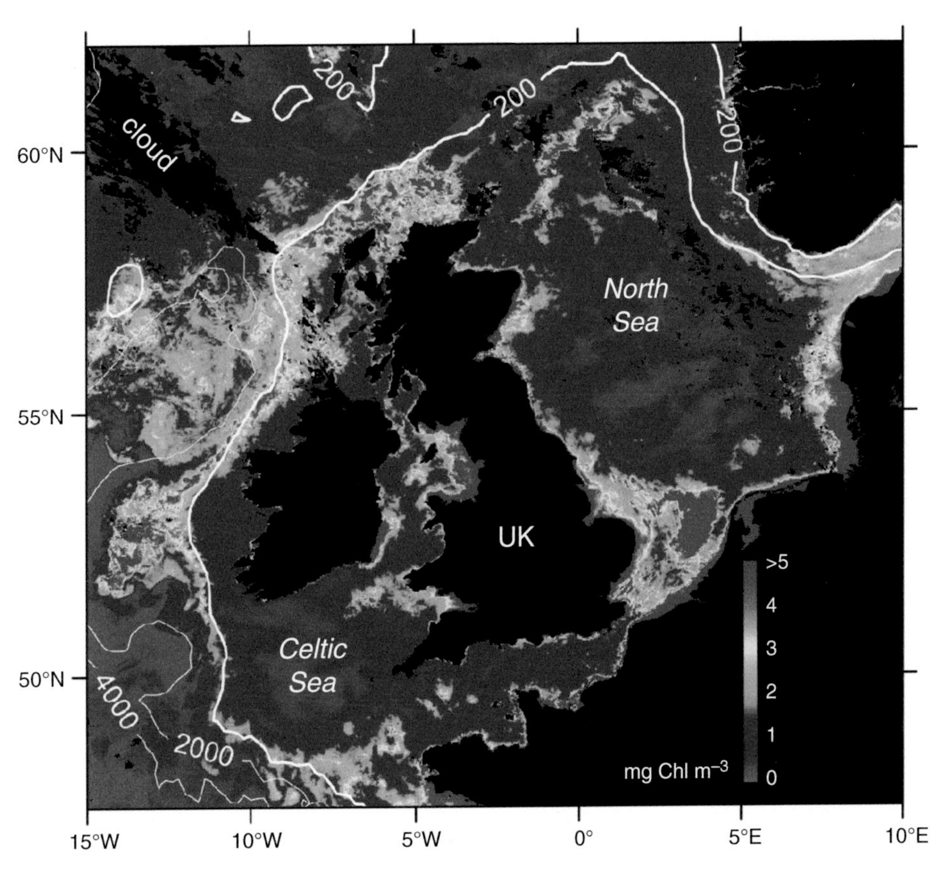

Figure 10.24 Satellite image of sea surface chlorophyll over the NW European shelf seas, taken May 30, 2004. Isobaths are contoured in white, with the 200-metre isobath marking the edge of the continental shelf. Image data courtesy NEODAAS, Plymouth Marine Laboratory UK. Bathymetry data from GEBCO.

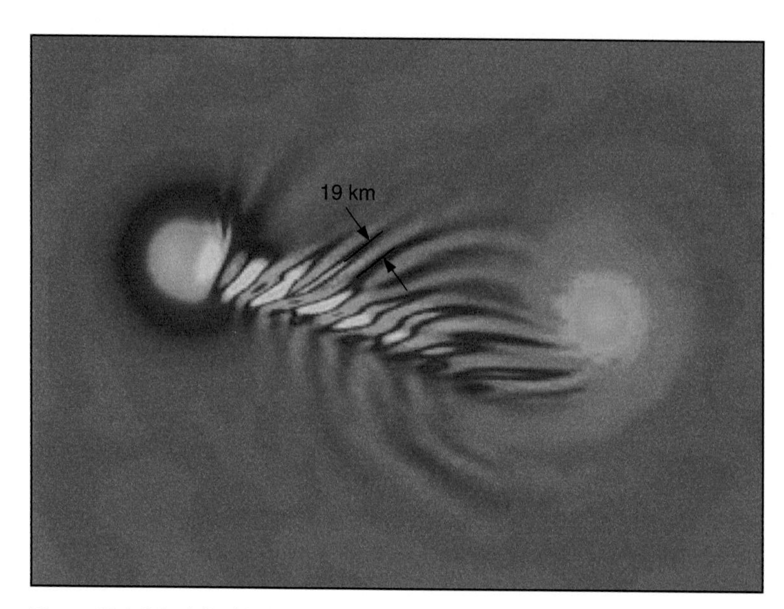

Figure 11.1 Modelled internal waves associated with two counter-rotating eddies shed in a stratified flow past an isolated seamount. The image is a plan view of the density perturbation away from the mean at a depth of 400 metres. The eddy on the left is rotating clockwise and is trapped over a seamount of 25 km diameter. The anticlockwise eddy on the right is being advected away by the mean flow, shedding the internal waves. From Blaise *et al.* (2010), courtesy of Springer.

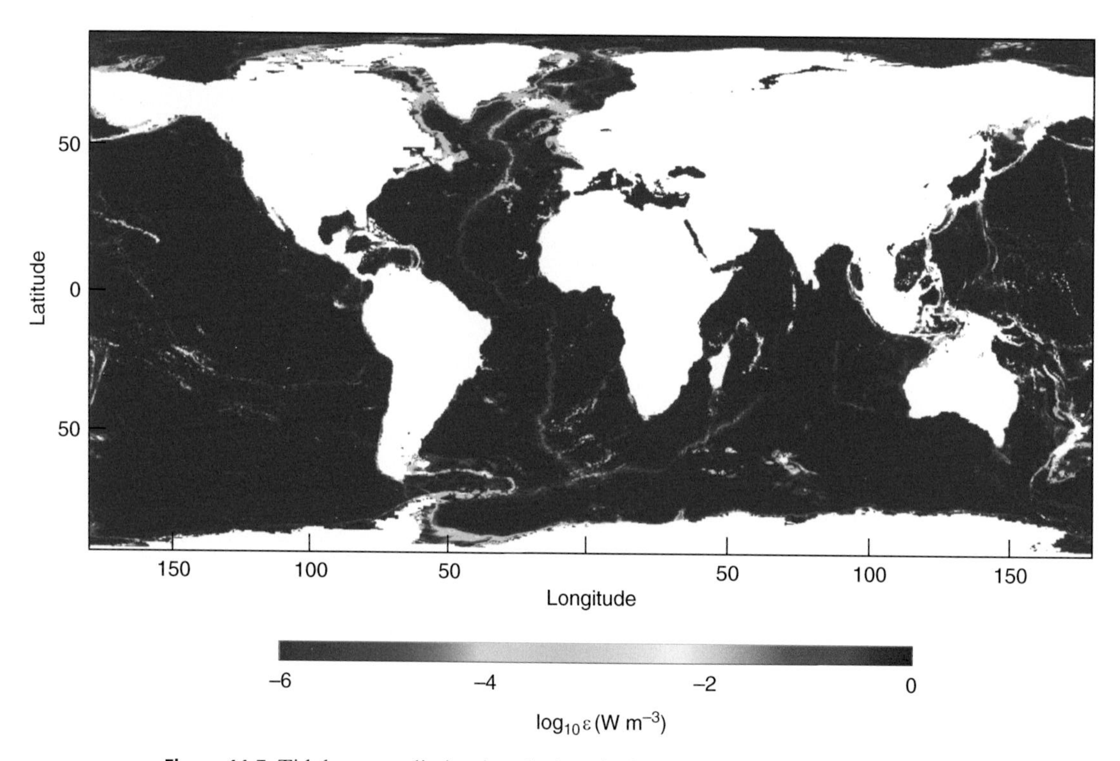

Figure 11.7 Tidal energy dissipation during the last glacial maximum. From (Green, 2010) with permission from Springer.

8 Tidal mixing fronts: their location, dynamics and biological significance

The tidal mixing fronts identified in Chapter 6 are the transition zones between areas which are vertically well mixed and those where weaker tidal stirring allows stratification to develop. In this chapter we will focus on these transition zones. Tidal mixing fronts have special properties which distinguish them from the mixed and stratified regimes that lie on either side of them. The large temperature gradients exhibited by the fronts are clearly apparent in satellite infra-red (I-R) imagery of the sea surface which provides a useful way of keeping track of the position of fronts and following their evolution. The large horizontal temperature gradients in the fronts also involve corresponding changes in density, and hence pressure gradients, which can result in jet-like flows along the fronts and a degree of cross-frontal flow. These frontal currents together with the rapid changes in water column stability which can occur in fronts and the consequent modification of light and nutrient availability in the frontal zone have important implications for primary production and higher levels in the food chain. In the final sections of the chapter we will consider these implications and examine the hypothesis that fronts are zones of significantly enhanced primary production, and assess the reasons for corresponding increases in activity at higher levels in the food chain.

8.1 Frontal positions from satellite I-R imagery

As we noted in Section 2.2.2, long wave energy radiated by the sea surface has a spectral peak at a wavelength $\lambda \sim 10\ \mu m$. This maximum in emission coincides with a minimum in atmospheric absorption by gases. Under cloud-free conditions, satellite I-R sensors can see though this 'window' and map the sea surface temperature (SST), by inversion of the Planck radiation law, with a resolution of a few tenths of a degree Centigrade.

Maps of SST obtained by satellites provide a powerful alternative test of the validity of the criterion for frontal positions (Equations 6.18, 6.23). In the early months of summer, the fronts are clearly manifest at the surface as rapid temperature

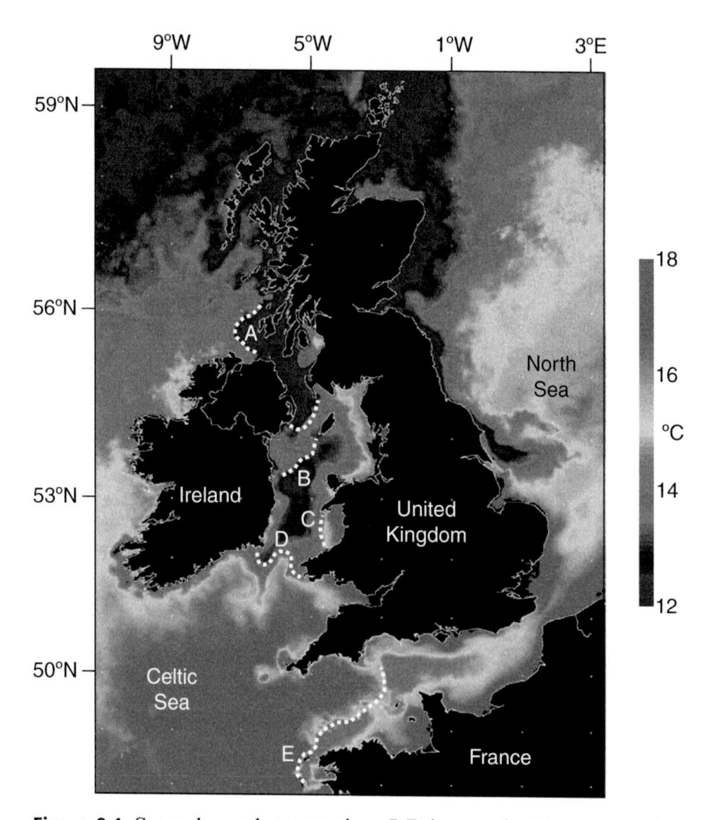

Figure 8.1 See colour plates version. I-R image showing sea surface temperature. The image is a composite between July 9 and 15, 2006. The main fronts to the west of the United Kingdom are marked with white dashed lines. A: the Islay front, B: Western Irish Sea front, C: Cardigan Bay front, D: St George's Channel front, E: Ushant and Western English Channel front. Image courtesy of NEODAAS, Plymouth Marine Laboratory, UK.

changes between the cool mixed water and the relatively warm surface water of the stratified region. Temperature differences across the fronts of ~2–3°C mean that they are readily observed from space in conditions of clear skies, as seen in the satellite image of the British Isles shown in Fig. 8.1.

Composites showing the variation in the positions of fronts in several images allow us to assess the variability of positions and the degree to which they conform to the map of our predictions. Compare the front information in Fig. 8.2a with the predictions based on $SH = \log_{10}\left(h/\overline{|\hat{u}|^3}\right)$ shown in Fig. 8.2b. The fronts occupy remarkably consistent positions close to a contour of $SH \approx 2.7$ in accord with the results of the analysis of the stratified-mixed transitions in Section 6.1. There is some variation in position, a large part of which is due to advection by the semi-diurnal current. After removal of this tidal displacement, we might expect to see some residual movement as a result of changes in u over the springs-neaps cycle in the tides and the variation of Q_i over the seasonal cycle. We will next examine the evidence for such adjustment and the processes involved.

8.2 Fortnightly and seasonal adjustment in the position of fronts

While a large part of the variability in the frontal positions evident in Fig. 8.2a is simply due to the to-and-fro of the tidal currents, a small component of the movement is attributable to the spring-neap cycle of tidal mixing. This has important implications for the biology of fronts. The physics of this spring-neap movement also provides us with a test of the theory of frontal positions which leads to further insight into the behavior of the fronts. We will now focus on how our simple theory initially fails when confronted by the spring-neap adjustment problem, and how we need to modify the theory to accommodate the spring-neap cycle and see what we learn from the extended theory.

8.2.1 The equilibrium adjustment

We found in Chapter 6 that the position of the fronts is controlled by the competition between buoyancy input by heating on the one hand and stirring by the tides and winds on the other. Since the tidal currents increase by a factor of ~2 between neap and spring tides, the stirring power, which is proportional to u^3, will be increased by $\sim 2^3 = 8$. We might expect the fronts to adjust their position in response to such large changes in stirring. If the change happened sufficiently slowly, there would be a continuous equilibrium between heating and stirring so that fronts

Figure 8.2 (a) Composite of frontal positions from clear images for the period May–September 1980, and (b) contours of SH from a numerical model showing the predicted frontal positions at $SH = 2.7$ (thick lines) and the extent of the transition zone (thin lines) defined by $\Delta SH = +/-0.5$. After Simpson and James, 1986, courtesy of the American Geophysical Union.

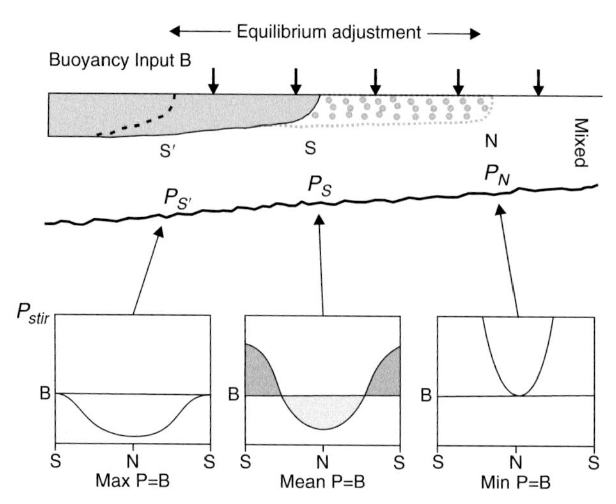

Figure 8.3 A schematic illustration of the spring-neap adjustment of frontal position. As the strength of tidal currents vary over the fortnightly cycle, the front tends to move to maintain the balance between buoyancy input B and stirring P. At the neaps position (N), B equals the minimum stirring at a point where the stirring power at mean tidal range is P_N. The corresponding springs position would be at S′ if there was time to establish a new balance in which B equalled the maximum stirring power at S′ but, in the time available (~7 days), the front retreats only as far as position S at which the *average* stirring over the S-N cycle matches the buoyancy. In practice the frontal movement is further restricted by the reduction of the efficiency of mixing in stratified conditions. (Note that P_N, $P_{S'}$ and P_S represent the stirring power at mean tidal range (M_2).)

would always be found at a point where the tidal stirring balances the buoyancy input (e.g. Equation 6.14):

$$\frac{ek\rho\overline{|\hat{u}|^3}}{3h} = \frac{\alpha g Q_i}{2c_p}. \tag{8.1}$$

Look at the schematic illustration in Fig. 8.3. As the currents increase from neaps to springs, a front would maintain the balance by moving to a position S′ where the *mean* stirring power is lower, i.e. to a higher value of SH. If we denote the mean stirring power due to M_2 at the neaps and springs equilibrium positions of the front by P_N and $P_S{}'$ respectively, then continuous equilibrium over the spring-neap cycle requires that:

$$\frac{P_N}{P_{S'}} \approx 2^3 = 8. \tag{8.2}$$

This means a shift between neaps and springs of $\Delta SH \sim 0.9$ which for typical frontal gradients would imply a displacement of the front by ~10–20 km.

8.2.2 Reasons for non-equilibrium: stored buoyancy and inhibited mixing

In practice, frontal positions are displaced much less than the equilibrium prediction, typically only 2–4 km (Simpson and Bowers, 1979; Simpson, 1981). The discrepancy arises for two reasons. First the adjustment time scale is much longer than the fortnightly cycle; physically this is because the mixing during the springs phase of the tidal cycle has to remove buoyancy accumulated during the neaps phase of the tide as well as mixing the new input of heat. This will slow the rate of adjustment of the front as spring tides approach, and so reduce the displacement distance of the front which will move only as far as S (see Fig 8.2). Allowing for this, we see that the spring tide position of the front is set by the condition that the *average* stirring power over the spring-neap cycle, P_S, matches the buoyancy input. We can formulate this condition and the corresponding one for neaps by neglecting wind mixing and assuming that the depth-mean tidal stream amplitude varies over the fortnightly cycle according to:

$$|\hat{u}| = \hat{u}_{M2}(1 + r_{SM}\cos\sigma t) \tag{8.3}$$

where $\sigma = 2\pi/T_{SN}$ is the angular frequency for the spring-neap cycle with period T_{SN} and $r_{SM} = \hat{u}_{S2}/\hat{u}_{M2}$ is a constant for a given location.[1] The balance at spring tide between buoyancy input and the average stirring, which defines the limit of the retreat of the front, is then given by:

$$\begin{aligned}\frac{\alpha g Q_i}{2c_p} &= P_S\frac{1}{T_{SN}}\int_0^{T_{SN}}(1 + r_{SM}\cos\sigma t)^3 dt \\ &= P_S\left(1 + \frac{3}{2}r_{SM}^2\right).\end{aligned} \tag{8.4}$$

The furthest advance of the front at neaps will be set by the balance for minimum stirring:

$$\frac{\alpha g Q_i}{2c_p} = P_N(1 - r_{SM})^3 \tag{8.5}$$

where P_N is the mean stirring power at the neaps frontal position. So for a constant heating rate and with $r_{SM} = 0.3$, the ratio of stirring power at the springs and neaps positions will be

$$\boxed{\frac{P_N}{P_S} = \frac{(1 + \frac{3}{2}r_{SM}^2)}{(1 - r_{SM})^3} = 3.3.} \tag{8.6}$$

This result implies much smaller displacement than for equilibrium adjustment ($\Delta SH \sim 0.5$) but still exceeds the observed displacement of fronts over the fortnightly cycle by about a factor of 2.

The second reason for over-estimating the adjustment distance lies with our assumption of a constant mixing efficiency e. This does not allow for the inhibiting of mixing

[1] For much of the NW European shelf seas, $r_{SM} \approx 0.3 - 0.4$.

by stratification. As we mentioned in Section 6.1, laboratory experiments on stirring (Turner, 1973; Linden, 1979) suggest that mixing efficiency decreases rapidly as stratification increases. This implies that, once stratification is established, it becomes harder to break it down and re-establish vertical mixing. So after the advance of the front in the period of weak stirring around neap tides, the efficiency of mixing is reduced in the newly stratified area so that the retreat of the front towards the following spring tide is reduced. This behaviour is also apparent in the seasonal cycle of stratification in the shelf seas; following the initial onset of stratification in the spring, stability continues to grow as the efficiency of mixing falls and robust stratification then continues throughout the summer and persists until convective overturn begins in late summer when surface heat exchange becomes negative (see Fig. 6.8a). Only rarely is the cycle re-started in the early weeks of stratification after an episode of strong mixing forces complete breakdown of the initial stratification (e.g. Fig. 6.12).

A summary plot of frontal adjustment over the springs-neaps cycle based on satellite I-R observations of frontal positions can be seen in Fig. 8.4, which compares the observed displacements of 400 frontal positions with theoretical models. The movements of several fronts on the European shelf, after correction for tidal oscillation, have been combined by plotting in a non-dimensional form of h/u^3 against the tidal range factor F which is the range of the tide divided by its value at neaps. The F values plotted are those applying two days prior to the observed position, a lag which gives the largest correlation between displacement and F. This lag indicates that the maximum displacement of the front into stratified waters occurs not at springs but ~ 2 days later after a sustained period of higher than average stirring. The form and magnitude of the small observed adjustment (~ 2–4 km for typical frontal gradients on the European shelf) is simulated reasonably well by models of

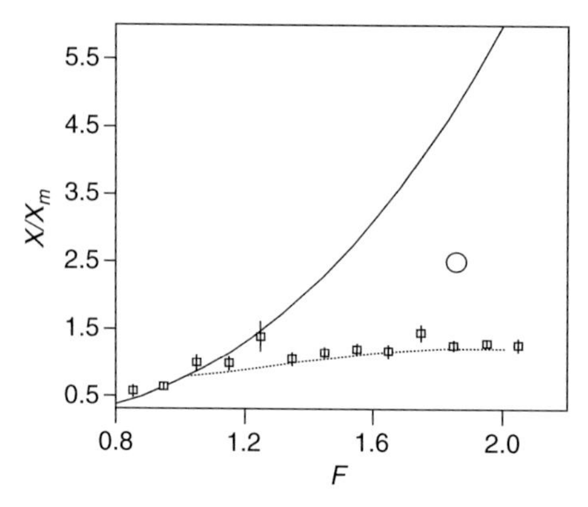

Figure 8.4 Summary of observed spring-neap frontal displacement data plotted against the tidal range factor F ($F = 1$ at mean neap tides). Data are plotted in non-dimensional form as X/X_m where $X = h/u^3$ is the value at the observed frontal position and X_m is the mean value for the frontal position. The continuous curve represents the equilibrium adjustment. The open circle shows the limit of adjustment over the fortnightly cycle when stored buoyancy is taken into account (Equation (8.6)) while the dotted curve is the result of model in which mixing efficiency varies with stratification. Adapted from Simpson, 1981, with permission from the Royal Society, London.

heating-stirring interaction which allow for stored buoyancy and the reduction in mixing efficiency (Simpson and Bowers, 1981; Sharples and Simpson, 1996). This type of variable-efficiency model has been successfully applied to simulate the observed movement of a tidal mixing front in the Bungo Channel of the Seto Inland Sea, Japan (Yanagi and Tamaru, 1990). Substantial adjustment of the front around Georges Bank in response to the springs-neaps cycle has also been observed in I-R imagery (Bisagni and Sano, 1993) and in mooring observations (Loder *et al.*, 1993), although in this case modulation of the tide is mainly monthly rather than fortnightly. In the Georges Bank frontal zone, sea surface temperature determined from I-R data shows an inverse relation to tidal range with a maximum correlation at a lag of ~3 days, while the mooring data show changes in vertical temperature differences which imply frontal displacements of as much as 5–10 km.

Following the onset of heating soon after the vernal equinox, the tidal mixing fronts advance rather rapidly to their summer positions and remain in almost fixed positions through the summer. This is partially due to the net heat supply increasing at its greatest rate at this time, but it also again reflects the reduction in the efficiency of mixing once stratification starts. Stratification, therefore, once initiated, grows rapidly and in most cases irreversibly. Although there are rather few studies of the development of stratification, which varies from year to year depending on weather conditions, it appears that the time scale for the fronts to reach their stable summer positions is of the order of 4–6 weeks (Pingree, 1975). Thereafter the mean position of the fronts remains very consistent until the autumn erosion of stratification as convective cooling sets in. The retreat of the fronts has been even less studied than the spring advance partly because the surface manifestation of fronts in SST generally disappears towards the end of the summer season.

Both the timing of the seasonal cycle and the variation of frontal positions are of great significance in relation to primary production, as we shall see in Section 8.7, but first we shall look more closely at the internal structure and flow field of the fronts.

8.3 The density field and the baroclinic jet

Earlier, in Section 3.7, we noted that the average non-tidal flows in shelf seas are often very weak. The strong horizontal density gradients across tidal mixing fronts provide a marked contrast to this, driving stronger mean flows and setting up significant mesoscale (scales of 10s of km) features. The strength of these flows, and their consistent appearance each year, make them ecologically important. We shall first look at the physics of flows in the vicinity of tidal mixing fronts and identify their generic features.

8.3.1 Expected flow from geostrophy

As we saw in Chapter 6 (Fig. 6.5) high resolution sections with an undulating CTD across tidal mixing fronts reveal the existence of strong horizontal gradients in

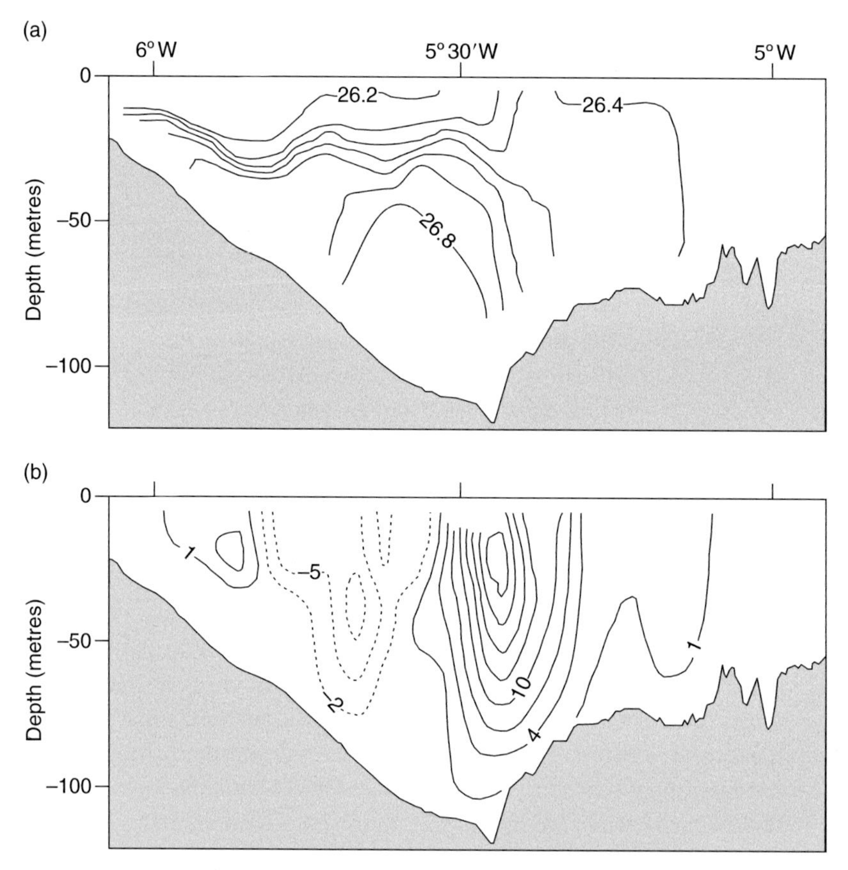

Figure 8.5 (a) A section across a tidal mixing front in the western Irish Sea in July 1996 (front B in Fig. 8.1) showing density contours as σ_t (kg m^{-3}). Density is derived from measurements from a CTD mounted on a towed Scanfish (Fig. 1.5). (b) Velocity normal to the section (cm s^{-1}) derived from the density field using the thermal wind Equation (8.7) and assuming the velocity is zero at bottom boundary. Adapted from Horsburgh *et al.*, 2000, with permission from Elsevier.

temperature, not just at the surface but throughout much of the water column. As we move from mixed to stratified waters, temperature generally increases in the surface layers but decreases in the bottom layers. Where salinity variations are small, density is mainly controlled by temperature and exhibits corresponding gradients of opposite signs in the surface and bottom layers, as seen in the example of a section across the front in the western Irish Sea shown in Fig. 8.5a. These changes in density cause pressure gradients with the isobars sloping across the front. If, as we might expect, the resulting flow field is in geostrophic balance with the pressure distribution (Equation 3.16), the current must be parallel to the front (i.e. normal to the pressure gradient).

Assuming that conditions are uniform in the direction of the front, which we choose to lie along the y axis, the velocity shear within the front will be given by the thermal wind relation (Equation 3.25) as:

$$\frac{\partial v}{\partial z} = -\frac{g}{\rho f}\frac{\partial \rho}{\partial x} \tag{8.7}$$

where the x axis is in the cross frontal direction and the v velocity component is parallel to the front. Equation (8.7) can be integrated upwards from the bed, where we assume $v = 0$, to determine the velocity field from density observations (e.g. see Section 3.3.3). Using the density section across the western Irish Sea front, Fig. 8.5b shows an example of the residual flow derived in this way. A relatively strong flow, referred to as the *frontal jet*, is seen to be concentrated in the high gradient region with a maximum speed of ~20 cm s^{-1} located at a depth of ~20 metres where the horizontal density gradient changes sign. Notice that it is the strong gradients in the lower part of the water column which make the larger contribution to the velocity shear and are responsible for the rapid increase in speed with height above the bed. This predominance of the gradients in the deeper water is commonly the case after midsummer when the surface gradients tend to weaken.

Box 8.1 Geostrophic balance in a tidal mixing front

Tidal mixing fronts have a characteristic density field in which stratified water lies adjacent to mixed water whose density is intermediate between the surface and bottom layers in the stratified region. In the schematic plot of typical density contours on a section normal to front shown in Fig. B8.1, density in the surface layers decreases in the x direction but this trend is reversed in the lower part of the water column (Fig. B8.1a). These density changes give rise to pressure gradients which are apparent in the slope of the isobars; isobars (dotted lines) have to be farther apart where the density is lower. The sea surface itself is an isobar which slopes towards the stratified side of the front where sea surface elevation is lower by an amount $\Delta\eta \approx \left(\frac{\overline{\rho}_s}{\rho_m} - 1\right)h$ where $\overline{\rho}_s$ and ρ_m are depth-mean densities in the stratified and mixed water respectively and h is the depth. $\Delta\eta$ is usually, although not always, positive (i.e. sea level is lower in stratified water); the density difference $\Delta\rho = \overline{\rho}_s - \rho_m > 0$ since the average temperature of the stratified water rises more slowly than that of the mixed regime.

In a steady state without friction (the geostrophic balance), the pressure gradient is balanced by the Coriolis force which acts at right angles to the flow (Fig. B8.1b). Hence the motion must be perpendicular to the pressure gradient and therefore parallel to the front (Fig. B8.1c). In the bottom layers, the isobars slope downwards with increasing x. This implies a pressure force in the x direction which means the force which acts at right angles to the flow must be directed out of the paper in the negative y direction (for the northern hemisphere). With increasing height above the bed, the slope of the isobars increases and, with it, the current speed until the level of the pycnocline is reached. Further up the water column, the density gradient reverses so that the slope of the isobars and the current speed are somewhat reduced

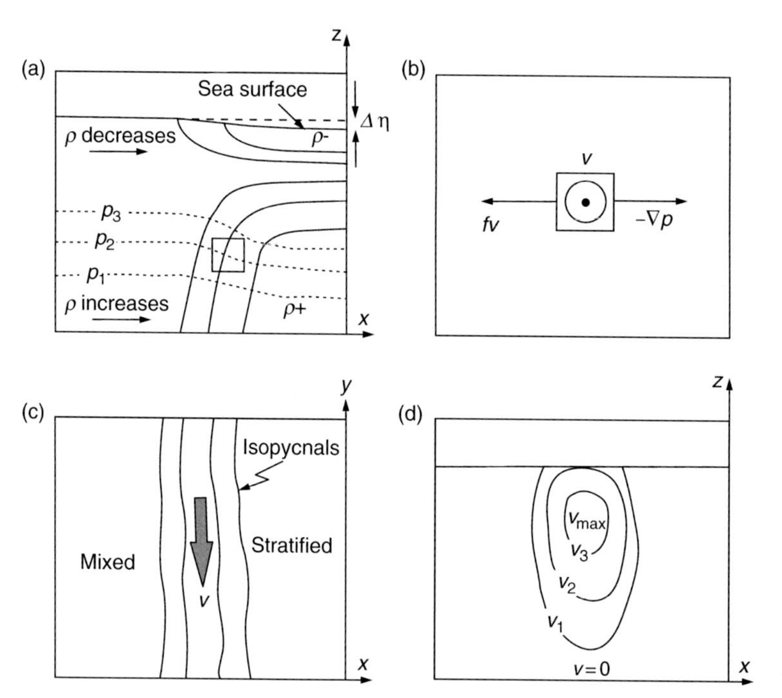

Figure B8.1 (a) Density (solid lines) and pressure surfaces (dotted lines) across a tidal mixing front; (b) the balance of forces within the box marked inside the front in (a); (c) plan view of the frontal jet; (d) vertical structure of the expected along-front jet.

as the surface is approached (Fig. B8.1d). Detailed calculations of the velocity from the density field assuming the geostrophic balance can be made with Equation (8.7).

8.3.2 Observed frontal jets

Patterns of flow similar to geostrophic predictions have been observed in tidal mixing fronts, and generally the predominant flow has been found to be parallel to the front. In some cases, the velocities inferred from the density field have been corroborated by direct measurements of the residual, non-tidal, flow by an acoustic Doppler current profiler (ADCP). In the example from the north eastern Celtic Sea shown in Fig. 8.6, the density and velocity fields have been observed on a section through a large meander in the front, crossing from mixed to stratified water and then back again, thus intersecting the front twice. The isopycnals in Fig. 8.6a have a pattern which is almost symmetrical about the centre of the strongly stratified region. This near symmetry is reflected in the inferred geostrophic flow in Fig. 8.6b with comparable jets flowing in opposite directions parallel to the front. These flows are confirmed by direct measurements with a shipborne ADCP (Fig. 8.6c) after removal of the barotropic tidal flow (Carrillo *et al.*, 2005). In this case, the mean flow was also measured

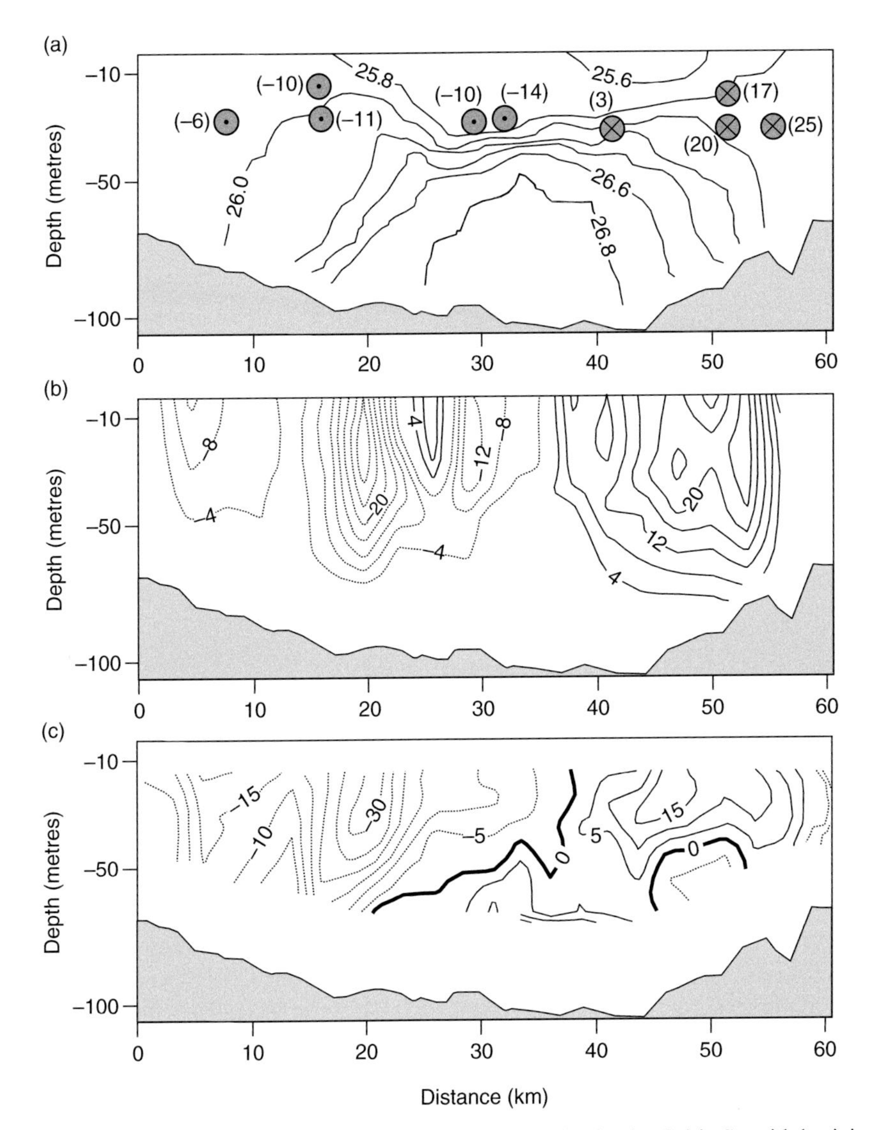

Figure 8.6 Section from west to east through a meander in the Celtic Sea tidal mixing front (front D in Fig. 8.1). (a) Density contours σ_t (kg m^{-3}) from an undulating CTD, along with measurements of velocity from drifter observations. Dotted circles are drifter velocities to the south (out of the page), crossed circles are velocities to the north (into the page). The position of the circles represents the depths of the drogues attached to the drifting buoys; numbers in brackets by the drifter positions are drift speeds in cm s^{-1}. (b) Residual velocity (cm s^{-1}) normal to the section inferred from the density using the thermal wind Equation (8.7) and assuming velocity is zero at the seabed. (c) As (b) but residual velocities derived from the shipborne ADCP with tidal flow removed. Adapted from Brown *et al.*, 2003, with permission from Elsevier.

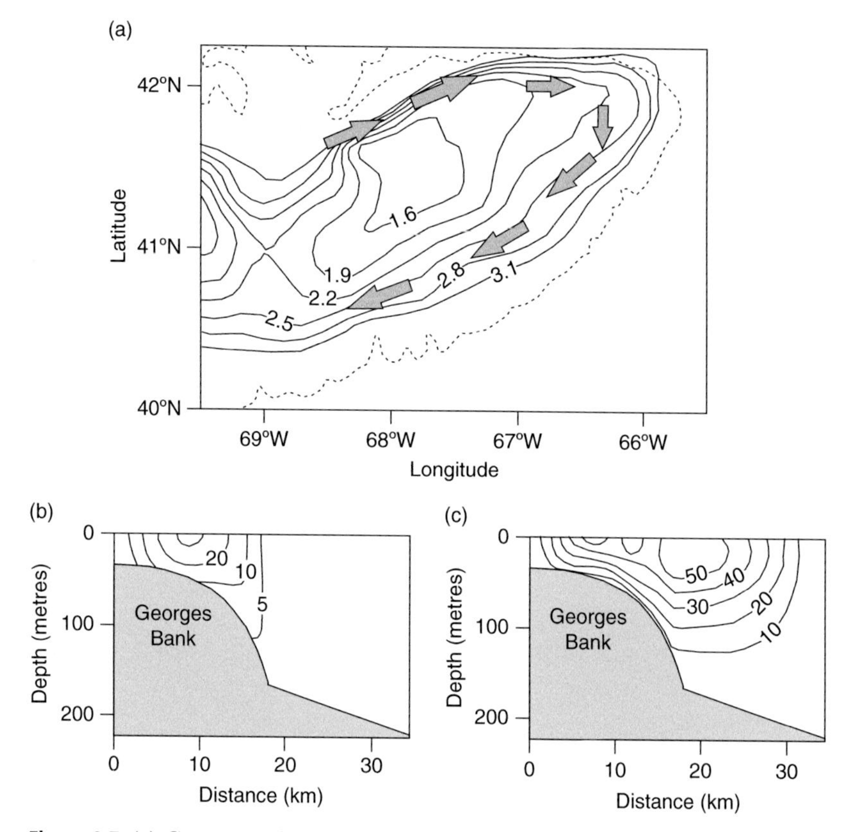

Figure 8.7 (a) Contours of *SH* over Georges Bank, indicating vigorous mixing over the bank contrasting with strong stratification off-bank. The dashed line is the 100-metre isobath. Arrows show the pronounced circulation around the bank which results from a combination of baroclinic forcing and tidal rectification. (b) and (c) show the along bank velocity (cm s^{-1}) on a cross-bank section over the north of the Bank in winter (b) and summer (c). Adapted from (Loder and Wright, 1985; Butman *et al.*, 1987), with permission from MIT Press and the *Journal of Marine Research*.

by a number of drifters deployed in the area and tracked by the Argos satellite system. The resulting estimates of the Lagrangian currents, illustrated in Fig. 8.6a, are in general agreement with the flow observed by ADCP and provide further confirmation of the flow pattern deduced from the density field

Where an isolated mixed region occurs in a surrounding stratified sea, a continuous tidal mixing front will border the mixed zone. Such a situation is formed on the north-western Atlantic shelf by the strong tidal mixing over Georges Bank. This results in complete vertical homogeneity on the bank throughout the year as we would expect from the distribution of *SH* shown in Fig. 8.7a. During the summer, as stratification develops in the deeper water, a tidal mixing front is established around the bank with a baroclinic jet as seen in Fig. 8.7b and c, with speeds up to ~50 cm s^{-1} around the bank (Loder and Wright, 1985; Butman *et al.*, 1987). In this case, the jet combines with a barotropic circulation around the bank due to tidal rectification which maintains a weaker flow around the bank in the absence of stratification during winter (Fig. 8.7b). In summer, the total flow within the jet has been estimated as $\sim 9 \times 10^5$ m^3 s^{-1} (0.9 Sv).

While rarely as intense as the flow around Georges Bank, the transports in frontal jets like those of Figs. 8.5 and 8.7 often exceed 10^5 m^3 s^{-1} (0.1 Sv) and in many cases are the dominant component of the summer circulation. This predominance during the summer regime of the mean flows set up by heating and stirring has been well demonstrated by observations of satellite tracked drogues which show a strong tendency to move normal to the density gradients and to circulate around strongly stratified regions (Horsburgh *et al.*, 2000; Brown *et al.*, 2003). Simulations with numerical models (Hill *et al.*, 2008) provide further confirmation of the large contribution from density driven flows which generally out-compete the influence of wind forcing during the summer. The closed circulation set-up over features such as Georges Bank have important impacts on the local biology, which will be explored in Section 8.6.3.

8.4 Baroclinic instability

So far we have envisaged the density driven flow to be steady and to have an approximately 2-D structure with a density field independent of y, i.e. not changing or varying only slowly in the along frontal direction. While some tidal mixing fronts appear to conform to this simplified picture, others do not and exhibit a convoluted form which is apparent in infra-red images. The image in Fig. 8.8 of the Ushant tidal mixing front off the coast of NW France shows strong convolutions along the front. These shifts in frontal position involve substantial displacements ($\sim$10 km) from the mean position of the front, and indicate the development of eddies through the process of *baroclinic instability* of the flow in the frontal jet.

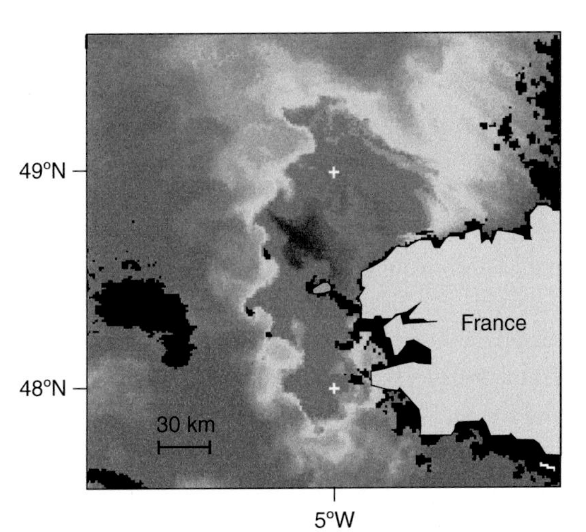

Figure 8.8 See colour plates version. Infra-red image of the TM front to the west of the Brittany peninsula, NW France. Note the complex structure of the boundary between warm stratified (orange, red) and colder mixed (green, blue) waters. Convolutions of the front have length scales in the range 10–40 km, indicating the instability of geostrophic flow with the development of eddies. Image courtesy of NEODAAS, Plymouth Marine Laboratory, UK.

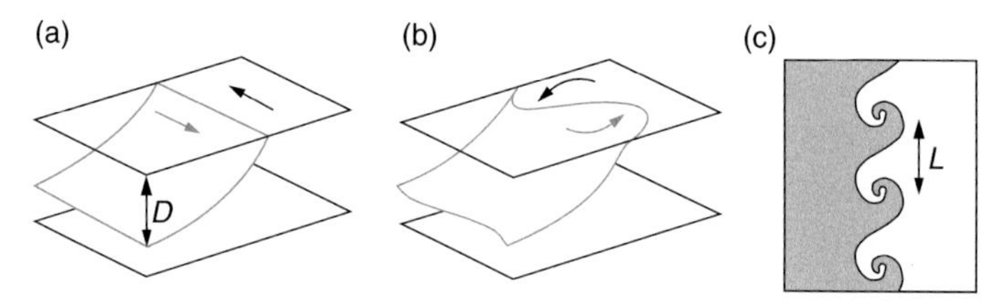

Figure 8.9 (a) Geostrophic flow along a boundary between stratified and mixed regimes, with surface mixed layer depth D. (b) Growth of sinusoidal perturbation in the flow. (c) Plan view of resultant baroclinic eddies, with length scale L.

The growth of a baroclinic instability is illustrated in Fig. 8.9. The process begins with an initial sinusoidal perturbation of the front (a meander) which grows as potential energy from the density field is converted into kinetic energy of a developing meander. The meander can eventually pinch off to form a cyclonic eddy. The kinetic energy of the eddy becomes available through the interchange of water parcels across the frontal boundary, a process which is most efficient when the exchange is along paths having half the slope of the initial isopycnal surfaces (Pingree, 1978). According to perturbation theory of baroclinic flow (Eady, 1949), the scale of the most rapidly growing instability and hence of the resultant eddies will be given by:

$$L \approx 4R_0' \approx 4\frac{\sqrt{g\frac{\Delta\rho}{\rho_0}D}}{f} \tag{8.8}$$

where R_0' (m) is the internal Rossby radius which can be determined from the density difference $\Delta\rho$ (kg m^{-3}) between surface and bottom layers in the stratified water and the thickness D(m) of the surface layer. For the example in Fig. 8.8, the observed values of $\Delta\rho$ and D indicate $L \sim 20$ km, which is of the same order as the observed scale of the eddy motions (20–40 km). The corresponding time scale for eddy development is ~2 days.

In principle, it should be possible to follow the evolution of developing eddy systems from sequences of I-R images, but in practice this approach is frustrated by the paucity of cloud-free images. It is also very difficult to follow the evolution of an eddy from a research vessel given the scales involved and the continuous displacement of the front by tidal advection. The mapping of temperature, salinity and phytoplankton distributions in an eddy by Pingree and co-workers (Pingree *et al.*, 1979) is one of the few such studies in the literature. Observation-based knowledge of the detail of eddy processes is therefore limited and we are forced to rely on theory and models to guide our understanding. Numerical models are able to simulate the main features of the instability process in fronts (James, 1988) while the laboratory experiments (Griffiths and Linden, 1981; Thomas and Linden, 1996) have illustrated the complexity of eddy fields in fronts which may include contra-rotating pairs of eddies. Some of the features in Fig. 8.8 are suggestive of such eddy pairs rather than individual cyclonic eddies (James, 1981).

It is not yet clear why some fronts are subject to frequent eddy development while others appear to be stable for much of the time. It may be that a sufficiently large gradient of tidal stirring at a front inhibits the growth of instabilities, but although there are a number of suggestive cases (e.g. the stable Islay front (Simpson *et al.*, 1979)), this idea remains a hypothesis to be tested. Where eddies do occur regularly, they clearly dominate the frontal zone and make a substantial contribution to the transfer of properties across the front.

8.5 Transverse circulation

Where fronts are not subject to frequent instability, we may still regard the front dynamically as having an approximately 2D, geostrophically balanced flow along the front direction. In the ideal frictionless case there will be no component of current normal to the front. In reality, friction will lead to some cross-frontal exchange. Including the frictional stresses in the equation of motion, we have the steady state version of Equation (3.13) with no external forces:

$$fv = \frac{1}{\rho}\left(\frac{\partial p}{\partial x} + \frac{\partial \tau_x}{\partial z}\right); \quad fu = -\frac{1}{\rho}\left(\frac{\partial p}{\partial y} + \frac{\partial \tau_y}{\partial z}\right). \tag{8.9}$$

Frictional stresses arise in the shear flow associated with the baroclinic jet and modify the along-front flow so that the geostrophic balance is no longer exact. The unbalanced component of the pressure gradient then acts to drive an *ageostrophic* component of flow normal to the front. Assuming the stresses are related to the velocity shear by a known eddy viscosity (Equation 3.32), the dynamical Equations (8.9) together with the continuity Equation (3.1) may be solved numerically for a given density distribution to determine the frontal flow. For realistic friction, the along-front jet is only slightly modified but friction does induce a significant cross-frontal circulation (James, 1978; Garrett and Loder, 1981). In Fig. 8.10 you can see an example, based on a model, of the circulation for the tidal mixing front at the edge of Georges Bank which shows two contra-rotating circulation cells combining to form a downwelling flow with a surface convergence and an upwelling movement in the mixed water close to the bank.

There is some evidence for both of these features occurring in fronts. Accumulations of surface-active material[2] and flotsam along lines parallel to the front are frequently observed in frontal zones in, or close to, the region of largest surface gradients, indicating a convergence of flow. At the same time, a slight minimum in surface temperature, which is often observed on frontal crossings at a point just prior to the sharp temperature rise (see Fig 6.5b for an example), suggests the influence of cold water upwelled from the lower layers of the stratified system. The direct measurement of these cross-frontal flows, which have magnitudes ~1 cm s^{-1}, poses a formidable challenge but the results from some dye tracer experiments (Chen *et al.*, 2008) provide further support for our conceptual picture of cross-frontal circulation.

[2] Surface active material refers to organic substances in a thin surface microlayer which reduces surface tension and damps capillary waves, often creating a visible slick.

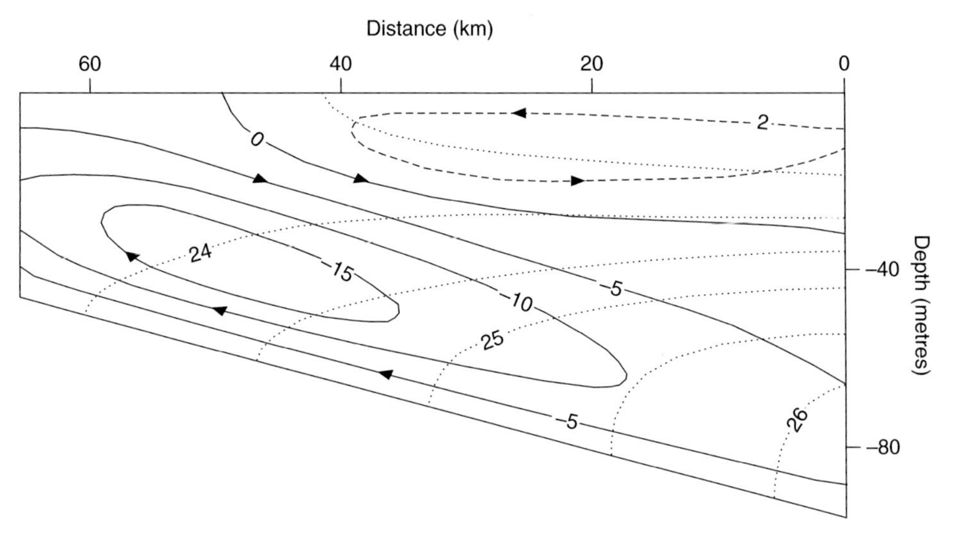

Figure 8.10 Section normal to a front, from a numerical model showing density contours (dotted lines) and streamlines (continuous and dashed lines) of transverse flow. Numbers on streamlines are the stream function in units of 10^{-2} m^2 s^{-1}. Transverse velocity in the midwater downward flow does not exceed 1 cm s^{-1}. From (Garrett and Loder, 1981) with permission from the Royal Society, London.

Physics summary box

- Tidal mixing fronts mark the transition point between regions that seasonally stratify and those that remain mixed. They are clearly seen in satellite images of sea surface temperature as places where surface temperature changes very rapidly.
- Where the tides provide the dominant source of mixing, the position of the fronts is well represented by a critical value of h/u^3, with h the water depth and u some measure of the strength of the tidal currents.
- Fronts change position due to advection by the tidal currents, and to changes in the amount of tidal mixing energy available over the spring-neap cycle. The former is a bulk movement of the front with the flow of water. The latter is an adjustment of the front's position, typically by 2–4 km, where the water within the region of the adjustment undergoes a fortnightly cycle in stratification and mixing.
- The mean flows associated with the front are illustrated in Fig. 8.11. Fronts are marked by a strong jet-like current. Looking from the stratified region toward the mixed region, this jet usually flows along the front to the left (northern hemisphere).
- Meanders in this along-front jet can become unstable, pinching off eddies which transfer water and its constituents across the front.
- Friction induces weak flows perpendicular to the front which involve a convergence of surface currents and a downwelling at the front. The convergence is often marked by a line of buoyant surface detritus along the edge of the front.

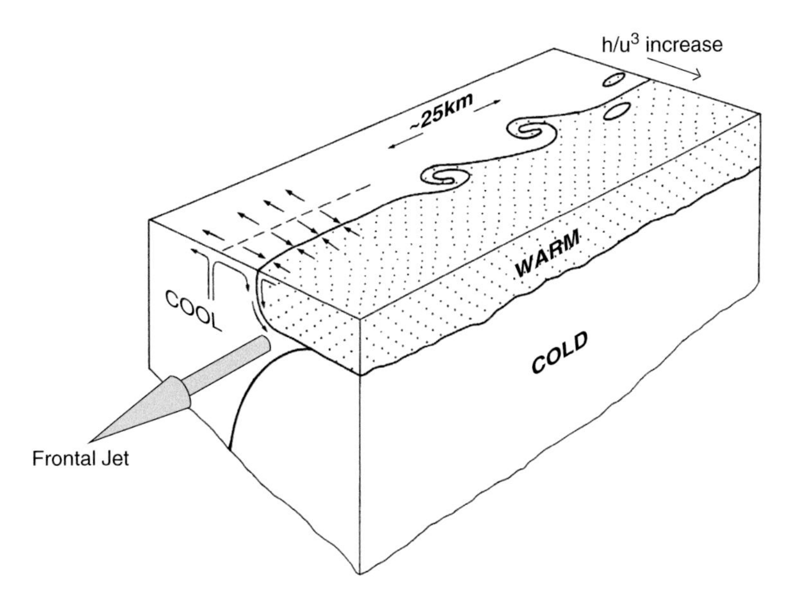

Figure 8.11 Schematic illustration of the structure and residual flows associated with a tidal mixing front. Adapted from (Simpson 1981) with permission from the Royal Society, London.

8.6 Frontal structure and biology

By now, you should appreciate that tidal mixing fronts, although defined in a straightforward way as the transitions between mixed and stratified systems, are complicated and, in some ways, rather subtle physical systems. At about the same time that the features were being recognised in the physical structure of shelf seas, it also became clear that the fronts were playing a role in the biochemistry (Pingree *et al.*, 1975; Fogg, 1988). In particular, sections through fronts indicated enhanced chlorophyll concentrations in the surface waters at fronts, as seen in the early data from the Ushant front in Fig. 8.12a. Satellite observations, such as the SST and ocean colour images in Fig. 8.12b, c, often show a surface manifestation of chlorophyll linked to the temperature gradient at the fronts in satellite images.

In addition to the large-scale shelf sea tidal mixing fronts, it was found that isolated islands in stratified regions could generate localised mixing due to the enhanced stirring brought about by the increased tidal currents in the shallowing water on the flanks of the islands. You can see this feature around the Isles of Scilly in the satellite images of Fig. 8.12 (Simpson *et al.*, 1982). Measurements of water properties on a section extending southwards from the Isles of Scilly, shown in Fig. 8.13, illustrate how the physical structure and the chlorophyll distribution around the islands have similar patterns to the larger scale shelf sea tidal mixing fronts. Note in Fig. 8.13c how the nitrate distribution is also correlated with the density structure of the front. Cold, sub-thermocline water contains the greatest concentrations of nitrate, locked below the thermocline when stratification began in

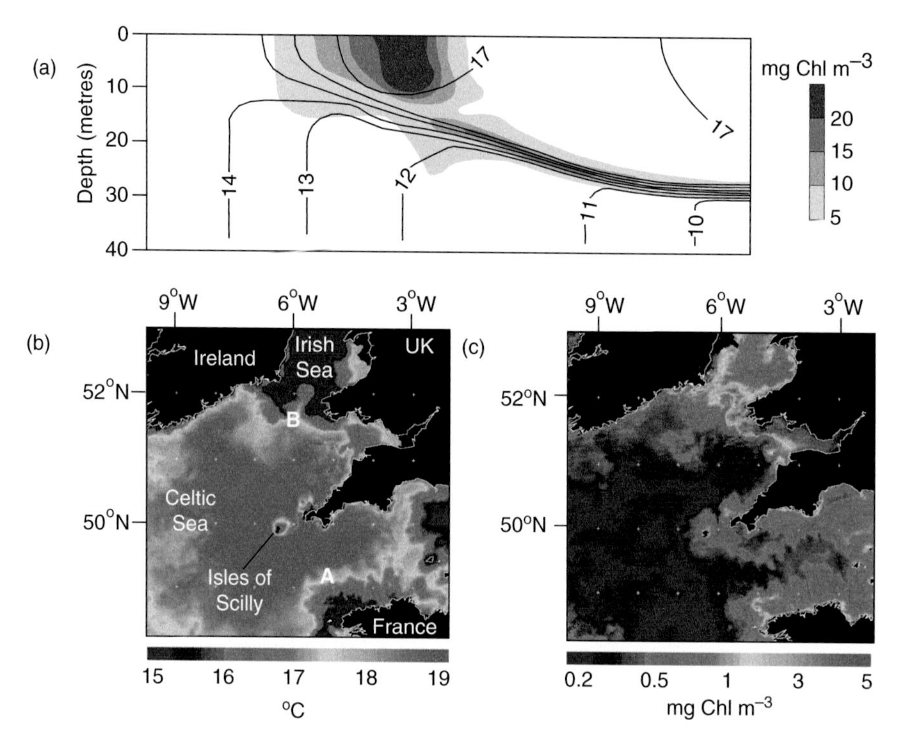

Figure 8.12 (a) See also colour plates section. A section of temperature (lines) and chlorophyll concentration (shaded) through the tidal mixing front off Ushant, NW France (from Pingree *et al.*, 1975, courtesy of *Nature*). The water depth was 100 metres; only the upper 40 metres has been plotted. (b) and (c) Satellite images of sea surface temperature and chlorophyll for the Celtic Sea. Images are composites over July 12–18, 2005, provided by NEODAAS Plymouth Marine Laboratory. Tidal mixing front A is the Ushant front across which the section in (a) was collected. Front B is the Celtic Sea front. Note also the 'island mixing front' and patch of mixed water surrounding the Isles of Scilly, along with an associated increase in sea surface chlorophyll.

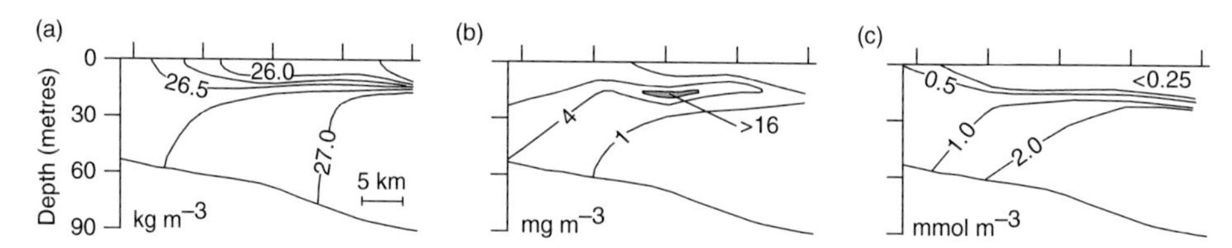

Figure 8.13 A section across the island mixing front around the Isles of Scilly in summer. (a) Density, showing the sharp pycnocline away from the islands, expanding to form bottom and surface fronts in the shallower water by the islands. (b) Chlorophyll concentration, indicating increased surface chlorophyll approaching the islands. (c) Nitrate concentration, with mixing close to the islands driving nitrate towards the surface. The tick marks along the upper axes indicate the positions of the sampling sites. Adapted from Simpson *et al.*, 1982, with permission from Elsevier.

spring. The warm surface water in the stratified region has almost no nitrate as a result of earlier uptake by the spring bloom. In the mixed region the water contains intermediate nitrate concentrations, maintained by the vigorous tidal mixing.

In the following sections we will detail the candidate mechanisms for this biochemical response to shelf sea and island tidal mixing fronts. We will then look at some of the links between fronts and higher trophic levels, and describe the role of the frontal physics. You will see that simply linking higher phytoplankton biomass through to the distribution of grazers and larger animals is not always justified by the observational evidence.

8.6.1 Enhancement of primary production at fronts

As the pattern of phytoplankton distribution at fronts became recognized, two hypotheses arose to explain the association between the phytoplankton and the temperature distribution. The weak, convergent current flow at the surface could simply be concentrating the phytoplankton at the front, or the phytoplankton enhancement could arise directly from *in situ* production due to some property of frontal dynamics aiding access to nutrients and light. Direct evidence that a frontal phytoplankton community actively grows came from the Georges Bank front (Horne *et al.*, 1989). The information on growth and nitrate uptake rates is summarized in Fig. 8.14. While the Georges Bank front has elevated carbon fixation rates, particularly when compared to the stratified region, the main contrast lies in the high *f*-ratio (see Section 5.1.6) and demand for nitrate at the front. The front is a site of elevated *new* primary production.

What physical processes supply nitrate to surface frontal waters, leading to this increase in new primary production? The schematic diagram in Fig. 8.15 summarises the biogeochemical environment and the possible nitrate supply mechanisms. There are four possible ways to transfer nutrients across and into a front: (1) increased vertical mixing allowed by the gradually weakening stratification at the front, (2) periodic mixing and re-stabilisation of the frontal region due to spring-neap adjustment, (3) eddy transfers across the front, and (4) fluxes associated with the friction within fronts (Loder and Platt, 1985).

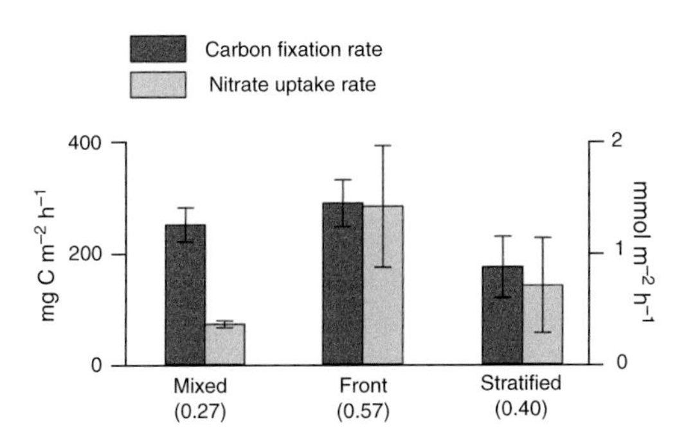

Figure 8.14 Carbon fixation rates and the nitrate uptake rates, integrated through the photic zone (upper 40 metres) at mixed, frontal, and stratified sites across the tidal mixing front on Georges Bank. Numbers in brackets are the *f*-ratios calculated as the ratio between nitrate and nitrate+ammonium uptake rates. Based on observations reported in Horne *et al.*, 1989.

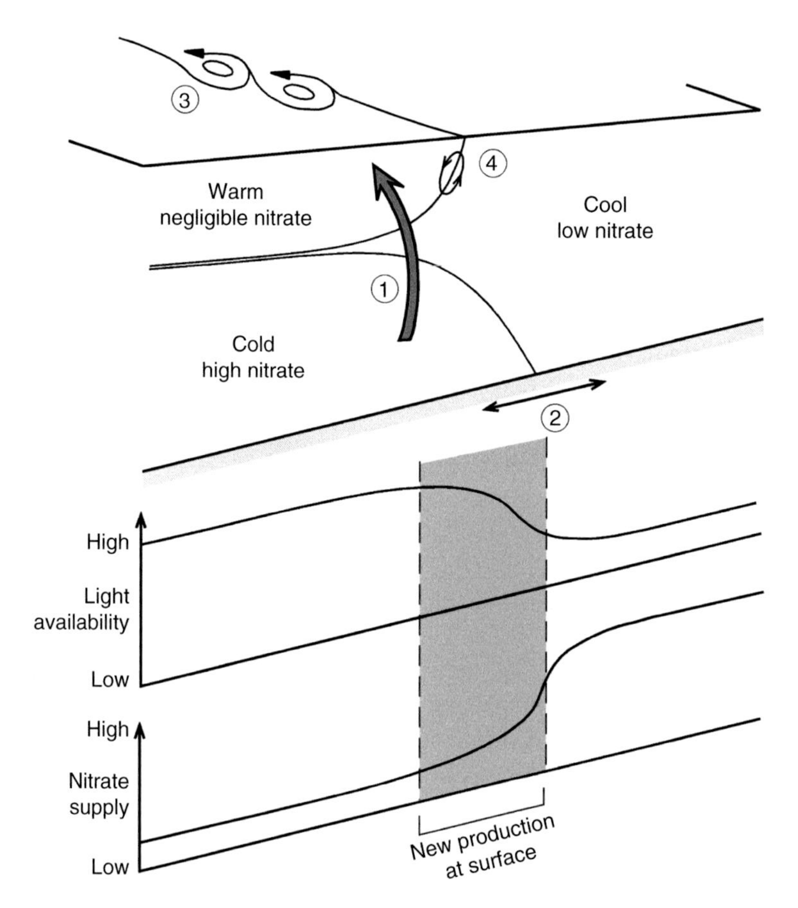

Figure 8.15 A schematic illustration of the changes in light availability and nitrate supply across a tidal mixing front, and the main mechanisms supplying frontal phytoplankton with nitrate. (1) increased vertical mixing of deep nitrate to the surface as the stratification weakens at the front; (2) the spring-neap adjustment of the front position; (3) eddy transfer of surface water across the front; (4) weak mixing of nitrate driven by interfacial friction within the front.

Increases in vertical mixing at the front will only support surface productivity and the accumulation of phytoplankton biomass as long as the rate of mixing is still low enough for near-surface residence times of phytoplankton cells to be sufficient to allow net growth. Figure 8.16a presents the results from a model incorporating only this process, which show that enhanced vertical mixing alone does indeed lead to a pattern of phytoplankton biomass within a front which is similar to observed biomass distributions and concentrations. Allowing vertical mixing at the front to respond to the spring-neap tidal cycle further enhances frontal production and biomass, and leads to fortnightly pulsing of surface biomass within the adjustment region, shown in Fig. 8.16b. This spring-neap modulation of frontal mixing has been

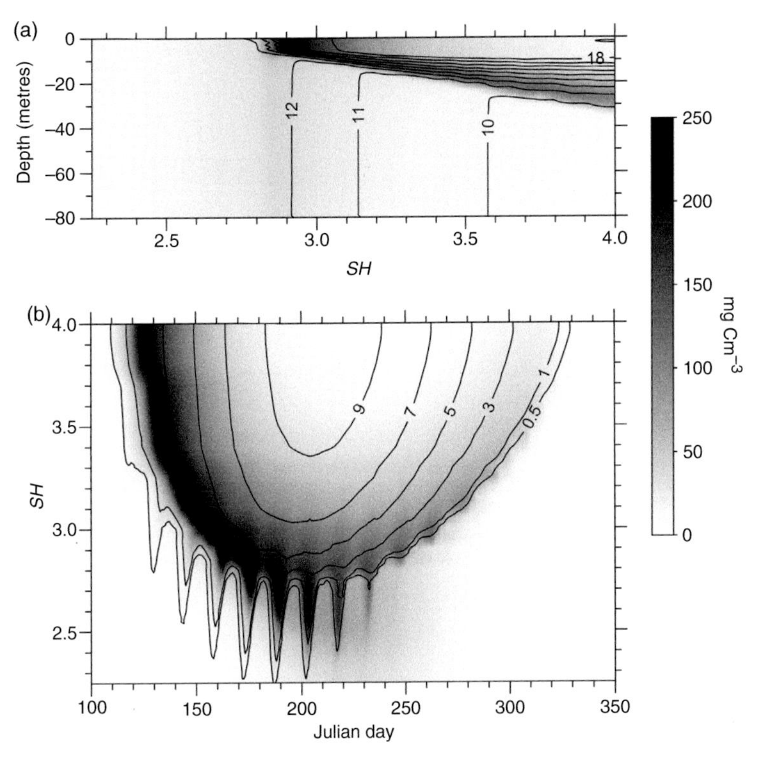

Figure 8.16 See colour plates version. Model predictions of (a) cross-frontal temperature (°C, line contours) and phytoplankton carbon (shaded), and (b) variations in stratification (surface-bottom temperature difference, °C, line contours) and surface phytoplankton carbon (shaded) between mid April and late November for different *SH*. In (a) only the M_2 tide was operating, so that phytoplankton within the front could only grow in response to changes in the vertical flux of nitrate at the front. In (b) both M_2 and S_2 tides operated to drive a spring-neap cycle in mixing. Adapted from Sharples, 2008, with permission from Oxford University Press.

proposed as a mechanism with the potential to raise annual productivity in frontal regions by 70%, compared to a front driven solely by the M_2 tide (Sharples, 2008).

The fact that this enhancement occurs within a narrow region at the frontal boundary has important implications for the horizontal resolution of shelf sea numerical models. There is some limited evidence from the western Irish Sea that such spring-neap frontal biomass pulses do indeed occur throughout summer (Richardson *et al.*, 1985). Further evidence for a spring-neap cycle comes from the Ushant front off northwest France, where estimates of spring-neap changes in the supply of nitrate to frontal waters have been shown to be consistent with changes in rates of phytoplankton nitrate uptake (Morin *et al.*, 1993).

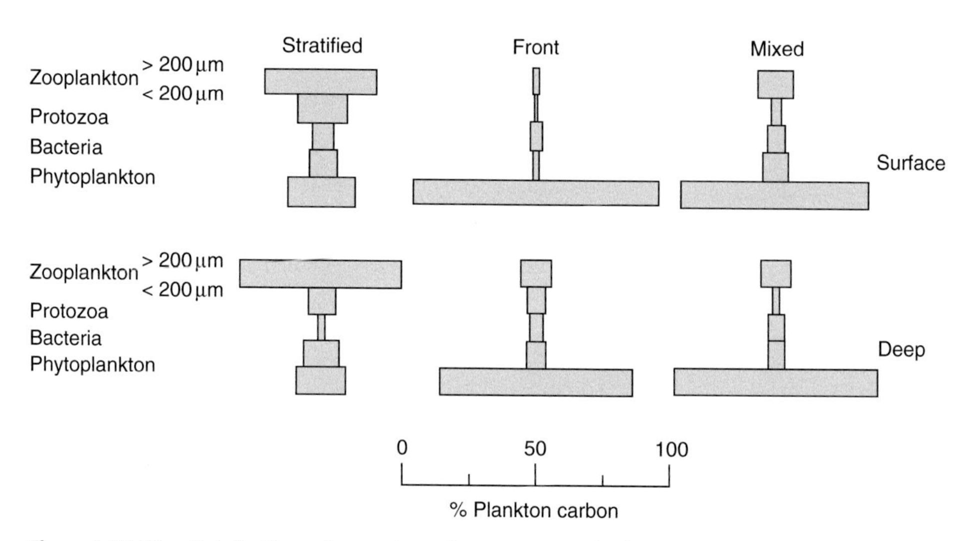

Figure 8.17 The distribution of organic carbon amongst plankton groups for surface (0–24 m) and deep (24–60 metres) waters in vertically mixed, frontal and stratified waters of the western English Channel. Adapted from Holligan *et al.*, 1984, courtesy of *Inter Research Journals*.

Calculations for typical fronts suggest that increased vertical mixing and the spring-neap adjustment could together supply 80% of the phytoplankton nitrate requirements, with the remaining 20% coming from eddy transfers and the transverse circulation (Loder and Platt, 1985). However, the details are likely to vary for different fronts. For instance, the front on Georges Bank may be fuelled by significant amounts of nitrate supplied by horizontal advection (Horne *et al.*, 1996). Modelling studies for Georges Bank indicate that much of the nutrient supply to the surface frontal waters is driven by short but intense mixing events during each semi-diurnal tidal cycle (Franks and Chen, 1996), with the steep slopes of the bank generating on-bank residual flows that will transport bottom layer nutrients to the front (Chen and Beardsley, 1998).

8.6.2 Zooplankton and tidal mixing fronts

Given such a clear increase in phytoplankton production and biomass within the surface waters of a front, we might reasonably ask if there is evidence of further impacts on the frontal ecosystem at higher trophic levels. Fig. 8.17 shows the average plankton distributions at a tidal mixing front in the western English Channel, indicating that the increase in phytoplankton carbon within the front emerges as the clearest signal (Holligan *et al.*, 1984). Phytoplankton represent 90% of total plankton carbon in the surface layer of the frontal zone, compared to 70% in the mixed water and 25% in the surface stratified water. There does not, however, appear to be any corresponding frontal response in the zooplankton biomass in Fig. 8.17.

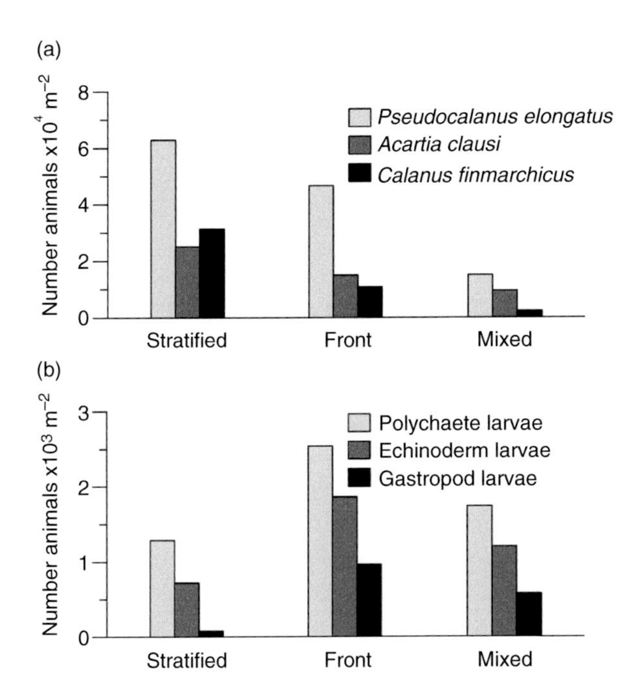

Figure 8.18 Distributions of the major zooplankton taxa across the tidal mixing front in the western Irish Sea, based on data in (Scrope-Howe and Jones, 1985). (a) The three most abundant taxa, illustrating a common trend of highest numbers in the stratified water, decreasing across the front into the mixed water. (b) The only three of the reported taxa that demonstrated an enhancement at the front. Note the different scales on the *y*-axes.

The lack of apparent zooplankton enhancement at a tidal mixing front is perhaps a little surprising; with such a strong response of the primary production and primary biomass to the front, we would expect the zooplankton to take advantage of an enhanced food supply. Is this lack of response a general property or does it apply to some zooplankton and not to others? To answer this question we can look at regions where the distributions of zooplankton species have been assessed in more detail. Fig. 8.18 summarises some of the data from a study of the seasonal changes in zooplankton across the western Irish Sea front (Scrope-Howe and Jones, 1985). There it was found that there was almost no consistent localised response of the zooplankton. In Fig. 8.18a you can see that concentrations at the front for most species were intermediate between those in stratified and mixed water. However, this well-resolved taxonomic study did show a small number of species to have marked increases in abundance at the front, for instance polychaete larvae, gastropod larvae, and the free-swimming stage of echinoderms (Fig. 8.18b). It is perhaps noteworthy that these are all larvae of benthic organisms. Similarly, observations have shown that some taxa of zooplankton exhibit peaks in abundance at the front on the north flank of Georges Bank (Wishner *et al.*, 2006), while for most others the frontal jet acts simply as a boundary between taxa (Butman *et al.*, 1987; Meise and Oreilly, 1996).

There is a good deal of evidence supporting a link between chlorophyll concentration (as a food supply) and the rate of copepod egg production (Uye and Shibuno, 1992; Runge *et al.*, 2006), so despite the lack of a clear signal in zooplankton biomass, we might expect tidal fronts to be sites of higher zooplankton production. For copepods, such a frontal increase in egg production has been observed in the

Inland Sea, Japan, but higher predation of copepod nauplii in the front was suggested as preventing this production following through to higher copepod concentrations (Uye *et al.*, 1992). Off northeast Scotland, copepod egg production and egg concentrations were found to be higher within a front, but successively older stages of the copepods showed less relation to the front (Kiørboe and Johansen, 1986).

8.6.3 Fronts and larger marine animals

While it would appear to be difficult to demonstrate a simple, clear picture of zooplankton enhancement at fronts, there are several unambiguous examples of predators of zooplankton being found to forage preferentially at tidal mixing fronts.

Basking sharks have been found to forage for zooplankton along a tidal front in the western English Channel (Sims and Quayle, 1998). The sharks were observed to concentrate their swimming close to sea surface slicks, indicators of surface current convergence, with foraging time being spent mainly in regions with higher zooplankton concentrations. Convergences associated with smaller scale tidal fronts around reefs, analogous to the island mixing fronts, have also been shown to accumulate a range of zooplankton. A variety of reef fish target the convergence zones to feed (Kingsford *et al.*, 1991). Seabirds have long been recognised as selectively feeding along fronts (Begg and Reid, 1996; Jahncke *et al.*, 2005). In the western Irish Sea, observations of seabird distributions have shown a correspondence between frontal structure and foraging behaviour and, in particular, an association of surface feeders with surface convergences (Durazo *et al.*, 1998). These studies indicate that the relatively fine scales of the surface flow convergence, rather than the broader scale of the phytoplankton distribution, could be providing the increased prey densities targeted by the sharks or seabirds, and that species can have different foraging locations depending on their foraging method and the structure of the front. Thus, tidal fronts are not necessarily important for foraging animals because they are sites of higher food production. They are also important because they make food accessible, either by concentrating zooplankton within convergence zones or by making zooplankton available on the thermocline closer to the surface.

Tidal mixing fronts have received particular attention in the context of commercially important fisheries. Spawning stocks of Atlantic herring have been shown to be in close proximity to tidal mixing fronts and the adjacent mixed water, with the mixed regions acting to prevent wider dispersal of larvae and determining the size of a stock (Iles and Sinclair, 1982) and also maintaining genetic identity of stocks (Sinclair and Iles, 1989). However, this concept of oceanographic retention controlling larval recruitment to a fish stock has been seen as contrary to observations of larval drift from spawning grounds (Cushing, 1986), and the idea that recruitment success depends on spawning being timed to coincide with a flush of suitable food (e.g. the match-mismatch hypothesis (Cushing, 1990)). It is possible that all three mechanisms may contribute to the full explanation of stock development, with the balance between them being dependent on species and environment.

At this stage we should note that there seems to be something of a paradox here. We have clear signals of primary production and phytoplankton responding to tidal mixing fronts, and at the other end of the food chain we have evidence of larger marine animals targeting the fronts. The evidence for the link provided by the zooplankton within this chain seems to be rather patchy. There are several possible reasons why a frontal response in zooplankton numbers may either be absent or, if it does occur, be difficult to observe. Zooplankton distributions are generally far patchier than phytoplankton, which makes observation of cross-frontal contrasts challenging; with limited sampling the patchiness is likely to dominate over any consistent horizontal distribution. Also, zooplankton respond to their environment at a far slower rate than the phytoplankton. The time between an increase in food supply, through to egg laying, successful hatching and maturation could be several days to a few weeks, compared to the relatively immediate response (a few days) of phytoplankton to increases in light or nutrients. This time lag will, to varying extents, de-couple zooplankton distributions from their physico-chemical environment (Abraham, 1998). For instance, we might expect any signal in frontal copepod production to gradually decrease from eggs, through to nauplii, through to adult copepod stages due to physical dispersion and the developmental time between egg and adult copepod, a pattern supported by observations off northeast Scotland (Kiørboe and Johansen, 1986; Kiørboe *et al.*, 1988). Other factors that need to be considered are the limited cross-frontal transports (Perry *et al.*, 1993), and the impact of different predation pressures affecting zooplankton on the mixed and stratified sides of the front (Meise and Oreilly, 1996). The examples of the basking sharks (Sims and Quayle, 1998), the reef fish (Kingsford *et al.*, 1991) and the seabirds (Durazo *et al.*, 1998) indicate the importance of accumulations of zooplankton at frontal convergences; these convergences can be very localized so that single 'frontal' sites chosen for study on the basis of the chlorophyll distribution could miss a region of higher zooplankton abundance. Detecting any frontal signatures in zooplankton requires many samples with fine horizontal resolution and with taxonomic resolution that can separate species and stages of species.

8.6.4 Fronts and fisheries

We conclude our discussion of the links between the physics and biology of fronts by looking in some detail at two examples of fisheries being supported by frontal systems – one where the front acts as a barrier to dispersion of larvae and another where both advection and retention associated with a tidal mixing front play roles in the recruitment of larvae to the adult stock.

(i) ***Nephrops norvegicus* and the western Irish Sea gyre**

Nephrops norvegicus (see photograph in Fig. 8.19), also known locally as the Dublin Bay prawn, is a demersal organism of considerable commercial importance in the

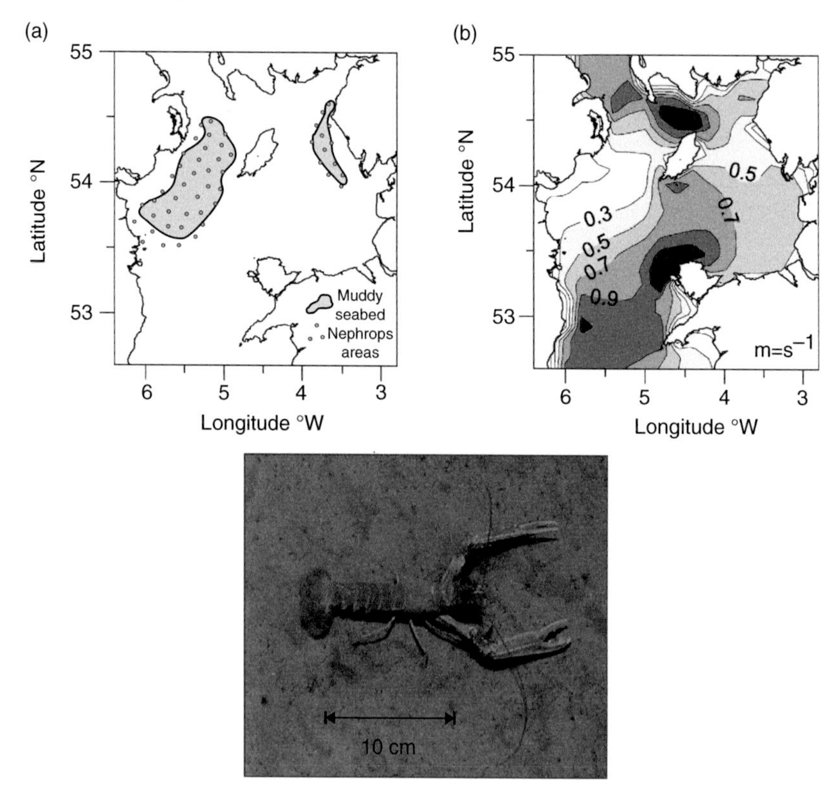

Figure 8.19 (a) The distribution of muddy seabed within the Irish Sea, along with the strongly correlated distribution of *Nephrops*. After White *et al.*,1988, courtesy Oxford University Press. (b) Amplitude of the M_2 tidal current speed (m s^{-1}) within the Irish Sea (model data provided by Sarah Wakelin, National Oceanography Centre, UK), showing the lower current speeds associated with the muddy regions. The photograph is of an adult *Nephrops*. (Photo by Inigo Martinez, Marine Scotland, Aberdeen.)

western Irish Sea. The adult *Nephrops* require a muddy substrate within which to make burrows. Fig. 8.19a shows the correspondence between the *Nephrops* distribution and the two main areas of muddy seabed in the Irish Sea (White *et al.*, 1988). Fig. 8.19b indicates that the muddy substrate corresponds to regions of weak tidal currents, with maximum M_2 tidal currents less than 0.5 m s^{-1}; this contrasts markedly with the central Irish Sea where currents can reach over 0.9 m s^{-1} and the seabed is predominantly sand or gravel. *Nephrops* larvae are free-swimming, so the question arises as to how these isolated populations are maintained, particularly as observations have shown that a strong southward buoyancy current along the Irish coastline in early spring acts to transport larvae away from the adult area (White *et al.*, 1988).

The weak tidal currents that are responsible for the muddy seabed are also weak enough to allow thermal stratification in summer. The western Irish Sea *Nephrops* distribution corresponds with a thermally stratified region in the west of the generally

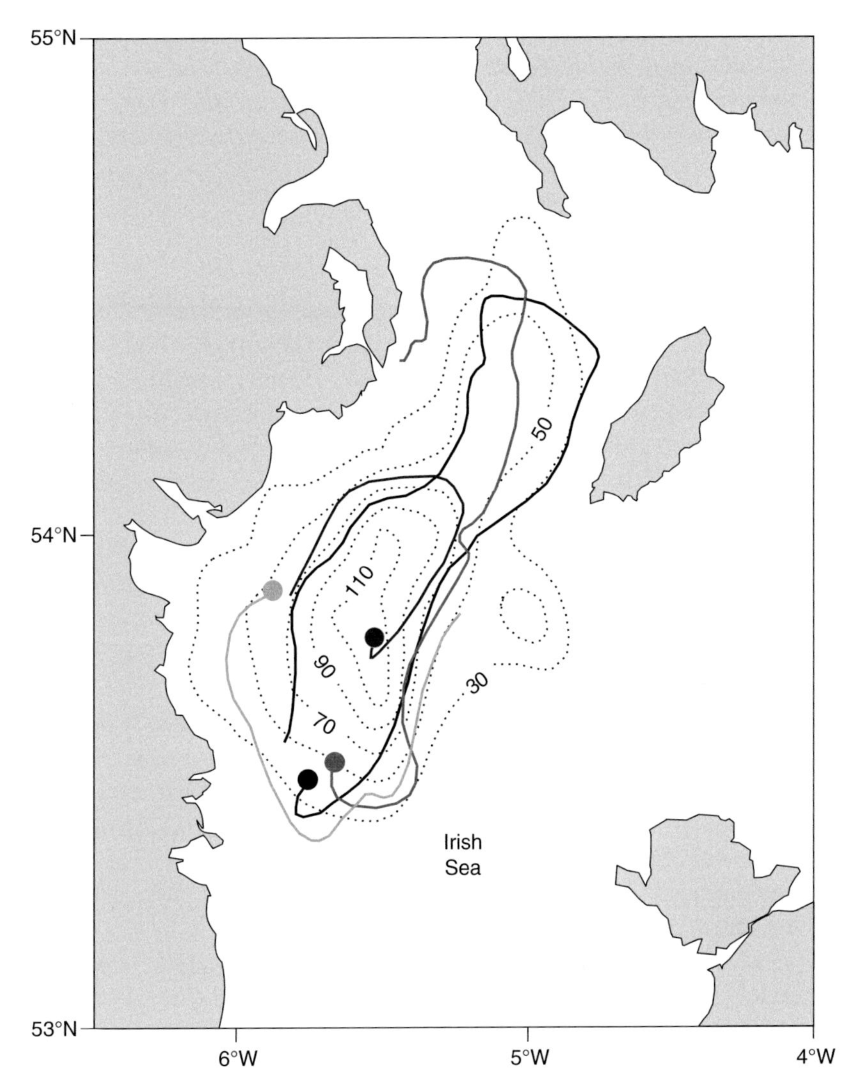

Figure 8.20 Trajectories of four satellite-tracked drifting buoys in the western Irish Sea, July 1995. The tracks are shown as bold lines, with the deployment position of each buoy shown as a filled circle. The dotted lines indicate the extent of stratification in terms of the potential energy anomaly, Φ (see Section 6.1.2). Adapted from Horsburgh *et al.*, 2000, courtesy of Elsevier.

very well mixed Irish Sea (see Fig. 8.5a). It is the dynamics of this isolated patch of stratified water, and particularly the cold pool of water trapped underneath the warm surface layer, that plays a vital role in the life of the *Nephrops*. The tracks of drogued, satellite-tracked buoys deployed in the region (Horsburgh *et al.*, 2000) are shown in Fig. 8.20. They follow a consistent anticlockwise flow, indicating the presence of a gyre circulation associated with the isolated patch of stratification. The gyre can be thought of as a continuous frontal jet circulating around the mud patch.

This circulation is believed to act as a retention mechanism for the *Nephrops* eggs and larvae, aiding the settlement of young *Nephrops* back onto the required muddy substrate (Hill *et al.*, 1997). It is probable that the 'leakiness' of this gyre, the inter-annual variability in the timing of spring stratification and the strength of the stratification, and the fishing pressure on the *Nephrops* population will combine to play important roles in the survival of *Nephrops*.

(ii) Fish stocks over Georges Bank

Georges Bank, off the northeastern United States, has long been recognised as an important feeding area for fish, particularly Atlantic cod and haddock, and has been commercially fished since at least the early nineteenth century, possibly much longer. Fish stocks generally need to maintain sufficient integrity for enough of the adults to meet up and spawn each year despite living within a dispersive marine environment. In 'choosing' a region for spawning, adult cod and haddock have to solve the problem of how their initially non-motile eggs and weakly motile larvae can reach the feeding area over the shoals of Georges Bank rather than being advected or dispersed elsewhere. Considerable research effort has been focused on understanding the links between the physical regime of Georges Bank and the life stages of the fish, primarily in response to the importance of the commercial fishing and the consequent pressures that the fish stocks have been experiencing.

The residual circulation around Georges Bank has already been shown (Fig. 8.7). Rectification of the barotropic tidal currents (see Sections 3.7 and 8.3.2) provides a clockwise circulation around the bank all year, but the development of the tidal mixing front separating mixed bank water from the surrounding stratification augments the flow considerably (e.g. from < 10 cm s^{-1} to > 20 cm s^{-1} along the southern flank of the bank (Loder and Wright, 1985)). During the summer stratified period far weaker, but as we shall see vitally important, cross-frontal residual flows have been observed and modelled (Lough and Manning, 2001; Chen *et al.*, 2003).

These two components of the flow, along-bank transport followed by weak cross-bank transfers, provide the fish eggs and larvae with their conveyor back to the shoals of Georges Bank. The pathway is shown in Fig. 8.21. Both cod and haddock spawn prior to and over the onset period of spring thermal stratification on the northeast end of Georges Bank. The residual along-bank flow then moves the eggs and subsequent larvae along bank isobaths towards the southwest, shown in Fig. 8.21a. This transport has been inferred from tracking the ages of eggs and larvae against the known background flows (Butman and Beardsley, 1987; Lough and Bolz, 1989). Observing the weaker cross-bank flows proved to be more difficult. A combination of residual flow set up by the tidal mixing front (Fig. 8.21b) and the vertical position of the larvae (either transported downward by the frontal downwelling, or actively swimming downward in the case of older larvae) is thought to lead to transport of larvae into the shoals over Georges Bank. The tidal mixing front thus plays a key role, partially by augmenting the along-isobath transport of larvae and then by providing a mechanism for cross-isobath transport.

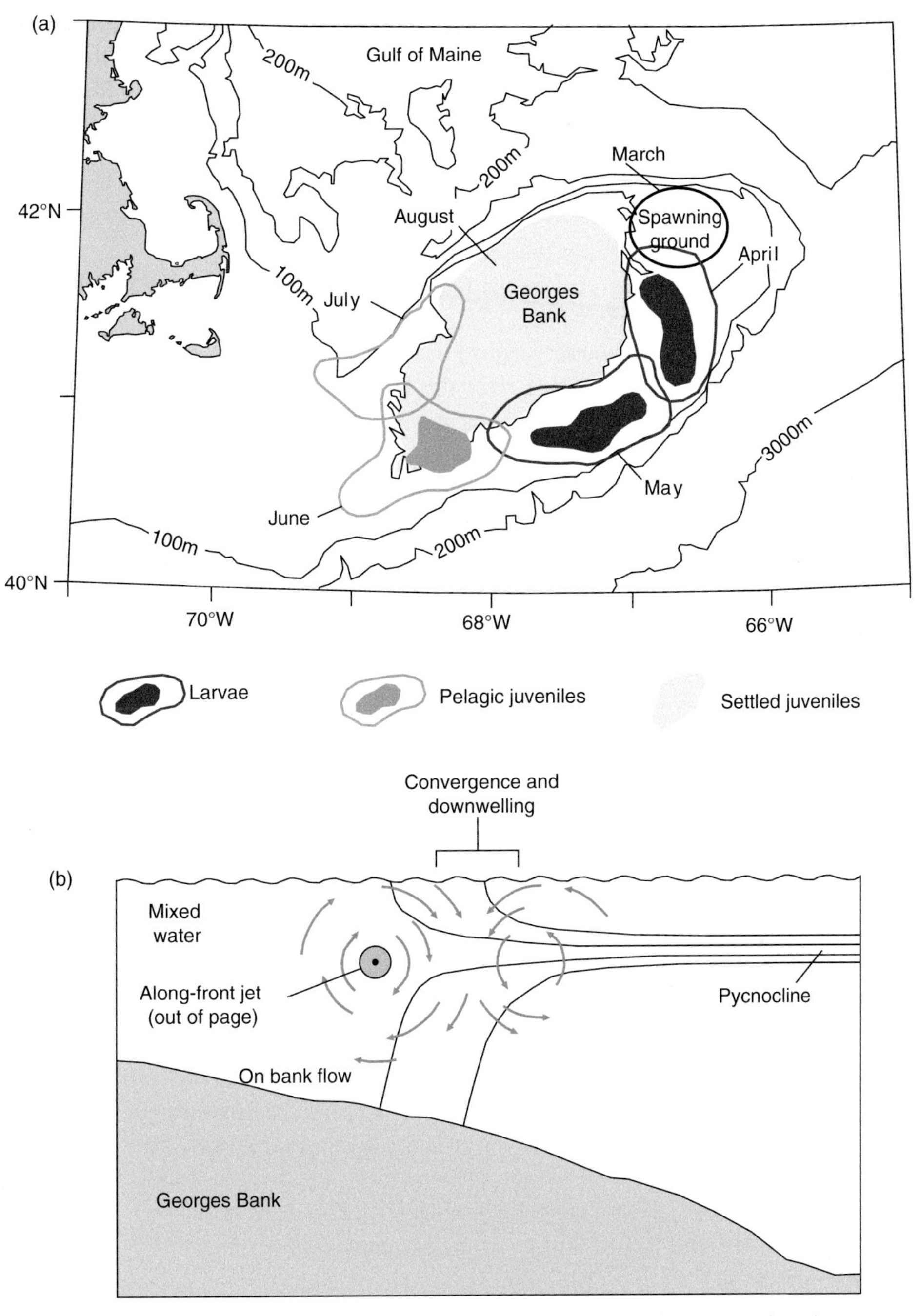

Figure 8.21 Route followed by fish larvae from the spawning ground to eventual settlement on Georges Bank (based on Lough and Manning, 2001, with permission from Elsevier). (a) The path followed by developing fish larvae within the along-bank residual jet (see Fig. 8.7a), transferring them from the spawning ground to eventual settlement on the Bank. (b) Schematic of the mechanism that transfers larvae from the jet onto the bank. Arrows indicate the directions of the residual flows within the front.

While the above summary suggests that the basic requirements of the fish stock are being provided by the tidal environment over Georges Bank, there are likely to be large inter-annual variations associated with changes to the density field by the transport of the nearby Scotian Shelf water (Townsend and Pettigrew, 1996).

The examples from the western Irish Sea and Georges Bank make no mention of the impact of frontal primary production on the survival of the fish stocks. Food must of course be important. In the case of *Nephrops* they are likely to be dependent on organic material reaching the seabed from the surface spring bloom and the SCM over the stratified region. For the Georges Bank, haddock and cod appear to use the mixed shoals over the Bank as the feeding area. The absolute food concentration is thought to be less than would normally be required, but the effective concentration is increased by the turbulent mixing over the bank leading to greater predator-prey contact rates (Werner *et al.*, 1996), a mechanism described in Section 5.2.3 (see Fig. 5.17). Tidal mixing fronts appear to be most important to larger organisms via the circulation which they induce, rather than the primary production that they support.

Summary

Shelf sea fronts, or tidal mixing fronts, mark the boundaries between regions that are always vertically well-mixed and those that stratify (thermally) during spring and summer. The positions of the tidal mixing fronts can be explained by assessing the relative impacts of the supply of heat through the sea surface and of mixing generated by the friction between tidal currents and the seabed. This leads to the prediction that fronts should follow contours of a critical value of h/u^3, with h the water depth and u some measure of the strength of the tidal currents. Analysis of frontal positions in ship surveys of temperature structure and in satellite images of sea surface temperature confirms the validity of this prediction. The density field at a front leads to the development of an along-front jet with a weak attendant cross-frontal circulation. For some fronts, the jet is unstable and this causes the development of baroclinic meanders and eddies, which can transfer water constituents between the front and the mixed water.

As well as marking the physical boundary between the biologically very distinct mixed and stratified regions, the fronts have their own biological characteristics arising from processes particular to the frontal zone. The enhanced growth of phytoplankton frequently observed at fronts results from a number of mechanisms, the most important of which are probably:

(1) A nutrient supply, greater than in strongly stratified regions, brought about by enhanced vertical mixing in the frontal zone.
(2) A fortnightly movement of the frontal boundary due to variation in stirring over the spring-neap cycle. This adjustment of the front involves the alternate mixing and re-stratification of a band, typically 2–4 km, in width extending along the front.
(3) Eddy transfers of properties across the front.

The evidence to support a simple link from the mixing, through the primary production, to the zooplankton and on up to the larger organisms is not strong. For zooplankton there is evidence for increases in reproduction rates (e.g. eggs produced per female) in response to the elevated food supply, but observations of increased densities of adults at fronts seem to be infrequent. This is probably because the combination of frontal circulation and the developmental time between zooplankton egg and a new breeding adult leads to a de-coupling between the primary producers and the adult zooplankton population. It should also be recognized that aggregation of zooplankton at a front could be occurring mainly within the small region of convergent flow, and could therefore have been missed by ship-based sampling procedures. Animals from fish larvae and juvenile fish, to basking sharks, and to seabirds have been seen to forage at fronts, particularly at the convergence zones where their zooplankton prey aggregate.

Fronts are also recognised as important for many commercially exploited fish stocks. The underlying reason may be linked, not so much to food supply at a front, as to the need for a mean flow from a spawning ground back to the adult feeding area, so that the free-drifting eggs and early stage fish larvae are not lost within the otherwise dispersive marine environment. The role of the baroclinic jets and the cross-frontal circulation in this context is illustrated in examples from the Irish Sea and Georges Bank.

FURTHER READING

Circulation and Fronts in Continental Shelf Seas, ed. J. C. Swallow, R. I. Currie, A. E. Gill and J. H. Simpson. The Royal Society, London, 1981.

Simpson, J. H., and J. Sharples. Does the Earth's rotation influence the location of the shelf sea fronts? *Journal of Geophysical Research*, **99**(C2), 3315–3319, 1994.

Franks, P. J. S. Phytoplankton blooms at fronts: patterns, scales and physical forcing mechanisms. *Reviews in Aquatic Sciences*, **6**(2), 121–137, 1992.

Loder, J. W., and T. Platt. Physical controls on phytoplankton production at tidal fronts. In *Proceedings of the 19th European Marine Biology Symposium*, 3–21, 1985, Cambridge University Press.

Problems

8.1. The density distribution on a section perpendicular to a shelf sea front located in latitude 50° N is shown in the figure below. Given that the currents are in steady geostrophic balance with the density field, estimate the speed and direction of the along-front flow in the region of strongest horizontal density gradients and present your result as a sketch of the velocity profile assuming that the velocity is zero at the seabed.

If the flow were to become unstable, what would you expect to be the dominant scale of the instability?

(Hint: From the figure estimate the horizontal gradient of density above and below a depth of 30 m. Use these values with the thermal wind equation to determine the vertical shear and hence the current profile.)

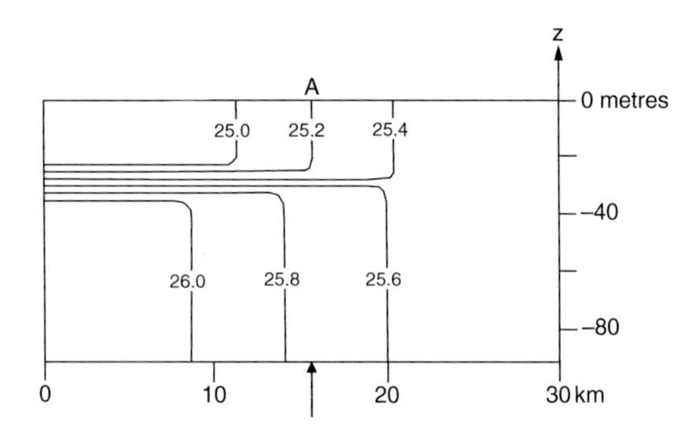

8.2. In a shelf sea region, where stirring is mainly due to tidal forcing, the potential energy anomaly $\Phi(y, t)$ can be represented as a function of $y = h/u^3$ and the time t which has elapsed since the spring equinox (March 21). If $\Phi_0(t)$ is the corresponding value of Φ which would occur in the absence of stirring, show that the efficiency of mixing e can generally be expressed as

$$e = \frac{y(\Phi - \Phi_0)}{C_d \rho t}$$

where k is the drag coefficient. Hence, show that e can be related to its value e_f at a TM front where $y = y_f$ and $\Phi \to 0$ by:

$$\frac{e}{e_f} = \left(1 - \frac{\Phi}{\Phi_0}\right)\frac{y}{y_f}.$$

On June 30, when $\Phi_0(t) = 205$ J m^{-3}, observations showed that $\Phi = 185$ J m^{-3} at a stratified position where $y = 2100$ m^{-2} s^{-3}. Compare the value of e at the stratified site with that at the front where $y_f = 400$ m^{-2} s^{-3} and suggest an explanation of your result.
(Hint: Start from an integrated form of the Φ equation (e.g. 6.24) and discount wind stirring. Use $C_d = 0.0025$ and $\rho = 1026$ kg m^{-3}.)

8.3. The kinetic energy input to baroclinic instability of fronts comes from the release of potential energy from the density field by the exchange of water particles across the front. In a section normal to a front, the isopycnals slope uniformly at an angle β to the horizontal. Show that the energy released by exchanging two particles of unit volume along paths inclined at an angle θ is given by:

$$\Delta \text{PE} = \frac{\partial \rho}{\partial z} g L^2 \tan\theta(\tan\beta - \tan\theta)$$

where $\partial\rho/\partial z$ is the vertical density gradient and L is the horizontal displacement of the particles. Hence show that the maximum energy release occurs when

$$\tan\theta = \frac{1}{2}\tan\beta$$

i.e. when the particles are exchanged along surfaces with half the slope of the isopycnals.

8.4. Satellite I-R data indicate that the displacement of TM fronts over the springs-neaps cycle corresponds to a change in the tidal mixing parameter of $\Delta SH \approx 0.25$. Use this information to determine the ratio of the amplitudes of the M_2 tidal stream at the extremes of the displacement cycle of a front occurring in a region of uniform bottom depth (h = constant). Hence estimate the extent of the displacement of the front in kilometres if the M_2 tidal stream amplitude u_{M2} varies as $u_{M2} = u_s(1 + \gamma x)$ where x is the distance normal to the front, u_s is the amplitude at the position of the front following springs where $x = 0$ and the constant $\gamma = 9 \times 10^{-5}\ \mathrm{m}^{-1}$.

8.5. Just after spring tides, the surface water on the mixed side of a TM front is found to have a nitrate concentration of 1 mmol m^{-3}. Noting any assumptions that you make, estimate:

(i) The mean rate of primary production ($\mathrm{g\,C\,m^{-2}\,d^{-1}}$) supported by the spring-neap frontal adjustment.
(ii) The contribution to the annual phytoplankton carbon fixation ($\mathrm{g\,C\,m^{-2}\,a^{-1}}$) made by the spring-neap adjustment.

8.6. Treat the instability scale at the front in Problem 8.1 as an estimate of the eddy size.

If the mixed water has a nitrate concentration of 2 mmol m^{-3}, estimate the total carbon that could be fixed when the eddy transfers nitrate into the stratified side of the front.

9 Regions of freshwater influence (ROFIs)

As we saw in Chapter 2, the input of buoyancy over much of the shelf seas is dominated by heating and cooling through the sea surface. There are additional exchanges of buoyancy at the sea surface through the processes of evaporation and precipitation, both of which modify the salinity of surface layers, but in temperate latitudes their contribution is generally small in comparison with heat exchange. In areas of the shelf adjacent to estuaries, however, freshwater discharge from rivers can make a dominant contribution to buoyancy input (Section 2.3) and maintain strong horizontal gradients of salinity.

In this chapter, we shall consider these Regions Of Freshwater Influence (ROFIs) (Simpson, 1997), the suite of processes which operate within them and the distinctive environment that results. In terms of physical processes, ROFIs have much in common with estuaries, but they differ insofar as estuaries are confined by land barriers and are generally of a smaller horizontal scale than that which is characteristic of ROFIs. In many cases, the circulation in an estuary is predominantly an along-channel flow while transverse motions are relatively small and the effects of the Earth's rotation can be neglected. We shall initially develop the theory of density-driven circulation for such a simplified, non-rotating system before proceeding to consider how rotation modifies the density-driven circulation in ROFIs. We shall next examine the way in which the density-driven circulation is involved in determining the water column structure of ROFIs in competition with stirring and then explore the implications for the biology and environmental health of ROFIs.

9.1 Freshwater buoyancy and estuarine circulation

The competition between buoyancy input and stirring in estuaries and ROFIs is somewhat analogous to that between surface heating and stirring discussed in Chapter 6, but there is an important difference. Whereas buoyancy input due to heat exchange is more or less spatially uniform, freshwater input from rivers enters through the lateral boundaries and is distributed by buoyancy-driven flows in a manner which is common to estuaries and ROFIs. This re-distribution of buoyancy

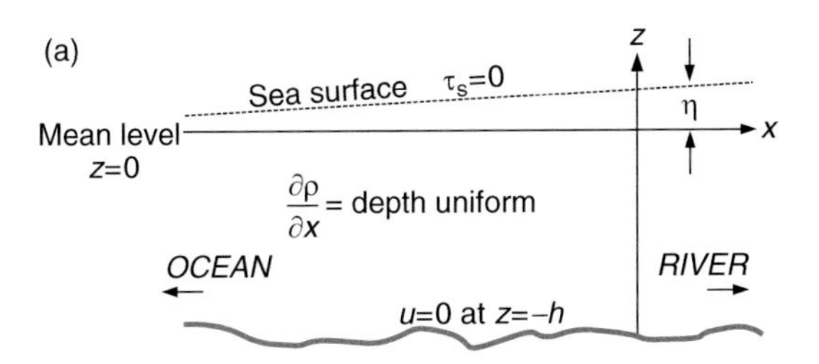

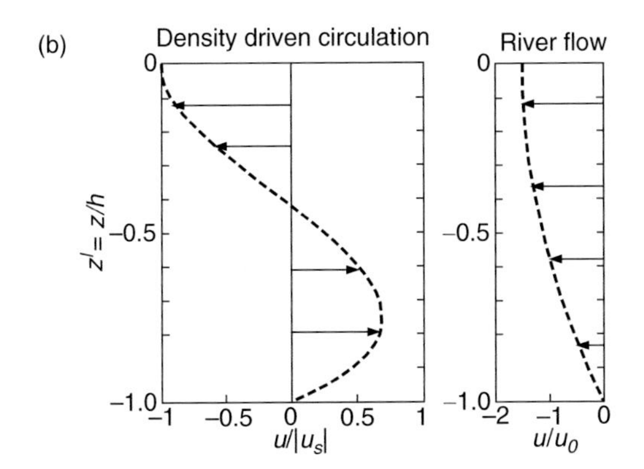

Figure 9.1 (a) Frame of reference used in the calculation of the residual flow using Equation (9.10). (b) The two components of the residual flow described by Equation (9.10).

makes the problem much more difficult than its heating-stirring counterpart and requires that we examine the spreading process in some detail. This we do, initially, by developing a simple model of the density-driven circulation and its interaction with tidal flow in a narrow estuary where, because of the smaller scale, we can neglect the effects of rotation.

9.1.1 The estuarine exchange flow

The flow of freshwater from a river into an estuary sets up and maintains horizontal gradients of salinity and hence of density. The density gradient implies a pressure gradient which drives a circulation. In a narrow estuary, where rotation is not important, we can determine the form of the circulation following a similar approach (Officer, 1976) to that of the classical calculation by Hansen and Rattray (Hansen and Rattray, 1966).

In this simple model of a narrow estuary, we assume that the horizontal density gradient is independent of depth and directed in the x direction which is parallel to the axis of the estuary, as illustrated in Fig. 9.1a. Note that this important assumption applies, not only in the case of a vertically mixed estuary, but also to estuaries with stratification if the vertical structure is uniform along the estuary.

We assume further that rotation can be neglected (Coriolis forces = 0) and that the flow is in steady state. Under these conditions, the momentum Equation (3.13) simplifies to a balance between the pressure gradient and the stress divergence, i.e.

$$\frac{\partial p}{\partial x} + \frac{\partial \tau_x}{\partial z} = 0. \tag{9.1}$$

Given that the pressure in such a steady flow will be hydrostatic (Equation 3.21), the pressure gradient can be written as:

$$\begin{aligned}\frac{\partial p}{\partial x} &= \frac{\partial}{\partial x}\left(\int_z^\eta \rho g dz\right) = \frac{\partial}{\partial x}\left(\int_z^0 \rho g dz + \int_0^\eta \rho g dz\right) \\ &= \int_z^0 \frac{\partial \rho}{\partial x} g dz + \rho_s g \frac{\partial \eta}{\partial x} = -gz\frac{\partial \rho}{\partial x} + \rho_s g \frac{\partial \eta}{\partial x}\end{aligned} \quad (9.2)$$

where, in the final step, the depth-uniform $\partial\rho/\partial x$ has been taken outside the integral and ρ_s is the density at the surface. Assuming (Equation 3.40) that the stress is related to the velocity shear by $\tau_x = -N_z\rho_0\frac{\partial u}{\partial z}$ where N_z is the eddy viscosity, the force balance can now be written as:

$$\frac{\partial p}{\partial x} = -gz\frac{\partial \rho}{\partial x} + \rho_s g\frac{\partial \eta}{\partial x} = -\frac{\partial \tau_x}{\partial z} = \rho_0 \frac{\partial}{\partial z}\left(N_z \frac{\partial u}{\partial z}\right). \quad (9.3)$$

If N_z is independent of depth and making the reasonable approximation that $\rho_s \approx \rho_0$ the reference density, this can be arranged to give:

$$\frac{\partial^2 u}{\partial z^2} = \frac{g}{N_z}\chi - \frac{g}{N_z}\xi z \quad (9.4)$$

where $\chi = \partial\eta/\partial x$ and $\xi = (1/\rho_0)\partial\rho/\partial x$. After integrating twice with respect to z, we have:

$$u(z) = \frac{g\chi}{2N_z}z^2 - \frac{g\xi}{6N_z}z^3 + Az + B \quad (9.5)$$

where A and B are constants of integration to be determined from the boundary conditions: (i) at the surface ($z = 0$) the stress $\tau_s = 0$ and (ii) at the bottom ($z = -h$) the velocity $u = 0$. Applying these conditions gives the constants as:

$$A = 0; \quad B = -\frac{g\chi h^2}{2N_z} - \frac{g\xi h^3}{6N_z} \quad (9.6)$$

so that the solution for u is:

$$u(z) = \frac{g\chi}{2N_z}(z^2 - h^2) - \frac{g\xi}{6N_z}(z^3 + h^3). \quad (9.7)$$

Note that an alternative bottom boundary condition, setting the bed shear stress to match the bottom stress given by a quadratic drag law $\tau_b = -k_b\rho|u_b|u_b$, yields similar results which you can see in a numerical simulation at the website. In a steady state, the gradients χ and ξ are not independent (Officer, 1976).

In a steady state, which you can see in a numerical simulation at the website, the gradients χ and ξ are not independent. They are related through the condition that the net transport must match the freshwater input R_w per unit width from the river, i.e.

$$R_w = \int_{-h}^{0} u dz = -\frac{g\chi h^3}{3N_z} - \frac{g\xi h^4}{8N_z} \tag{9.8}$$

which means that the surface slope χ must be related to ξ according to

$$\chi = -\frac{3}{8}\xi h - \frac{3N_z R_w}{gh^3} \tag{9.9}$$

and the velocity profile, which we will write in terms of the fractional depth $z' = z/h$, becomes:

$$\begin{aligned} u(\zeta) &= \frac{g\xi h^3}{48N_z}(1 - 9z'^2 - 8z'^3) - \frac{3}{2}\frac{R_w}{h}(z'^2 - 1) \\ &= u_s(1 - 9z'^2 - 8z'^3) \qquad - \frac{3}{2}u_0(z'^2 - 1) \\ &\quad \text{density-driven} \qquad\qquad \text{river flow.} \end{aligned} \tag{9.10}$$

The resulting circulation is shown in Fig 9.1b, and consists of a component driven by the density gradient ξ with surface velocity seaward of $u_s = g\xi h^3/48N_z$ and an average downstream flow of $u_0 = R_w/h$ due to the river discharge R_w (in units of $m^2 s^{-1}$). The density-driven component is landwards below $z = -0.42h$ with a maximum speed of $0.69u_s$ at $z = -0.75h$. This flow will interact with the density gradient to move lower salinity water seaward in the surface layer and denser, more saline water landward in the lower layer and thus induce stratification.

It should be remembered that the form of the estuarine circulation which we have derived applies to estuaries which may be vertically well mixed or stratified, the only requirement being that the horizontal gradient of density is independent of depth. The velocity profile represented by Equation (9.10) was used by Hansen and Rattray to derive a corresponding steady state salinity profile and from the combination of the two profiles they established a widely used system of estuarine classification (Hansen and Rattray, 1966).

9.2 Density-driven circulation in a ROFI: rotation and coastal currents

The arguments of Section 9.1 for the density-driven circulation in an estuary can be extended to a ROFI system where the horizontal flow is no longer restricted to one dimension by the topography of the estuary. The response to density forcing is then free to take a two-dimensional form. The increase in horizontal scale, previously set by the width of the estuary, means that we must now take account of the Coriolis forces. We will first look at what happens to the low salinity surface outflow as it leaves the confines of the estuary, and then re-visit the estuarine exchange problem but this time including the Coriolis term.

9.2.1 Coastal buoyancy currents

As low salinity water from the estuary enters the shelf sea on an open coast, the Coriolis force tends to deflect it back towards the coast and to establish a flow parallel to the coast. In the northern (southern) hemisphere, such *coastal currents* or river plumes move with the land boundary to the right (left) of the direction of flow, i.e. in the direction of propagation of a Kelvin wave (see Section 3.6.2). Except near the equator, where $f \rightarrow 0$, the influence of the Earth's rotation frequently dominates, and discharge from an estuary results in a current flowing along the coastal boundary. This flow transports the buoyancy input away from the estuary source and distributes it into the ROFI where it is mixed into the water column at a rate depending on the level of stirring imposed by the tides and wind stress.

A well-defined example of such a density-driven current is the Norwegian Coastal Current (NCC) which flows northwards along the Norwegian coast with a transport of up to 3 Sv (3×10^6 m^3 s^{-1}). This current is mainly driven by a large input of brackish water which enters from the Baltic Sea via the Skagerrak (the 'estuary' in this case). The flow of this low salinity water along the coast is clearly evident in Fig. 9.2, which shows the salinity distribution of the northern North Sea and a salinity section normal to the coast (Mork, 1981; Rodhe, 1998). The current, which is ~75 km wide, flows mostly over the deep water (>200 metres) of the Norwegian trench which lies close to the coast. Because the water depth generally exceeds that of the current, the flow is scarcely influenced by frictional stresses at the seabed so that, in the absence of wind stress, we would expect the current to be in geostrophic balance. To a first approximation this seems to be the case, although the dynamics are complicated by instability of the current. Satellite imagery (Mork, 1981) shows that the flow can be unstable and develops large meanders and eddies with scales of ~80 km. This baroclinic instability is essentially the same as that which we found can occur in the frontal jets of TM fronts (Section 8.4) and which has been studied in laboratory experiments (Griffiths and Linden, 1981). Although the eddies resulting from the instability induce some horizontal mixing which tends to broaden the current, the large water depth means that the brackish water is far from the stresses at the bottom boundary and is only slowly mixed downwards. This limited vertical mixing, together with some reinforcement of the flow by other freshwater sources along the coast, allows the flow to remain as a distinct low-salinity feature for ~1000 km along the Norwegian coast.

More usually, the depth near the coast is shallower and the coastal current interacts with the seabed generating significant frictional stresses which modify the flow. This is the situation along the Dutch coast in the North Sea, shown in Fig. 9.3a, where the estuarine discharge from the Rhine, carrying the large freshwater input from the river (~2000 m^3 s^{-1}), is deflected to the right as it mixes with the ambient water of the North Sea and establishes a persistent, density-driven current flowing northwards along the coast. This Rhine ROFI is located in water of depth 10–20 metres and exhibits a salinity deficit reaching $\Delta S \sim 4$ near the coast. It extends ~30 km from the coast and, in the along coast direction, more than 200 km into

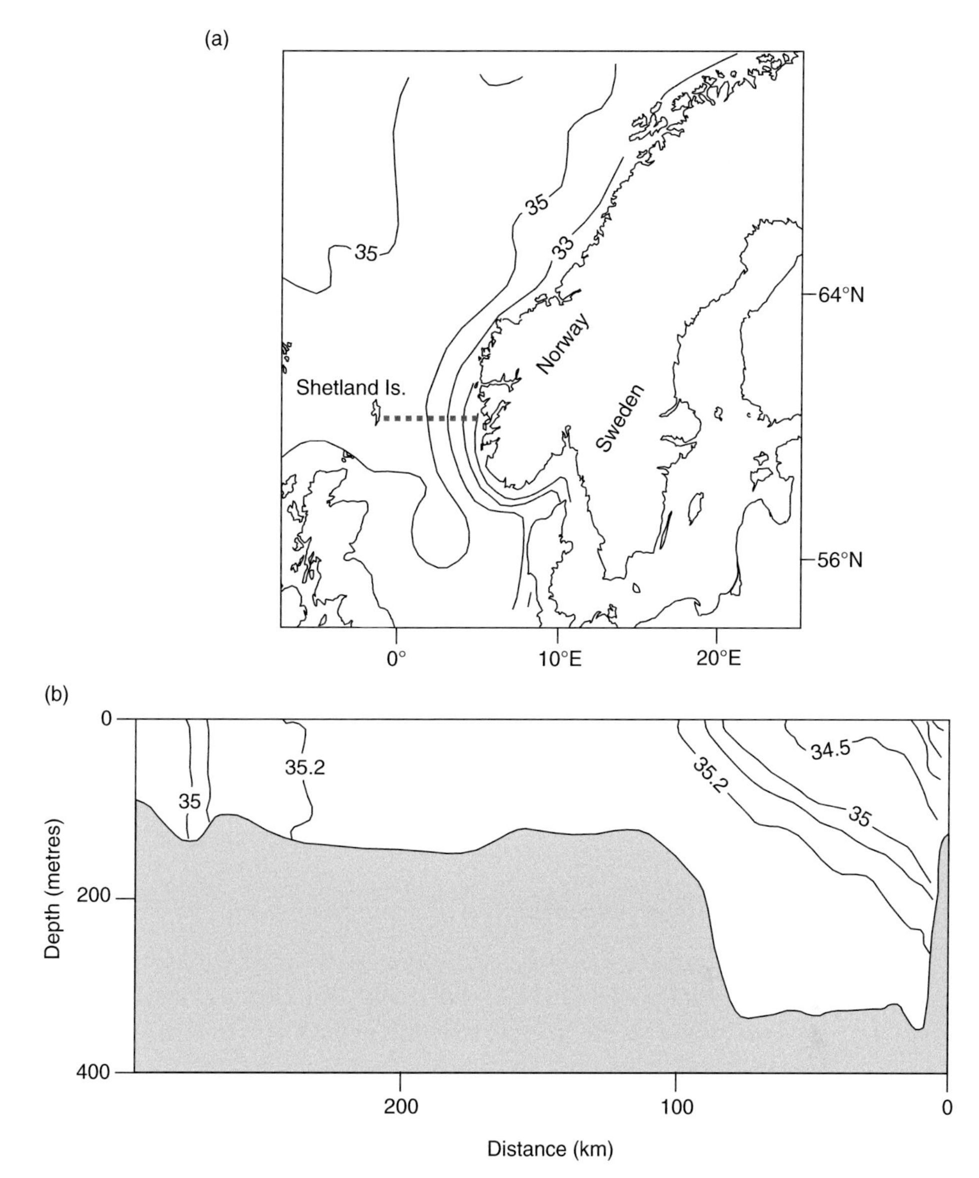

Figure 9.2 The Norwegian Coastal Current: (a) Average surface salinity for the northern North Sea in May showing low salinity water flowing northward along the Norwegian coast. After (Mork, 1981); (b) Salinity distribution in February on a vertical section (grey dashed line shown in (a)) from Shetland to the Norwegian coast. The coastal current, flowing into the page, can be seen over the deep water of the Norwegian Trench. After (Rodhe, 1998), courtesy of the Royal Society, London.

the German Bight where other freshwater inputs contribute to maintaining the salinity deficit of the flow.

Figure 9.3b shows that the freshwater input from the Rhine also delivers large quantities of nutrients whose influence dominates in the coastal current, as can be

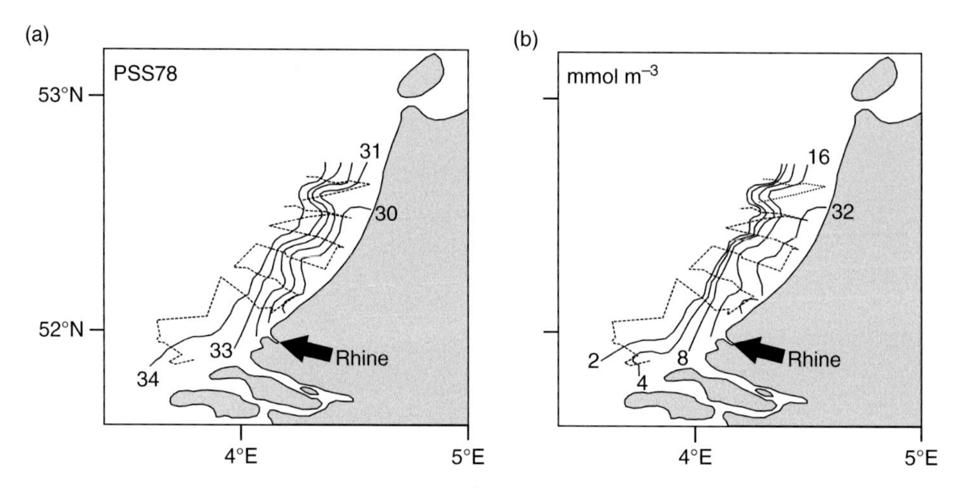

Figure 9.3 The surface distribution of (a) salinity and (b) nitrate off the coast of the Netherlands, as a result of the Rhine inflow. The dotted line is the course followed by the research vessel during sampling. Figure courtesy of D. J. Hydes, National Oceanography Centre, UK.

seen in the nitrate distribution. Nitrate levels in the coastal current are more than an order of magnitude greater than in the North Sea outside the ROFI and, as the plots show, there is a close correlation of salinity and nitrate. This high nutrient loading is an important feature of many ROFIs, and we shall return to consider its impact on the ROFI environment in Section 9.8.3.

9.2.2 Residual flows in a ROFI

Let us now have a look at the solution to the estuarine circulation problem, but this time including the effects of the Earth's rotation. As well as the influence of the Coriolis force we might expect the flow in the ROFI to be affected by frictional forces as in the estuarine case. The simplest model which allows for the combined influences of rotation and friction is that of Heaps (Heaps, 1972) in which it is assumed that the flow is forced by a depth-uniform horizontal gradient in the x direction which is normal to the coast, as shown in Fig. 9.4. Conditions do not vary in the along coast direction so that derivatives in y are zero (i.e. $\partial/\partial y = 0$) and frictional stresses in x and y are again controlled by the eddy viscosity. With these assumptions the equations of motion (Equation 3.13) for a steady state with hydrostatic pressure (Equation 9.2) become:

$$
\begin{aligned}
fv - \frac{1}{\rho_0}\left(\frac{\partial p}{\partial x} + \frac{\partial \tau_x}{\partial z}\right) &= fv - g\frac{\partial \eta}{\partial x} + gz\frac{1}{\rho_0}\frac{\partial \rho}{\partial x} + N_z\frac{\partial^2 u}{\partial z^2} = 0 \\
-fu - \frac{1}{\rho_0}\frac{\partial \tau_y}{\partial z} &= -fu + N_z\frac{\partial^2 u}{\partial z^2} = 0.
\end{aligned}
\tag{9.11}
$$

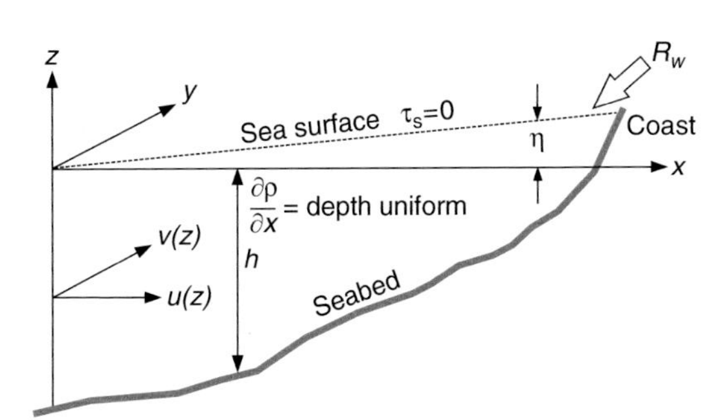

Figure 9.4 Frame of reference for the analytical solution of Heaps (Heaps, 1972) to density-driven flow influenced by the Earth's rotation.

Continuity requires that the net transport normal to the coast is balanced by the net input of river water R_w per unit length of the coast:

$$\int_{-h}^{0} u dz = R_w \tag{9.12}$$

and the solution is closed by boundary conditions, which are:

(i) At the surface ($z = 0$) the stresses $\tau_x = \tau_y = 0$;
(ii) At the bottom ($z = -h$) stresses follow a linear drag law: $\tau_x = -k_b^{'}\rho u_b$; $\tau_y = -k_b^{'}\rho v_b$ where u_b and v_b are the velocity components at the seabed.

The mathematical details of Heaps' analytical solution are rather intricate and lead to algebraically complicated expressions for the velocity profile. The equations are, however, readily solved numerically (see Heaps' and ROFI modules at the website). The essential result shown in Fig. 9.5a is an along-gradient component of circulation, analogous to the estuarine circulation, plus a component of flow perpendicular to the gradient which increases with height above the bed. With an appropriate choice of the eddy viscosity ($N_z = 2.5 \times 10^{-3}$ m^2 s^{-1}), the Heaps theory is seen to be in fair accord with the observed structure of the mean sub-tidal flow in the Rhine ROFI; look at the correspondence between the Heaps solution and the ADCP data in Fig. 9.5a. The persistence and the offshore extent of the flow are illustrated in Fig 9.5b, which shows a 16-day average of the surface flow measured by HF radar (Souza, Simpson, *et al.*, 1997) together with the surface salinity contours. Inside the region of salinity deficit (S< 34.5) almost all vectors indicate flow to the north-east with speeds up to ~0.15 m s^{-1}.

The balance between the alongshore and cross-shore components of the circulation is set by the ratio of the frictional forces to the Coriolis force, known as the *Ekman number* $Ek = N_z/(fh^2)$. For large friction (high *Ek*) the flow reverts to an estuarine circulation. By contrast for low *Ek*, the flow is controlled by a geostrophic balance and is a pure coastal current parallel to the isopycnals, in which the sea surface slopes upwards to the coast according to:

$$\frac{\partial \eta}{\partial x} = -\frac{h}{\rho_0}\frac{\partial \rho}{\partial x} \tag{9.13}$$

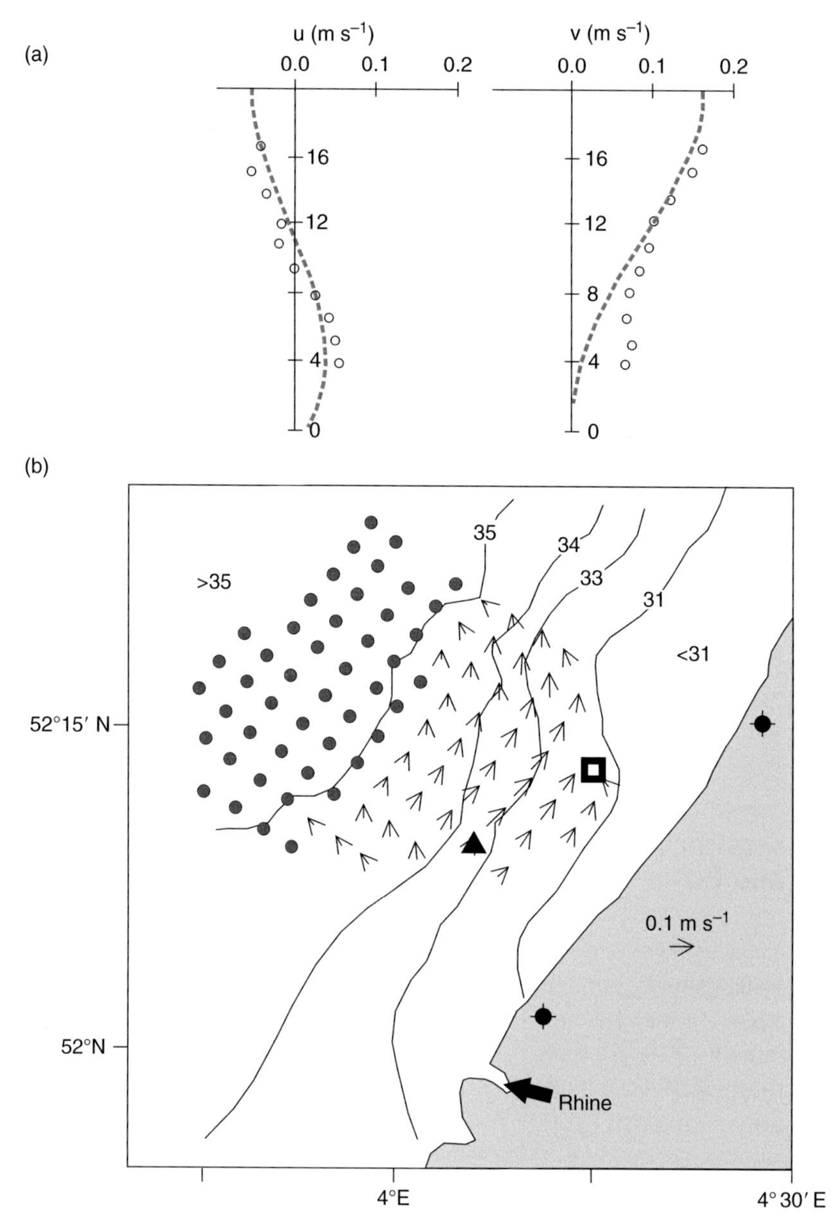

Figure 9.5 (a) The Heaps solution applied to the Rhine ROFI. The open circles represent long-term average current profiles versus height above bed (m), measured by an ADCP moored close to the seabed. The dashed lines are the prediction of the Heaps solution with $N_z = 2.5 \times 10^{-3}\ \mathrm{m^2\ s^{-1}}$. The ADCP was situated at the position marked □ in (b). (b) Surface flow in the Rhine ROFI, measured by HF radar and averaged over 16 days, after Souza *et al.*, 1997, courtesy *Journal of Marine Research*. Dashed lines are salinity isolines. ⌖ show the positions of the radar antennae. The position marked ▲ is that of a mooring used to collect the data shown in Fig. 9.11.

so that the alongshore surface velocity is just:

$$v_s = \frac{1}{f\rho_0}\frac{\partial p}{\partial x} = \frac{g}{f}\frac{\partial \eta}{\partial x} = -\frac{gh}{f\rho_0}\frac{\partial \rho}{\partial x}. \tag{9.14}$$

In such a frictionless flow with cross-shore density gradients typical of the Rhine ROFI, the alongshore surface flow would be $v_s \sim 0.3$ m s^{-1}. This is about twice that observed, while at the same time there is an observed cross-shore flow of $u \sim 0.05$ m s^{-1}; both of these observations indicate the influence of friction.

The presence of a persistent and pronounced alongshore flow under density forcing is characteristic of many ROFIs, but not all conform to this pattern all the time. Wind influence on along-shelf flow can alter the speed and the structure of the plume of low salinity water as it moves along the coast. For instance, the coastal current flowing southwards from the mouth of the Chesapeake Bay, in the eastern United States, has been found to be the sum of buoyancy- and wind-driven currents, with moderate winds affecting the vertical structure and width of the buoyant plume (Lentz and Largier, 2006). The wind stress and ambient flows may, at times, outcompete the density forcing and may even reverse the coastal flow. An extreme example is the plume of the Amazon, a ROFI notable for the huge freshwater discharge and for it being free of Coriolis influence at 0° latitude. There, the outward spreading of the low salinity plume is strongly influenced by the North Brazil Current (Geyer *et al.*, 1996). We will return to the influence of wind forcing on ROFIs in Section 9.6.

9.3 Stratification control: circulation versus stirring in estuaries and ROFIs

The tendency of the estuarine circulation is to promote stratification by moving lighter water over heavier water. This will be opposed by stirring due to the tidal flow and wind stress. In many estuaries and ROFIs, tidal stirring predominates, so we shall focus initially on the competition between the stratifying effect of the estuarine circulation and the mixing induced by the tidal flow. We will initially look at the detail of this competition in an estuary, the simplest case, where the density-driven flow is 2D and can be assumed to follow the results of Section 9.1 for a non-rotating estuarine system. Then we will go on to describe observations of the buoyancy-stirring interaction in ROFIs.

9.3.1 The buoyancy-stirring competition in an estuary

Let us take an approach that is analogous to the description of the heating stirring competition that we described in Chapter 6. As with the heating-stirring problem, we shall utilise the potential energy anomaly, Φ, as the state variable for stratification. From Equation (6.1) we have for the time derivative of Φ:

$$\frac{\partial \Phi}{\partial t} = \frac{g}{h}\int_{-h}^{0}\left(\frac{\partial \hat{\rho}}{\partial t} - \frac{\partial \rho}{\partial t}\right) z\,dz. \tag{9.15}$$

If flow is in the x direction only and we assume that density changes due to advection are dominant, then the advection diffusion equation (Section 4.3.3) simplifies to

$$\frac{\partial \rho}{\partial t} = -u\frac{\partial \rho}{\partial x}. \tag{9.16}$$

On taking the vertical average and again assuming $\partial\rho/\partial x$ is independent of z, we have also that:

$$\frac{\partial \hat{\rho}}{\partial t} = -\hat{u}\frac{\partial \rho}{\partial x} \tag{9.17}$$

where $\hat{u}$ is the depth mean velocity. Substituting from Equations (9.16) and (9.17) into (Equation 9.15) gives us:

$$\frac{\partial \Phi}{\partial t} = \frac{g}{h}\frac{\partial \rho}{\partial x}\int_{-h}^{0}(u-\hat{u})z\,dz = gh\frac{\partial \rho}{\partial x}\int_{-1}^{0}(u-\hat{u})z'\,dz'. \tag{9.18}$$

This equation allows us to evaluate the change in Φ due to advection by any velocity profile $u(z)$. We shall first use it to determine the stratifying effect of the density-driven component of the estuarine circulation of (Equation 9.10):

$$u(z') = u_s(1 - 9z'^2 - 8z'^3). \tag{9.19}$$

On substituting into Equation (9.18) and integrating, we have:

$$\left(\frac{\partial \Phi}{\partial t}\right)_{Est} = \frac{1}{320}\frac{g^2h^4}{N_z\rho_0}\left(\frac{\partial \rho}{\partial x}\right)^2. \tag{9.20}$$

If this stratifying effect is opposed only by tidal stirring, we have an analogous competition to that between heating and stirring (Equation 6.17). In the present case, stratification will be maintained or increase only if

$$\begin{aligned} \left(\frac{\partial \Phi}{\partial t}\right)_{Est} &\geq e\frac{k_b\rho_0\hat{u}^3}{h} \\ \text{or} \quad \frac{1}{320}\frac{g^2h^4}{N_z\rho_0}\left(\frac{\partial \rho}{\partial x}\right)^2 &\geq \frac{4e}{3\pi}\frac{k_b\rho_0\hat{u}_{M2}^3}{h} \end{aligned} \tag{9.21}$$

where e is, again, the efficiency of tidal mixing and the final step involves averaging the stirring over the tidal cycle. If we use an empirical formula for the eddy viscosity $N_z = 3 \times 10^{-3}\hat{u}_{M2}h$ (Bowden *et al.*, 1959), we can further simplify our criterion for the development of stratification to give the condition on the density gradient for stratification to develop as:

$$\frac{1}{\rho_0}\frac{\partial \rho}{\partial x} \geq \frac{\sqrt{0.4ek_b}}{g}\left(\frac{\hat{u}_{M2}}{h}\right)^2 \sim 2.1 \times 10^{-4}\left(\frac{\hat{u}_{M2}}{h}\right)^2 \tag{9.22}$$

using $k_b = 0.0025$ and $e = 0.004$. This criterion for the development of water column stability is somewhat analogous to Equation (6.17), but there is an important difference. In the heating-stirring problem we could reasonably expect to be able to measure the heating buoyancy source, Q_i. In Equation (9.22) we need to know the horizontal density gradient, which is itself a complex result of the interaction between the buoyancy source (R_w, the riverine supply of fresh water) and the buoyancy spreading and mixing within the estuary. The horizontal density gradient is really a part of the 'solution' rather than the external forcing. Nevertheless, if density gradient data are available, this criterion can be a useful guide to the development of water column stability in the estuarine environment.

9.3.2 Observations of stratification in ROFIs

In a ROFI, the x component of the circulation illustrated in Fig. 9.5a acts in the same way as the estuarine circulation to promote stratification in competition with the effects of stirring by wind and tide. Changes in the balance between stratifying and mixing occur on a variety of time scales from seasonal through monthly and fortnightly down to one day or shorter. Seasonal variations in the supply of buoyancy may arise from annual cycles of freshwater discharge. In tropical regions which experience monsoonal cycles, the variation in freshwater discharge may be extreme so that most of the annual input of buoyancy occurs in a short, wet season. Since the annual cycle of heat exchange in the tropics is generally small compared with midlatitudes, the monsoonal spate of freshwater may dominate the seasonal cycle. This is the case, for example, in the Gulf of Thailand (Stansfield and Garrett, 1997; Simpson and Snidvongs, 1998), where the wet season input of buoyancy from the Mekong and other rivers between July and October induces strong haline stratification over much of the Gulf. This is slowly eroded by tidal and wind stirring, but with significant stratification persisting at least until February in areas of weaker tidal stirring. We will return to the interesting questions relating to tropical ROFIs in Chapter 11.

In temperate latitudes, the seasonal cycle of runoff is usually less pronounced (for instance see Fig. 2.9 for the monthly average discharge of the Rhine), so that the changes in stratification are dictated mainly by variations in wind and tidal stirring. The predominance of shorter time scales in the variation of stratification has been well documented for the ROFI in Liverpool Bay. Figure 9.6a shows how the region is bounded by the coasts of northwest England and north Wales where a marked offshore gradient in density is maintained by freshwater input from the Dee, the Mersey and several other rivers along the northwest coast of England. The year-long time series of the surface to bottom density stratification from moored instruments shown in Fig. 9.6b illustrates the way that stratification develops and breaks down on a variety of short time scales. Short episodes of strong stratification ($\Delta\rho \sim 1$–2 kg m^{-3}) occur throughout the year, interspersed with episodes of strong mixing. There is no clear evidence of an annual cycle. This pattern contrasts sharply with the cycle in areas subject to thermal stratification, shown using model data in Fig. 9.6c. As we have seen in Chapter 6, this involves a switch to a stably stratified water column in the spring which persists, usually without break, until the autumn overturn.

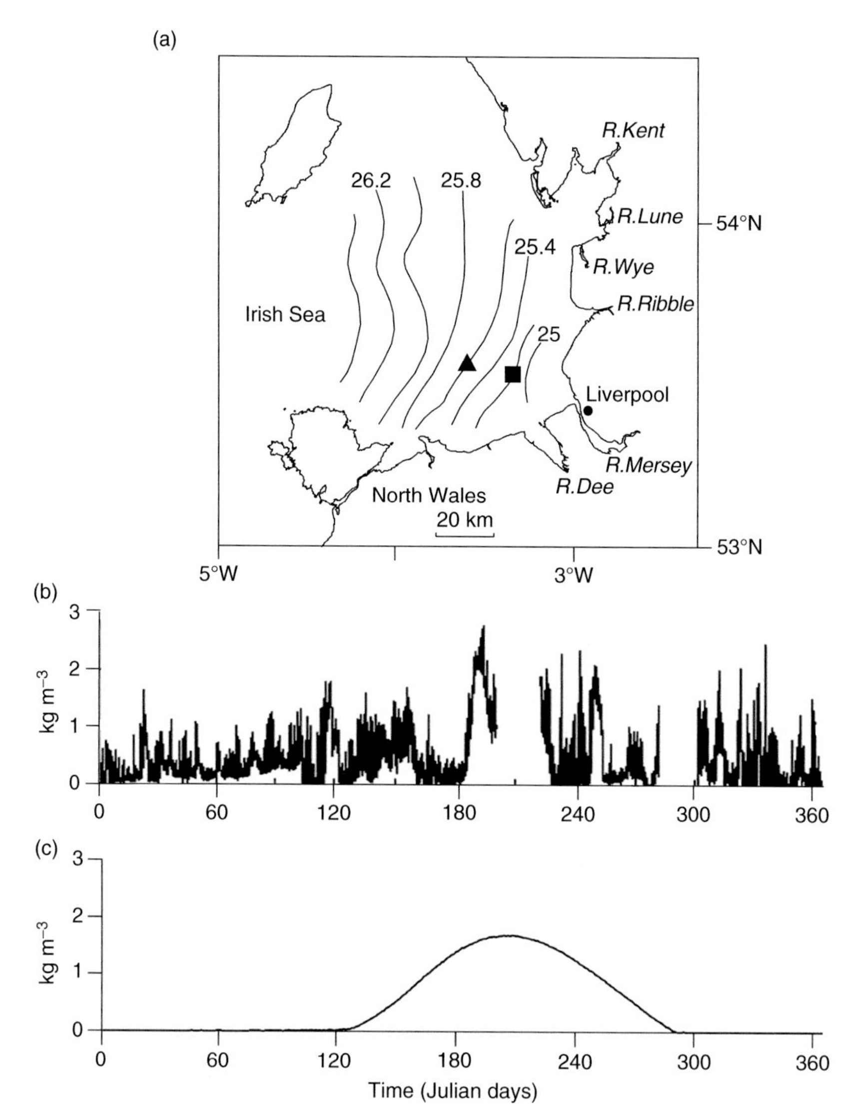

Figure 9.6 (a) A typical map of the distribution of surface isolines of σ_t (kg m^{-3}) in the Liverpool Bay ROFI, (Sharples and Simpson, 1993) courtesy of the Estuarine Research Federation. (b) A year-long time series of the difference between the near-bed and surface density, $\Delta\rho$(kg m^{-3}), measured by a mooring in Liverpool Bay. The position of the mooring is marked as ■ in (a) (from Verspecht *et al.*, 2009, courtesy of Springer). (c) The typical annual progression of $\Delta\rho$ in the nearby Celtic Sea, where stratification is attributable to seasonal changes in surface heat fluxes. Based on the model of Sharples, 2008. In (a) ▲ marks the mooring position for the data in Fig. 9.7.

Much of the variation of the stratification in the Liverpool Bay ROFI is due to irregular changes in runoff from the rivers and in wind stirring, but closer inspection reveals that there is also an important periodic component associated with changes in tidal stirring. Remember how we saw in Chapter 8 that the position of the tidal

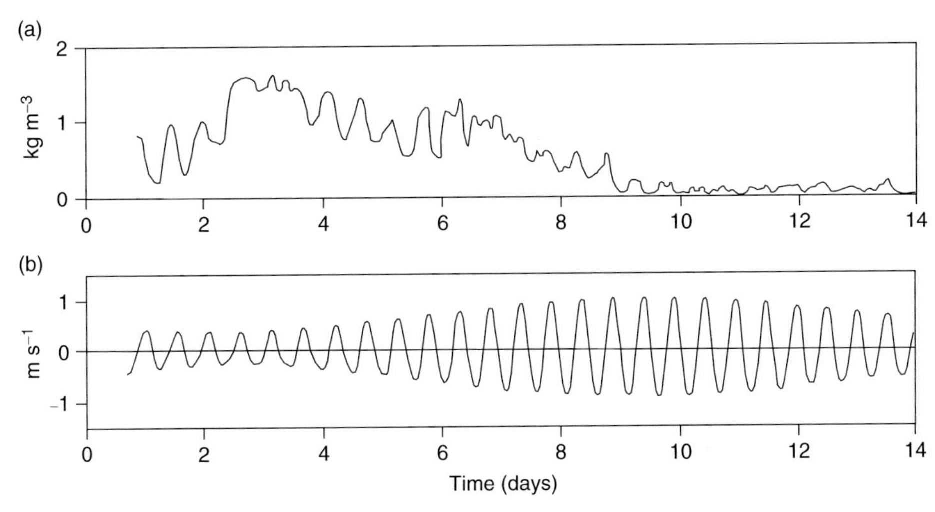

Figure 9.7 (a) Bottom-surface density difference over a neap-spring transition in Liverpool Bay. (b) Depth-averaged tidal currents in Liverpool Bay. Data were collected by the vertical array of current meters shown in Fig. 1.7. The position of the mooring is marked in Fig. 9.6. Adapted from Sharples and Simpson, 1993, with permission from the Estuarine Research Federation.

mixing fronts, which indicates a balance between heating and stirring influences, adjusts over the spring-neap cycle as the tidal currents change. The condition for maintaining freshwater-driven stratification (Equation 9.22) allows for a similar adjustment of the balance as the springs-neaps tidal cycle imposes a regular cycle of change in the stirring. Figure 9.7 shows how this modulation of stirring is frequently observed in a fortnightly cycle of stratification and mixing. During the period around neap tides (day 1–2), reduced stirring has allowed the development of pronounced stratification with surface-bottom density difference $\Delta\rho \approx$ 1–2 kg m^{-3}. As the tides increase towards spring tides (day 9–10), greater stirring results in a complete breakdown of stratification $\Delta\rho \approx 0$ kg m^{-3}. This pattern can be complicated by monthly modulation of the tides due to the N_2 tidal constituent which results in one springs-neaps cycle in each month being stronger than the other (Sharples and Simpson, 1995). The onset of the stratification as stirring decreases is rapid, resulting in the propagation of a front westward through the bay to about 4° W until a geostrophic balance is established.

Box 9.1 Density currents versus stirring in a tank experiment

The interaction between horizontal buoyancy spreading and vertical mixing has been explored in a revealing series of laboratory studies (Linden and Simpson, 1988) based on a modified version of the classical lock-exchange experiment. In the original experiment, a rectangular tank is equipped with a central 'lock gate' which separates two miscible fluids of different density (usually brine and freshwater with different

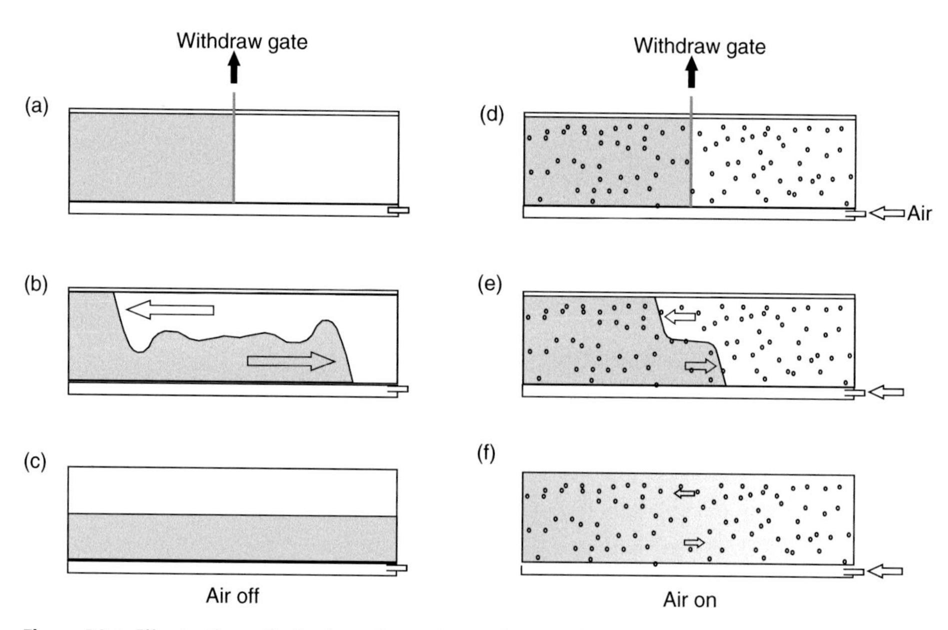

Figure B9.1 Illustration of a 'lock exchange' experiment without (left) and with (right) turbulent mixing generated by bubbling air through the base of the tank. (a, d) Initial laboratory tank set-up, with high density water (grey) separated from low density water (white) by a barrier. (b, e) The barrier is removed, allowing two-way exchange. (c, f) As the experiment progresses, for the case of no mixing the fluid eventually reaches a resting state with stable stratification.

colours added) as seen in Fig. B9.1a. When the gate is removed, a density-driven flow develops with the light fluid moving over the heavier fluid to create a stratified 2-layer flow, shown in Fig. B9.1b. The leading edge of the light water at the surface (termed the 'gravity head') and that of the heavy water at the bed both advance at a speed of $\sim \sqrt{g'D}$ where $g' = g(\Delta\rho/\rho_0)$ is the reduced gravity and D is the layer depth. When the two gravity heads reach opposite ends of the tank, the flow becomes more complicated as momentum is reflected back into the tank at the boundaries. Eventually, the motion dies out due to frictional damping, leaving a stably stratified water column (Fig. B9.1c) with the two layers separated by the small amount of water of intermediate density produced by the mixing which has occurred during the experiment.

In the modified version of the experiment, the tank is fitted with a false bottom which allows air bubbles to be blown into the tank, illustrated in Fig. B9.1d. When the air supply is switched on, the rising air bubbles induce vertical mixing which opposes the development of the density current and greatly reduces the horizontal transport of buoyancy (Fig. 9.8e, f). If the airflow is switched off, the density gradients accelerate the fluid and re-establish stratified flow. Although the experiments are conducted without rotation, there is a clear analogy with buoyancy stirring competition in the ROFI regime and they serve as a useful model, helping us to visualise the processes operating in the full-scale system. One version of the experiment has revealed the interesting effect of varying the modulation period of the stirring. Switching between bubbles on for T seconds and then off for the same period results in a horizontal exchange which increases with T even though the net stirring remains the same. This result reflects the time required to establish the density current regime. The fortnightly period of the springs-neaps modulation is much longer than the time required to accelerate the density current but, as we shall see in Section 9.4, a process at the much shorter semi-diurnal period can play a major part in controlling stratification.

9.4 Tidal straining

In addition to the spring-neap cycle in stability, the time series of Fig. 9.7 also indicates that in the Liverpool Bay ROFI there is a strong component of variation at the semi-diurnal tidal frequency. This semi-diurnal periodic component of stratification, sometimes referred to as Strain-Induced Periodic Stratification (SIPS), is the result of an interaction between the semi-diurnal tidal current and the horizontal density gradient in a process referred to as *tidal straining* (Simpson *et al.*, 1990). This is a common process generating stratification in estuaries and ROFIs, which is distinct from the more enduring stratification induced by the estuarine circulation. We can investigate its effects again by utilising the energetics of buoyancy supply and mixing.

9.4.1 The straining mechanism

Figure 9.8 illustrates the process for the case where the tidal current is a standing wave oscillation parallel to the density gradient. At the start of the ebb, we shall assume that the water column is fully mixed so that the isolines of salinity (and thus density) are all vertical. As the ebb flow develops, friction at the bed causes the velocity to decrease towards the bed with the result that lighter, lower salinity water near the surface is moved seawards faster than heavier water near the bed. The result of this differential advection is the development of stable stratification which continues until the end of the ebb flow.

On the flood, the shear in the velocity profile is reversed so now heavier water is advected landwards faster than water near the bed. Stratification is progressively removed and, in the absence of mixing, the system would return to a fully mixed state at the end of the flood. If, however, there has been significant mixing by the tidal flow during the ebb, the process will not be symmetrical between ebb and flood and complete vertical uniformity will occur before the end of the flood when the isohalines will be vertical. Thereafter, the straining will tend to force heavier water over lighter and produce an unstable condition which will result in convection and vigorous mixing.

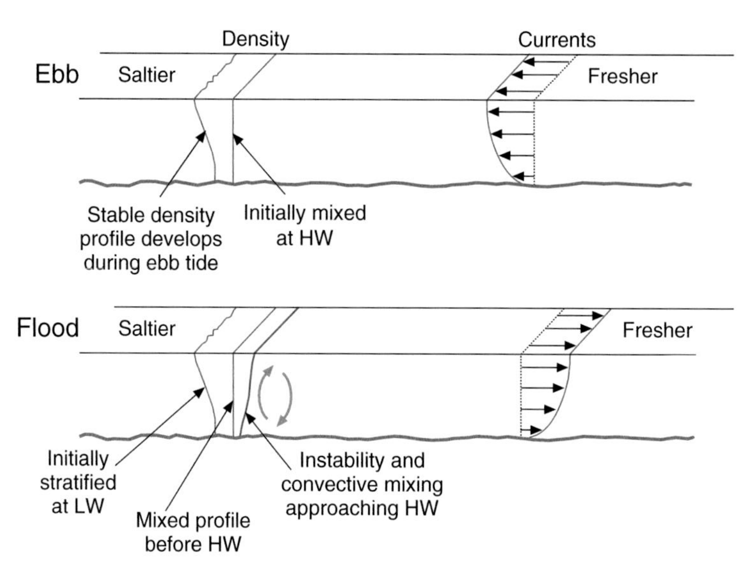

Figure 9.8 Straining of the density profile in a ROFI regime where the tide is in the form of a standing wave. (a) During ebb flow shear in the tidal current moves lighter water over denser thus promoting stratification. (b) In the following flood flow, the current shear is reversed and stratification is reduced. Tidal and wind mixing during ebb tides can mean that flood currents 'over strain' the density profile, leading to instability and convective mixing.

We can analyse the contribution of tidal straining to the evolution of stratification by using Equation (9.18) with a tidal velocity profile of the form:

$$u(\zeta) = \hat{u}(a - bz'^2) \tag{9.23}$$

which, with $\hat{u}$ the depth-mean current and the constants $a = 1.15$ and $b = 0.425$, is a good approximation to observed profiles of tidal flow (Bowden and Fairbairn, 1952). On substituting, we have:

$$\begin{aligned}\left(\frac{\partial \Phi}{\partial t}\right)_{St} &= gh\frac{\partial \rho}{\partial x}\int_{-1}^{0}(u-\hat{u})z'dz' = gh\hat{u}\frac{\partial \rho}{\partial x}\int_{-1}^{0}(a-bz'^2-1)z'dz' \\ &= gh\hat{u}\frac{\partial \rho}{\partial x}\left(\frac{b}{4}-\frac{(a-1)}{2}\right) = 0.031gh\hat{u}\frac{\partial \rho}{\partial x}.\end{aligned} \tag{9.24}$$

This is the source term which produces the observed periodic component of stratification. The condition for SIPS to occur is that the average positive contribution of $\partial\Phi/\partial t$ on the ebb should exceed the effects of tidal mixing, i.e.

$$\boxed{\frac{2}{\pi}\left(0.031ghu_2\frac{\partial \rho}{\partial x}\right) > \frac{4}{3\pi}\frac{ek_b\rho_0\hat{u}_{M2}^3}{h}} \tag{9.25}$$

which reduces to:

$$\frac{1}{\rho_0}\frac{\partial \rho}{\partial x} > \frac{2}{3}\frac{ek_b}{0.031g}\left(\frac{\hat{u}_{M2}}{h}\right)^2 \cong 2.2\times 10^{-5}\left(\frac{\hat{u}_{M2}}{h}\right)^2. \tag{9.26}$$

This condition on the density gradient required for SIPS to occur is of the same form as that for the occurrence of estuarine stratification (Equation 9.22) but with the important difference that the numerical constant is an order of magnitude lower. SIPS therefore occurs more readily than enduring estuarine stratification. We will now look more closely at tidal straining and some of its more subtle physical effects using two examples from the NW European shelf.

9.4.2 Tidal straining in Liverpool Bay

Liverpool Bay provides a good example of a case where the tide is a standing wave, and the major axis of the tidal ellipse is aligned with the horizontal density gradient. For the region where the data shown in Fig. 9.7 was collected, the depth-mean M_2 tidal stream amplitude is $\hat{u}_{M2} \sim 0.6\,\mathrm{m\ s^{-1}}$ in a water depth of $h = 35$ metres. Using Equation (9.26), the gradient required to produce significant periodic stratification is $> 6.5 \times 10^{-9}\,\mathrm{m^{-1}}$. The observed horizontal gradient was $1/\rho\ \partial\rho/\partial x \sim 5 \times 10^{-8}\,\mathrm{m^{-1}}$ which would indicate that periodic stratification should occur, as observed, for much of the springs-neaps cycle.

An illustration of the evolution of stratification over two tidal cycles in response to tidal straining is shown in Fig. 9.9, which is based on regular hourly profiles from a research vessel at a station in the Liverpool Bay ROFI. Stratification of both salinity, shown in Fig. 9.9a, and temperature, in Fig. 9.9b, can be seen to develop during the ebb phase of the tide. This produces stable density stratification which reaches a maximum around low water slack, as seen in Fig. 9.9c. This stability is eroded during the following flood flow so that complete vertical mixing is evident for some time before high water (in Fig. 9.9c high water occurred at 0400, while vertical homogeneity was reached at about 0200). During this latter part of the flood, the straining mechanism is operating to produce unstable density gradients in the water column as denser water is moved over less dense water by the shear in the tidal current. In this situation, sometimes termed 'over-straining', potential energy from the instability of the water column can induce convective motions which enhance turbulent mixing. The colours in Fig. 9.9c show the resulting impact on measurements of turbulent dissipation, which is revealed as a marked contrast in turbulence between the ebb and flood half cycles (Rippeth *et al.*, 2001). During the ebb, the development of stratification inhibits turbulence in the upper part of the water column, while late in the flood high dissipation levels are seen to extend throughout the water column to the highest level of observation (~6 metres below the surface). This cycle of dissipation, which can also be seen repeated in the second tidal cycle, contrasts sharply with the situation where horizontal gradients are small when a regular M_4 cycle of dissipation with equal intensity on ebb and flood is observed (see Section 7.2.1).

9.4.3 Tidal straining in the Rhine ROFI

The Rhine ROFI serves as a second, contrasting example of the action of tidal straining. The important difference between the two cases is that the tidal regime

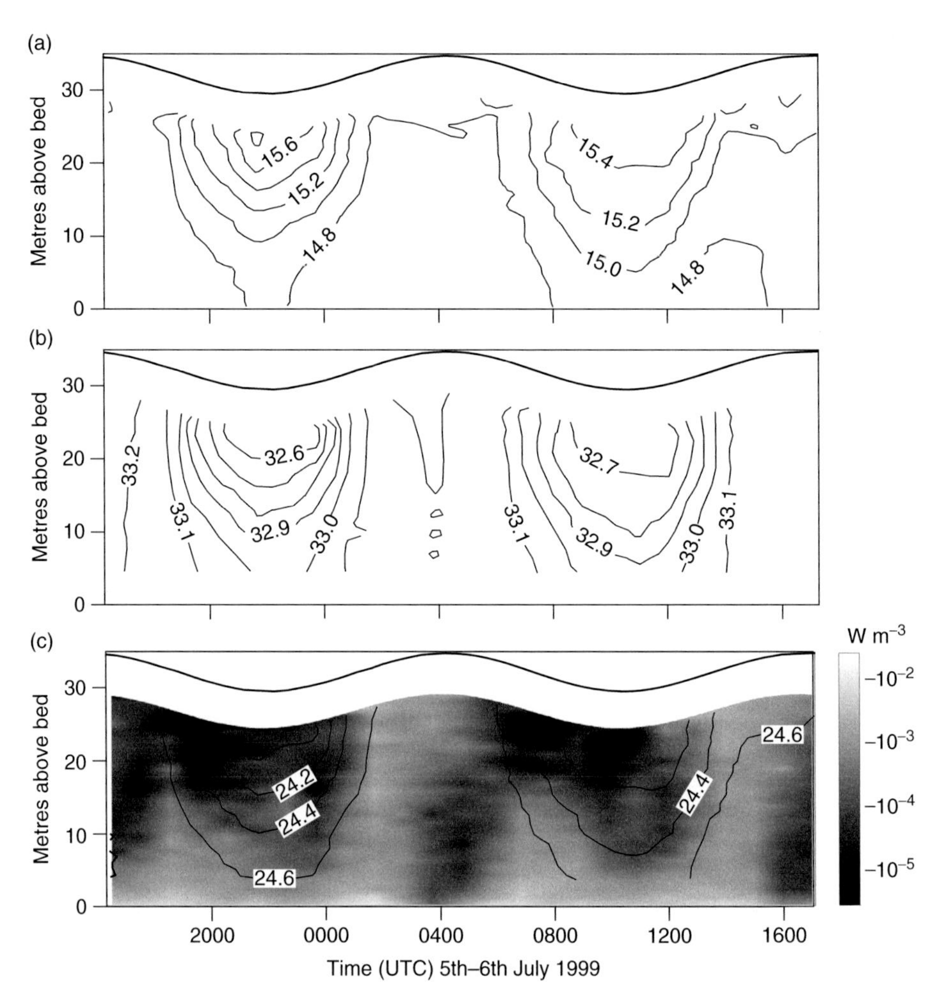

Figure 9.9 See colour plates version of (c). Two tidal cycles of (a) temperature, (b) salinity, and (c) σ_t (kg m^{-3}, lines) and turbulent dissipation (W m^{-3}, shaded) measured from a vessel anchored in Liverpool Bay. The vessel position was the same as that of the mooring data in Fig. 9.7. Adapted from Rippeth *et al.*, 2001, with permission from the American Meteorological Society.

along the Dutch coast is dominated by a progressive Kelvin wave, in which the velocity is in phase with the elevation, i.e. maximum flow to the north-east occurs close to high water. This is the opposite of what we have seen in the standing wave regime of Liverpool Bay, where slack water coincides with HW since the velocity is $\sim 90^\circ$ out of phase with elevation. It also means that in the Rhine ROFI, the major axis of the tidal current ellipse is aligned with the coast, orthogonal to the horizontal density gradient; we might think that this should reduce any potential for tidal straining as the dominant tidal shear is in the wrong direction.

As in Liverpool Bay, there is a continuous competition in the Rhine ROFI between the stratifying influence of freshwater and the stirring effects of tidal flow and wind

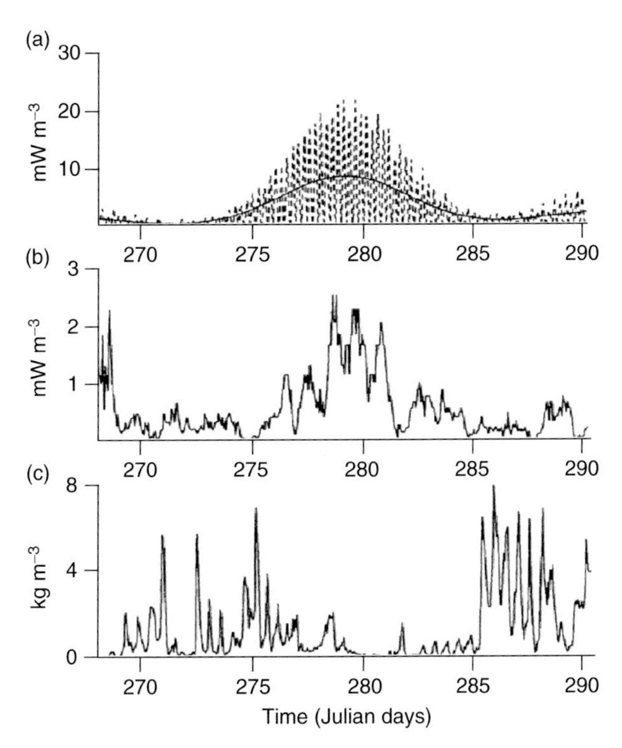

Figure 9.10 Time series of mixing contributions and stratification in the Rhine ROFI Adapted from Souza and Simpson, 1997, courtesy of Elsevier. (a) Depth-mean tidal stirring power (dashed line is instantaneous, solid line is tidally averaged) from a mooring in the Rhine ROFI. (b) Wind stirring power during the mooring deployment. (c) The bottom-surface density difference measured by the mooring at the position shown in Fig. 9.5.

stress. Stratification again tends to vary with the springs-neaps cycle in the tides, but episodes of wind stirring also make a substantial contribution to vertical mixing. In Fig 9.10 you can see how vigorous tidal stirring at a spring tide can be reinforced by an episode of wind stirring, together resulting in virtually complete vertical mixing following spring tides (e.g. see the reduction in stratification shown in Fig. 9.10c around the spring tide at day 279). By contrast, following neaps, the water column becomes stratified with a bottom-surface density difference > 5 kg m^{-3} at times. It is also clear that there is a strong semi-diurnal periodicity evident in Fig. 9.10c. The obvious mechanism behind this semi-diurnal periodic stratification is again tidal straining. But how does the straining operate if the major axis of the tidal flow is parallel to the coast, i.e. orthogonal to the horizontal density gradient?

The normally weak cross-shore tidal shear must be enhanced during periods of stratification. In order to understand how this occurs, we need to recognise how the form of the tidal ellipse responds to changes in water column stability (Visser *et al.*, 1994; Souza and Simpson, 1996). Remember that the tide along the Dutch coast takes the form of a Kelvin wave travelling to the north-east along the coast. In Fig. 9.11a, measurements made by HF radar show that when the water column is well-mixed, the tidal ellipses are close to being degenerate (i.e. rectilinear); current meter data in the region indicate that this is the case at all depths. Conversely, Fig. 9.11b, corresponding to the stratified, low mixing period around neap tides, shows that the surface tidal ellipses become more circular with the current vector

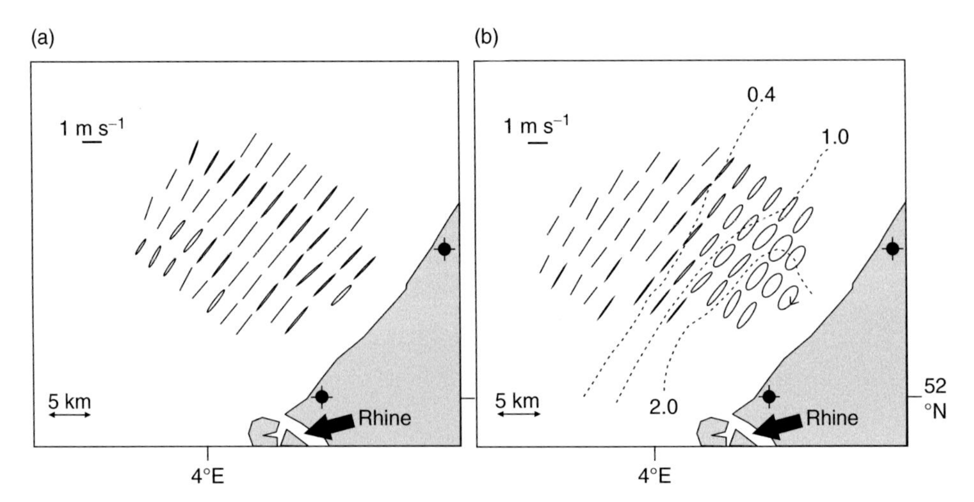

Figure 9.11 HF radar data of surface currents in the Rhine ROFI (Visser *et al.*, 1994), showing the change in tidal ellipse structure between (a) a vertically-mixed situation at spring tides, and (b) a stratified period at neap tides. In (b) the dotted contour lines show stratification as the salinity difference ΔS between 10m and 1m depth. ◆ indicate the positions of the radar antennae. With permission from Elsevier.

rotating in a clockwise direction. At the same time, current meter measurements reveal that the ellipses in the bottom layer acquire an anticlockwise form so that top and bottom layers rotate in opposite directions.

It is this vertical contrast in ellipticity which is responsible for the strong cross-shore shear which acts on the cross-shore density gradient to induce the M_2 cycle of stratification (Simpson and Souza, 1995). We show how this works schematically in Fig. 9.12. At time $t = 0$, it is low water when, in a Kelvin wave, there is peak reverse flow, i.e. the current is directed to the south-west, there is no cross-shore flow and the water column is assumed to be mixed. Quarter of a tidal period later ($t = T/4$), the alongshore flow is zero, but there is now a pronounced two-layer flow in the cross-shore direction due to the difference in ellipticity between the surface and bottom layers. This cross-shore shear flow acts on the density gradient to induce stratification by moving lighter water near the coast over heavier water further offshore. Maximum stratification occurs at high water ($t = T/2$) when the alongshore flow is maximal to the north-east and the cross-shore flow is zero. After the next quarter cycle ($t = 3T/4$), the cross-shore flow is now the reverse of that at $t = T/4$ and straining removes the stratification so that complete vertical homogeneity is restored after a full cycle ($t = T$).

Figure 9.12 also indicates that the cross-shore straining will lead to upwelling at the coast in the first half of the tidal cycle and downwelling in the second half. The occurrence of upwelling has been noted by the appearance of a band of cold water near the coast detected in I-R imagery of the region by de Boer *et al.*, 2009.

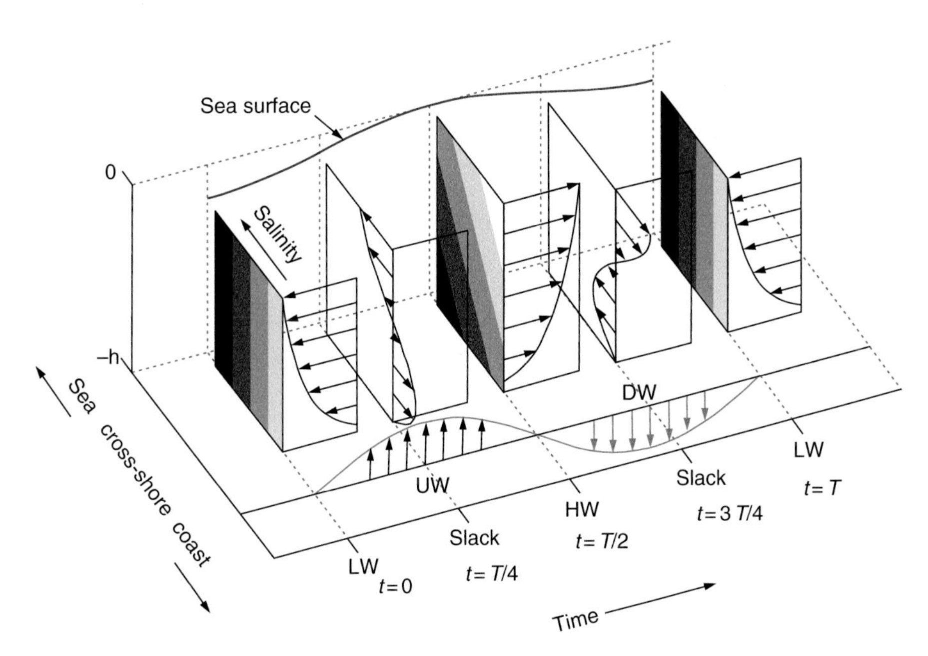

Figure 9.12 Schematic illustration of the action of counter-rotating surface and bottom layer tidal current ellipses in developing strong cross-shore shear and subsequent tidal straining of the horizontal density gradient. The cross-shore density gradient and vertical density structure are illustrated by the shading (black is the densest water); maximum stratification occurs at HW. UW and DW show periods of upwelling and downwelling respectively. See main text for details. Based on de Boer *et al.*, 2009, with permission from Elsevier.

9.5 Modulated non-tidal transports in ROFIs

Density-driven circulations are important because they provide a mechanism for net transport through an estuary or ROFI against the background of the limited net movement caused by the often much larger tidal flows. More generally, we are interested in any non-tidal (or *residual*) flow that can lead to net transport; such transports are often important to the biochemistry, ecology and water quality of estuaries and ROFIs.

We have seen already in the Linden-Simpson experiments and in the theory that the steady state density-driven circulation in both estuaries and ROFIs responds to changes in vertical mixing. Reduced N_z, which controls the vertical mixing of momentum, increases the circulation. For instance, in the non-rotating case (Equation 9.10) there is an inverse dependence of density-driven current speed on N_z, and we know that N_z will respond to the semi-diurnal tide (e.g. low N_z at slack water) and to the spring-neap cycle (e.g. tidally averaged N_z will be lower at neap tides). Hence we might expect the density-driven circulation to increase at slack water and during neap tides (e.g. Sharples and Simpson, 1995; Ribeiro *et al.*, 2004).

There is a further mechanism for the generation of depth-varying residual circulation on semi-diurnal time scales which arises through the interaction between stratification, mixing and the tidal flows. For example, in a standing wave tidal regime, the increasing stratification and reduced vertical mixing on the ebb tend to decouple the surface layer, which then moves faster in the offshore direction. On the flood the increased vertical mixing reduces the vertical shear in the flow, thus increasing the shoreward movement of the bottom layer relative to the surface. Both ebb and flood half cycles, therefore, act to increase residual circulation, in the same direction as the density-driven circulation, relative to that which would occur in the absence of straining. This enhancement of the circulation by tidal straining, identified by Jay and Musiak (Jay and Musiak, 1996), has been recognised in observations (Stacey *et al.*, 2008) and demonstrated in a numerical models (Prandle, 2004).

Modulated residual flows in estuaries and ROFIs will have impacts on the flushing or retention of material. As an example, consider the asymmetry in vertical turbulent mixing as a result of tidal straining and its effect on the transport of suspended sediments. We saw in Section 9.4.2 how the stratification which develops on the ebb in a standing wave regime leads to a severe damping of turbulence. On the flood, the straining acts to erode stratification and, therefore, to promote turbulence. Towards the end of the flood, the water column can become vertically mixed and the straining mechanism then tends to produce instability and convection which will further enhance turbulence. The result is a pronounced variation in vertical mixing with much higher diffusivity, K_z, on the flood than on the ebb. This contrast in vertical mixing means that flood flow is able to maintain much higher levels of sediment in suspension than on the ebb, thus favouring the transport of sediments in the direction of the flood. The operation of this asymmetry in the transport of suspended sediments has been observed in the York River estuary (Scully and Friedrichs, 2007).

9.6 The influence of wind stress

So far in this chapter, we have concentrated on the interaction between freshwater input and the action of the tides in stirring the water column and straining the density field. In many ROFI systems, tidal influence predominates for much of the time, but episodes of strong winds can modify the ROFI regime especially in cases where the tides are weak. In extreme cases like the ROFI of the Rhone River debouching in to the Mediterranean Sea, where the tidal flow is almost negligible, wind stress becomes the dominant mechanical forcing (Reffray *et al.*, 2002). In conditions of light winds, the Rhone discharge forms a plume which flows to the west along the French coast under the control of the buoyancy forcing and the Coriolis force. At times of strong winds, however, the surface stress out-competes the buoyancy forcing and, depending on the wind direction, the Rhone plume may be driven away from the coast or even reversed, so that the Rhone ROFI regime is highly variable (Forget and Andre, 2007).

This influence of both the magnitude and direction of the wind stress on the flow and structure is apparent in other ROFIs even where the tidal influence is substantial. In a narrow estuary where the flow is constrained by the boundaries, wind stress operating in the seawards direction tends to strengthen the estuarine circulation and, hence, to increase stratification. Conversely, a landward component of wind stress

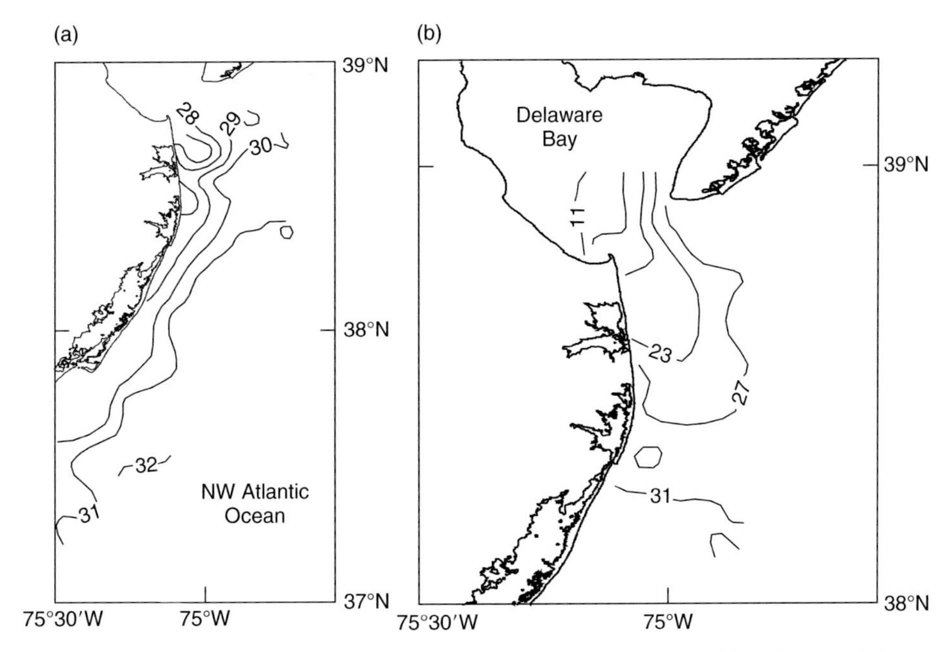

Figure 9.13 Surface salinity in the ROFI outside the Delaware River. (a) Following light winds from the northeast (downwelling favourable); (b) Following strong winds from the southwest (upwelling favourable). Figure provided by Rich Garvine, based on Munchow and R.W.Garvine, 1993.

acts to reduce the circulation and diminish stratification. This 'wind straining' mechanism (Scully *et al.*, 2005) also operates in ROFIs (e.g. Verspecht *et al.*, 2009), but in a rather different way since the flow is constrained horizontally only by the coast. In deep water at some distance from the coast, we know that the Ekman transport (Section 3.5.1) will be directed at right angles to the wind stress, so that it is the alongshore wind component which is responsible for generating the coast-normal transport and the consequent effects on the structure of the ROFI.

As an example, Fig. 9.13 shows the influence of wind stress on the ROFI generated by the Delaware River on the east coast of the United States. Observations (Munchow and Garvine, 1993) shown in Fig. 9.13a have revealed that coast-parallel winds which are downwelling favourable (i.e. in a direction of the coastal current) constrain the low salinity waters close to the coast and reinforce the along coast jet. Conversely upwelling favourable winds tend to arrest the alongshore flow and force offshore flow with a consequent increase in the width of the ROFI, as shown in Fig 9.13b. Similar responses to wind direction have been demonstrated for the Columbia River discharge onto the NW shelf of the United States (Hickey *et al.*, 2005). In shallow water, the cross-shore component of wind may also drive upwelling and downwelling motions and hence influence the structure of the coastal jet as has been shown for the Rhine ROFI (Souza and Simpson, 1997).

It should also be remembered that the wind generates surface waves which can contribute to mixing near the surface through the turbulence which is induced by wave breaking. Moreover, if the water is shallow enough, waves, especially long period swells, may have orbital motions which extend down to the seabed and induce bottom stresses which promote vertical mixing. To separate the effect of stirring by the surface wind stress and the contribution from wave motions is difficult and requires independent wind and wave data from local direct measurements, as in an extreme shallow case studied by Wiles, *et al.*, 2006.

9.7 Modelling the physics of ROFIs

It should be clear from the forgoing sections that ROFIs are host to a diverse range of interacting physical processes. While we have seen that analytical models can represent the key aspects of the individual processes, numerical methods are needed if we are to try and simulate the full complexity of ROFI behaviour and thus test the extent of our understanding of the way they work. Because of the range of processes involved, their interactions and the multiple dynamic feedbacks involved, ROFIs constitute a demanding challenge to the skill of numerical models.

While the ultimate goal must be prognostic full physics 3D models, much progress to date has been made with 1D simulations at a point, using a turbulence closure scheme similar to the approach described in Chapter 7. In addition to specified forcing by tides and winds, it is also necessary, in a 1D ROFI model, to prescribe the horizontal gradients of density which must be determined from observations. If a full suite of forcing is available, these 1D models can give a satisfactory first order simulation of the evolution of density structure and flow in a ROFI (Sharples and Simpson, 1995) and can reproduce the enhanced dissipation (e.g. as seen in Fig. 9.9c) resulting from tidally forced convection (Simpson *et al.*, 2002). The focus in 1D models on vertical exchange processes makes them a useful test-bed for evaluating turbulence closure schemes. A flexible 1D model framework for comparing different closure schemes is available in the General Ocean Turbulence Model (GOTM; see: http://www.gotm.net/index.php).

A 3D full physics model involves much greater freedom than is the case of a 1D model, as it is forced only by inputs at the boundaries (i.e. tidal elevations and freshwater input at the lateral boundaries, heat and wind stress at the surface). To properly determine the evolution of the system, a 3D model must accurately represent the vertical mixing of buoyancy and momentum at each point in the model domain. Any deficiency in calculating the mixing rates here will cause errors in the horizontal density gradients which will modify the circulation and hence the velocity shear. This in turn will cause further misrepresentation of the vertical mixing. The effect of tight feedback loops of this kind is particularly acute in ROFIs where such a complex suite of processes is operating.

In spite of the problems here, 3D models have been applied, with some success, to test our understanding of the way ROFIs work (Whitney and Garvine, 2006). A detailed model of the Rhine region (de Boer *et al.*, 2006), for example, gives a convincing demonstration of ROFI behaviour including the development of cross-shore straining, periodic stratification and upwelling along the coast during periods of reduced vertical mixing.

Physics summary box

- ROFIs are distinctive regions of the shelf seas where the freshwater input from rivers constitutes a buoyancy source which is comparable in magnitude to that of surface heating and cooling further offshore.
- In contrast to surface heating, this buoyancy input is not distributed uniformly over the surface but enters at one or more localised estuarine sources at the coast.
- The buoyancy input drives a circulation in which the low salinity water emerging from the estuary is deflected by the Coriolis force to the right/left in the northern/southern hemisphere and, in the absence of friction, forms a current flowing parallel to the coast.
- The flow structure is complicated by frictional effects which induce an additional two-layer flow normal to the coast and parallel to the density gradient.
- This estuarine-type flow moves low salinity water over heavier water further offshore and thus tends to stratify the water column in competition with stirring by the tides and winds.
- Variations in freshwater input, tidal stirring and wind stirring shift the balance in this competition so that ROFIs experience large changes in stratification on a variety of time scales in contrast to other areas of the shelf seas.
- When tidal stirring predominates, its regular variation over the springs-neaps cycle may induce a fortnightly switching between stratified and fully mixed conditions.
- The interaction between shear in the tidal currents and the density gradient leads to periodic variations in stratification at the M_2 tidal frequency through the process of tidal straining.
- The effect of tidal straining is subtly different for ROFIs with progressive and standing wave tidal regimes.
- Tidal straining can act to augment density-driven circulation and sediment transport in estuaries and ROFIs.

9.8 Biological responses in estuaries and ROFIs

There are three aspects of ROFIs that we will consider when we look for biological responses. First, the horizontal density gradient in a ROFI tends to establish a vertically sheared mean flow. We might expect any near-surface material in the water to be advected offshore and alongshore. Nearer the seabed, things are likely to drift shoreward, possibly even returning to the estuary. We have already seen the importance of persistent flows to the maintenance of some shelf animal populations (Section 8.6.3); mean flows in ROFIs and estuaries can play a similar role. Second, cycles in mixing and stratification have a great impact on phytoplankton growth, which we have seen in the context of the annual spring bloom (Section 6.3.1) and the modulation of primary production by the spring-neap tidal cycle at shelf sea fronts (Section 8.6.1) and within the shelf sea seasonal thermocline (Section 7.3.2). ROFIs can exhibit stratification-mixing cycles ranging from the semi-diurnal and spring-neap tidal cycles, through to seasonal responses to changes in freshwater supply. We might ask if the phytoplankton are seen to respond to any or all of these? Finally, a ROFI is a coastal zone where nutrients and pollutants borne by the freshwater will have their first impact on the shelf sea system. The physics of the ROFI regime is crucial in determining where these inputs go and what will be their impact on the ROFI ecosystem.

9.8.1 Density-driven flows: export and return

Any buoyant or actively surface-dwelling organism in an estuary or ROFI will tend to experience a net movement out of the estuary and off- and alongshore in the coastal zone due to the density-driven circulation. Conversely, anything living down near the seabed will tend to be moved back towards the coast, possibly re-entering the estuary. You can imagine that a combination of these two density-driven mean flows could be exploited by an organism in order to remain roughly in one place or to limit its dispersion. All an organism needs is some capacity for controlling its vertical position and hence which component of the density-driven flow it experiences.

Organisms which can swim vertically have the potential to react on short time scales to their environment, by using different stages of the tidal cycle to maintain their horizontal position. Utilisation of the much slower density-driven exchange tends to occur over time scales longer than tidal, with changes in the depth of an organism occurring at the very low frequency associated with different life stages (e.g. the ontogenic behaviour of some mesozooplankton introduced in Section 5.2.1). We have already seen the patterns of low salinity water outside Delaware Bay on the northeast shelf of the United States (Fig. 9.13). Several organisms have been recorded apparently taking advantage of the exchange flows between the Bay and the shelf. For instance early stage fiddler crab larvae, which are buoyant, are thought to be flushed from the estuaries of Delaware Bay out onto the shelf. Later stages are found near the seabed on the shelf, where the return flow of the density-driven circulation brings them back into the Bay (Epifanio *et al.*, 1988). This ontogenic shift in the depth distribution of larval stages, combined with later stages being found further offshore, is illustrated in Fig. 9.14. Notice how the later stages are found in the seaward sampling stations outside the Bay (Fig. 9.14a, b), while at the same time they have a more even, vertical distribution compared to the surface bias of the earlier stages (Fig. 9.14c).

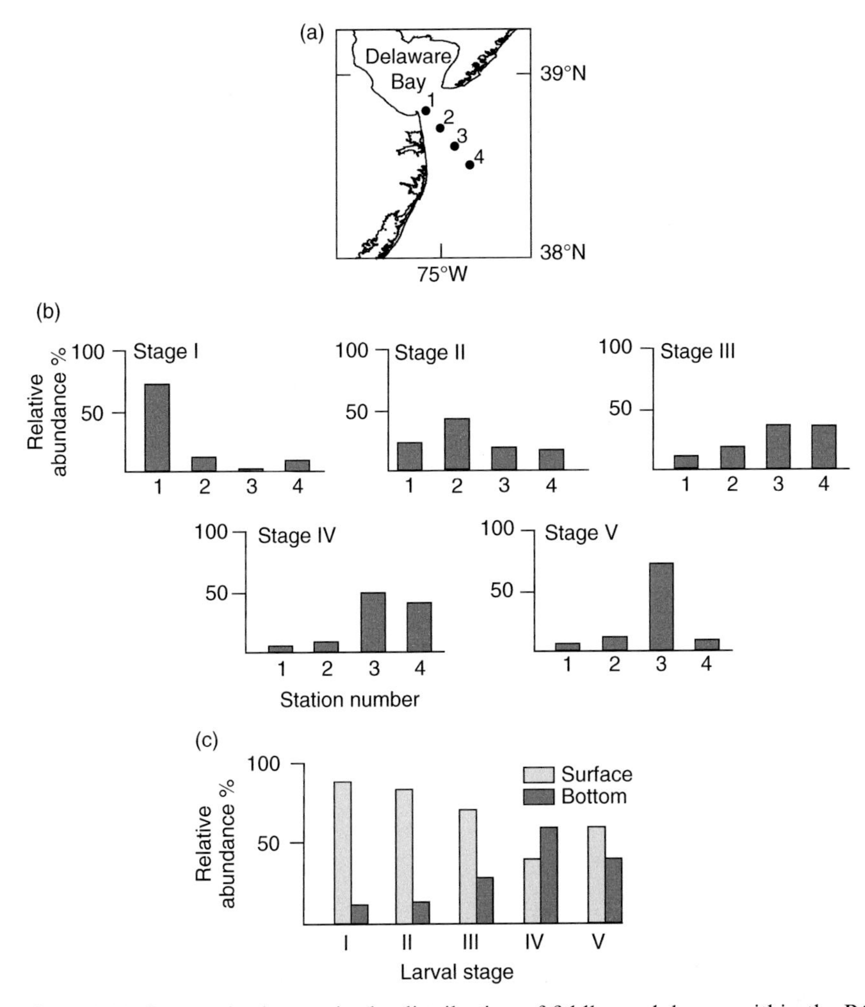

Figure 9.14 Ontogenic changes in the distribution of fiddler crab larvae within the ROFI at the entrance to Delaware Bay. (a) The 4 sampling stations from the Bay mouth seaward. (b) Changes in the location of larval stages, expressed as relative abundance of each stage at each sampling position. (c) Changes in the vertical distributions of the different larval stages. Adapted from Epifanio *et al.*, 1988, with permission from Inter-Research.

Alongshore transport of phytoplankton has also been linked to coastal buoyancy currents, often with a focus on nuisance species (e.g. Franks and Anderson, 1992). A dramatic example is that of the odyssey thought to be undertaken by the dinoflagellate *Prorocentrum mariae-lebouriae* in Chesapeake Bay, shown in a series of along-Bay surveys of salinity and *Prorocentrum* cell numbers in Fig. 9.15. The surface bloom of *Prorocentrum* reaches the mouth of the Bay in winter, driven by the seaward (southward) surface flow (Fig. 9.15a). Circulation between the mouth of the Bay and the adjacent shelf, combined with weak stratification, leads to the cells being distributed through the water column. Figure 9.15b shows that, as stratification

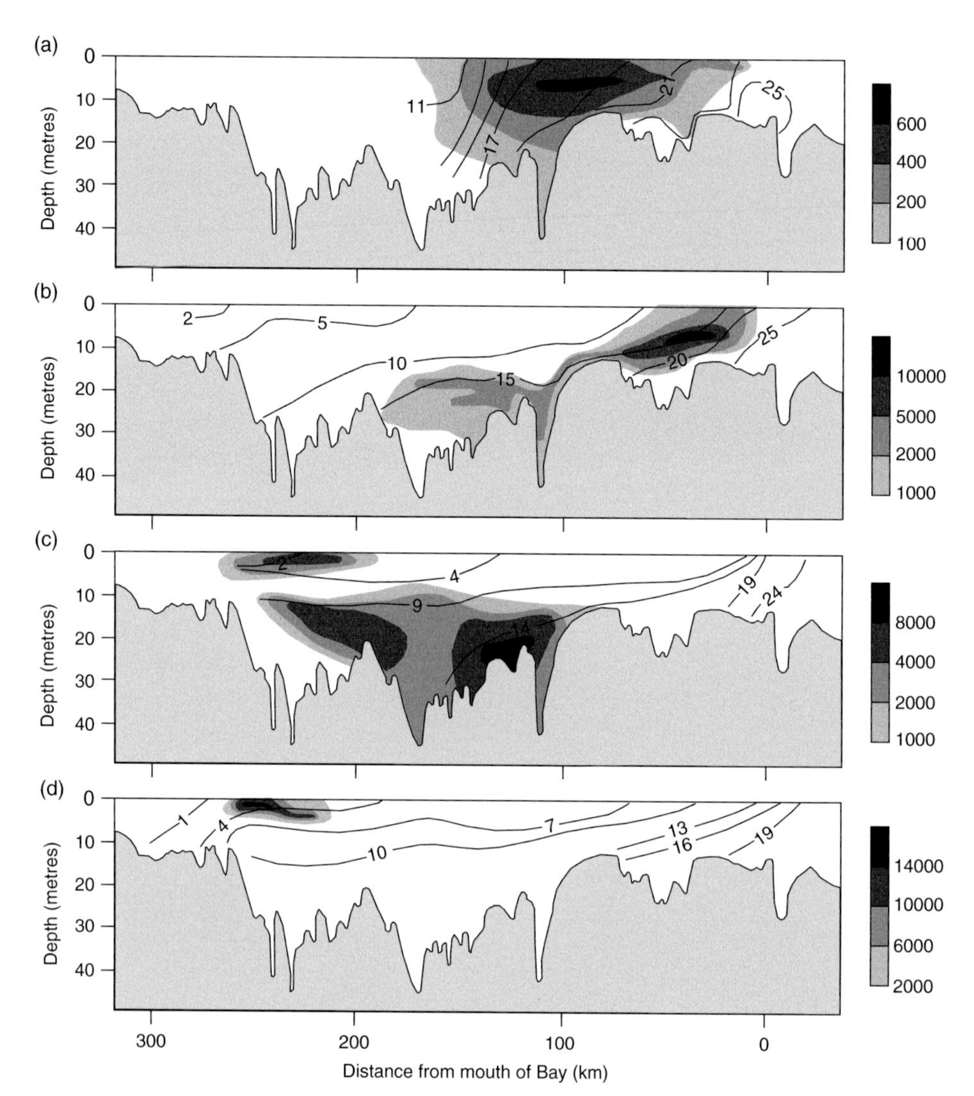

Figure 9.15 Seasonal progression of stratification and the position of the dinoflagellate *Prorocentrum mariae-lebouriae* along a transect running from the head of Chesapeake Bay to seaward of the Bay mouth, after Tyler and Seliger, 1978, courtesy of ASLO. Lines are contours of σ_t (kg m^{-3}), concentrations of *Prorocentrum* are shaded (cells ml^{-1}). Note that the contour intervals are different for each panel. (a) mid-winter, (b) late winter, (c) late spring, (d) mid-summer.

increases due to higher river flows in late winter, a population of *Prorocentrum* is found trapped in the bottom layer where it can be moved northward under the influence of the near-bed density-driven flow. Towards the end of spring, shown in Fig. 9.15c, cells are located in the upper part of the Bay, later forming surface blooms through the summer (Fig. 9.15d) which can then be advected

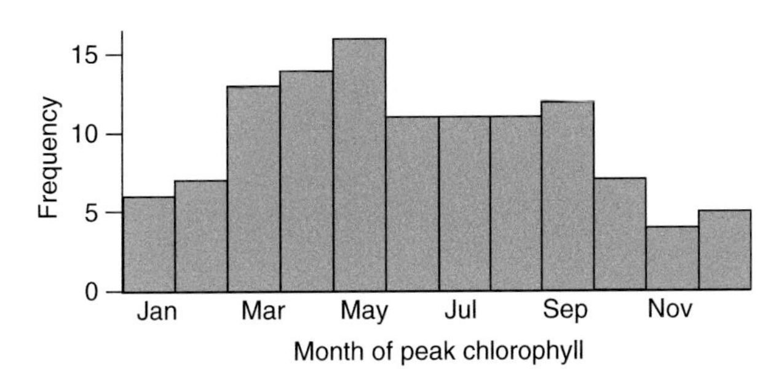

Figure 9.16 Frequency of the timing of peak chlorophyll concentration from an analysis of data from northern temperate coastal regions. After Cloern and Jassby, 2008, courtesy Wiley-Blackwell.

southward by the density-driven surface flow to complete the circuit. We might expect that the length of time spent in the dark during the bottom-layer journey from the mouth to the head of the Bay would be challenging for the phytoplankton. However, experiments carried out at the time of the surveys showed that the *Prorocentrum* cells could remain viable for 4 weeks in the dark (Tyler and Seliger, 1978).

9.8.2 Responses to cycles in stratification

Given the importance of stratification in coastal waters as a driver of increases in surface primary production, we might expect that the regular cycles of stratification in a ROFI will have consequences for phytoplankton growth. Certainly it is recognised that the seasonal cycles of primary production in freshwater-influenced coastal waters look very different from the cycles in the open shelf seas (Cloern and Jassby, 2008). An analysis of the timing of peak chlorophyll concentrations during the year for a variety of northern temperate coastal regions, many of which are influenced by freshwater inputs, is illustrated in Fig. 9.16. While there is some tendency towards favouring spring, peak chlorophyll can occur at any time of the year with a notable broad response through the whole spring-summer period. This pattern is very different from the dominance of spring growth seen away from the coast in temperate shelf seas. In conducting this analysis, Cloern and Jassby noted that the often unpredictable nature of the timing of coastal blooms will have impacts on grazing communities; unlike the open shelf seas, the coastal consumers do not have the relative convenience of a consistent annual event in the arrival of food.

While coastal primary production is often viewed as unpredictable, we have seen that many ROFIs do have stratification-mixing cycles with regular periodicity. Is there evidence of the primary producers responding to these cycles?

Consider first the spring-neap cycles in ROFI stratification. We know that phytoplankton in the open shelf seas can respond on this time scale, so do they within ROFIs? In Fig. 9.17, data from Puget Sound, a fjord on the coast of the northwest United States, provides an example of a degree of correlation between the spring-neap tidal cycle, the strength of stratification, primary production

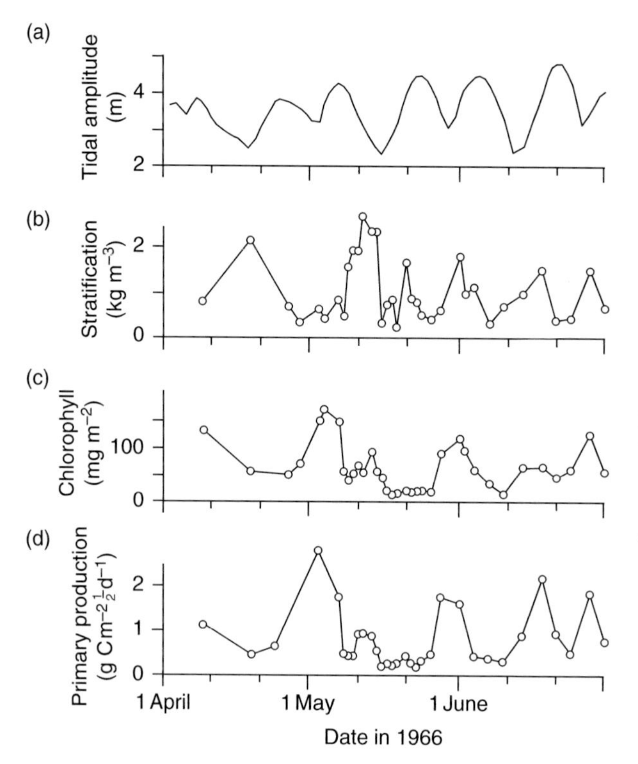

Figure 9.17 Variability of stratification and primary production during spring and summer 1966 in Puget Sound. (a) Tidal amplitude, indicating the spring-neap variability. (b) Stratification, expressed as the density difference between a depth of 25 metres and the sea surface. (c) Chlorophyll, integrated from the surface to the 1% light level. (d) Rate of primary production (carbon uptake) integrated from the surface to the 1% light level. Based on (Winter *et al.*, 1975) with permission from Springer.

rates and the concentrations of surface layer chlorophyll. Note the responses to the sequence of neap tides between May 14 and June 25. Similar behaviour in surface chlorophyll responding to a spring-neap cycle has been reported in the York River, a tributary of Chesapeake Bay (Sin *et al.*, 1999), within San Francisco Bay (Cloern, 1991), and in the coastal plume of the Fraser River (Harrison *et al.*, 1991).

We should not be surprised that phytoplankton growth can respond to stratification-mixing cycles driven by the spring-neap tidal cycle in ROFIs. With cell doubling times of typically 1 to 2 per day, the timing of the spring-neap cycle is well suited to driving such a growth response. But what about tidal straining? Are there examples of primary production responding to a modulation of stratification on a semi-diurnal time scale? This is a difficult question to address, not least because, as we saw in Chapter 5, the usual method of measuring primary production rates requires sample incubations over several hours. One tantalising example comes from Liverpool Bay, with measurements taken over one tidal cycle in a region we now know to experience tidal straining. Observations of both autotrophic and heterotrophic urea assimilation rates (shown in Fig. 9.18a), alongside knowledge of the cycling of stratification (Fig. 9.18b), show increases in assimilation rates correlating with the breakdown and subsequent re-establishment of stratification either side of a mixed period.

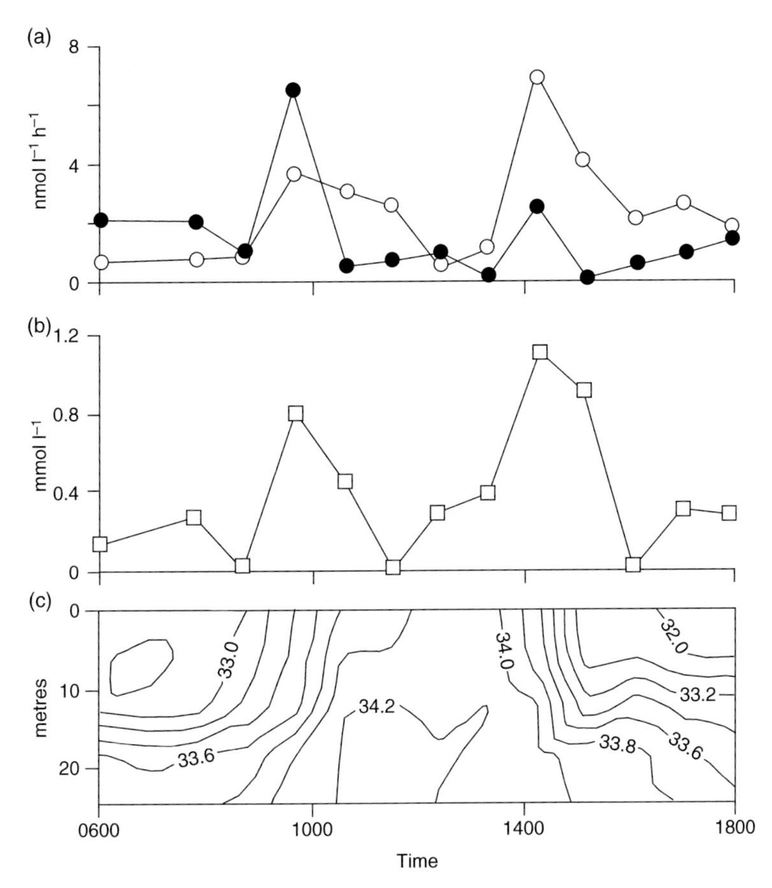

Figure 9.18 Response of primary producers to semi-diurnal changes in stratification in Liverpool Bay, UK. (a) Uptake rates of urea-nitrogen for autotrophic (○) and heterotrophic (●) phytoplankton at a depth of 10 metres. (b) Concentration of urea-nitrogen at a depth of 10 metres. (c) Time-depth series of changes in salinity. Times are BST (UTC + 1 hour) on May 4, 1977. High water was at 1230 BST and it was a spring tide. Based on Floodgate *et al.*, 1981, with permission from *Nature*.

Originally these data were interpreted in terms of the tidal advection of a front back and forth past the anchored research vessel (Floodgate *et al.*, 1981). Subsequent re-analysis of the salinity changes suggested that, rather than advection of a front, the cycle of stratification and mixing was being driven by tidal straining (Simpson *et al.*, 1991). The data in Fig. 9.18 could indicate that the phytoplankton nutrient uptake (and potentially growth) was responding to the straining-induced changes in stratification. Alternatively, while a front may not have been a consistent feature of the region, horizontal patchiness in urea uptake could have been advected past the research vessel. In Problem 9.3 you can think about how you might interpret the data in Fig. 9.18, and what additional data you would collect, in order to decide between these two hypotheses. Modelling studies indicate that semi-diurnal stratification will not trigger phytoplankton blooms (Lucas *et al.*, 1998), though in shallow waters the changes in light driven by tidal variations in sea level could be important (Lucas *et al.*, 1999).

9.8.3 Impacts of riverine material in ROFIs

Rivers play a global role in delivering material important to biogeochemical cycling into the ocean. Silicate is, as we saw in Chapter 5, a vital nutrient for diatoms, and the most important source of silicate (as silicic acid) to the ocean is via rivers and estuaries (Treguer *et al.*, 1995). Considerable attention has been focused on the large fluxes of organic carbon down rivers and through estuaries. If coastal and shelf primary production simply recycles this carbon, rather than requiring further carbon from the inorganic pool, then the shallow seas could be net sources of CO_2 back to the atmosphere. The physics of estuaries and ROFIs controls the residence times of silicate and organic carbon in the near shore zone, which affects their modification, utilisation and subsequent dispersion and supply beyond the coasts. Many estuaries are found to be net sources of CO_2 to the atmosphere, but evidence collected over wider shelf regions suggests that this effect is constrained close to the coast and that shelf seas play an important part in the removal of CO_2 from the atmosphere. We will discuss that further in Chapter 10, when we deal with carbon export across the shelf edge.

Estuaries have long been recognised as sources of both industrial and agricultural pollutants, since they are used either deliberately or accidentally to transport unwanted substances into the ocean. In the past there was an implicit assumption that pollutants input to the coastal zone were quickly dispersed and diluted through the ocean. More recently it has become recognised that the physics and biology of the coastal seas undermine the concept of the ocean as a limitless diluter; dispersion away from the coastal zone can be weak, with plenty of opportunity for pollutants to influence the biochemistry and cause problems.

An example off the coast of NW Europe is the radioactive discharge from the Sellafield nuclear power plant and waste reprocessing facility on the northwest coast of England (Prandle, 1984). The radio-isotope caesium-137 (^{137}Cs), discharged by the Sellafield plant into the Irish Sea, is distributed by a coastal buoyancy current round the north of the UK and then southward into the North Sea (see Fig. 3.17). The tendency for ^{137}Cs to attach to fine sediment particles leads to the onshore near-bed current in this freshwater-influenced coastal region, driving some of the ^{137}Cs back into the estuaries (Mudge *et al.*, 1997). The concentrations of ^{137}Cs are far below anything likely to be harmful, but the discharges have provided a serendipitous tracer of coastal oceanography and also a way to estimate recent sedimentation rates in estuaries (Mamas *et al.*, 1995).

Remember in Chapter 5 that we noted the large amount of atmospheric nitrogen that is fixed to support agriculture; humanity has doubled the global rate of N_2 fixation. Once fertilisers have been applied to farmland, rain leaches much of it into rivers where it is transported to the sea. Note that primary production in freshwater is generally limited by phosphate, so this supply of nitrogen to the rivers is able to reach the coast relatively unutilised. Such runoff of agricultural nitrogen, along with the use of rivers and estuaries for sewage disposal, can have dramatic biochemical consequences within ROFIs. See, for instance, the high nitrate concentrations in the Rhine ROFI shown in Fig. 9.3. The high nutrient load being supplied to the surface layer of a stratified ROFI can drive high rates of primary production, resulting in *eutrophication*. These large supplies of nitrogen into estuaries and coastal regions are frequently implicated as a cause of harmful algal blooms (HABs) of species of phytoplankton that are toxic to other organisms (Anderson *et al.*, 2002). The majority of phytoplankton that exhibit toxicity are dinoflagellates (Smayda, 1997), with the stratification induced by the freshwater thought to play a role in the bloom development. Exactly why excess nutrients should trigger blooms of toxic species is far from clear; the most successful prediction strategies tend to be regionally specific statistical models that draw on correlations between past blooms and environmental conditions, or real-time monitoring of the phytoplankton community.

Regardless of whether or not the phytoplankton are toxic, the subsequent degradation of larger amounts of phytoplankton when they sink down into the bottom layer generates a high oxygen demand by the heterotrophic bacteria as they oxidise the organic material. The stratification that supported the surface primary production provides a barrier to oxygen transport from the surface layer down into the bottom waters, and so bottom waters can become seriously de-oxygenated (e.g. Weston *et al.*, 2008). This can have important consequences for bottom water marine life (Pihl *et al.*, 1991), substantially reducing the biodiversity of the ecosystem. A dramatic example of bottom water de-oxygenation as a result of eutrophication is the 'dead zone' off the coast of Louisiana and Texas, USA (Rabalais *et al.*, 2002). Nitrogen supplied by the Mississippi and Atchafalaya rivers drives surface phytoplankton production within the stratified coastal buoyancy current as it flows westward along the northern boundary of the Gulf of Mexico. The resulting degradation of the settling phytoplankton material in the bottom layer of the stratified water column leads to a large area of low oxygen concentrations (termed *hypoxic*, with oxygen concentrations < 2 mg l^{-1}), since the pycnocline prevents easy replenishment of dissolved oxygen from the surface waters and atmosphere. The survey of the region in Fig. 9.19a shows an area of about 17,000 km^2 experiencing hypoxia. The limitation of the hypoxic waters to the lower part of the stratified water column is clear from Fig. 9.19b. The reduction in oxygen leads to large changes in bottom water marine life (Rabalais *et al.*, 2002). Animals that have sufficient swimming ability to control their position vertically within the water column (e.g. fish, demersal invertebrates such as shrimp) attempt to leave the hypoxic region. The colloquial term of dead zone largely arises from a failure to catch demersal fish and shrimp in the area. Animals that have less ability to leave the region of low oxygen gradually become stressed and, if concentrations of oxygen drop low enough, die. For instance, starfish and brittle stars are stressed when oxygen drops below about 1.5 mg l^{-1}, and are found dead when there is less than 1 mg l^{-1}. Fauna living within burrows in the

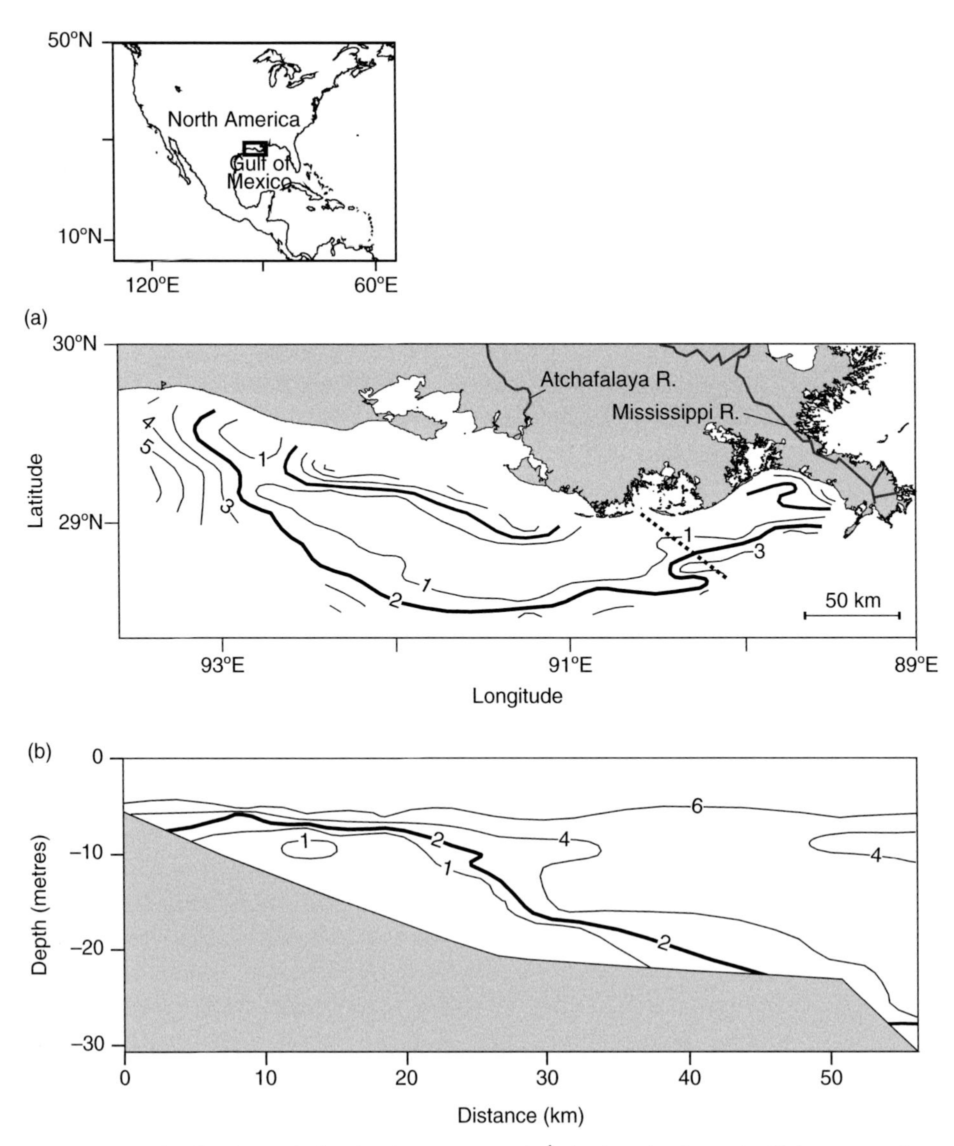

Figure 9.19 Distributions of dissolved oxygen (mg l^{-1}) in the 'dead zone' off the coast of Louisiana and Texas. (a) Bottom water dissolved oxygen in mid-summer 1993. (b) A section across the shelf (dotted line in (a)) showing the low dissolved oxygen limited to the deeper water. This section was taken in August 1990 during upwelling favourable winds, which have brought low oxygen water shoreward into shallower areas. Adapted from Rabalais *et al.*, 2002, courtesy of *Annual Reviews*.

sediments become stressed at 1 mg l^{-1}, and die when oxygen falls below about 0.5 mg l^{-1}. The scale of this problem is such that a task force, led by the U.S. Environmental Protection Agency, has been set up to plan and implement the reduction of nitrogen loading on the Mississippi; considering the size of the Mississippi catchment, which covers 40% of the entire USA, achieving an adequate reduction in nutrient loading will be a major challenge.

Summary

ROFIs are host to a wide range of physical processes; they combine many of the characteristics of both the central shelf and estuarine regimes in a dynamic system of considerable complexity. They occur where the flux of freshwater from one or more estuaries inputs significant amounts of buoyancy to a coastal region. The resulting horizontal gradients of density tend to drive a circulation analogous to that which occurs in an estuary, but the flow in a ROFI is modified by the Coriolis forces which tend to generate coast parallel currents moving in the direction of Kelvin wave propagation (i.e. with the coast to the right of the flow direction in the northern hemisphere). The competition between the stratifying effect of the freshwater buoyancy and stirring by wind and tide is more subtle than that between heating and stirring (explored in Chapter 6) because the buoyancy is injected through specific sources at the coast rather than being distributed uniformly over the sea surface. Variations in tidal stirring over the spring-neap cycle, along with more irregular changes in wind stirring and freshwater input, produce a highly variable regime which can change from strongly stratified to vertically well mixed on a variety of time scales. Periodic variations in stratification may occur on a semi-diurnal time scale in response to tidal straining, a process which results from the interaction between vertical shear in the tidal currents and the horizontal density gradient. The form of the interaction depends on whether the tide is predominantly a standing or a progressive (Kelvin) wave. Wind stress may be important in both cases, both by adding extra stirring and by driving the along-coast flow towards, or away from, the land boundary.

Given this dynamic complexity, and particularly the range of time scales on which buoyancy and mixing processes compete, it is not surprising that primary production in ROFIs does not operate with the seasonal consistency that we see in deeper shelf sea regions. Rather than a single bloom of surface productivity in spring, production in ROFIs can peak at any time of the year. Spring-neap modulation of stratification in many ROFIs forces similar periodicity in phytoplankton growth. There may be evidence that phytoplankton uptake of nutrients can respond to the faster semi-diurnal stratification of tidal straining, but the rapid switching between mixed and stratified sites is unlikely to result in a measurable response in phytoplankton biomass.

An ecologically important feature of the physics of ROFIs is the mean, density-driven circulation set up by the horizontal density gradient. Outflow at the sea surface, and shoreward flow near the seabed, can be exploited by organisms that release larvae into the water column but then require some method for transport of older, developed larvae back towards the adult habitat. The slow density-driven flows (O(cm s^{-1})) and the typical horizontal scale of ROFIs are often well suited to the ontogenic time scales of the young of many estuarine animals. This is an area of active research, particularly in the context of understanding how the physical environment supports commercially important coastal fisheries.

As the zone of first impact of whatever contaminants are carried by estuarine waters, the water quality in many ROFIs is monitored closely. The idea of the coastal ocean being a limitless receiver of unwanted domestic, industrial and agricultural effluents has now been superseded by the knowledge that the physics of these regions tends to limit dispersion into the wider coastal ocean. Contaminants that adsorb onto sediments will be transported by the onshore near-bed flow back into the source estuary, or into estuaries downstream in the coastal buoyancy current. The often high nutrient loads of estuarine water can, in stratified ROFIs, drive high surface primary productivity and are often seen to correlate with growth of toxic species of phytoplankton. As that organic material eventually sinks into the bottom layer and begins to be consumed by bacteria, high biological oxygen demand can severely deplete the bottom waters of dissolved oxygen, with major impacts on coastal organisms.

ROFIs are host to a complex set of processes, making them perhaps the most difficult of all the regimes of the shelf seas to predict. While we have a first order understanding of many of the processes involved, combining them in models to fully simulate their physical and biological behaviour remains a major challenge.

FURTHER READING

de Boer, G. J., *et al.* SST observations of upwelling induced by tidal straining in the Rhine ROFI. *Continental Shelf Research*, **29(1)**, 263–277, 2009.

Hill, A. E. Buoyancy effects in coastal and shelf seas, In: *The Sea*, vol. **10**, ed. K. H. Brink and A. R. Robinson, John Wiley & Sons, Chichester, 21–62, 1998.

Hypoxia in the Gulf of Mexico: http://www.gulfhypoxia.net/.

Problems

9.1. Along the Dutch coast at latitude 52.25° N, the density increases outwards from the coast with a depth uniform gradient of 1.5×10^{-4} kg m^{-4}. Estimate the cross-shore surface slope and hence the alongshore surface current at a point where the water depth is 18 m assuming that the flow is in steady geostrophic balance and that there is zero pressure gradient at the bed. Indicate the form of the velocity profile and estimate the net transport per unit width along the coast. Compare your answer with (i) the result of integrating the thermal wind (Equation 3.28), and (ii) the steady flow obtained numerically with the ROFIZ Simulation Module using a realistic value of the eddy viscosity (say $N_z = 0.02$ m^2 s^{-1}).

9.2. In a ROFI where the tide is predominantly a standing wave, the water column is well mixed at high water. If the subsequent ebb flow has a peak velocity of

$0.65 ms^{-1}$, estimate the water column stability Φ at low water which would result from tidal straining given that the density gradient along the principal axis of the tide has a magnitude of 2.1×10^{-4} kg m^{-4} and the depth is 25 m. what would be the reduced value of Φ if tidal stirring is included?
(Hint: Assume the velocity profile has the form given in (Equation 9.23) and integrate Equation (9.24) over the ebb half cycle. Use Equation (9.25) to estimate the contribution of tidal stirring with $e = 0.004$; $k = 0.0025$.)

9.3. The data from the Liverpool Bay anchor station shown in Fig. 9.18c looks very similar to that shown in Fig. 9.9c, suggesting that rather than advection of a front, the data in Fig. 9.18c reflect the semi-diurnal process of straining-induced stratification. If this is true, there are interesting implications for the biochemistry of a ROFI arising from the observations shown in Fig. 9.18a, b.
Write down any arguments you can think of that support the 'frontal advection' and the 'SIPS' hypotheses as explanations for the physical and biochemical data presented in Floodgate *et al.*, 1981. Do you favour one or other of the hypotheses, and why? Are there other data that you would collect to allow you to separate the possible explanations?

9.4. At high water a ROFI has a typical vertically mixed chlorophyll profile with concentration 1 mg Chl m^{-3}. Using reasonable values for the net phytoplankton growth rate, estimate:

(i) The change in near-surface chlorophyll in response to one semi-diurnal cycle of SIPS.
(ii) The change in near-surface chlorophyll in response to a neap-tide stratification period (similar to that in Fig. 9.7).

Specify what assumptions you make.

9.5. Crustacean larvae in the surface waters of a ROFI need to develop to a final stage where they can then swim downward towards the onshore mean flow of the bottom water. Depending on rainfall, the horizontal density gradient in the ROFI ranges between 1×10^{-7} and 1×10^{-6} m^{-1}, depth is about 40 metres and the typical offshore extent of the ROFI is 50 km.

What limits do the physical characteristics of the ROFI place on the developmental time scale of the larvae?
(Hint: Use the offshore component of the mean ROFI circulation in Heaps, 1972, and assume $N_z = 5 \times 10^{-3}$ m^2 s^{-1}.)

10 The shelf edge system

At the outer edge of the shelf is a region where the gentle slopes of the shelf give way to the much steeper topography of the continental slope, and bottom depths rapidly increase down to the abyssal plains of the deep ocean. On average, the depth at which this slope transition occurs is about 130 metres, but this varies through the world's oceans. Off NW Europe the shelf edge is at a depth of 200 metres, while at high latitudes the shelf edge is deeper, typically 400–500 metres around Antarctica and off Greenland. Because the topography is steep, with slopes as large as 1:10, the transition between shelf and the deep ocean is usually limited in extent (~50 km). It is in this rather narrow region that the very different regimes of the shelf and the deep ocean adjust to each other. In this chapter, we shall consider how this adjustment occurs and how it controls the important exchanges between shelf and deep-ocean. We will look at wind-driven upwelling, the most studied process linking the physics and the ecology of the shelf edge which, in many parts of the world, supports important stocks of plantivorous fish. We shall also consider the upwelling (and downwelling) driven by the bottom Ekman layer of along-slope flows, and the consequences for nutrient supply to, and organic material export from, shelf seas. We will describe the density contrasts that develop in winter between temperate shelf seas and the adjacent ocean that can lead to downslope cascades of shelf seawater and its constituents. Finally, we consider the role of the internal tide, a prominent shelf edge process which strongly influences the biochemistry and is important in relation to commercial fisheries.

10.1 Contrasting regimes

The deep ocean and shelf regimes differ radically in a number of respects, illustrated in the schematic in Fig. 10.1. Most obviously they differ in the depth of the water column, on average 3.8 km for the deep ocean as opposed to 0–200 metres for the shelf, a difference which has important consequences for the relative extent of

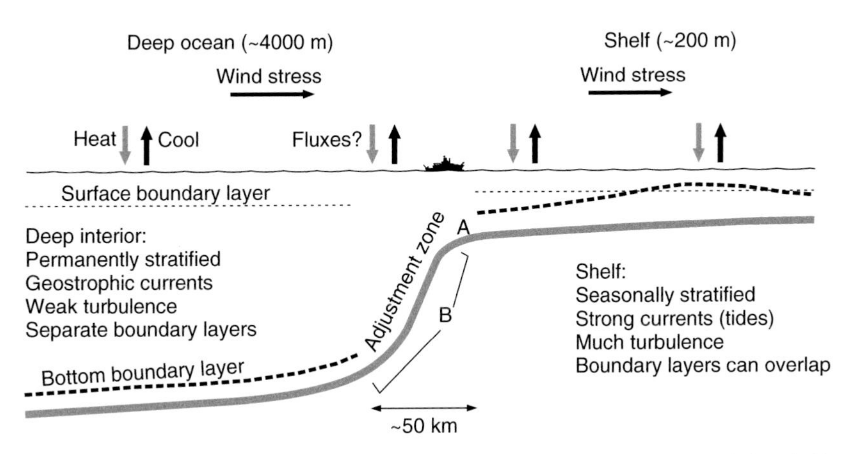

Figure 10.1 The very different dynamical regimes of the deep ocean and the shelf sea must adjust to each other in the rather narrow region of the slope. This mutual adjustment involves dynamical processes which contribute to the specialised environment of the slope. A is the shelf edge, the region marked B is the continental slope. The ship silhouette (RRS *James Cook*, length 100 m) is in proportion to the 200m depth of the illustrated shelf edge, but remember that the horizontal scale is substantially shortened.

boundary influences. We have seen that the frictional boundary layers induced at the surface and bottom occupy a large proportion of the water column in the shelf seas. These boundary layers are frequently energetic with high levels of turbulence; they may overlap and promote complete mixing through the water column. By contrast, in the deep ocean, the boundary layers occupy only a very small fraction of the total depth. Moreover, the bottom boundary layer is generally much weaker in the deep ocean because the barotropic tidal currents are typically 1–2 orders of magnitude smaller than on the shelf. In between the relatively thin surface and bottom boundary layers of the deep ocean is a large interior region where conditions are isolated from boundary stresses and where the flow varies rather slowly and is normally close to geostrophic balance. Turbulence in the deep ocean, between the boundary layers, is generally very weak; on average the typically vertical diffusivity is only about $1 \times 10^{-5}\ \mathrm{m^2\ s^{-1}}$ (Munk, 1966; Munk and Wunsch, 1998). There is considerable interest in the role of steep seabed topography in the deep ocean increasing turbulence in the ocean interior (e.g. Ledwell *et al.*, 2000; Garabato *et al.*, 2004), but even in these regions the strength of the turbulence is substantially lower than that which we find in shelf seas.

At the shelf edge, these very different dynamical regimes exist in rather close proximity, separated only by the narrow region of the continental slope. Within the width of the slope, the differences between the regimes have to be reconciled. The processes involved in this mutual adjustment create a rather special shelf edge regime which, as we shall see, has profound effects on the biology over a range of trophic levels. In particular the flow field and the mixing occurring at the shelf edge are of crucial importance in relation to the transfer of nutrients between the open ocean and the shelf seas and the fixation of carbon and its transfer as organic carbon across the slope into the deep ocean.

10.2 Bathymetric steering and slope currents

An important and fundamental aspect of the flow at the shelf-deep ocean boundary is that the steep topography of the slope, in combination with the Earth's rotation, imposes a particular discipline on the currents. This constraint on the flow, which is known as the Taylor-Proudman theorem after its discoverers, will now be derived from the equations of motion following the approach of Brink (Brink, 1998).

10.2.1 The Taylor–Proudman theorem

We start by assuming that the flow is steady and in geostrophic balance so that from Equation (3.16) we have:

$$fu = -\frac{1}{\rho}\frac{\partial p}{\partial y}; \quad -fv = -\frac{1}{\rho}\frac{\partial p}{\partial x}. \tag{10.1}$$

We are aiming at understanding how geostrophic flows might be influenced by a sloping seabed, which could lead to flows converging or diverging. So we want to know how geostrophic flows change horizontally, while at the same time being constrained by flow continuity. Differentiating Equations (10.1) with respect to x and y respectively and combining with the continuity Equation (3.1), we have

$$\boxed{-\left(\frac{\partial u}{\partial x} + \frac{\partial v}{\partial y}\right) = \frac{\partial w}{\partial z} = 0} \tag{10.2}$$

i.e. the vertical velocity is everywhere independent of depth. Since $w = 0$ at the surface ($z = 0$), it follows that $w = 0$ throughout the water column. This condition that $w = 0$ everywhere means that the flow has to be parallel to the isobaths; cross-isobath flow would require a component of flow upward or downward. We can express this formally as

$$-\underset{\sim}{u}.\nabla h = w = 0 \tag{10.3}$$

i.e. the flow vector $\underset{\sim}{u}$ at the bed must be perpendicular to ∇h and hence parallel to the isobaths.

In addition, for a steady state and neglecting diffusion, Equation (4.35) for the transport of density becomes

$$u\frac{\partial \rho}{\partial x} + v\frac{\partial \rho}{\partial y} + w\frac{\partial \rho}{\partial z} = 0. \tag{10.4}$$

Since the flow is geostrophic, we can use the thermal wind balance (Equation 3.30) to write the density gradients in terms of the velocity shear to give:

$$\frac{\partial \rho}{\partial x} = -\frac{\rho_0 f}{g}\frac{\partial v}{\partial z}; \quad \frac{\partial \rho}{\partial y} = \frac{\rho_0 f}{g}\frac{\partial u}{\partial z}. \tag{10.5}$$

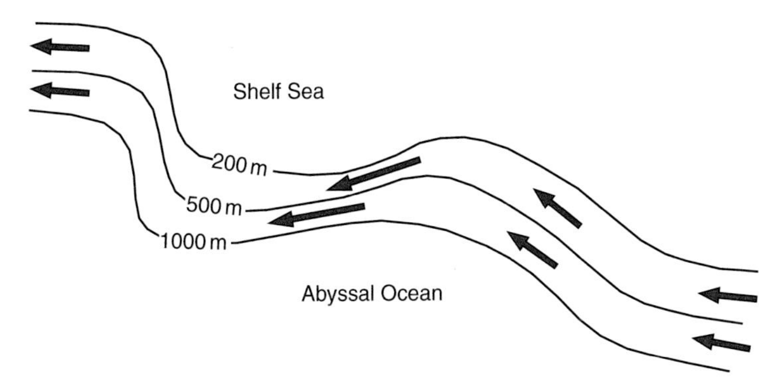

Figure 10.2 The Taylor–Proudman theorem. The speed of the current may vary with depth but the direction must be parallel to the isobaths at all levels. Notice that the current increases where isobaths are closer together. Bathymetric steering of the flow acts to inhibit cross-slope exchange. Deviations from isobath-parallel flow may occur due to (i) unsteady flow (e.g. in the oscillatory tidal flow), (ii) energetic flow in which the Rossby number is large (> 0.1) and (iii) friction in the boundary layers.

Substituting in 10.4 and setting $w = 0$ we have:

$$-u\frac{\partial v}{\partial z} + v\frac{\partial u}{\partial z} = \underset{\sim}{u} \times \frac{\partial \underset{\sim}{u}}{\partial z} = 0 \tag{10.6}$$

which means that the velocity shear is parallel to the velocity, i.e. the flow direction is the same at all levels and parallel to the isobaths, although the current speed may vary with depth.

This result, that the flow is topographically steered at all levels, constitutes a fundamental constraint on steady geostrophic flow. Because of the particularly steep topography over the continental slope, there is a strong tendency for this constraint to operate there and for the flow to follow the bottom contours along the slope, as illustrated in Fig. 10.2. Notice how the flow speed increases where the isobaths are closer together. Cross-isobath flow can only occur where one or more of the assumptions of the Taylor–Proudman theorem are broken and the flow is no longer geostrophic. This will be the case if (i) the flow is unsteady, or (ii) it is so vigorous that the non-linear terms in the dynamical equation become significant and violate the linearising assumption, or (iii) frictional effects are important as is the case in the boundary layers near the surface and the seabed.[1]

10.2.2 Slope currents

The dominating role of topographic steering is apparent in the pronounced 'slope currents' which occur on many continental shelves and transport large volumes

[1] In Chapter 3 we mentioned how the commonly held idea that the Gulf Stream bathes NW European shelf seas in warm water is flawed because of the very weak mean flows seen in shelf seas. The strong inhibition of cross-isobath flow in the Taylor-Proudman theorem provides another reason for dispelling this misunderstanding.

of water parallel to the shelf edge with relatively little movement across the slope. The slope current along the Hebridean shelf edge, which was the focus of an extensive study of processes at the shelf edge,[2] provides a good illustration of the general character of these currents. The bathymetry of the region is shown in Fig. 10.3a. Each of two sections of the along-slope flow in Fig. 10.3b,c are based on a series of repeated traverses across the shelf over a period of 13 hours with a shipborne ADCP (Souza *et al.*, 2001). After removal of the tidal signal by a least squares analysis (Simpson *et al.*, 1990), a strong jet-like current is seen to flow along the slope with speeds approaching 0.3 m s^{-1} in the core of the current. The close similarity of the structure at the two sections, which are separated by about 55 km along the slope, indicates the spatial consistency of the current. The flow is largely barotropic (i.e. almost depth independent), is ~15–25 km wide and has a maximum over the steep slope close to the 400-metre isobath. The longer-term existence of the slope current is shown in Fig. 10.3d using a 25-day average of the slope current in the summer derived from ADCP and current meter moorings. This shows a similar structure but with a lower maximum flow speed. The currents meters also showed a very weak cross-shore flow (<0.02 m s^{-1}). Below the wind mixed layer (~100-metre depth) the current vector; $\underset{\sim}{v}$ is remarkably consistent. We can quantify this by defining a 'steadiness factor',

$$St = \frac{|\overline{\underset{\sim}{u}}|}{\overline{|\underset{\sim}{u}|}} = \frac{vector\ mean}{scalar\ mean} \tag{10.7}$$

with $St = 1$ for a flow with constant direction. The slope current in Fig. 10.3d has $St \sim 0.8 - 0.9$, illustrating the steering effect of the slope bathymetry. For flow on the shelf and off in the ocean, St is typically <0.5. The total transport of this slope flow in summer is ~1 Sv (i.e. 1×10^6 m^3 s^{-1}) and increases to ~2 Sv in winter.

The Hebridean slope current forms part of a continuous slope transport which extends along the European shelf edge northwards from the Bay of Biscay to the Norwegian Sea. The existence of this current has been inferred from measurements at locations along the shelf edge and from simulations with numerical models like that shown in Fig. 10.4a (Pingree and Cann, 1991). The continuity of the slope current and the constraining effect of bathymetric steering are also convincingly illustrated by the paths of drifting drogue buoys. Figure 10.4b shows the results of an experiment in which clusters of drogues were released at three closely adjacent but contrasting locations: one on the shelf, one over the slope and a third in deep water at the bottom of the slope. Those released in deep water tend to move in large eddies which are the characteristic features of the deep ocean circulation, while those on the shelf exhibit weak and variable motion. By contrast, those over the slope show consistent and rapid movement northwards at speeds of 0.1–0.3 m s^{-1}.

[2] The NERC-funded Shelf Edge Study (SES) 1994–96. This programme yielded much new information about shelf edge processes, but it also made us appreciate the intensity of the fishing effort which is focussed on the shelf edge and slope. Long term observations were frequently disrupted through irresponsible fishing by vessels trawling up moored equipment.

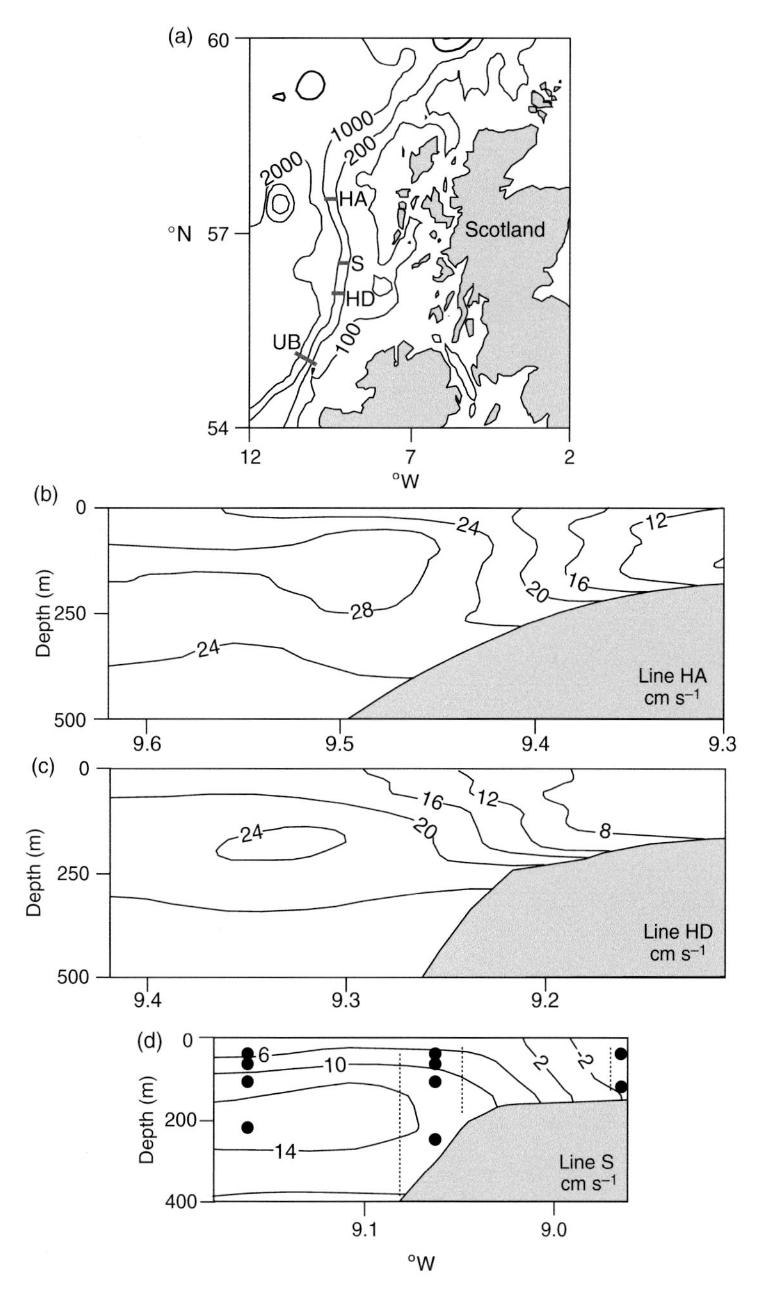

Figure 10.3 (a) Slope topography (contours in metres) and location of the current sections on the Malin and Hebridean slope during the LOIS-SES project; (b) Snapshot of the residual flow on section HA determined in summer 1996; (c) As (b) but for section HD; (d) The slope current, averaged over a 25-day period (August 12–September 6 1995) on the S section, using data from moored ADCPs and current meters. In (d) positions of upward-looking ADCPs are marked with vertical dashed lines, and the locations of individual current meters are shown with black circles. Adapted from Souza *et al.*, 2001, with permission from Elsevier.

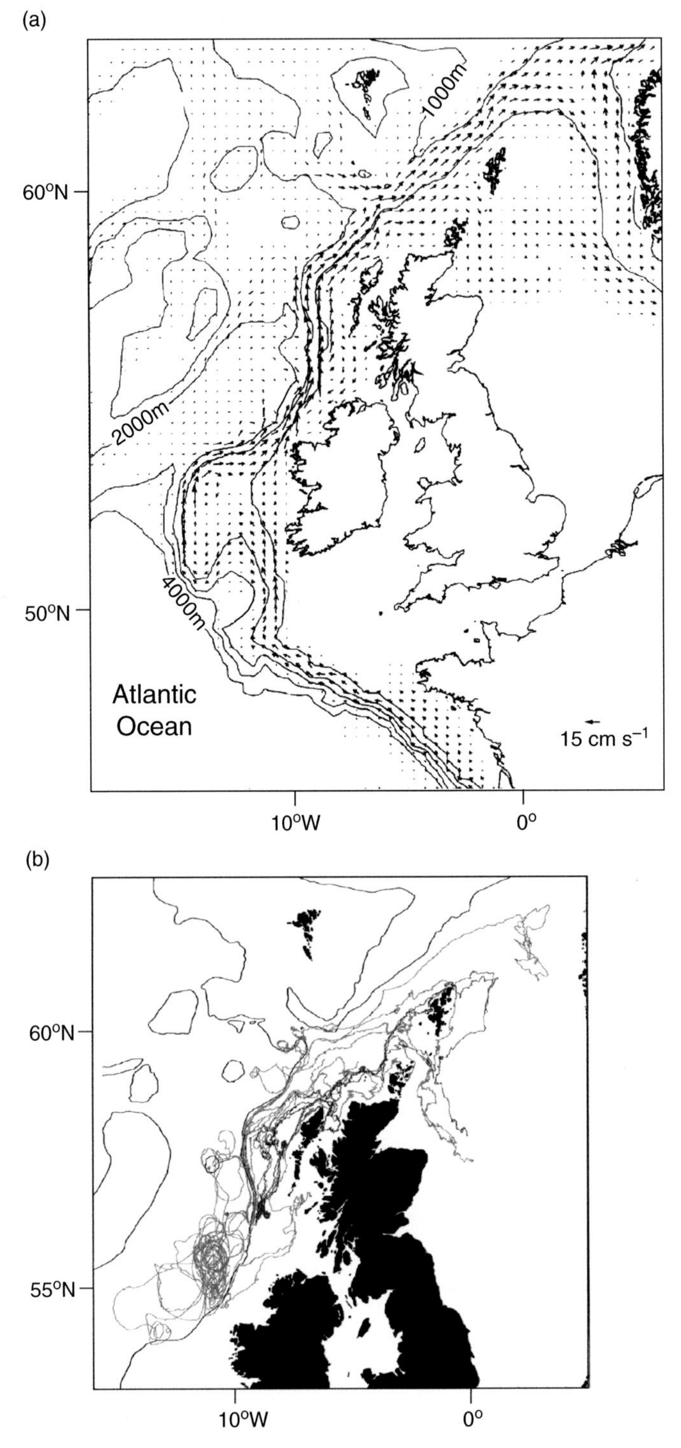

Figure 10.4 See colour plates version for (b). (a) Flow along the NW European shelf edge derived from a barotropic numerical model forced by a meridional density gradient equivalent to a downward slope of the sea surface of 10^{-7} (1 cm in 100 km) in the deep water. The model

10.2.3 Forcing of slope currents

The influence of bathymetric steering is manifest in the occurrence of pronounced currents over continental slopes in many parts of the world ocean. While they are all constrained to flow parallel to the topography, they are forced by a number of different mechanisms.

On the western margins of the ocean, the mean wind-driven circulation over an ocean basin is concentrated in the relatively narrow jet of a Western Boundary Current. The westward intensification of the boundary currents is the result of the large-scale oceanic gyre circulation flowing under the influence of the latitudinally varying Coriolis forcing, as first shown by Henry Stommel (Stommel, 1948). The topography of the shelf edge and slope provides a region for trapping the boundary flow, and potentially altering its strength and width (e.g. note again the schematic in Fig. 10.2). In other words, the current is primarily forced by much larger-scale processes, but it forms close to the shelf edge simply because the geostrophic oceanic flow will be constrained by the Taylor-Proudman theorem in the region of steep topography. This is the case for the Gulf Stream as it flows through the south and mid Atlantic Bights before it leaves the slope boundary at Cape Hatteras. The Kuroshio is similarly constrained for part of its passage along the continental slope off Japan, as is the East Australia current as it flows southwards off the eastern margin of Australia and the Brazil Current off southeast South America.

At the eastern boundaries of the ocean basins, the wind-driven basin circulation is much broader and weaker and other mechanisms are involved in driving slope currents. In the case of the eastern margin of the North Atlantic, the strong and persistent flow which we have seen flowing along the continental slope of northwest Europe has its origin in the mutual adjustment of shelf and oceanic regimes to the meridional density gradient, via a process known as JEBAR (Joint Effect of Baroclinity and Relief). In the North Atlantic, as in all other ocean basins, there is a poleward increase of the density in the upper ~ 1 km of the water column, mainly due to the decrease in temperature with latitude. Typically this gradient has a value $\frac{\partial \rho}{\partial y} = O[10^{-4}]$ kg m^{-3} km^{-1}.

We can understand how the meridional density gradient interacts with the steep bathymetry of the slope by developing a highly simplified model (Pingree, 1990). Consider an ocean of uniform depth H adjacent to a slope and shelf where the depth contours are oriented north-south along the y axis, as shown in Fig. 10.5. The density is assumed to be independent of depth and a function of y only, while the

Caption for Figure 10.4 (*cont.*) predicts a pronounced flow over the continental slope which is almost continuous from the Bay of Biscay northwards to the northern North Sea. The current tends to be strongest over the upper part of the slope and to increase northwards as far as 60° N. After (Pingree and Cann, 1991); (b) Tracks of satellite-tracked drifting buoys released during the SES programme. Surface floats were attached to high drag drogues to follow the currents at 50 m depth (Burrows *et al.*, 1999). Clusters of drogues were released at three locations: one on the shelf (blue), one over the slope (green) and a third in deep water at the bottom of the slope (red). Figures courtesy of Elsevier.

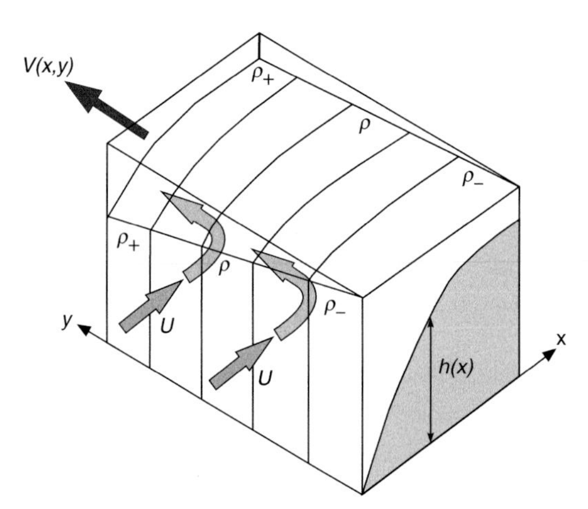

Figure 10.5 A schematic illustration of the density field and flow driven by the JEBAR process along the shelf slope. The figure is drawn for the northern hemisphere. Adapted from Hill, 1998, courtesy John Wiley & Sons.

depth $h(x)$ is a function of x only. The steady state, vertically integrated equations of motion for frictionless, linear flow can then be written:

$$-fV = -g\frac{\partial\eta}{\partial x}h; \quad fU = -g\frac{\partial\eta}{\partial y}h - \frac{g}{2\rho_0}\frac{\partial\rho}{\partial y}h^2 \tag{10.8}$$

where U and V are the vertically integrated transports in x and y. Again, we want to know how this flow will vary horizontally while maintaining flow continuity, so we differentiate these equations with respect to y and x respectively and combine with the continuity equation

$$\frac{\partial U}{\partial x} + \frac{\partial V}{\partial y} = 0 \tag{10.9}$$

to give the meridional surface slope as:

$$\frac{\partial\eta}{\partial y} = -\frac{h}{\rho_0}\frac{\partial\rho}{\partial y}. \tag{10.10}$$

For the shelf ($h = h_s$), and the deep ocean ($h = H$), the surface slopes will be:

$$\left(\frac{\partial\eta}{\partial y}\right)_{h_s} = -\frac{h_s}{\rho_0}\frac{\partial\rho}{\partial y}; \quad \left(\frac{\partial\eta}{\partial y}\right)_H = -\frac{H}{\rho_0}\frac{\partial\rho}{\partial y}. \tag{10.11}$$

Since $H \gg h_s$ the downward slope with latitude will be greater over the deep ocean and, in consequence, the difference between sea surface height over the shelf and that over the ocean will increase with latitude, as is illustrated schematically in Fig. 10.5. This difference will result in a geostrophic current parallel to the isobaths. At the same time, the dynamical balance in the y direction implies a geostrophic transport in deep water towards the slope of:

$$U = \frac{gH^2}{2\rho_0 f}\frac{\partial\rho}{\partial y}. \tag{10.12}$$

As it reaches the slope, this flow turns to the north and joins the meridional current, which is increasing with latitude as the zonal difference in height increases.

The JEBAR forcing mechanism has been explored theoretically by Huthnance (Huthnance, 1984) in a more realistic model which includes the effects of friction. He showed that the current should be barotropic and that its magnitude v_s will vary across the slope according to

$$v_s = \frac{g}{2\rho_0 K_b'} h(H - h) \frac{\partial \rho}{\partial z} \tag{10.13}$$

where k_b' is a linear drag coefficient. This picture of the slope current is confirmed by numerical simulations of flow driven by JEBAR (Pingree, 1990; Blaas and de Swart, 2000) and is consistent with observations. In the case of the slope current along the European shelf edge, the meridional gradients do not vary greatly over the annual cycle, so we might expect that the current, if driven by JEBAR, would not exhibit large differences between winter and summer. This appears to be the case although, as mentioned above, there is some significant strengthening of the flow in winter when wind forcing becomes more significant.

Note the difference between currents driven by the JEBAR mechanism, which are 'true' slope currents in the sense that the steep topography of the slope plays a key role in the forcing mechanism, and flows forced by other mechanisms which are concentrated and steered along the slope. In addition to the steering of western boundary currents mentioned above, slope currents can appear as a result of other forcing mechanisms. Off the western coast of North America, for instance, the California Current, which is seasonally forced by upwelling-favourable winds, involves a southward-flowing jet on the shelf close to the shelf edge (Winant *et al.*, 1987). Another example is the Benguela Current, the eastern boundary flow in the South Atlantic. The western side of the current tends to be relatively poorly defined, but the eastern side is concentrated as an upwelling-driven frontal jet constrained by the slope bathymetry (Nelson and Hutchings, 1983; Peterson and Stramma, 1991).

10.3 Cross-slope transport mechanisms

In constraining currents to be parallel to the isobaths, topographic steering also inhibits cross-slope transfer as is apparent from the tracks of the buoys deployed in the slope current shown in Fig. 10.4b. We have already noted, however, that there are a number of ways in which this constraint can be circumvented when the assumptions of the Taylor-Proudman theorem are not satisfied. We now consider some of the more important candidate mechanisms contributing to transfer across the isobaths.

10.3.1 Wind stress in the surface boundary layer

In the surface layers, the action of large frictional stresses imposed by the wind can invalidate the geostrophic balance and induce substantial cross-slope transfer. We found in Section 3.5 that a steady wind in the x direction exerting an along-slope stress of τ_w, in deep water and away from lateral boundaries, induces a vertically integrated flow at right angles to the wind stress given by the Ekman transport relation (Equation 3.43) as:

$$V = \frac{\tau_w}{\rho_0 f}. \tag{10.14}$$

If the wind is blowing parallel to the shelf edge in the northern hemisphere with the shelf to the left of the wind direction, the transport will be offshore and normal to the bathymetry. For $\tau_w = 0.1$ Pa (wind speed $W \sim 7.5$ m s^{-1}), this cross-shore transport will be $V \sim 1$ m^3 s^{-1} per metre of shelf edge. As we saw in Section 3.5.3, continuity requires that such an offshore Ekman transport in the surface layer must be balanced by onshore flows in the lower part of the water column with an upwelling flow at the ocean boundary. When the upwelling-favourable wind switches on, it takes some time for the system to adjust and to establish the nearbed onshore flow of water. As an example, consider the time series of alongshore winds and the response of the lower water column collected by one of us during work at the shelf edge of northeast New Zealand, shown in Fig. 10.6. Pulses of upwelling-favourable wind stress (Fig. 10.6a) often correlate with onshore near-bed mean flow (Fig. 10.6b) and reductions in near bed temperature (Fig. 10.6c) indicating the on-shelf transfer of cooler, deeper water. The most obvious upwelling events in Fig. 10.6 occur at the end of May, end of June, and following the July 19. A more detailed correlation analysis of the time series indicates that about 44% of the near bed cross-slope current variability and 65% of the near bed temperature variability can be explained by changes in the along-slope wind stress, with the currents lagging the wind by 37 hours and the temperature lagging the wind by 65 hours. The New Zealand shelf is relatively narrow (the mooring that provided the data in Fig. 10.6 was about 35 km from the coast), but localised transfers of deeper water across a shelf edge also occur on wider shelves (Johnson and Rock, 1986).

At this stage it is worth noting that the front between the cold, upwelled water and the warmer shelf surface water is not generally seen to be a stable feature. The alongshore geostrophic current set up on the warm side of the front generates mesoscale frontal instabilities, and potentially eddies, analogous to those seen at the shelf sea tidal mixing fronts (Section 8.4). The structure of the instabilities in upwelling systems has been observed to take the form of a number of pronounced filaments of surface water streaming offshore to beyond the shelf edge at intervals along the coast set by prominent features in the coastal topography (Haynes *et al.*, 1993).

Upwelling of deep water from seaward of the shelf break has important consequences for the supply of nutrients to the shelf seas, while the wind-driven offshore

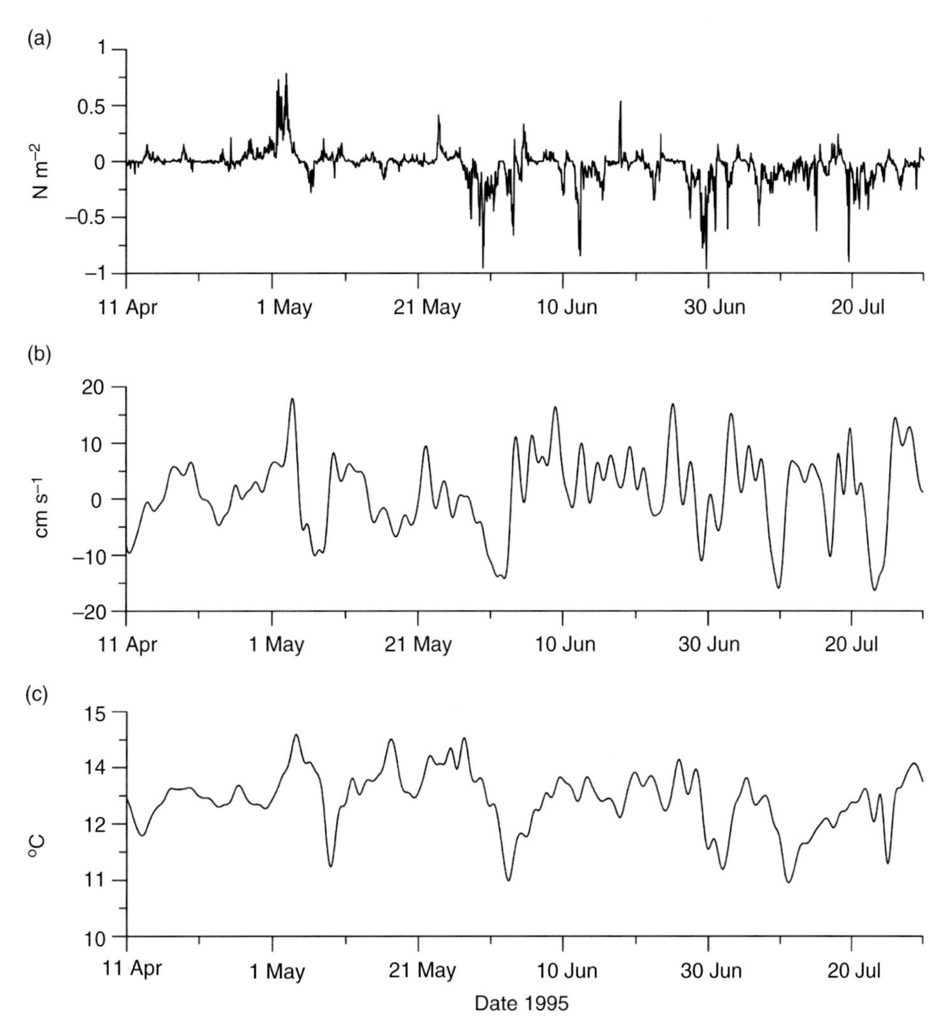

Figure 10.6 (a) Alongshore component of surface wind stress off northeast New Zealand during the austral winter, 1995. Negative wind stress is to the southeast and upwelling favourable; (b) Cross-shelf edge component of currents measured by a current meter 10 metres above the seabed at the shelf edge (position: 35° 30.22' S, 174° 54.77' E). The currents have been low-pass filtered to remove tidal variability; negative flow is on-shelf; (c) Near-bed temperature measured by the current meter, low pass filtered to remove tidal variability.

surface flow, with associated eddies and filaments, can export surface material off the shelf. Also, remember that if the wind direction is reversed then surface water piles up against the coast, generating a surface slope that will push deep shelf water towards the shelf edge and the open ocean. We then have a mechanism for exporting deep shelf water and importing surface oceanic water. We shall explore the biogeochemical and ecological consequences of these transports later in this chapter.

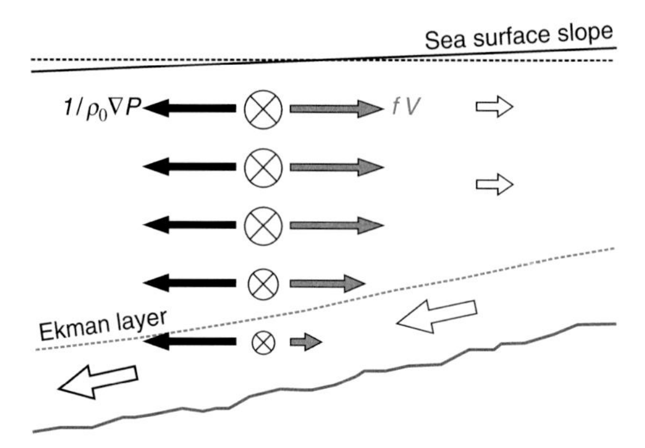

Figure 10.7 Cross-shelf downwelling circulation induced by a slope current flowing northward (northern hemisphere). Crossed circles indicate the flow into the page. Black arrows indicate the direction and strength of the pressure gradient, grey arrows the Coriolis force.

10.3.2 The bottom boundary layer

Just as in the surface layer, frictional stresses at the bottom boundary contravene the assumption of geostrophy and allow an important cross-isobath flow forced by a slope current. As we have seen in Section 10.2.2, slope currents are often approximately barotropic and thus their influence extends to the seabed over the slope. The effect of bed friction on an along-slope flow is shown in Fig. 10.7. Frictional effects in the bottom boundary layer slow the current so that the Coriolis force near the bed is reduced and there is, therefore, an unbalanced pressure component which drives a down slope flow (large white arrows in the bottom Ekman layer in Fig. 10.7). Continuity requires a return flow towards the shelf in the upper part of the water column (small white arrows).

The example of Fig. 10.7 is analogous to the slope current along the Hebridean shelf edge west of Scotland. Flowing with the shelf to the right of the current direction, the near-bed current is deflected to the left of the direction of the slope current, i.e. in the downslope direction (look at the cross-shelf sea surface slope in Fig. 10.5). If the stress is related to the barotropic current velocity $\hat{v}$ by a quadratic drag law $\tau_b = k_b \rho_0 \hat{v}^2$, the net transport in the cross-slope direction has a magnitude of:

$$T_b = \frac{\tau_b}{\rho_0 f} = \frac{k_b \hat{v}^2}{f}. \tag{10.15}$$

At latitude 58°N with $\hat{v} = 0.2\,\mathrm{ms}^{-1}$ and $k_b = 0.0025$, the net transport across the slope is $T_b \simeq 0.8\ \mathrm{m}^2\ \mathrm{s}^{-1}$. In a steady state, continuity will require that this cross-slope flow is balanced by a compensating flow in the upper part of the water column, as in Fig. 10.7. In view of the consistency of the slope current, this bottom Ekman-layer flow provides an important mechanism for the transfer of material between the shelf and the deep ocean, sometimes referred to as the *Ekman drain*. If the along-slope current is in the opposite direction (shelf to the left of the current direction), transport in the bottom boundary layer will be upslope (northern hemisphere). This is the case along part of the east coast of North America, where the Gulf Stream flow is northwards over the slope. The resulting upslope transport in the bottom boundary layer, with a compensating offshore flow in the upper part of the water column, is thought to contribute significantly to the supply of nutrients to the shelf (Condie, 1995). In both cases the transport in the bottom Ekman layer is the result of bed friction, and so the bottom Ekman layer will be turbulent. This turbulence will result in a mixed near-bed layer, the observation of which is a useful indicator of near-bed cross-slope flow.

Transport in the surface and bottom Ekman layers may reinforce or oppose each other, but their average effect can be a major contribution to cross-slope exchange. For the European shelf region between 54° N to 57 N, the net flux, forced by surface and bottom boundary layer motions for the period 1960–2004, has been estimated from a full-physics model as a net downwelling transport of 1.75 $m^2\ s^{-1}$ (Holt and Proctor, 2008). The corresponding horizontal transports along the ~350 km length of this shelf break sector amount to $\sim 1.75 \times 350 \times 10^3 = 0.6$ Sv on to the shelf in the upper layers and off-shelf near the bed.

10.3.3 Cascading

It has long been suspected (Cooper and Vaux, 1949) that near-bed transport across the slope in mid- and high latitudes may also be forced by density differences between the shelf and slope produced by intensive cooling of shelf water during the winter months. Because of the shallower depth, shelf water tends to cool more rapidly than water over the slope and hence becomes denser than the slope water towards the end of the winter cooling period. If the shelf water density exceeds that of the adjacent slope by $\Delta\rho$, it will acquire a negative buoyancy b:

$$b = \frac{(\rho_{shelf} - \rho_{slope})}{\rho_0} g \simeq -\frac{\Delta\rho}{\rho_0} g = -g'. \tag{10.16}$$

Such dense water sitting at the top of the slope would be unstable and tend to flow down the slope in an intense density current or *cascade*. The downslope flow will be deflected by the Coriolis force and be diverted into an along-slope flow, but frictional stress in the bottom boundary layer undermines geostrophy and ensures that there will be a downslope component of flow even if bed topography is uniform in the along-slope direction. Figure 10.8 illustrates the force balance for steady flow of a layer of thickness h on a uniform slope with slope δ relative to horizontal. The downslope component of gravity, $g' \sin \delta$, which is driving the flow, is opposed by the frictional stresses F_v and the Coriolis force. For the forces acting in the plane of the slope, the balance in the direction of the flow can be written as

$$g' \sin\delta \sin\theta = F_v \tag{10.17}$$

where θ is the angle between the flow direction and the isobaths. In the direction normal to the flow, the balance is given by:

$$(f \cos\delta) V_g = g' \sin\delta \cos\theta, \tag{10.18}$$

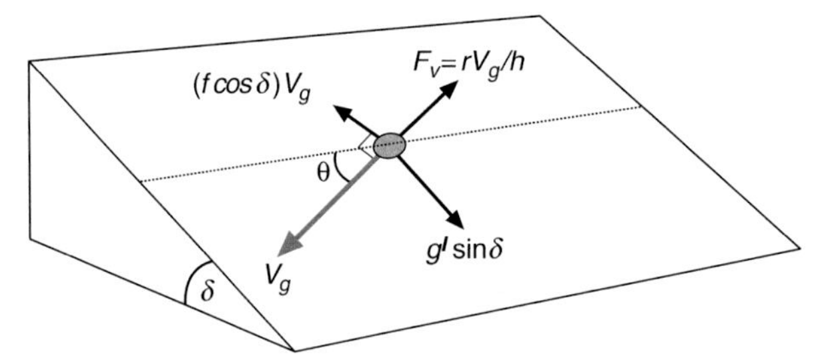

Figure 10.8 Forces acting in the plane of the slope during a steady flow of a layer of thickness h down a uniform slope at a speed V_g. The gravitational force due to the excess density of the cascading flow acts down the slope and is balanced by the Coriolis force $(f \cos \delta)$ V_g acting to the right of the flow (northern hemisphere) and a drag force F_v induced by friction at the top and bottom boundaries of the flow.

where V_g is the speed of the flow, which is assumed to be vertically uniform in the cascading layer. In deriving Equations (10.17) and (10.18) we have neglected a small contribution from the horizontal component of the Earth's rotation. The frictional term F_v represents the combined frictional stresses at the top and bottom of the cascading layer. If we assume that F_v can be represented by a linear drag law of the form $F_v = r\rho V_g$, then the flow direction and speed are given by:

$$\tan\theta = \frac{r}{fh\cos\delta} \simeq \frac{r}{fh}; \quad V_g = \frac{g'h\sin\delta}{\sqrt{f^2h^2\cos^2\delta + r^2}} \simeq \frac{g'h\delta}{\sqrt{f^2h^2 + r^2}} \tag{10.19}$$

where the final steps are valid approximations when $\delta < 15°$. Bowden (Bowden, 1960) first applied this relation to show that the flow of dense water along the sides of the Iceland-Faroes ridge was descending at an angle of 5–10° relative to the isobaths.

The results in Equation (10.19) apply to flow on a slope which is uniform in the along-shelf direction. Where the along-slope topography is punctuated by canyons running down the slope, the density current will tend to concentrate in the canyons whose walls can support the component of the pressure gradient normal to the flow which is needed to balance the Coriolis force in a purely downslope flow. This is illustrated in Fig. 10.9, which is based on the results of a high-resolution 3-dimensional numerical model of cascading down an idealised canyon (Chapman and Gawarkiewicz, 1995). A section down the central axis of the canyon, in Fig 10.9a, shows the higher density water moving down the slope. Looking in the direction of the downslope flow at a section across the canyon (Fig. 10.9b), the effect of the Coriolis force is clear as the core of the flow is displaced towards the right-hand wall of the canyon (northern hemisphere).

We might expect cascading events forced by differential cooling to occur infrequently, probably not more than once per year at the end of the winter cooling period. They will, moreover, be short-lived features since the cascade flow will rapidly eliminate the density instability. Such infrequent events of short duration at the shelf edge are difficult to observe. Fig. 10.10 shows one of the few recorded examples – an event which occurred at the Malin shelf edge west of Ireland near the end of winter

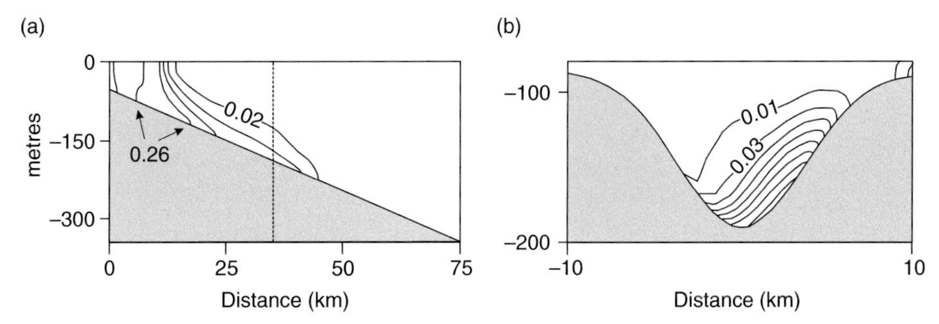

Figure 10.9 Model results of a canyon steering a downslope cascade of dense shelf water. (a) The excess density (kg m^{-3}) along a transect down the central axis of the canyon; (b) A section across the canyon, 30 km from the coastal boundary, of the downslope flow of the density current. Adapted from Chapman and Gawarkiewicz, 1995, courtesy of the American Geophysical Union.

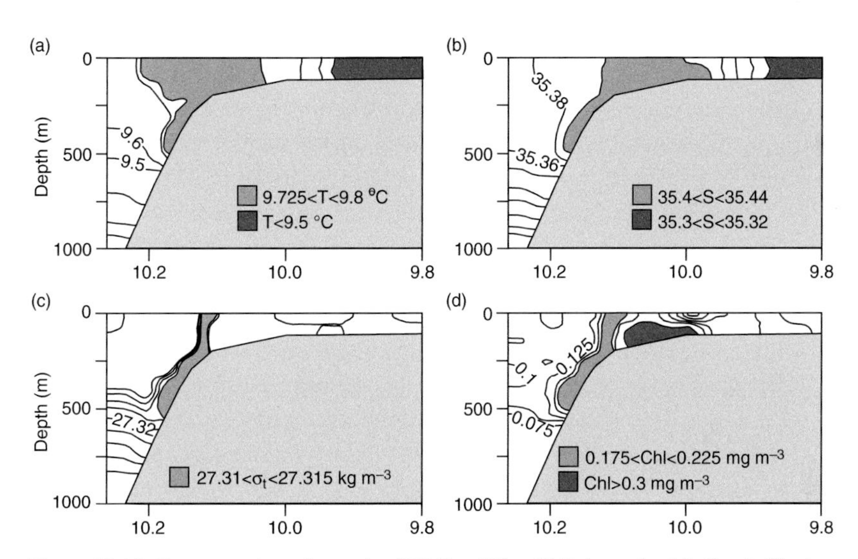

Figure 10.10 Cross-section along the UB line (Fig. 10.3a) on the Malin shelf edge and slope during a winter cascade. (a) Temperature (°C), (b) Salinity (PSS78), (c) Density (σ_t, kg m^{-3}), and (d) Chlorophyll (mg m^{-3}). Shaded areas in (c) and (d) indicate the descending plume of water with a slight density excess carrying chlorophyll from the shelf down to 500 m on the slope. Adapted from Hill *et al.* (1998), courtesy of *Journal of Marine Research.*

cooling in February 1996 (Hill *et al.*, 1998). At this time, winter convection penetrated down to depths of ~500 metres in the adjacent Atlantic while on the shelf, convective mixing was limited to the water depth of <200 metres. Fig. 10.10a shows that the temperatures on the shelf away from the slope were consequently significantly lower than in the upper layers of the deep ocean, while in Fig. 10.10b you can see that the cooler shelf water also had slightly higher salinity. This combination of higher salinity and colder temperature resulted in high-density shelf water, which was unstable. The subsequent cascading of the water down the upper slope can be seen in

Fig. 10.10c. In this case, the biogeochemistry of the descending plume of water provided tracers that confirmed the shelf origin; for instance Fig. 10.10d shows that the plume had a distinctive chlorophyll signature.

We can apply the dynamical balance (10.19) to these observations at the Malin shelf edge. Taking parameter values of: $h = 50$ m; $f = 1.19 \times 10^{-4}$ s^{-1}; $r = 7.5 \times 10^{-4}$ m s^{-1}; $g' = 10^{-4}$ m s^{-2}; $\delta = 4.5°$; we find: $\theta \simeq 7°$; $V_g \simeq 0.07$ m s^{-1}. This indicates an almost geostrophic flow of ~7 cm s^{-1} inclined at a small angle of 7° to the isobaths, which implies a downslope speed of 0.8 cm s^{-1} and a corresponding downslope transport of ~0.4 m^2 s^{-1}. In this case, therefore, cascading appears to induce a contribution to cross-slope transport which is of the same order, and in the same direction, as that due to the Ekman drainage forced by the slope current. Cascading due to differential cooling, however, operates only for short periods (a few days) so its average transport will not generally compete with the more persistent influence of Ekman drainage.

As the cascading water descends, it tends to mix with the overlying slope water so that its excess density $\Delta\rho$ decreases and the downslope flow will diminish as the cascade reaches an equilibrium level (~500 metres in the Malin shelf example) where its density matches that of the ambient water. More elaborate models of the cascading process allow for the entrainment of ambient water into the descending plume and, in principle, can predict the equilibrium depth, although uncertainties remain about the friction and entrainment parameters in such models.

Cascading is not limited to differential cooling at the shelf edge. An alternative forcing mechanism is the production of high-salinity water on the shelf in arid regions which are subject to high evaporation as we saw in Section 2.3.4 for Spencer Gulf, South Australia. Cascading in the Gulf, down the slope to ~250 metres depth, commences at the end of the summer season and, over a period of about 3 months, drains the high-salinity water accumulated in the Gulf during the previous summer (Lennon *et al.*, 1987). Cascades are potentially an important source of shelf water to the deeper ocean, transporting heat, salt and carbon. A global inventory of 61 confirmed cases of cascading has led to a tentative estimate of a cascade transporting between 0.05 and 0.08 Sv per 100 km of shelf edge, or 0.5–0.8 m^3 s^{-1} per metre length of slope (Ivanov *et al.*, 2004). One final point to consider is that, in lower latitudes, the advent of large-scale desalination plants to alleviate water shortages in arid regions will lead to substantial enhancement of salinity. Such plants could therefore generate local cascading effects which may significantly modify the local shelf environment.

10.3.4 Meandering and eddies

If the flow along the slope encounters abrupt changes in the direction of the isobars, forcing the current to meander and go round 'tight corners', the non-linear terms in the dynamical equations may become significant and the flow will not be geostrophic. In setting up the geostrophic balance, we assumed that the non-linear terms ($v\frac{\partial u}{\partial y}$ etc.) could be neglected so that $\frac{Du}{Dt} = \frac{\partial u}{\partial t}$. This linearising assumption is not valid when the flow involves large radial accelerations. A useful criterion for the breakdown of geostrophy in this way is based on the *Rossby number*, *RN*, which represents the ratio of inertial and Coriolis forces. Flow is judged to deviate significantly from geostrophy when

$$RN = \frac{V}{fL} > 0.1 \qquad (10.20)$$

where V and L are velocity and length scales characterising the flow. High Rossby number flow can result in incursions of the slope current on to the shelf as, for example, in the deflection of the Kuroshio into the Sea of Japan and the East China Sea by a tight bend in the topography where $RN \simeq 0.3$ (Hsueh *et al.*, 1996). Irregularities in the topography can also generate eddies and meanders and, hence, increase cross-frontal transfer. Such influence is exemplified by the effect of a pronounced topographic feature, the Charleston Bump, in promoting cross-stream perturbations in the Gulf Stream as it flows as the slope current along the south and mid-Atlantic Bight of eastern North America (Oey *et al.*, 1992; Miller, 1994).

10.4 The internal tide

Cross-slope flow can occur when the time scale of the motion means it cannot be regarded as steady. The oscillatory motions of the tides are an important case of such unsteady motion. The barotropic tide, which originates in the deep ocean and flows onto and off the shelf every tidal cycle, varies sufficiently rapidly to be exempt from the geostrophic constraint. The motion is reciprocating so it does not involve direct volume transport across the slope, but in stratified waters these barotropic tidal flows can lead to large internal tidal waves travelling along the pycnocline. These internal tidal waves propagating onto continental shelves can often generate substantial turbulence and thus promote vertical mixing. In combination with these mixing processes, which are irreversible, tidal flow may result in some net transport of scalar properties across the slope. Another cross-slope transport mechanism due the internal tides results from solibores, non-linear internal waves, which are a by-product of the internal tide. We will next examine in some detail the processes involved in the generation of the internal tide and the associated wave motions.

10.4.1 Generation of internal motions

The generation of the internal tide at the edge of the continental shelves is a globally ubiquitous phenomenon with important implications for the biogeochemistry. Energy is fed into the internal tide by the interaction of the barotropic tidal flow over the slope with water column stratification. The essential mechanism, which is illustrated for uniform two-layer stratification in Fig. 10.11, involves the displacement of the stratified water column up and down a steep slope. During the flood,

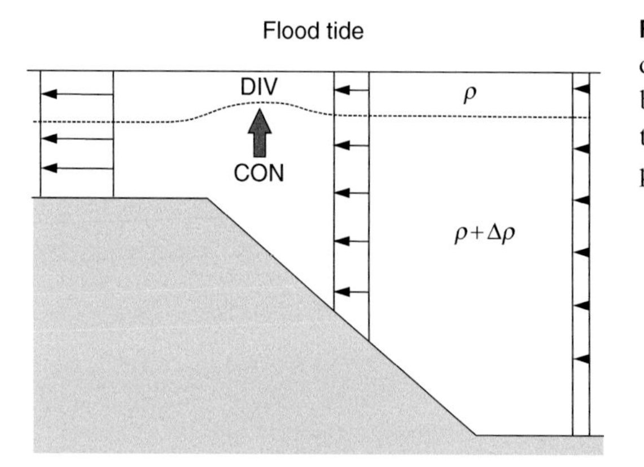

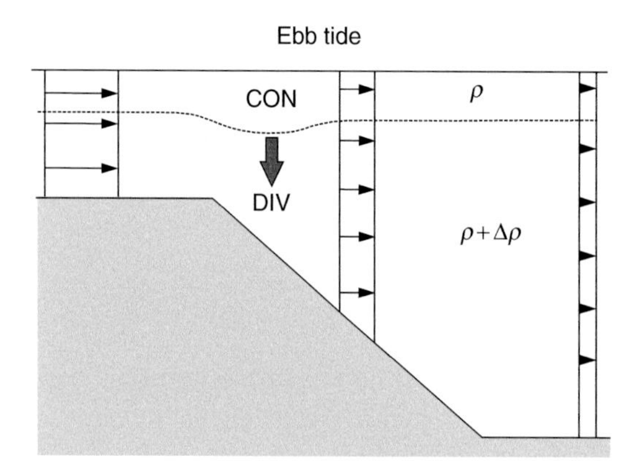

Figure 10.11 Convergent (CON) and divergent (DIV) flows generated by the barotropic tide as it flows onto and off the shelf, causing oscillation of the pycnocline.

depth-uniform, on-shelf flow results in an elevation of the pycnocline as the stratified column is 'squeezed' in the decreasing water depth. Conversely, during the ebb, the pycnocline is depressed. This periodic forcing of the pycnocline results in the generation of waves of tidal frequency which propagate both away into the deep ocean and onto the shelf. Initially the waves are predominantly semi-diurnal in character but, as they propagate away from the source, they take the form of an internal hydraulic jump or shock and evolve into a number of waves of shorter period and persistent profile (Holligan *et al.*, 1985; Pingree and Mardell, 1985).

10.4.2 Propagation and evolution of the waves

Observation of large amplitude waves of this kind propagating over the slope and on to the shelf have been reported in many shelf edge regions. Figure 10.12 shows some examples from the west and east coasts of North America (Loder *et al.*, 1992; Pineda, 1999) (Fig. 10.12 a,b), New Zealand (Sharples *et al.*, 2001a) (Fig. 10.12c) and the Celtic Sea (Sharples *et al.*, 2007) (Fig. 10.12d). The data from these four different regions show broadly the same types of pattern. The tidal-period oscillation of the

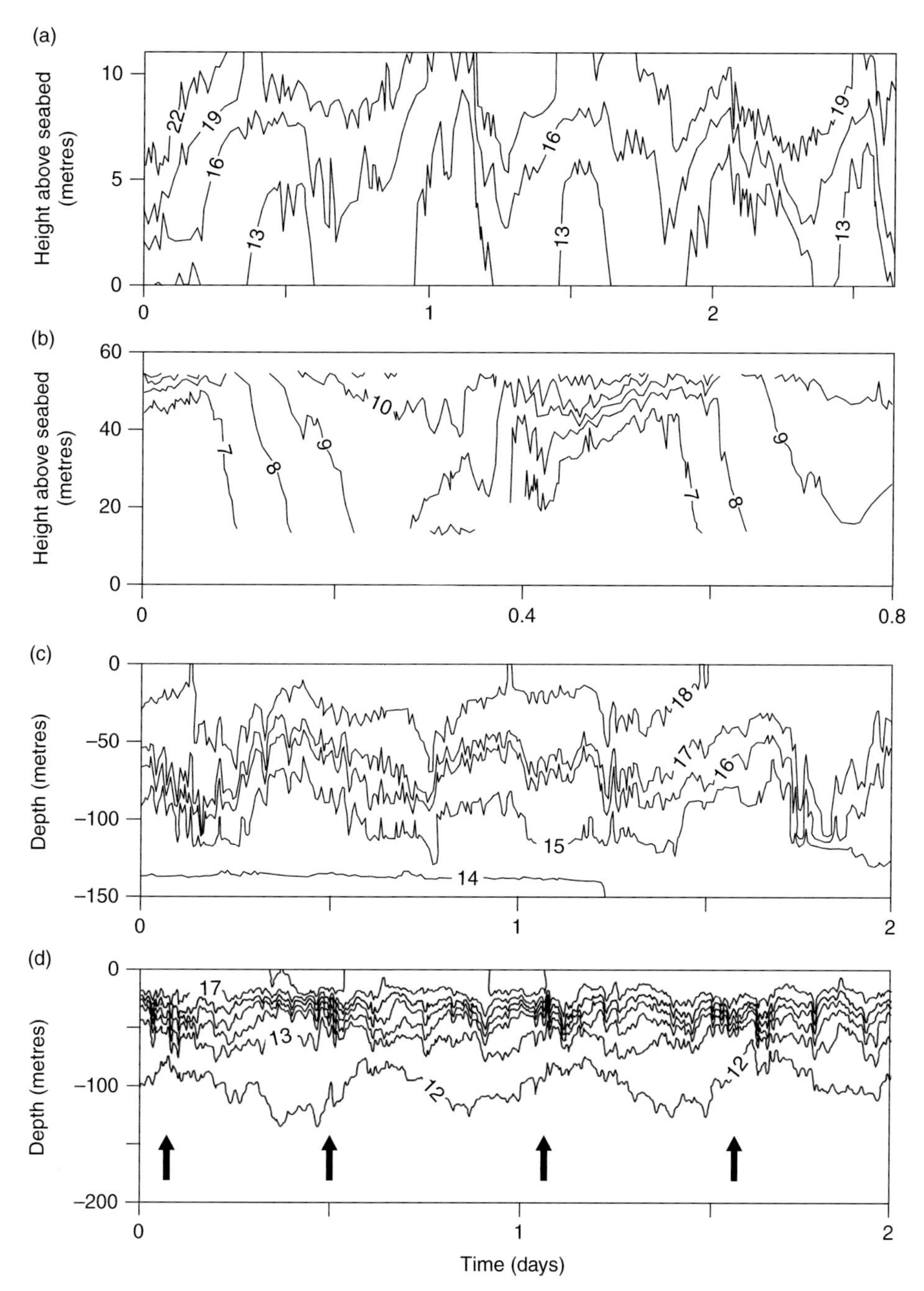

Figure 10.12 Examples of internal tide observations from moored vertical arrays of temperature loggers. (a) The shelf off southern California, adapted from (Pineda, 1999) courtesy of ASLO; (b) The northeast edge of Georges Bank, from (Loder *et al.*, 1992) courtesy the American Geophysical Union; (c) The shelf edge off northeast New Zealand (Sharples *et al.*, 2001a); (d) The shelf edge of the Celtic Sea (Sharples *et al.*, 2007). In (d) arrows mark packets of energetic high-frequency internal waves occurring once each tidal cycle.

isotherms is clear in all cases. This is the main internal tidal wave propagating into the shallower coastal or shelf water. Superimposed on this main wave are much shorter period oscillations of the isotherms, typically with a period of 10–30 minutes. Particularly energetic packets of these waves are evident in Fig. 10.12d. These shorter internal waves are generated as the main internal tidal wave begins to be influenced by the shoaling seabed. As these waves propagate, they generate an oscillatory baroclinic flow field which at the sea surface leads to convergences and divergences. To help visualise the motion involved, see Section 4.2 and particularly Fig. 4.8. This surface flow pattern can, under the right wind conditions, cause alternating patterns of rough and calm water which can be detected by radar. Images taken by Synthetic Aperture Radar (SAR) can allow the tracking of internal waves over large distances (Fu and Holt, 1984), and can provide information on the source of the often complex patterns of internal waves seen over the shelf edge. Figure 10.13a shows an example off the northeast shelf of New Zealand (Sharples *et al.*, 2001a), with two main wavefronts crossing onto and across the shelf separated by about 15 km. Each wavefront has a train of parallel 1–2 km wavelength internal waves behind it. Under the right sea conditions these smaller internal waves can be seen using standard marine x-band radar. The image in Fig. 10.13b shows a snapshot of the radar screen aboard the RRS *Charles Darwin* in summer 2005 during the cruise when we collected the data shown in Fig. 10.12d. Being aware of the passage of the internal waves past the ship turned out to be very useful, as we were able to keep our turbulent dissipation measurements going and catch the high mixing rates associated with these packets of waves.

We can see in Fig. 10.14 the development of a packet of these internal waves in data from near the Hebridean shelf edge (Small *et al.*, 1999). The internal tide there is observed to propagate on to the shelf with a phase speed $c \sim 0.4$ m s^{-1} and a wavelength ~18 km. As it does so, it evolves into a group of waves of shorter period which follow an initial depression of the isotherms. This development, which is due to non-linear interactions in the wave motion, is captured in a series of 'snapshot' sections (Fig. 10.14 a–d) obtained by towing a vertical string of thermistors through the advancing wave field at a vessel speed of ~2.2 m s^{-1} which is large compared with c. Over a period of ~6 hours, the initial depression of the isotherms is seen to acquire a following group of waves with wavelength ~500 metres and a crest-to-trough displacement of ~30 metres.

The relatively energetic orbital velocities associated with these waves are shown in Fig. 10.15 using data collected by an upward-looking moored ADCP on the Hebridean shelf. The waves generate flows of up to 50 cm s^{-1} with vertical velocities ~10 cm s^{-1}. Large amplitude waves of this kind exhibit rectified flows O[0.02] m s^{-1} and can thus contribute significantly to net transport across the slope. When the barotropic motion was removed, the waves shown in Fig. 10.15 were seen to sustain an off-shelf bottom layer transport of 5 m^2 s^{-1} for about 1.5 hours (Inall *et al.*, 2001).

The internal tide has been described as only 'quasi-periodic' and most long-time series show indications of intermittency in the internal tidal behaviour. It is important, therefore, to remember that while internal tidal motions are clearly

(a)

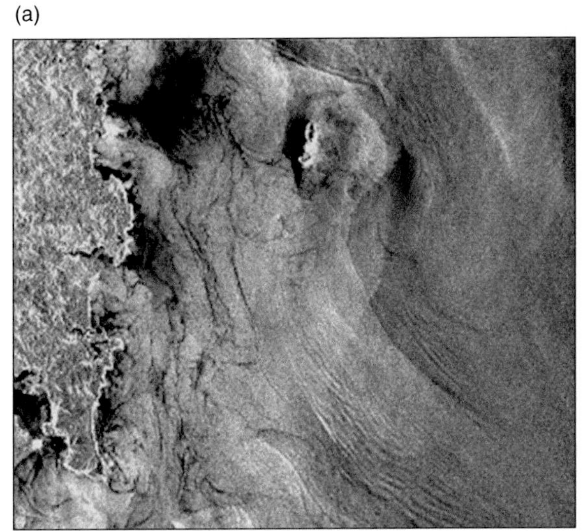

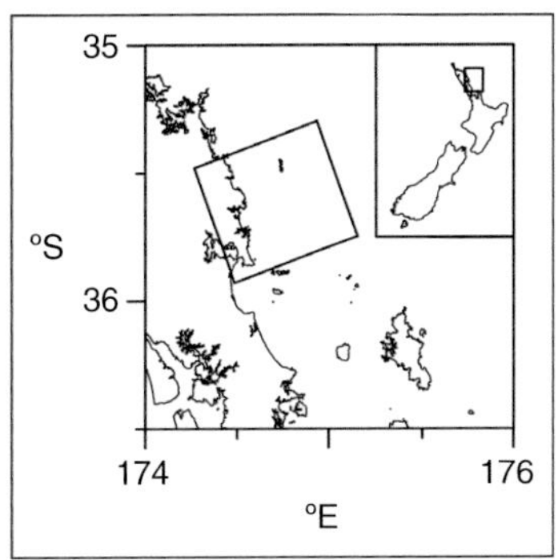

(b)

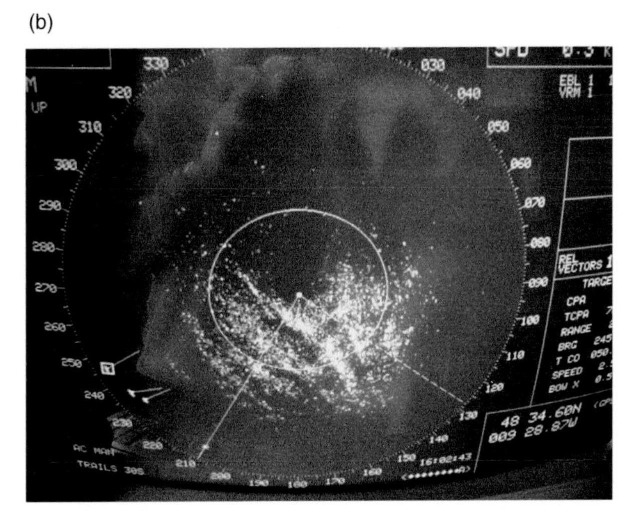

Figure 10.13 (a) Synthetic Aperture Radar (SAR) image off the northeast New Zealand shelf; the map to the right shows the area of the image. From Sharples *et al.*, 2001a, courtesy the American Geophysical Union; (b) A photograph taken of the radar screen aboard RRS *Charles Darwin* during a research cruise to the shelf edge of the Celtic Sea in July 2005. The 3 targets in the left of the screen in (b) are fishing vessels, approximately 6 km from the RRS *Charles Darwin*.

driven by the barotropic tide, they are not always in a consistent phase relation with it. To date, there seems to be no general explanation for the intermittency and spatial variability of the baroclinic tides, although it may be related to the low phase velocity of the internal modes which allows the barotropic tidal flow and other currents to significantly modify the propagation of the tide. Changes in the stratification will also act to modify the propagation speed away from the

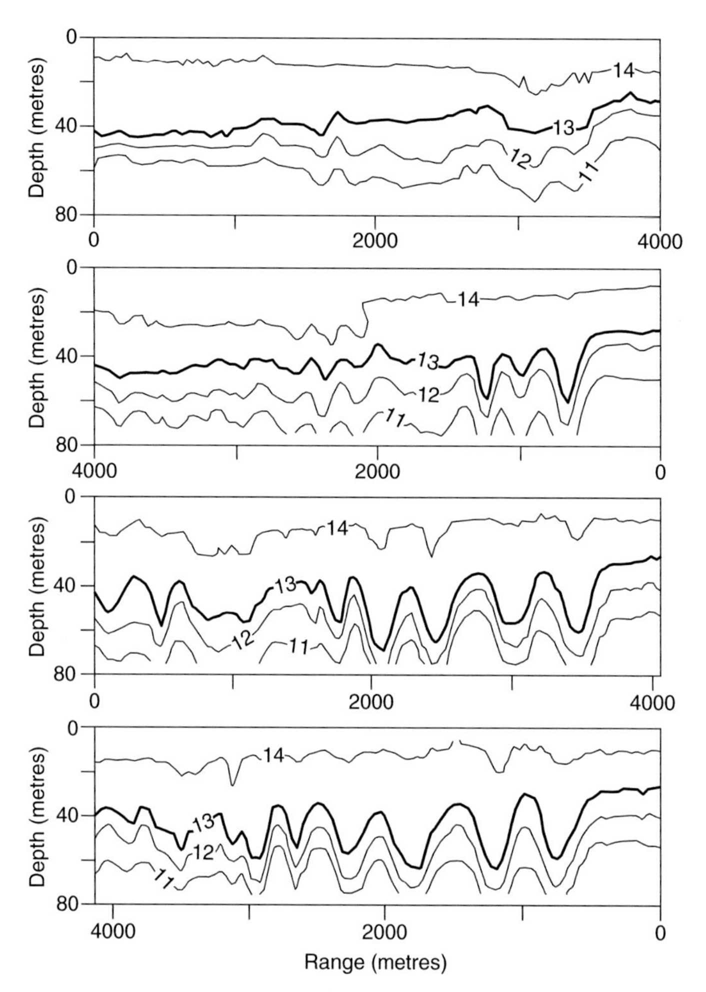

Figure 10.14 A sequence of sections showing the development of a packet of internal waves over the Hebridean shelf, observed using a towed vertical array of temperature loggers. Starting with the top panel the entire sequence took about 6 hours to complete, during which time the waves propagated a distance of ~9 km towards the shelf relative to the water. From Small *et al.* (1999), courtesy of the European Geosciences Union.

generation zone, and so, further on the shelf, the observed phase of the internal motions relative to the local barotropic tide will vary.

10.4.3 Enhanced mixing at the shelf edge

Although less consistent than the barotropic tide, internal tidal motions contain significant amounts of energy which they transport at the group velocity $U_g = 0.5$ to $1 \times c$ (see Section 4.2.3). It has been suggested that this energy flux of the internal tide on to the shelf may be responsible for driving mixing in the pycnocline of stratified areas remote from the shelf edge (Pingree and Mardell, 1985). While there

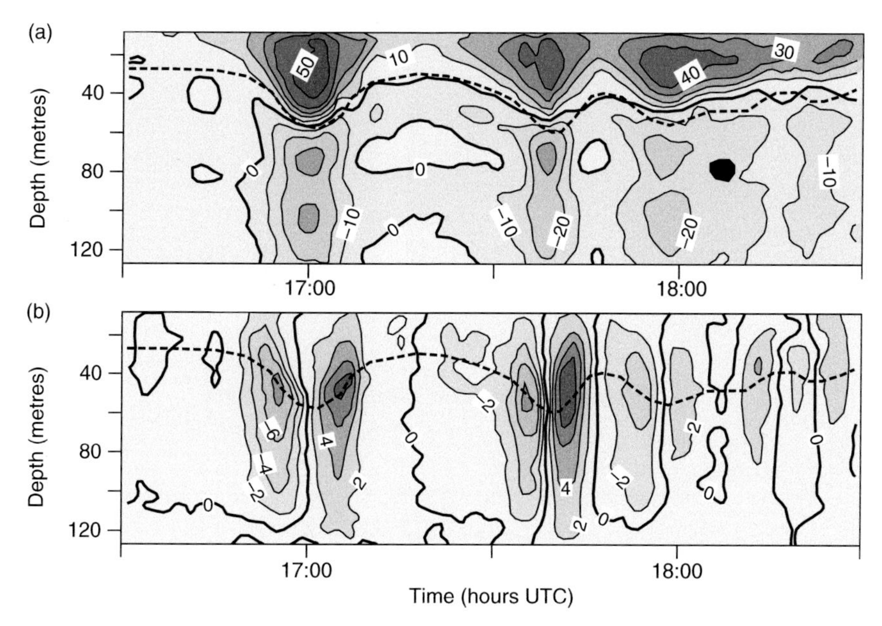

Figure 10.15 See colour plates version. (a) Horizontal and (b) vertical velocities (cm s^{-1}) observed by a moored ADCP for a packet of waves travelling on to the Hebridean shelf. The dashed grey line indicates the vertical displacement of a particle initially at a depth of 25 m. Note the opposing horizontal flows in the top and bottom layers and the large up and down vertical velocities as each trough passes the mooring. Image courtesy of Mark Inall, Scottish Association for Marine Science.

is evidence for such 'action at a distance' for waves travelling *away* from the shelf (Holligan *et al.*, 1985), recent observations indicate that most of the energy of the internal tide propagating on to the shelf is dissipated within 2–3 wavelengths of the semi-diurnal internal tide which is a distance of ~50–75 km for the Celtic Sea shelf (Green *et al.*, 2008). In part, this rapid energy loss from the internal waves results from their interaction with the local barotropic tide on the shelf which effectively increases the bottom frictional drag and augments dissipation.

The fact that most of the internal tidal energy is dissipated close to the source implies that there should be intensified internal mixing near the shelf edge. At the shelf edge of the Celtic Sea, this enhanced mixing is apparent as a band of cooler water along the shelf edge. The satellite image of sea surface temperature in Fig. 10.16a illustrates this, while an early set of observations from a towed, undulating CTD shown in Fig. 10.16b indicates the spreading of the pycnocline isopycnals over the shelf edge as a result of the internal mixing. It is worth noting here the contrasting effects of boundary-driven (e.g. from bed friction of surface wind stress) and internal wave mixing. Boundary stresses produce mixed boundary layers and tend to sharpen the pycnocline between the surface and bottom layers. Internal mixing operates directly on the pycnocline that support the internal waves, and so acts to smear out the pycnocline.

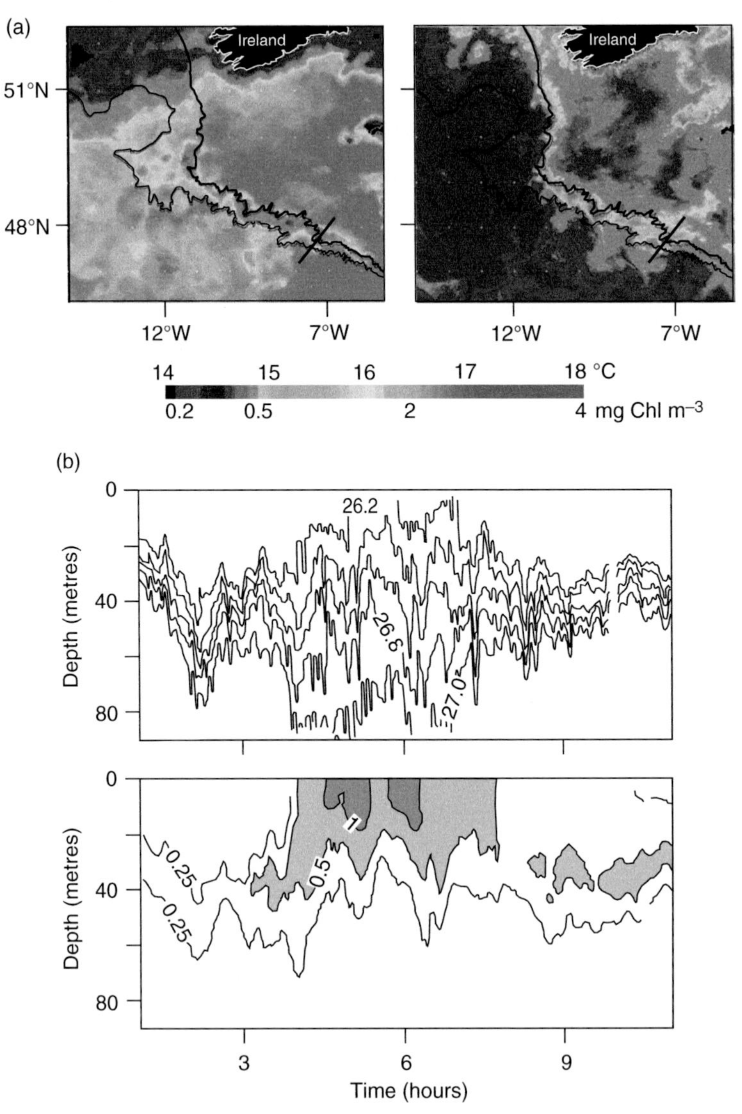

Figure 10.16 See colour plates version of (a). (a) Composite images of the Celtic Sea, June 13–19, 2004 courtesy of NEODAAS, Plymouth Marine Laboratory, UK. The left panel is sea surface temperature collected by AVHRR, the right panel is sea surface chlorophyll taken by MODIS. The bold contour line marks the 200 metre isobath at the shelf edge, the thin contour is the 2000 metre isobath. (b) A cross-shelf edge section of density (kg m^{-3}, upper panel) and chlorophyll (mg m^{-3}, lower panel) observed using a Batfish towed, undulating CTD system along the bold line marked in (a), adapted from Pingree *et al.* (1982), with permission from Elsevier. The research vessel speed was about 8 knots (15 km h^{-1}).

The two satellite images of sea surface chlorophyll and sea surface temperature in Fig. 10.16a show a band of locally increased chlorophyll coincident with the region of cool water, while the section in Fig. 10.16b illustrates the sub-surface chlorophyll maximum either side of the shelf edge being mixed upward at the shelf edge and reaching high surface concentrations. This link between the physics and the chlorophyll, and the broader ecological consequences, will be explored further in Section 10.9.1.

Physics summary box

- The very different regimes of the deep ocean and shelf seas meet and adjust to each other over the continental slope which is generally a narrow region of steep topography.
- Steady, geostrophic flow over a steep slope on a rotating earth is constrained to be parallel to the isobaths at all depths. This 'bathymetric steering' tends to trap currents (e.g. the Gulf Stream) over the slope and prevents them from moving onto the shelf.
- The steep slope topography and the latitudinal density gradient combine through the JEBAR mechanism to drive slope currents.
- Slope currents, forced by JEBAR, have a large barotropic (depth-independent) component so that their influence extends to the seabed over the upper part of the slope.
- Bathymetric steering inhibits cross-slope exchange, but some cross-slope flow can occur when the geostrophic constraint is broken, notably in the top and bottom boundary layers where friction is an important term in the dynamics.
- Friction at the seabed can undermine the geostrophy of an along-slope flow. Persistent downslope flow in the bottom boundary layer of a steady slope current at the eastern boundary of the ocean provides an important mechanism for the transfer of material from the shelf to the deep ocean (the Ekman drain). Upslope flow (upwelling) in the bottom boundary layer can occur under western boundary currents as they impinge against the shelf slope.
- Episodic downslope flow may occur infrequently due to 'cascading' after winter cooling helps to increase the density of shelf water so that it exceeds that of deeper water on the slope.
- In the surface boundary layer, wind forcing may drive cross-slope transports with consequent upwelling and downwelling at the shelf edge as well as on the shelf.
- The barotropic tide forces stratified water up and down the slope, which acts as a wave maker and generates an internal tide which propagates as a series of internal waves on to, and away from, the shelf.
- Most of the energy transported by these waves is dissipated close to their source and results in a region of increased mixing in mid-water with a consequent reduction in sea surface temperature which has been detected in I-R imagery.

10.5 The biogeochemical and ecological importance of the shelf edge

We have seen that the relatively steep seabed slope at the shelf edge is fundamental to limiting the transfers of water and its constituents between the shelf seas and the open ocean. Along-slope flows are far stronger than cross-slope transfers. This has meant that, to some extent, it has been possible to treat the shelf and the deep ocean separately. However, while the cross-slope physical flows are weak, the cross-slope biogeochemical gradients can be very strong so that biogeochemical fluxes between the open ocean and the shelf seas can be substantial. The high rates of primary production in shelf seas require adequate supplies of the macronutrients nitrate, phosphate and (for the diatoms) silicate (see Section 5.1.6). We saw in Chapter 9 that agricultural and wastewater runoff into rivers can supply large quantities of nutrients, particularly nitrate, but their effects are seen predominantly close to the coast. Taking the shelf seas as a whole, biogeochemical budgets suggest that 80–90% of the nitrogen and 50–60% of the phosphate that is required by shelf sea export production are supplied to the shelf seas from the deep ocean (Liu *et al.*, 2010). Silica, in contrast to nitrogen and phosphorus, is mainly supplied through the weathering of rocks on land and subsequent transfer down rivers into the shelf seas (Treguer *et al.*, 1995); processes at the shelf edge, therefore, control the flux of silicate from the shelf seas to the open ocean (Liu *et al.*, 2010). As well as being the gateway controlling the fluxes of nutrients between the shelf seas and the open ocean, the shelf edge is often a site of increased biological production which is implicated in supporting distinct ecosystems that contrast with both the open ocean and the shelf seas.

10.6 Upwelling, nutrient supply and enhanced biological production

The upwelling of deep oceanic water by persistent alongshore winds (see Sections 10.3.1 and 3.5.3), and the resulting supply of nutrients and enhancement of coastal primary production is probably the most studied aspect of the biological impacts of physical processes along the edges of the continental shelves. Along the narrow shelves of Peru, northwest North America, northwest and southwest Africa, this upwelling supplies nutrients across the continental shelf edge into the shallow coastal water, leading to enhanced primary production. These archetypal, wind-driven upwelling regions are well represented in the literature (see our suggestions for further reading at the end of this chapter). Here we will briefly address the biological consequences of wind-driven upwelling, focusing on the time scales of responses of different components of the ecosystem to changes in the alongshore winds. We will also consider how nutrients can be supplied by upwelling which is driven through the bottom Ekman layer of along-slope flows (see Section 10.3.2), particularly in regions where a western boundary current is forced to flow along a continental slope.

10.6.1 Wind-driven upwelling and biological response

Nutrient supply and the phytoplankton response

In clear sky conditions, the elevated chlorophyll concentrations arising from upwelling-favourable winds are often apparent in satellite images. For instance, comparison of concurrent images of sea surface temperature and sea surface chlorophyll, off the Iberian Peninsula (Fig. 10.17a), illustrates the clear link between the cool upwelled water and the enhanced phytoplankton concentration along the entire Portuguese shelf. Looking in more detail at the southern Portuguese shelf, shown in Fig. 10.17b, we can see a filament and eddy of upwelled water containing elevated chlorophyll. Such eddies and filaments, which, as we noted in 10.3.1, are correlated with topographic irregularities along the coastline (Haynes *et al.*, 1993), are known to be important for the export of material off the shelf (Alvarez-Salgado *et al.*, 2001). Figure 10.18 shows a cross-slope section off the Iberian Peninsula

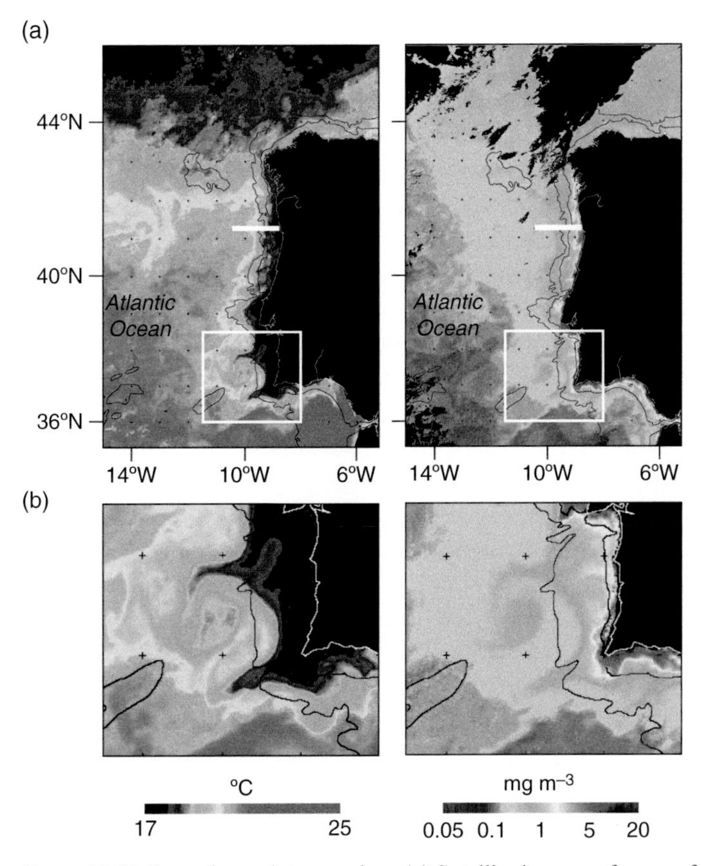

Figure 10.17 See colour plates version. (a) Satellite images of sea surface temperature (left) and sea surface chlorophyll (right) off the Iberian Peninsula. Both images are composites taken over July 25–27, 2007; (b) Detail of the boxes marked in (a) showing two large filaments of upwelled water containing chlorophyll extending off-shelf. Images from NEODAAS, Plymouth Marine Laboratory, UK.

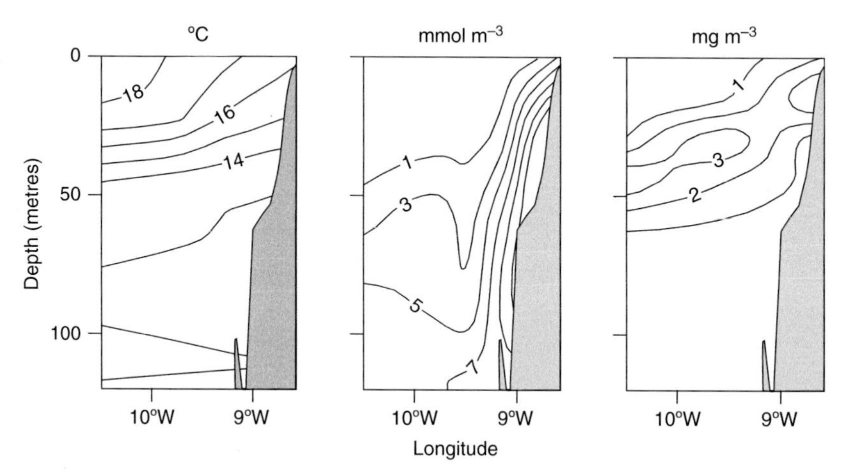

Figure 10.18 Observations of temperature (left), nitrate (centre) and chlorophyll (right) along a transect line (white line in Fig. 10.17a) perpendicular to the Portuguese coast, collected during July 1994. Adapted from Ballestero and Boxall (1997), courtesy of ESA/Earthnet Online.

(Ballestero and Boxall, 1997) which illustrates how the cool, nitrate-rich water is raised towards the sea surface in upwelling over the very narrow shelf, and produces an increase in surface and sub-surface chlorophyll at the coast.

We need to remember that upwelled water, while high in nitrate, will initially have a low phytoplankton concentration; it brings with it a seed stock of phytoplankton that will utilise the nitrate and grow as the water begins to experience the higher light levels at the sea surface. The availability of light for phytoplankton growth will depend on how stratification develops in the upwelled water once it starts to respond to the near-surface heating. The combination of the developing physical structure and the nutrient assimilation and growth rates of the phytoplankton will lead to some delay between the initial wind event and the subsequent response of the different species within the seed phytoplankton community. This delay in biogeochemical response has been observed in the Peruvian upwelling system by carrying out experiments on the phytoplankton nutrient uptake rates next to a series of drogued buoys that were used to track recently upwelled parcels of water (MacIsaac *et al.*, 1985). Note in Fig. 10.19 that there is a delay of ~2 days in the uptake of nitrate, with the maximum nitrate uptake occurring about 5 days after the upwelling event brings the phytoplankton and nitrate-laden water up into the euphotic zone. Chlorophyll was also seen to peak, but later again than the maximum in nitrate uptake; chlorophyll later decreased as the phytoplankton became nutrient limited. The whole sequence from arrival of upwelled water at the sea surface to depletion of nitrate and the reduction of biomass took 8–10 days, and was limited to a narrow region of 30–60 km from the coast. The phytoplankton population during the tracked upwelling event was dominated by diatoms. This is a common observation in upwelling regions (e.g. off the Iberian Peninsula (Huete-Ortega *et al.*, 2010), during monsoon-driven upwelling in the Arabian Sea (Barber *et al.*, 2001), in the

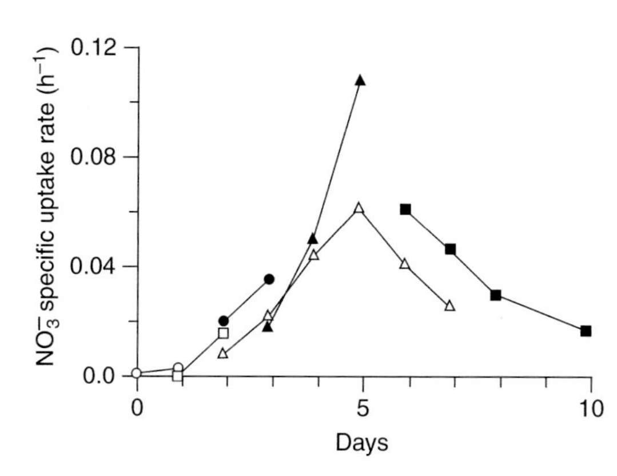

Figure 10.19 The specific uptake rate of nitrate in recently upwelled water off the coast of Peru. The different symbols are for data collected by different drogue buoys, with all of the measurements put onto a common time base with 0 days the time when upwelled water first reached the euphotic zone. After MacIsaac *et al.* (1985), with permission from Elsevier.

Californian upwelling (Wilkerson *et al.*, 2006)), often involving an upwelling-driven diatom bloom being succeeded by smaller flagellates and dinoflagellates as the upwelling relaxes (Smayda and Trainer, 2010; Mitchellinnes and Walker, 1991). This initial bloom of diatoms, responding to the high nutrient concentrations, is an example of the opportunistic nature of these relatively large, fast-growing cells, as discussed in Chapter 5.

In an environment where the upwelling-favourable winds are more infrequent (e.g. the time series in Fig. 10.6), vertical turbulent mixing will be a relatively more important part of any biogeochemical response. A short burst of winds may bring the pycnocline closer to the sea surface, but as the winds relax the pycnocline will return towards its previous depth. In order to produce a significant response in the primary production, the necessary net transfer of nutrients into the photic zone requires mixing across the pycnocline (Hales *et al.*, 2005).

Secondary production and fish

So far we have focused on the autotrophic community of photosynthesising phytoplankton which require the upwelled nutrients. Populations of heterotrophic phytoplankton are often much larger (Joint *et al.*, 2001), and they can have an important grazing impact on the diatoms (Kiørboe *et al.*, 1998). However, as we described in Chapter 5, these microzooplankton grazers respond on the same time scales as those on which the photosynthetic phytoplankton respond to the supplies of nutrients and light. How do the larger grazers, such as copepods, fish larvae and the planktivorous fish, which have longer response times to changes in their environment, cope with the variability of food supply?

Copepods usually dominate the zooplankton communities in upwelling systems. They have life cycles with intervals of several days to weeks, depending on species, between eggs being produced, hatching, and adulthood. This delay between egg production and the appearance of the adults will decouple copepod numbers from the conditions that triggered their initial production. We have noted similar decoupling in the context of the lack of a clear copepod response to shelf sea fronts in

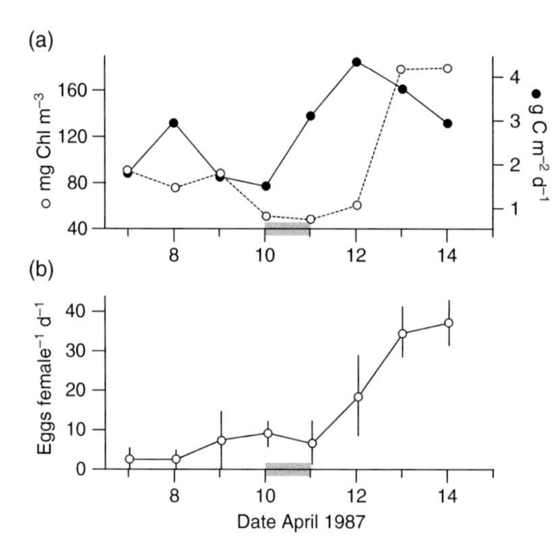

Figure 10.20 Production of phytoplankton carbon and copepod eggs in the Benguela upwelling system. The shaded bar on the date axis indicates when upwelled water arrived at the fixed sampling station. a Phytoplankton growth rate (solid circles) and phytoplankton biomass (open circles); (b) Copepod fecundity (egg production rate per female). Adapted from Armstrong *et al.* (1991), with permission from Elsevier.

Section 8.6.3. The earliest stage of the copepod life cycle can be linked to upwelling. Consider the example in Fig. 10.20 from the Benguela upwelling system off the coast of southwest Africa, which shows the initial pulse of phytoplankton growth, the subsequent increase in phytoplankton biomass and a correlated increase in the copepod eggs produced per female. Longer term cycles in copepod biomass and community composition will be the result of adding together a sequence of such upwelling responses (Peterson and Keister, 2003). We can view longer-term variability in copepod populations as a damped response to changes in the upwelling region, driven by inter-annual changes in, for instance, the mean winds, the number of upwelling wind events and the broader oceanographic environment.

In general, larger phytoplankton cells are more efficient at transferring energy up though the food chain (Legendre and Rivkin, 2002), and, in terms of fisheries, we noted in Chapter 5 that it is recognised that fish larvae and, planktivorous fish target food particles larger than about 5 μm in size (Cushing, 1995). We can see, therefore, why diatom-producing upwelling regions support some of the world's largest commercial fish stocks. Fish species with the highest biomass in upwelling systems are plankton-eating fish, e.g. anchovies, sardines and mackerel.

Environmental variability and fisheries in upwelling regions

The transfer of energy from sunlight through to the protein harvested by fishing vessels in upwelling areas is through a very short food chain, with the important result that the success of the fisheries is very responsive to changes in the primary production. Inter-annual changes in the oceanographic environment, for instance as a result of changes to the wind or large-scale basin-wide changes in ocean currents, can modulate the growth of the phytoplankton and very rapidly (i.e. the same year or 1 year later) affect the success of the fish stocks. The clearest example of a basin-wide change in the environment having an impact on fisheries is that of the El Niño

Southern-Oscillation (ENSO). These ENSO events, which occur every few years, involve a large-scale change in the atmospheric pressure gradient between the south-east Pacific and Indonesia. The change in pressure gradient is accompanied by a reduction in the strength of the trade winds which are normally responsible for maintaining a shallow thermocline along the equatorial and subtropical coast of South America and a deeper thermocline in the west Pacific. Weaker trade winds allow the thermocline to relax, becoming deeper on the South American coast so that any coastal upwelling brings relatively warm, nutrient-poor water to the surface. There is a strong link between the years of positive temperature anomalies at the Peruvian coast and reductions in the catch of anchovies (Escribano *et al.*, 2004).

The ENSO is a large-scale phenomenon that we can see has impacts on fisheries because it alters the dynamics of upwelling over a long period; i.e. in a positive ENSO phase all upwelling events will be poor at supplying nutrients to the sea surface because of the consistent change in the depth of the thermocline. There are also ways in which the frequency of short-term wind events within an otherwise upwelling-favourable system could affect the survival and growth of fish. Within the California upwelling system observations have shown that first-feeding anchovy larvae have a better chance of survival if they have access to high densities of large dinoflagellates within the sub-surface chlorophyll maximum (Lasker, 1975). While the winds are important for supplying nutrients to the shelf, episodes of calm weather are required for this layer of dinoflagellates to form. Strong winds (e.g. during an upwelling event) disperse the layer and substantially reduce the food available to the larvae. The necessary 4-day period of winds < 5 m s^{-1} has been termed a 'Lasker Event'.[3]

Upwelling and bottom water hypoxia

So far we have concentrated on the impacts of nutrient upwelling on the growth of phytoplankton and subsequent responses of higher trophic levels, including commercially important fish stocks. There are of course other routes that the organic carbon fixed by the phytoplankton could take. Remember the case of the Louisiana 'Dead Zone' (Fig. 9.19), where supplies of nutrients from rivers led to increased surface phytoplankton growth, but also to reduced oxygen concentrations in the bottom waters as the sinking organic material decayed. Upwelling systems exhibit analogous behaviour. Figure 10.21a, b illustrate a clear link between dissolved oxygen (DO) and organic carbon distributions over the Oregon shelf and shelf edge. Organic carbon is fixed in the surface water in response to upwelled nutrients, but flocculates and sinks towards the seabed, with hypoxic conditions developing in the bottom water as a result of the decay of this sinking organic material. This development of low oxygen bottom waters is a seasonal feature of the western American shelf, correlated with the seasonal upwelling-favourable winds which prevail in the summer months. The typical annual cycle in the vertical distribution of oxygen concentration mid-shelf off Washington, northwestern United States, is shown in Fig. 10.21c. Note how the

[3] After Reuben Lasker who suggested this condition for larval survival.

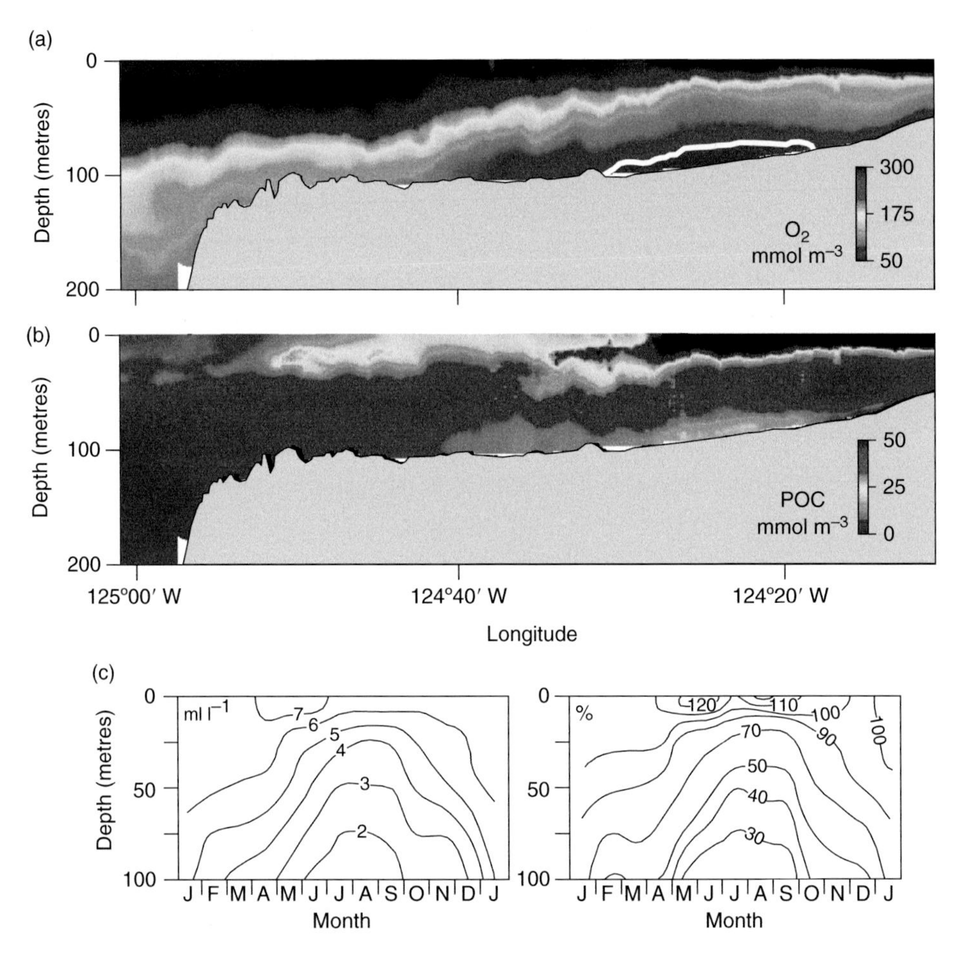

Figure 10.21 See colour plates version of (a) and (b). (a) Transect of dissolved oxygen concentration across the shelf and shelf edge off Oregon. The bold white contour line marks the extent of hypoxic bottom water; (b) Transect of particulate organic carbon (POC) across the Oregon shelf and shelf edge; (c) Annual cycle of the dissolved oxygen concentration (left) and % saturation (right) on the Washington shelf. (a) and (b) are high resolution observations collected in May 2001 during a period of upwelling-favourable winds. After Hales *et al.*, 2006, courtesy of the American Geophysical Union. (c) is adapted from Landry *et al.* (1989), with permission from Elsevier.

surface water achieves super-saturated D.O. conditions in summer, due to enhanced primary production in response to the upwelling of nutrients. Below the productive surface layer the oxygen concentration becomes hypoxic (D.O. saturation between 1% and 30%). While upwelling of deep water from off the shelf edge can introduce water with low oxygen concentrations, the hypoxia in the bottom waters of the California upwelling system in summer is attributed to the decay of organic carbon previously fixed in response to nutrients supplied by upwelling (Connolly *et al.*, 2010).

Figure 10.22 See colour plates version. Photograph of a coccolithophore bloom in Jervis Bay, east Australia, following the upwelling of nutrients driven by a filament of the East Australia Current impinging against the shelf edge (Blackburn and Cresswell, 1993). Image provided by Susan Blackburn, CSIRO, photographer Ford Kristo, Australian National Parks. Reproduced with permission from CSIRO Publishing.

An upwelling favourable wind event does more than input nutrients over the shelf edge. During the upwelling winds, surface water will be transferred off-shelf, and so export the carbon fixed in the surface waters as a result of earlier upwelling. The filaments mentioned earlier in this section are one mechanism for efficient transport of surface waters off the shelf, with rates of transport measurable using satellite imagery (Gabric *et al.*, 1993). Off Monterey Bay, part of the California current system, a strong correlation between surface primary production and the export of particulate organic carbon (POC) into the deep water of the continental slope has been seen in normal upwelling years (Pilskaln *et al.*, 1996). The same study also showed that in El Niño years, with very low nutrient upwelling, the correlation was substantially reduced as the phytoplankton community shifted from large diatoms to much smaller picoplankton and flagellates.

10.6.2 Biological response to upwelling under along-slope flow

Apart from upwelling forced by wind stress, we have seen (Section 10.3.2) that the along-slope flow of a western boundary current trapped against the slope topography can drive upwelling through the bottom frictional Ekman layer. Clear evidence of a biogeochemical response to this process is limited. An example of a biological response attributed to such a current-driven upwelling event has been recorded off the east coast of Australia. In this case, the East Australian Current, the southwest Pacific equivalent of the Gulf Stream, develops large meanders and filaments which can impinge against the shelf slope. Figure 10.22 shows a dense bloom of coccolithophores in Jervis Bay, 130 km south of Sydney, triggered by nutrients that arrived in the coastal water as such a filament impinged against the upper continental slope (Blackburn and Cresswell, 1993).

10.7 Shelf edge ecosystems driven by downwelling slope currents

While upwelling of nutrients from the deep ocean is an obvious cause of increased biological activity at the ocean margins, there are other possible mechanisms promoting enhanced growth and biomass. Indeed, edges of continental shelves where downwelling predominates are also often found to be sites of elevated phytoplankton and zooplankton, and to be regions important to commercial fisheries.

The potential of a downwelling-favourable slope current to support shelf edge productivity can be seen, for example, at the Mid Atlantic Bight and in the Bering Sea. The shelf edge of the Mid Atlantic Bight has increased phytoplankton biomass and production (Marra *et al.*, 1990) thought to be supported by nutrients originating from remineralised organic material in the bottom shelf water. The nutrient source was confirmed in this case using a freon tracer. The dynamics behind this nutrient source were identified using an elegant numerical model experiment that was aimed at understanding the formation of the shelf break front in the region (Gawarkiewicz and Chapman, 1992). The cross-shelf component of the pathway followed by particles originating in surface shelf water is shown in the schematic in Fig. 10.23. Nearshore surface water on the shelf is drawn downward and then offshore to feed the bottom Ekman layer of the along-shelf current. Seaward of the shelf edge the bottom boundary layer detaches from the seabed, and the shelf water is drawn upward along the sloping isopycnals into the photic zone over the upper slope.

By contrast the extensive shelf edge of the Bering Sea, with the downwelling-favourable Bering slope current and the Alaska Stream, draws nutrients from the open ocean. Primary and secondary production rates within the 'Green Belt' along the shelf edge are 60% greater than the adjacent shelf and more than 250% greater than the adjacent deep ocean, with marked correlations with commercially important fish stocks and the distributions of seabirds, whales and seals (Springer *et al.*, 1996). Using salt as an independent tracer of fluxes, this high productivity at the Bering Sea shelf edge is thought to be supported by slope current baroclinic eddies above the

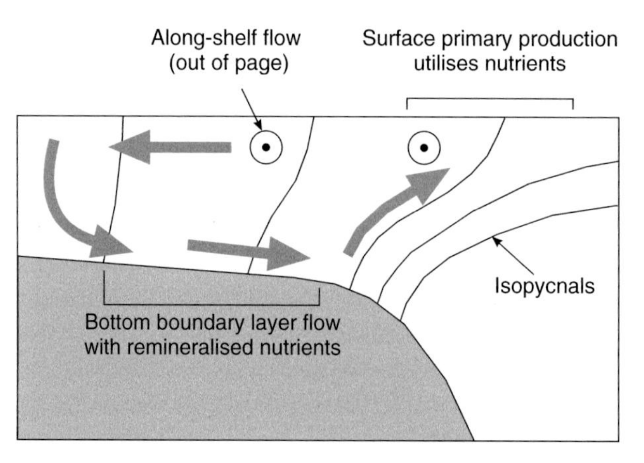

Figure 10.23 Schematic representation of how the along-shelf flow drives nutrients from shelf bottom water up to the photic zone over the shelf edge of the Mid Atlantic Bight.

bottom boundary layer driving a flux of deep ocean nitrate toward the shelf. Elevated diapycnal mixing at the shelf edge front then mixes the nutrients upward to the photic zone (Coachman and Walsh, 1981). These eddy fluxes of nutrients act down the nutrient gradient, which in the case of the Bering Sea is particularly strong due to iron-limitation of adjacent oceanic primary production resulting in High Nutrient Low Chlorophyll (HNLC) conditions. Advection of bottom shelf water within the bottom boundary layer of the along-shelf edge currents plays an important role here by supplying much of the iron required by the shelf edge primary production (Aguilar-Islas *et al.*, 2007). The mechanism responsible for increasing the diapycnal mixing over the shelf edge, mixing both the macronutrients and the iron upward at the shelf edge, has not yet been identified.

10.8 Internal tides, mixing and shelf edge ecosystems: the Celtic Sea shelf edge

We next focus on the effects of the internal tide on shelf edge biology, using the Celtic Sea as a case study. Studies of the internal tide in this region began with the work of Pingree, Mardell and Holligan in the early 1980s, with the early recognition of the role that the breaking internal waves had on the nutrient and chlorophyll distributions (Pingree and Mardell, 1981; Holligan *et al.*, 1985). Subsequently our own detailed measurements of the physical and chemical environment and the phytoplankton community response have shown how the across-shelf edge gradients in vertical turbulent mixing are pivotal in supporting a distinct ecosystem.

10.8.1 Nutrient supply and primary production at the shelf edge

The impact of the internal tide on the shelf edge biogeochemistry has already been anticipated in Fig. 10.16, where we saw a marked correspondence between the region of cool water mixed upward by the breaking internal tide and a band of increased concentrations of sea surface chlorophyll. Figure 10.24 shows that this is a regional phenomenon, with the band of elevated shelf edge chlorophyll reaching from northern Biscay, round the west of Ireland, and up to the north of Scotland – a distance of about 1500 km. This consistent pattern of shelf edge chlorophyll is strikingly similar to the ‘Green Belt’ of the Bering Sea. The mixing of cool water upwards at the Celtic Sea shelf edge during summer stratification is expected to be associated with a flux of nutrients, and the band of cool water has been observed with locally increased concentrations of surface nitrate of ~0.5 mmol m^{-3} (Pingree and Mardell, 1981). This concentration of nitrate is small compared to the concentrations we might expect below the pycnocline in a summer stratified shelf region, but there is an important contrast with the surface layer away from the shelf edge where nitrate is generally undetectable using the usual analysis techniques used for shelf sea work (i.e. < 0.05 mmol m^{-3}).

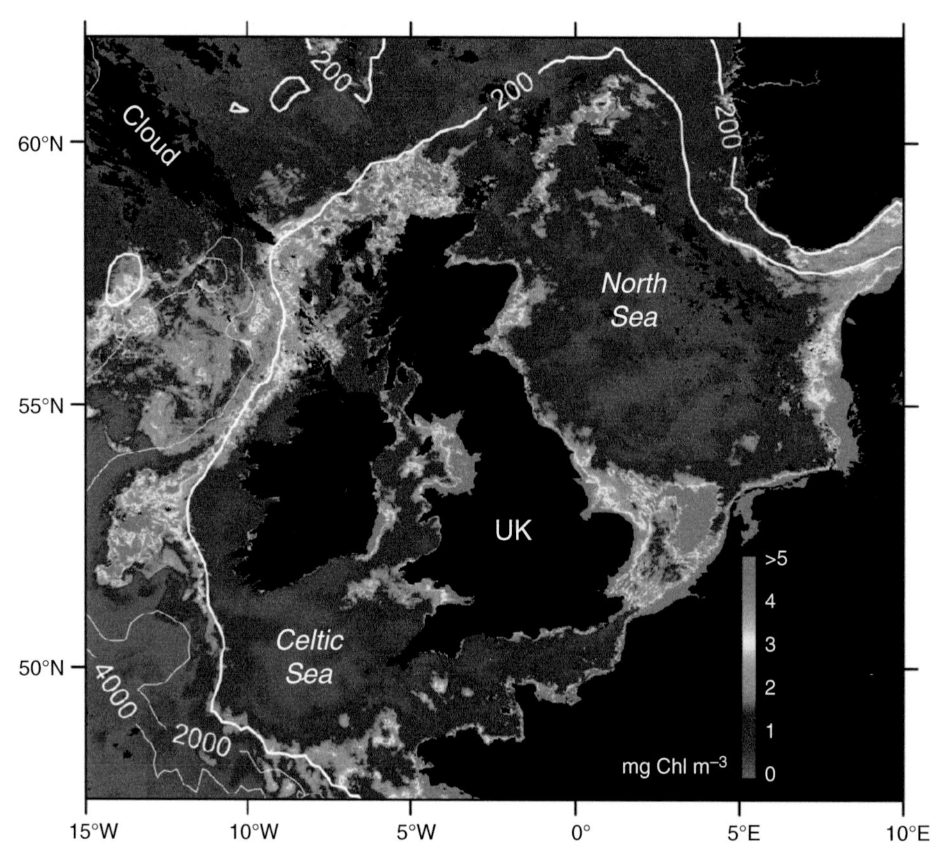

Figure 10.24 See colour plates version. Satellite image of sea surface chlorophyll over the NW European shelf seas, taken May 30, 2004. Isobaths are contoured in white, with the 200-metre isobath marking the edge of the continental shelf. Image data courtesy NEODAAS, Plymouth Marine Laboratory UK. Bathymetry data from GEBCO.

Direct measurements of the vertical turbulent transport of nitrate from the deeper water towards the surface at the Celtic Sea shelf edge have indicated a strong dependence on the contrasts in mixing by the internal tide over the spring-neap cycle (Sharples *et al.*, 2007). By combining turbulent dissipation measurements and observations of the vertical nitrate gradient, neap tide nitrate transport was found to be about 1.3 mmol m^{-2} d^{-1}, similar to the flux measured into the base of the shelf sea thermocline away from the shelf edge (Sharples *et al.*, 2001b; Rippeth *et al.*, 2009). At spring tides, however, the nitrate flux was found to increase to ~9 mmol m^{-2} d^{-1}, attributable to packets of high-frequency internal waves (e.g. such as those shown in Fig. 10.13b). Moreover, while at neap tides the surface nitrate concentration was below detection limits, at spring tides surface waters were found to have ~0.5 mmol m^{-3} of nitrate, suggesting that the flux at spring tides exceeded the phytoplankton community's capacity to assimilate it.

Mixing nutrients towards the sea surface, and the apparent increase in sea surface chlorophyll seen in observations such as in Fig. 10.16 and Fig. 10.24, suggests that the shelf edge should be found to be a region of significantly enhanced primary production. However, we need to be careful in making the leap from biomass to production. The mixing that supplies the nutrients to the surface will also be redistributing the chlorophyll, so the surface chlorophyll could simply be the result of upward turbulent transport of chlorophyll from within the thermocline. Also, while mixing might be thought to promote phytoplankton growth by supplying nutrients, it will also hinder photosynthesis by disrupting the light regime experienced by the phytoplankton. Remember also that when interpreting satellite images we have no information in the images about what is happening beneath the sea surface. Measurements of carbon fixation rates at the Celtic Sea shelf edge at neap tides have been found to be 400–800 mg C m^{-2} d^{-1}, and 300–600 mg C m^{-2} d^{-1} at spring tides (Sharples *et al.*, 2007). In both cases the range shows the effect of typical cloudy and sunny days, while the lower rates at spring tides could be an indication of the effect of increased mixing on the light received by the phytoplankton. These rates are about a factor of 2 greater than the rates of carbon fixation on the adjacent shelf and in the open ocean (Sharples *et al.*, 2009).

10.8.2 Phytoplankton community gradients across the shelf edge

While the shelf edge primary production rates are increased above those on either side, a more dramatic contrast is seen when we consider the detailed characteristics of the shelf edge productivity. Measurements of nutrient assimilation in the region have shown that the shelf edge band of elevated chlorophyll is associated with an *f*-ratio of about 0.7, compared to typically <0.3 away from the shelf edge (Joint *et al.*, 2001), indicating that the shelf edge primary production increase is the result of new production responding to the supply of deep nitrate. So how is this injection of new nutrients affecting the phytoplankton community structure?

A detailed picture of the chlorophyll distribution and the phytoplankton community structure across the Celtic Sea shelf edge is shown in Fig. 10.25, based on data collected on a research cruise aboard the RRS *Charles Darwin* in 2005 (Sharples *et al.*, 2009). This was the same cruise that collected the data in Fig. 10.12d. Fig. 10.25a shows a subsurface chlorophyll peak associated with the thermocline either side of the shelf edge. At the shelf edge the effects of internal tide mixing are evident, with a broader thermocline and cooler, high chlorophyll water outcropping at the sea surface corresponding to the typical surface satellite imagery of temperature and chlorophyll (e.g. Fig. 10.16). Figure 10.25b shows that when considering all of the chlorophyll in the upper 100 metres of the water column there is no clear evidence that the shelf edge has higher phytoplankton biomass than anywhere else on the section. Interesting contrasts begin to appear when we consider what the chlorophyll is comprised of. Figure 10.25c shows that the shelf edge phytoplankton population contains proportionally more diatoms than either the shelf or the open ocean, and that 60% of the chlorophyll is held within phytoplankton cells greater than 5 μm size

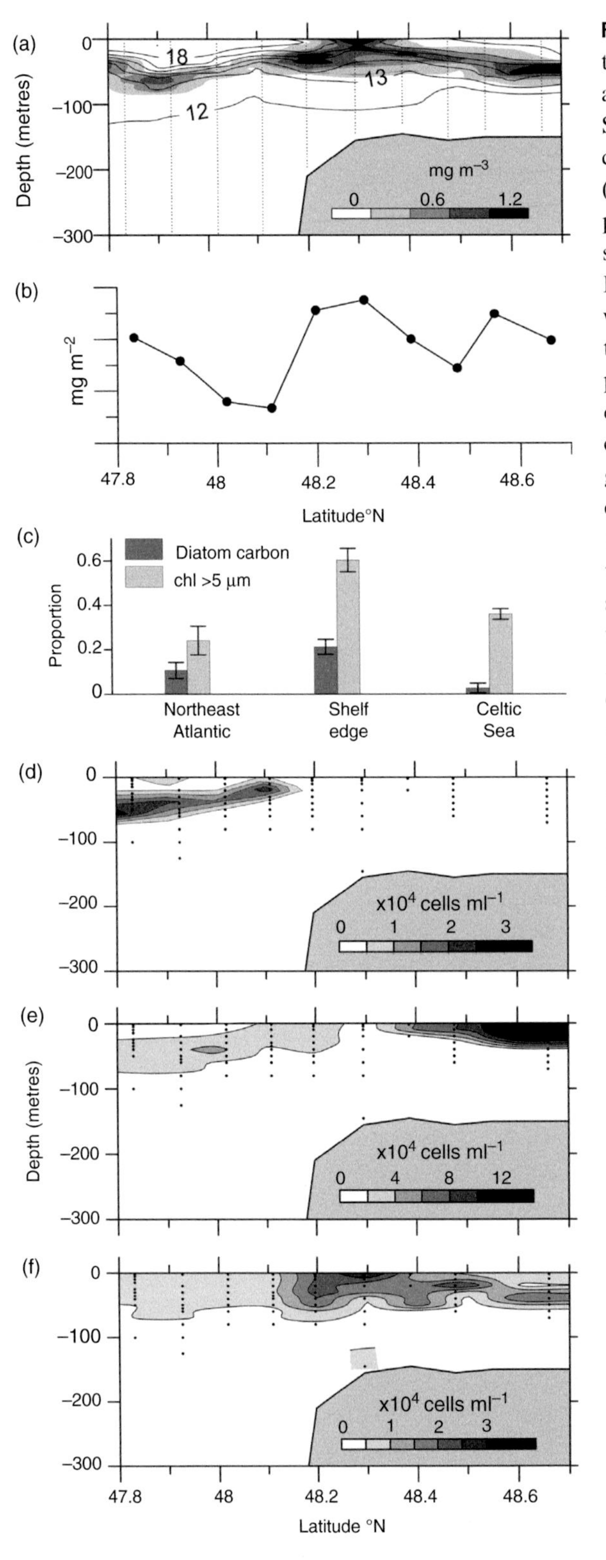

Figure 10.25 Observations of the phytoplankton community across the shelf edge of the Celtic Sea. (a) Temperature (line contours) and chlorophyll *a* (shaded). Vertical lines mark the positions of the CTD profiles; the shelf edge is at latitude 48.2°N; (b) Depth-integrated chlorophyll *a* within the upper 100 metres along the section; (c) The proportion of phytoplankton carbon within the diatoms, and the proportion of chlorophyll *a* in cells of size greater than 5 μm, from samples collected through the upper 100 metres at the two ends of the CTD section and at the shelf edge; (d)–(f) Distributions of the cyanobacterium *Prochlorococcus*, the cyanobacterium *Synechococcus*, and picoeukaryotes. In (d), (e) and (f) the dots mark the positions of the water samples used in the flow cytometer analysis. Adapted from Sharples *et al.* (2009), courtesy of the American Geophysical Union.

compared to 25% and 35% in the open ocean and on the shelf. In Fig. 10.25d–f a detailed analysis of the phytoplankton community based on data from a flow cytometer shows that the open ocean thermocline was dominated by small cyanobacteria. The smallest were *Prochlorococcus* at about 0.5–0.7 μm size, and were mainly found in the open ocean. *Synechococcus*, larger at 0.8–1.5 μm, had their maximum concentrations in the surface layer of the shelf sea. Within the shelf sea thermocline, and particularly at the shelf edge in a region with very few cyanobacteria, small eukaryotes (2–5 μm size) were found in large numbers.

The horizontal and vertical gradients in the phytoplankton community shown in Fig. 10.25 appear to be consistent with the changes in the vertical supply of nitrate. Remember from Chapter 5 that small cells are particularly well adapted to life in nutrient-deplete environments, such as the thermocline of the open ocean where nitrate fluxes are around 0.1 mmol m^{-2} d^{-1}, and the surface layer of the shelf sea where nitrate is limiting and the phytoplankton are almost entirely dependent on ammonium as the nitrogen source. At the base of the shelf sea thermocline, where the nitrate flux is greater at 1–2 mmol m^{-2} d^{-1}, the community begins to shift to larger cells. In the region with the largest supply of nitrate, at the point on the shelf edge where the internal tide breaks, we find the largest phytoplankton. However, the contrasts in nitrate supply can only be a part of the explanation. In Chapter 5 we noted that small cells will always have an advantage in the competition for nutrients; the growth of larger cells probably requires a relative reduction in the grazing pressure (Section 5.2.4). Preliminary evidence at the Celtic Sea shelf edge has shown that the numbers of small copepods do not appear to exhibit a response at the shelf edge (Sharples *et al.*, 2009). However, the complete picture of how the phytoplankton community contrasts develop in this region has yet to be determined, though the marked gradients in ecology and physics in the region perhaps provide an ideal environment to assess the dynamics of the phytoplankton and grazer communities.

10.8.3 A possible link to the fish at the Celtic Sea shelf edge

There are many examples worldwide of high zooplankton biomass on ocean margins that are also important fishing grounds (see, for instance, the Bering Sea 'Green Belt' (Springer *et al.*, 1996), off Vancouver Island (Mackas *et al.*, 1997), the Mid Atlantic Bight (Cosper and Stepien, 1984), the shelf edge of southeastern Australia (Young *et al.*, 2001) and the Patagonian shelf (Sabatini and Colombo, 2001). All of the research we have been involved in along the shelf edge of the northwest European continental margin has been carried out against a background of high fishing activity, targeting spawning stocks of mackerel, horse mackerel, blue whiting, hake, monkfish and megrim. Regional data on the distribution of fish eggs and larvae show that the shelf edge is a site for spawning fish along most of the enhanced chlorophyll band of Fig. 10.24 (Ibaibarriaga *et al.*, 2007). The persistent, northward slope current along the northwest European continental margin is implicated in the transport of the larvae back to the adult grounds (Reid, 2001), analogous to the role of advection around Georges Bank (Fig. 8.21). The phytoplankton community arising from the internal tide mixing may also be playing a role in supporting the fish. We saw in Chapter 5 the importance of food particles > 5 μm for fish larvae, and Fig. 5.16 illustrates evidence for small fish larvae ingesting phytoplankton. This suggests that it is the existence of a large-celled phytoplankton community, rather than the enhanced level of primary production, which is important to the first-feeding fish larvae (Sharples *et al.*, 2009).

As well as the commercially important fish stocks, the northwest European shelf edge is known to be a site favoured by basking sharks (Sims *et al.*, 2003), and to be an important boundary in the distributions of whales and dolphins (Weir *et al.*, 2001; Kiszka *et al.*, 2007). Such correspondences are very common globally. And it is not just the pelagic biology that follows the shelf edge. The cold water coral *Lophelia pertusa* is found extensively around the continental slope of northwest Europe, fuelled by labile organic material within large phytoplankton cells and faecal pellets raining down from the surface waters (Kiriakoulakis *et al.*, 2004). This implicates the distinct phytoplankton community of the shelf edge as a candidate food source for the corals, though more research is needed. Distinct shelf break and slope epibenthic communities have also been found off Greenland and in the Mediterranean (Mayer and Piepenburg, 1996; Colloca *et al.*, 2004).

10.9 Exporting carbon from continental shelves

The previous sections focused largely on the supply of nutrient-laden water from seaward of the shelf break by upwelling or mixing, and the biogeochemical and ecological responses to that nutrient supply. We will now consider how physical processes at the shelf edge lead to transports from the shelf to the ocean, particularly addressing the problem of the export of carbon from the productive ocean margins into the deep sea.

Carbon export from the shelf seas has been the subject of several major national research programmes. The large-scale, multi-disciplinary research effort required to address biogeochemical fluxes across the shelf edge were first used in the Shelf Edge Exchange Processes (SEEP I) and SEEP II programmes carried out on the east coast of the United States in the 1980s. The SEEP research aimed to test the hypothesis that up to 50% of a shelf seas' annual primary production could be exported into the adjacent deep ocean (Walsh, 1991; Biscaye, 1994). The hypothesis was eventually rejected, with observations showing that only about 6% of the shelf primary production was exported off the shelf and into the open ocean (Falkowski *et al.*, 1994). However, while small this percentage still meant that the carbon exported off the shelf exceeded the carbon delivered to the shelf down rivers, and so indicated that the shelf was net autotrophic (or, in other words, the shelf was a net sink for atmospheric CO_2). Part of the challenge of the SEEP projects, and subsequently other shelf edge exchange research (e.g. the Ocean Margin Exchange project (OMEX) (Wollast and Chou, 2001), the Kuroshio Exchange Project (KEEP) (Wong *et al.*, 2000), and the Shelf Exchange Study (LOIS-SES) (Hill *et al.*, 1998) was to determine what physical processes lie behind this carbon export.

Recall from Chapter 5 (Section 5.2.1) that the shelf edge is important to carbon export as it controls the link between the productive shelf seas and the continental slope sediments and deep ocean where carbon can be sequestered for long time scales. Even in the absence of any biological activity, carbon can be exported as DIC if cool, high-latitude water containing high concentrations of DIC becomes sufficiently dense to sink below the depth of the ocean's winter mixed layer.

10.9.1 Exporting carbon from upwelling systems: coastal filaments and canyons

Upwelled water will contain both nutrients and DIC. As we noted in Chapter 5, if the C:N has the Redfield ratio we might expect there to be no net export of carbon arising from the increased primary production driven by the nitrate. The physics of an upwelling system can, however, act to convert DIC to organic carbon and then export the fixed carbon before it can be remineralised. Indeed, overall the current best estimate is that the eastern boundaries of the world's oceans, including all of the major upwelling systems, are overall net exporters of carbon and weak sinks for atmospheric CO_2 (Jahnke, 2010). We will now look at some of the physical processes in upwelling systems that can export carbon.

Off Iberia and off northwest Africa the offshore transport mechanism provided by upwelling filaments (described in Section 10.3.1, and shown in Fig. 10.17) provides a conduit for dissolved organic material, including DOC, into the subtropical gyre of the northeast Atlantic (Alvarez-Salgado *et al.*, 2007) and potentially providing an important source of organic nutrients to the oligotrophic subtropical gyre (Torres-Valdes *et al.*, 2009). Also, the bathymetric features that play a large part in the development of upwelling front instabilities and filaments concentrate the deposition of POC. A remarkably high-resolution study of the Nazaré Canyon off the Iberian upwelling region (Masson *et al.*, 2010) has shown how the canyon tends to have larger sedimentation rates than the adjacent slope, and that seafloor sediments in the canyon have higher organic carbon content than the slope sediments. However, while we may link the observation of organic carbon deposition with the primary production triggered by the upwelling, the study also showed that the canyon was not a deposition site for the products of pelagic carbon fixation. Canyon sediments were found to be derived from terrigenous material from the shelf, while slope sediments contained more organic carbon fixed within the overlying water column. So, off the narrow upwelling shelf of Iberia, it appears that the filaments export carbon fixed in response to the upwelling, while the canyons hold onto and sequester carbon originating from the shelf and the rivers.

10.9.2 Seasonal downwelling and a role for the shelf thermocline

Off the California shelf (Hales *et al.*, 2005) the high nitrate concentrations of the upwelled water, which are completely utilised by photosynthesis, lead to the surface waters absorbing CO_2 from the atmosphere. Here it is the seasonality of

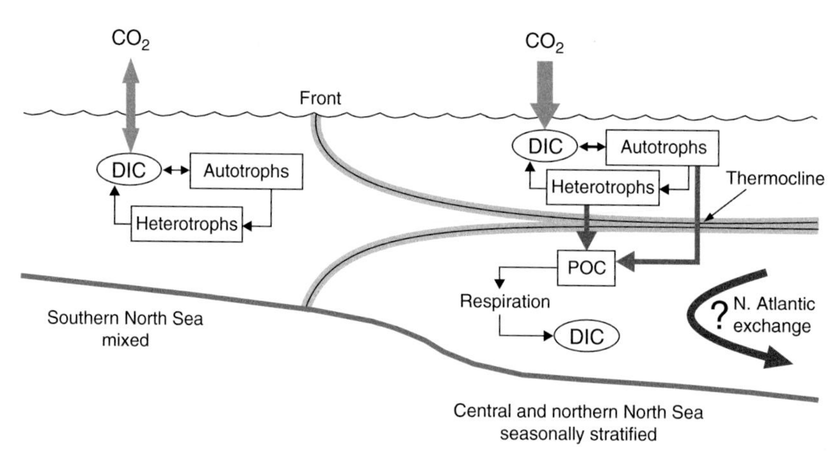

Figure 10.26 Schematic illustration of the role of the thermocline in exporting carbon from the central and northern North Sea (adapted from (Thomas, Bozec, *et al.*, 2004)). The question mark by the N. Atlantic exchange signifies uncertainty in how much of the deep water in the northern North Sea can be exported before winter convection allows CO_2 out-gassing to the atmosphere.

the upwelling that plays an important role in carbon export, as the DIC generated by respiration in the bottom waters of the shelf is exported off-shelf when the upwelling relaxes and downwelling winds occur in winter. This behaviour of the Californian system can be understood by considering the example we have already shown in Fig. 10.21. Remember that the export of the POC below the thermocline, followed by bacterial degradation, led to significant reductions in the bottom water dissolved oxygen concentration. These low oxygen conditions are maintained by the overlying thermocline preventing mixing with the surface layer and replenishment via contact with the atmosphere. Bacterial breakdown of organic carbon in the lower layer will generate inorganic carbon, which will also be kept away from contact with the atmosphere by the thermocline. Thus organic carbon is removed from the surface layer by sinking, and then the thermocline acts to trap increasing concentrations of DIC ready for export by the later downwelling conditions.

The Californian shelf is very narrow, but this trapping role of the shelf sea thermocline has been highlighted as the key to carbon export from the much more extensive shelf of the North Sea (Thomas *et al.*, 2004). The North Sea is broadly split into two distinct biogeochemical regimes. The southern North Sea is shallow, which combined with the strength of the tides results in a vertically mixed water column all year. The central and northern areas of the North Sea are deeper, which allows the archetypal temperate seasonal cycle of thermal stratification and mixing (e.g. see Section 6.1). The consequences of this partitioning on the storage, release and export of carbon are illustrated schematically in Fig. 10.26. Primary production in the mixed region draws CO_2 from the atmosphere, but as POC is cycled through the heterotrophs and remineralised to increase DIC, the water column is able to ventilate excess CO_2 back to the atmosphere. In the stratifying

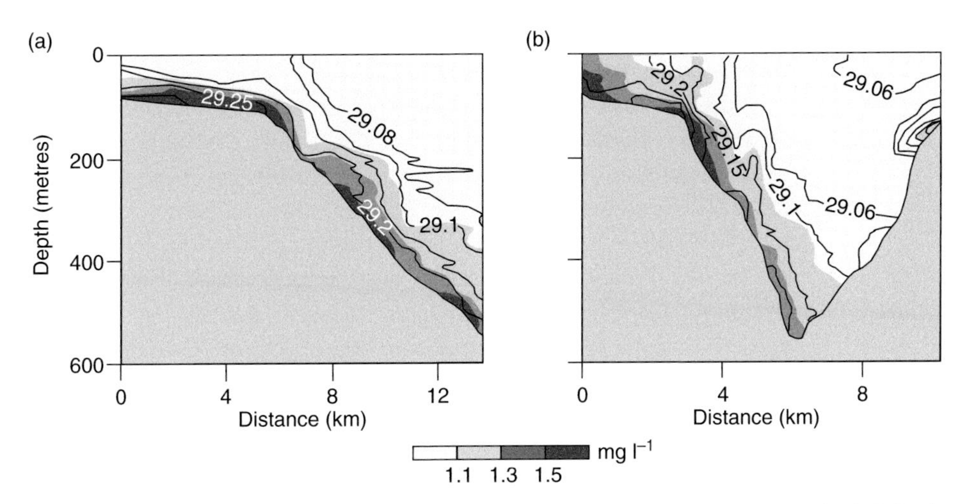

Figure 10.27 Sections of density (lines) and suspended sediment concentration (shaded) in the Cap de Creus canyon, Gulf of Lions. Adapted from Canals *et al.* (2006), with permission from *Nature*. (a) is a section down the middle of the canyon, (b) is a cross-canyon section looking into the canyon.

region, however, the products of the spring bloom and the summer subsurface primary production can sink into the deeper water where they are respired to DIC. As with the Californian example, the DIC is trapped below the thermocline and so cannot vent excess CO_2 to the atmosphere. Exchange of this deep water with the North Atlantic across the shelf edge of the North Sea can then export the DIC. However, the amount of this carbon that can be exported to the Atlantic is fundamentally controlled by the rate at which the deep layer of the North Sea can be transported across the shelf edge. There is a limited window of opportunity for exchange with the Atlantic to remove the deep layer DIC, because as soon as autumnal and winter convective overturning begin on the shelf the excess DIC can begin to equilibrate with the atmosphere. Problem 10.5 at the end of this chapter gives you the opportunity of exploring the likely net carbon export across this shelf edge boundary.

10.9.3 Ekman drain and winter cascading

In the example of the Californian shelf the main mechanism for removing the deep water from the shelf into the adjacent Pacific Ocean was the shift in regional wind patterns to downwelling-favourable winds. For the North Sea, with a generally westerly airflow, the same physics could be contributing to off-shelf transport. The North Sea, and indeed much of the shelf edge of the northwest European continental margin, is also affected by the pervasive flow of the slope current (see Section 10.2.3). Flow through the bottom Ekman layer of this slope current is downslope (e.g. Section 10.3.2), and it has been noted to generate consistent downwelling flows in models of the European margin (Holt *et al.*, 2009). This Ekman drain is analogous to the upwelling under a western boundary current, but provides a means to export carbon rather than supply nutrients.

The LOIS-SES experiment has provided us with a direct observation of a transfer of organic carbon down the slope (see Fig. 10.10). The biogeochemical importance of such cascades of cool, relatively dense shelf water has been described for the case of the East China Sea (Tsunogai *et al.*, 1999). There the cooling shelf water takes up CO_2 from the atmosphere, adding to the organic carbon fixed during the preceding productive months. The net effect is an annual carbon export from the East China Sea down the adjacent slope equivalent to 35 g C m^{-2} a^{-1}, a process named the *continental shelf pump*. Incorporating the process into a global circulation model led to a suggestion that continental shelves could export about 0.6 Gt of carbon every year (Yool and Fasham, 2001), which is about 30% of the annual anthropogenic supply of carbon to the atmosphere. Dramatic evidence of the channelling of such seasonal, cooling-driven cascades of dense water containing significant amounts of carbon is shown in Fig. 10.27 for the Gulf of Lions in the western Mediterranean. Note the pattern of down-canyon flow (Fig. 10.27a) and the Coriolis-driven tendency for the cascade to be displaced to the right against the canyon wall (Fig. 10.27b), analogous to the modelled cascade shown in Fig. 10.9. Current speeds within a series of these observed cascades typically reached 70–80 cm s^{-1}. The cascade flows have a significant effect on the morphology of the canyon seabed and contain large amounts of suspended sediment. At the end of the cascading season chlorophyll fluorescence was seen down at depths >2000 metres; the overall contribution of the cascades to carbon export from the shelf area of the Gulf of Lions was estimated to be 50 g C m^{-2} a^{-1}.

Summary

We have seen in this chapter that the edges of the continental shelf are host to a rich variety of physical processes which mediate important biochemical exchanges between the shallow, productive shelf seas and the adjacent deep ocean. Two physical processes in particular exert strong controls on the environment: (1) the steep bathymetry of the narrow slope region constrains geostrophic current to flow parallel to the isobaths, thus preventing them from crossing the slope and thereby impeding exchange between shelf and deep ocean. Some exchange across the slope is still possible through mechanisms which evade the constraint of bathymetric steering, notably in the surface and bottom boundary layers where frictional stresses drive substantial cross-slope flows; (2) the barotropic tidal flow over the slope acts as a generator of internal wave motions by forcing a stratified water column up and down the slope.

Surface wind stress and wind-induced upwelling, which brings nutrients in to the photic zone, plays a key role in driving greatly enhanced primary production in some shelf edge regions, such as the Benguela system off the SW coast of

Africa which supports large fish catches based mainly on plankton-eating fish. Upwelling and enhanced primary production can also be induced by along-slope flow, though the evidence here is far more limited. Perhaps rather surprisingly, downwelling may, in some cases, be favourable to plankton growth. Both upwelling and downwelling can result in a transport of fixed carbon from the shelf edge region out into the deeper waters of the open ocean, and it is possible that the generation of organic material in upwelling regions acts as a source of organic nutrients to the nutrient-impoverished oligotrophic gyres. We also saw that the thermocline in the outer shelf waters often plays a role in the export of carbon, by trapping DIC remineralised from sinking organic material. The relative timing between the processes driving cross-shelf edge transports of the deeper shelf water and the seasonal breakdown of shelf stratification is important, as winter mixing on the shelf will re-establish contact between the high DIC water and the atmosphere.

As the internal tide and shorter internal waves propagate towards and onto the shelf, they induce vertical mixing which broadens the thermocline, cools the surface water and promotes the nutrient supply to support enhanced plankton growth over the slope and the adjacent shelf. While there is often a focus on this increased production at the shelf edge, the case study of the Celtic Sea illustrates how significant contrasts in the species structure of the phytoplankton community are set up across the shelf edge and correlate with the horizontal differences in vertical supply of nitrate. The existence of large-celled phytoplankton species at the shelf edge is not a response solely to the observed enhanced supply of nutrients, but is also conjectured to require decoupling between the phytoplankton and their grazers. The very distinct contrasts across the shelf edge may make the region an ideal study site for assessing the links leading to phytoplankton community biodiversity. The importance of the shelf edge for commercial fisheries may be linked to this shift in community structure, as might the often highly diverse benthic ecosystems found on the continental slope.

The water at the outer edges of temperate and high latitude shelf regions can be prone to substantial cooling in winter, increasing its potential to absorb atmospheric CO_2 and also making it sensitive to density-driven cascades down the continental slope. Fluxes of inorganic and organic carbon in such flows represent an important export of carbon, and high sedimentation rates within canyons make them sites of long-term burial of POC.

It is perhaps remarkable how a narrow band around the edges of the ocean basins acts as such a fundamental control on the productivity of the shelf seas and on the export of carbon to the deep sea, This narrow band is also a region of marked horizontal gradients in ecosystem structures. Understanding, and correctly simulating, the physics of the shelf edge is vital because of the region's role in controlling crucial biogeochemical fluxes and supporting large commercial fisheries.

FURTHER READING

Carbon and Nutrient Fluxes in Continental Margins: A Global Synthesis, K. K. Liu, *et al.*, editors. Springer-Verlag, Berlin Heidelberg, 2010.

HERMES Hotspot Ecosystem Research on the Margins of European Seas (www.eu-hermes.net/intro.html)

Huthnance, J. M. Circulation, exchange and water masses at the ocean margin: the role of physical processes at the shelf edge. *Progress in Oceanography*, **35(4)**, 353–431, 1996.

The Dynamics of Marine Ecosystems: Biological-Physical Interactions in the Oceans, by Kenneth H. Mann and John R. N. Lazier, Blackwell Publishing, 2006. Chapter 5.

Problems

10.1. A barotropic current flows poleward along the continental slope between the 200-metre and the 500-metre isobaths at latitude 58°N. If, on the 15 km section normal to the slope between these isobaths, the average speed of the current is 0.23 m s^{-1}, estimate the difference in sea surface height across the current. Assuming that the bottom depth increases uniformly between 200 metres and 500 metres, estimate the total transport of the current. Assess the feasibility of monitoring the transport in such a slope current by satellite altimeter observations of sea surface height (r.m.s. uncertainty ~1–2 cm).

10.2. The thickness of the bottom Ekman layer, D, below a steady geostrophic flow, speed V_s, can be estimated from the empirical relation:

$$D = 0.4u_\star/f$$

where $u_\star = (\tau_b/\rho)^{1/2}$ is the friction velocity and $\tau_b = k_b \rho V_s^2$ is the stress at the bottom boundary. Use this relation to show that the observed value of D in a slope current can be used to estimate the bottom stress, the transport T_b normal to the slope in the bottom boundary layer and the speed of the along-slope current, V_s.
Temperature measurements in a slope current at latitude 49°N indicate a uniform mixed bottom boundary layer of thickness $D \approx 45$ metres. Estimate τ_b, T_b and V_s given that $k_b = 0.0025$.

10.3. An upwelling-favourable along-shelf wind of 12 m s^{-1} drives water from below the thermocline onto a narrow temperate shelf of width 20 km. The sub-thermocline water has a nitrate concentration of 10 mmol m^{-3}.

(i) Estimate the rate of primary production that could be supported by the upwelling.
(ii) What speed would a slope current have to have in order to drive the same biological production?

10.4. Two moorings are used to estimate an internal tidal wave energy decrease of 5 J m^{-3} between them, with the wave taking 14 hours to pass from one mooring to the next. Typical vertical profiles of density and nitrate concentration between the moorings are:

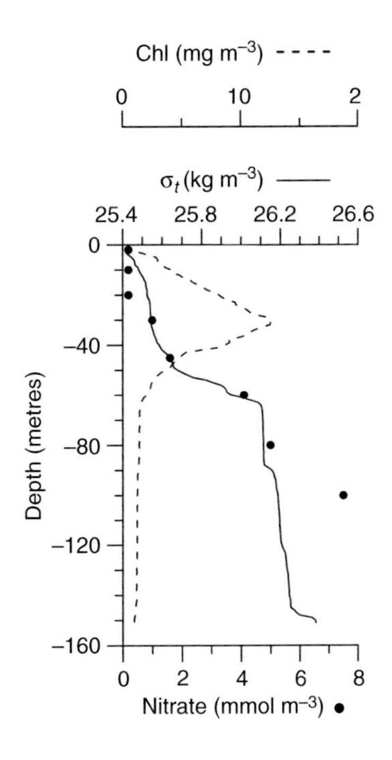

Estimate the flux of nitrate (mmol m^{-2} d^{-1}) into the base of the subsurface chlorophyll maximum driven by internal tidal mixing.

10.5. Assuming the bottom layer of the North Sea has a mean thickness of 50 metres, and using Fig. 10.24, estimate the volume of the bottom layer of the seasonally stratifying central and northern North Sea.
Thomas, *et al.*, 2004 observed that 13 mmol m^{-3} of DIC is stored in the bottom water of the stratified North Sea as a result of surface layer primary productivity. Use the estimate of annual bottom layer exchange with the North Atlantic in Huthnance, *et al.*, 2009 to calculate the potential export of DIC from the North Sea to the North Atlantic.
Discuss why your answer differs from that estimated by Thomas and co-workers.

11 Future challenges in shelf seas

In this final chapter we shall try to provide a perspective of the science of shelf seas and an indication of some of the important challenges which remain. In looking to the future of the subject, we shall highlight the need to move from temperate latitude shelf seas, which have been the focus of most research to date, into the shelf seas of the Arctic and the tropics. In both these areas, the suite of physical processes controlling the shelf sea environment and its biogeochemistry is substantially different from that operating in temperate latitudes. As well as looking at the prospects for future research in these areas, we shall also consider new ideas on the role of the shelf seas in the global ocean system and their putative influence on climate change since the last ice age. But first, we shall try to identify the big questions which remain in relation to the scientific understanding of the mid latitude shelf seas.

11.1 Remaining puzzles in the temperate shelf seas

As we have seen in previous chapters, the dominant physical processes controlling the environment of the shelf seas in temperate regions have been identified. The different shelf sea regimes have been defined in terms of the particular processes which dominate them, and the interaction of these processes has, in several cases, been simulated at least to first order in numerical models.

The level of physical understanding of the regimes is generally sufficient to provide a sound basis for biogeochemical studies, and considerable progress has been made particularly in the elucidation of the factors controlling primary production. The roles of stratification and reductions in vertical turbulent mixing in controlling the spring bloom of phytoplankton are well established. Our understanding of how layers of phytoplankton grow within pycnoclines, driven by modification of the vertical turbulent exchange in regions of strong vertical gradients in light and nutrients, has developed over the past two decades. Sub surface primary production can now be regarded as of equal importance to the spring bloom in terms of the supply of organic material to the rest of the shelf sea ecosystem.

At trophic levels above the primary production, there is some understanding of how the patchiness in shelf sea ecosystems, from zooplankton up to commercially important fish stocks, is linked to the physical environment. While phytoplankton growth is obviously required to provide the basis of food for higher trophic levels, there are many instances where the physics plays a more direct role, for instance in modifying prey distributions, having an effect on the availability of prey or by providing vital transport pathways between spawning grounds and feeding areas.

While progress has been made in understanding the temperate shelf seas and their ecosystems, much remains to be done in investigating the detail of process interactions and in refining models of the different shelf sea regimes. We shall now consider some examples of outstanding research questions relating to the temperate shelf seas which provide challenges for future studies.

11.1.1 Mixing in the pycnocline

In Chapter 7, we explained the importance of the rather small amount of mixing which occurs in the seasonal thermocline and drives nutrients upwards to fuel primary production in the subsurface maximum. Three candidate mechanisms for supplying the stirring power for this mixing have been identified, namely: (1) internal wave motions, (2) wind-driven inertial oscillations and (3) peaks in boundary-forced mixing in the springs-neaps cycle. The environment characteristics which determine the relative contributions in different parts of the shelf seas are not fully understood.

In principle, it should be possible to simulate the effects of inertial oscillations with existing models provided good-quality, high-resolution wind data are available. Recent studies have pointed to the importance of the shear induced across the thermocline by the rotating shear vector. Peaks in the magnitude of the bulk shear (velocity difference between top and bottom layers) occur when the shear vector aligns with the wind stress and/or the bottom stress vector (Burchard and Rippeth, 2009). These shear 'spikes' induce a temporary lowering of the Richardson number which may be sufficient to induce mixing. Accurate modelling of such intermittent bursts of mixing is likely to be a severe challenge, and therefore a good test, for turbulence closure schemes.

Internal tides and waves present an even more difficult challenge. As we noted in Chapter 10, most of the large energy input to the internal tide at the shelf break is dissipated close to the source (typically within ~75 km of the slope) and does not influence the inner regions of broad shelf seas where locally generated internal waves are the alternative candidate power source. In principle, it should be possible to simulate the generation and propagation of these waves but, as discussed in Chapter 7, present-day large-scale models of shelf seas have insufficient horizontal resolution to represent components of the internal waves, which can have wavelengths of 1 km or less. An alternative approach would be to use the theory of wave generation in stratified flow over variable topography to estimate the local power source which, together with assumptions about horizontal scale of propagation before dissipation, could be used as a basis for parameterisation of the mixing effect.

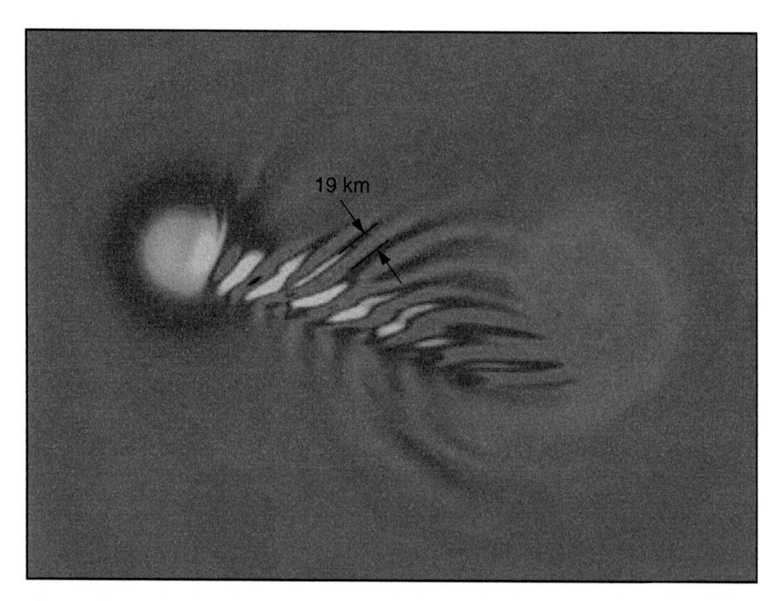

Figure 11.1 See colour plates version. Modelled internal waves associated with two counter-rotating eddies shed in a stratified flow past an isolated seamount. The image is a plan view of the density perturbation away from the mean at a depth of 400 metres. The eddy on the left is rotating clockwise and is trapped over a seamount of 25 km diameter. The anticlockwise eddy on the right is being advected away by the mean flow, shedding the internal waves. From Blaise *et al.* (2010), courtesy of Springer.

There is ongoing work with fine resolution models of internal wave generation by stratified flow over topography (e.g. Xing and Davies, 1998). Unstructured model grids now allow model resolution to be concentrated in regions where small scale waves are generated (Lai *et al.*, 2010). The example in Fig. 11.1 shows the remarkable detail that this enhanced resolution permits, allowing simulation of the patchiness of internal waves. However, the high resolution of these models, combined with the need to remove the hydrostatic assumption when dealing with large amplitude, short wavelength internal waves, makes them computationally expensive; computing power needs an order of magnitude shift before these direct approaches can be used as models of whole shelf systems. This high-resolution modelling is another possible route to designing parameterisations suitable for regional numerical models.

11.1.2 Towards the fundamentals of biogeochemistry and phytoplankton ecosystems

You may have noticed an important difference, which we mentioned in Chapter 5, in the characteristics of the physical and biological processes that need to be understood in order to quantify the behaviour of shelf seas. Physical oceanography is rooted in a fairly small set of fundamental equations. If you have the equations of motion and continuity, and use the hydrostatic assumption, it is possible to derive most of the basic features of ocean circulation.[1] In contrast to the physics, the equations commonly used to describe the fundamentals of primary production, such as the shape of the growth-radiation response of the phytoplankton or the uptake of nutrients, are empirical. While we understand the fundamentals of the biochemistry of photosynthesis and the molecular diffusion of nutrients to a cell that underpin these descriptions, there is a real challenge in formulating this basic biochemistry on a more fundamental level (Falkowski *et al.*, 2008). Progress is being made here, for instance by quantifying the interplay between the costs and benefits of different acclimation strategies in phytoplankton (Geider *et al.*, 2009), so that there appears to be a real prospect of significant advances in our models of primary production on the horizon. This is clearly of importance to the modelling of phytoplankton globally, not just in the shelf seas. However, the marked horizontal gradients in phytoplankton ecosystem structures that we see in the shelf seas, driven by the contrasts in physical forcing (e.g. Fig. 10.25), provide a particular challenge at a location which can be reached without several days steaming in a research vessel.

Such a change in the approach to describing phytoplankton biochemistry could revolutionise our models of phytoplankton communities. Most phytoplankton ecosystem models in regular use are based on a formulation of phytoplankton growth that can be recognised in the pioneering work of Gordon Riley (Visser and Kiørboe, 2006). In a development of the earlier single species models, the phytoplankton ecosystem is now often split into a number of *phytoplankton functional types* (PFTs), e.g. diatoms, dinoflagellates, cyanobacteria, flagellates. Each PFT has a single representative 'species' in the model with driving parameters taken from laboratory cultures or observations. Additional complexity is added in attempts to describe, for instance, different grazer types, recycling of organic material, and coupling between the biochemistry of the water column and the benthos. While such models have been successful in replicating the typical structures of present-day phytoplankton communities (e.g. Kiørboe, 1993; Denman and Gargett, 1995), there are important limitations. The phytoplankton species are 'hard-wired' to their PFT representatives, and so the modelled ecosystem lacks the capacity to adapt to shifts in its environment in ways that we expect the real ecosystem to do. Also, the complex biochemical interactions and the parameter values that control them are often lacking a sound observational basis. We would argue that such models cannot address the subtleties of the diversity of the phytoplankton community, and yet it is likely that shifts in phytoplankton diversity in response to physical processes are a more powerful driver of whole ecosystem biodiversity than simply assessing the role of physics in determining phytoplankton carbon fixation. This is an area of overlap between the biological oceanography and the fisheries oceanography that has high potential reward: in regions where gradients in the physics appear to correlate with the distribution of higher trophic levels, such as commercially important fish, we need to move on from simply linking physics to the production of organic carbon and

[1] Of course, this fundamental rigour in the physics still has the problem of how to deal with turbulence and friction, which is an area in which we are still reliant on parameterisations.

instead begin to determine the response in the diversity of the phytoplankton community and the quality of the food that they provide.

An alternative model approach is to simplify much of the biochemistry back to a complexity that we can support with data, and instead focus the computational power on simulating several tens or hundreds of species (Follows *et al.*, 2007), with parameter ranges set by what we know is biogeochemically possible, providing a continuum of phytoplankton through the functional types. The resulting ecosystem structure then emerges out of competition between the species within the physical and chemical environment, in a way which is more flexible in its ability to respond to changes in the physical forcing. Combining this approach with more fundamental descriptions of cellular processes is a potential direction in modelling phytoplankton ecosystems. At the same time, the role of a diverse grazer population in providing top-down constraints on the structure of the phytoplankton community, as described in Chapter 5, would add further strength to our attempts at simulating physical forcing of the ecosystem. Linking the emergent phytoplankton approach (Follows *et al.*, 2007) with a size-based view of grazing (e.g. Baird, 2010; Banas, 2011) has considerable potential.

11.1.3 Organism size and the scales of fluid motion

Can you imagine life as a 10 μm cell living in a turbulent fluid, or a 2 mm copepod trying to catch that 10 μm organism? Recall from Chapter 5 that molecular viscosity plays an enormous role in determining nutrient supply to the cell and, if motile, its ability to move through the fluid and interact with food particles. This is currently an area of fruitful research (e.g. Karp-Boss *et al.*, 1996; Visser and Kiørboe, 2006; Langlois *et al.*, 2009), providing some fundamental, physical understanding of growth and grazing in pelagic food webs. In many cases this knowledge is able to reveal the mechanics that control observed traits, such as motility or foraging strategies. However, these new results have not yet been applied to inform the most widely used ecosystem models. In future model development, close attention to the guiding principles of how different sized particles are constrained by the physics of the surrounding fluid could lead to more realistic descriptions of how organisms are allowed to operate within the food web.

If we move from single cells and small zooplankton, through to fish larvae and larger fish, fluid viscosity loses its controlling influence, but the medium to large scale (10s cm to several metres) motion of vertical turbulent transport, along with the structure of mean flows (e.g. estuarine flows, tidal rectification, frontal convergences), then become the major environmental control. We have seen several examples of how mean circulation plays a vital role in an organism's life cycle, as for example in the larval transport by the frontal jet around Georges Bank (Fig. 8.21) or density-driven flows in the ROFI outside Delaware Bay which ensure the survival of fiddler crabs (Fig. 9.14). These are relatively large scale, persistent flows. The challenge is perhaps how to make compatible measurements of smaller scale, more ephemeral flows (e.g. the mesoscale variability of frontal convergences) and the distribution and behaviour of organisms within them. To some extent, this is a technical problem, one that we often come across in interdisciplinary research. The properties of the sea that allow us to describe the physics of the ocean are often easily measured (e.g. salinity and temperature, well-resolved vertical profiles of currents). Making parallel measurements of the biology, particularly as the size of an organism increases and its number density in the ocean decreases, requires some innovation. The emergence of low-powered digital cameras, autonomous vehicles and combinations of acoustic techniques (e.g. vessel-mounted ADCPs and zooplankton/fish echo sounders) are all areas where new insights into physical drivers of ecosystems are possible.

11.1.4 Observations and numerical models

In addition to the technical challenges in linking physical and biological observations at the right scales, there is a similar challenge associated with model development and our ability to provide suitable observational data for validation. Look again at the image of internal waves generated by flow past a seamount in Fig. 11.1. How would you try to observe that in the real ocean? We could acquire images from satellite-borne radar (e.g. Fig. 10.13a), but the sensitivity of such imagery to sea state and the orbital characteristics of the satellite are not amenable to gaining any insight into the temporal evolution of the waves, which would be vital for any sensible model validation. Also, we would really want to know the vertical density structure underneath the sea surface rather than just the surface manifestation of the waves viewed by the satellite sensor. Perhaps a survey with an autonomous vehicle would do the job? Assume the vehicle can travel at 1 m s^{-1}, and that we want to survey an area about 10 km $\times$ 10 km. Spacing our survey tracks 1 km apart would lead to a total survey length of 109 km, which would take about 1 day 6 hours to complete. The internal wave field is likely to modify substantially over that time, so our survey could not be viewed as synoptic. A research vessel operating acoustics (e.g. Fig. 1.6) and (for internal waves closer to the sea surface) towing an undulating vehicle (Fig. 1.5) at 8 knots could get round the survey in 4 hours: still not synoptic, but getting much closer to what we might want to provide the modellers with. Models deal with noise-free, clear, synoptic (and seductive) fields of any data type that the modeller wants to see. Observationalists work from slow research vessels, in a real, noisy environment, and cannot measure many of the modelled parameters at scales compatible with the physics. Particularly in interdisciplinary oceanography, not just in the shelf seas, this is an area of significant challenge; combining models and observations together in a meaningful and insightful way which goes beyond simply working on the same project with sets of different tools and different approaches.

11.2 Regional questions

There remain many interesting topics for research in those regions of the shelf seas globally that have received less research attention compared to the temperate shelf regions on which our work, and this book, have mainly focused. These other regions

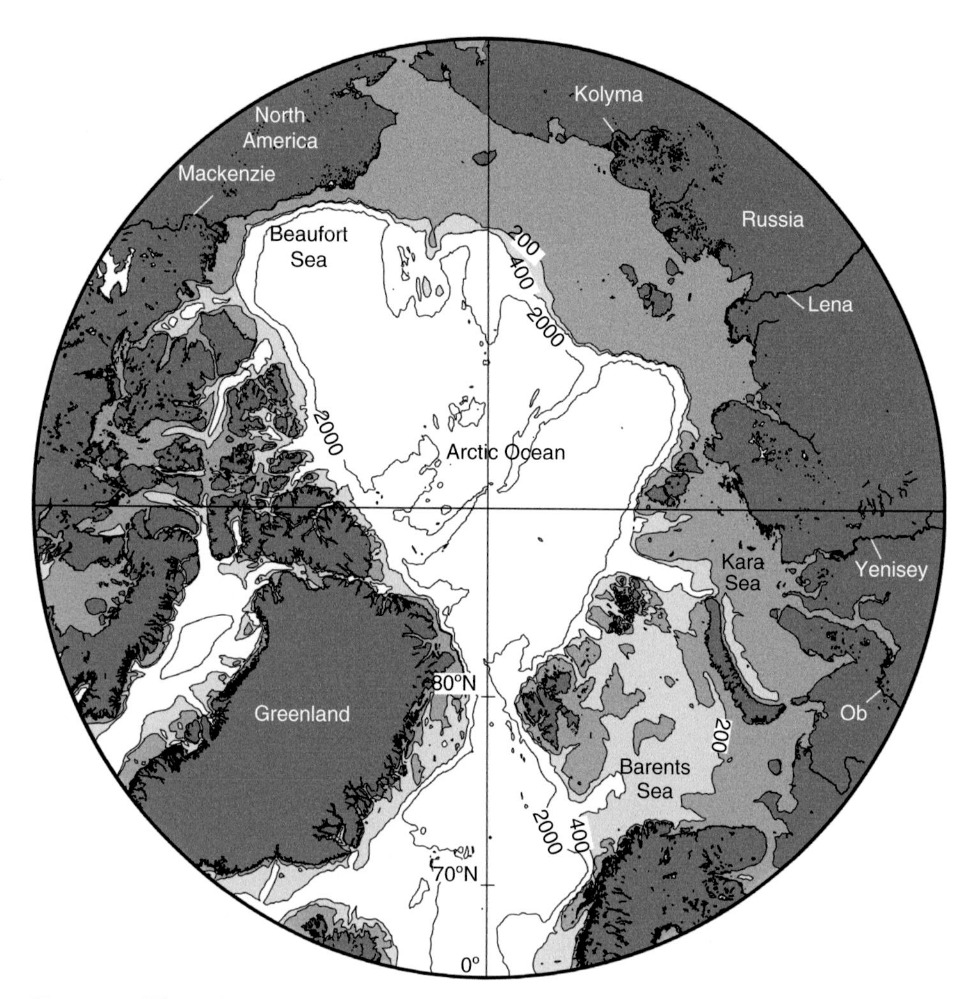

Figure 11.2 The major rivers supplying freshwater to the shelf seas around the Arctic Ocean. Isobaths are marked at 200, 400 and 2000 metres. Coast data based on (Wessel and Smith, 1996), figure provided by Clare Postlethwaite, National Oceanography Centre, UK.

have generally received limited research effort either because the environment is difficult to work in or because the bordering nations have other priorities for their investment.

11.2.1 Arctic shelf seas

The Arctic is host to very large areas of shelf which up to now have been little studied. You can see in Fig. 11.2 that approximately 40% of the area of the Arctic Ocean is occupied by shelf seas, much of it in the very broad shelf regions north of Russia where the shelf extends out as much as 1000 km from the coastline before it reaches the deep water of the Arctic basin. Much of this shelf area is relatively

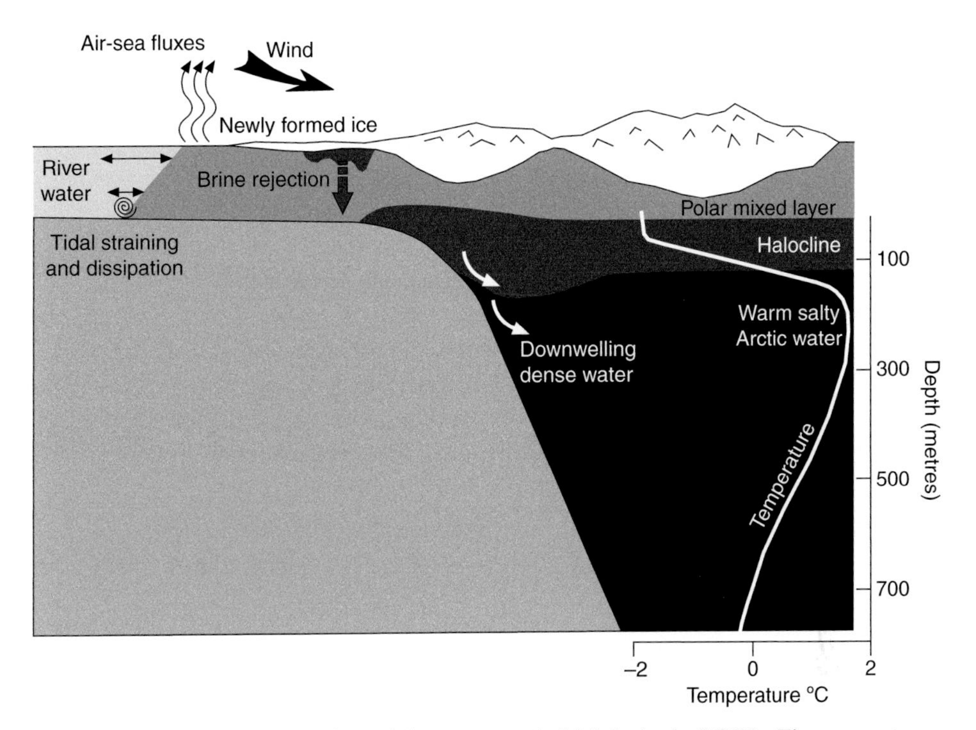

Figure 11.3 Schematic illustration of the processes in high latitude ROFIs. Figure courtesy of Yueng-Djern Lenn, Bangor University, UK.

shallow (50–100 metres depth) and is influenced by massive inflows of freshwater, mainly from the Russian rivers, although the Mackenzie River makes a substantial contribution on the Canadian side of the Arctic. The total annual freshwater input averages to ~60 000 $m^3\ s^{-1}$ which is almost 10 times that of the Columbia River in northwestern North America, and ~30 times that of the Rhine flowing into the North Sea. This inflow is concentrated in the summer period with a large peak in June and minimal flow during the winter months when the rivers are frozen.

Much of the Arctic shelf therefore qualifies as a ROFI but with major differences from temperate systems which are highlighted in Fig. 11.3. The most obvious difference is the seasonal ice cover which prevails over much of the shelf for a large fraction of the annual cycle. In recent years, the retreat of the ice in the summer has increased rapidly, thus extending the area of shelf which is subject to seasonal switching between ice cover and an open free surface interacting directly with the atmosphere. During ice-free periods, radiation inputs to the water column will increase surface warming and primary production, while wind forcing will exert a more direct and important influence on vertical mixing. In our warming world, earlier ice melt can lead to shifts in the timing of primary production (Sarmiento *et al.*, 2004; Kahru *et al.*, 2011).

The change to a free surface will also modify the barotropic tidal flow by removing the upper frictional boundary layer associated with ice cover. Tidal currents are not large over most of the Arctic shelf (<0.5 m s^{-1}), though they contain a high proportion of the kinetic energy of the flow, contributing to mixing in the boundary layers. Tides can also be locally important in increasing ice formation by opening leads and polynyas, so providing more area for ice formation and influencing the amount of dense water created (Hannah *et al.*, 2009; Postlethwaite *et al.*, 2011). As well as exerting a major influence on water column stability, the large freshwater input to the Arctic shelf seas will also bring with it substantial inputs of nutrients, organic carbon and other components from terrestrial sources. A further radical difference from the temperate case is the process of *brine rejection* which occurs during the freezing of salt water. As ice is formed mainly from freshwater, the salt left behind remains in the ambient seawater and increases its salinity and hence its density. The importance of this process and the potential of the heavy water formed to cascade into deeper water was first recognised by the explorer Fridtjof Nansen (Nansen, 1906).

Understanding the impact of these additional processes on the structure and circulation of the Arctic shelves and the response of the biological system presents a demanding but interesting challenge for the future. The practical difficulties of working in these high latitude regions are substantial but there is considerable motivation. The large areas of the Arctic shelf which are free from ice during the summer are postulated to make a large contribution to carbon fixation (Bates, 2006), with the potential to change markedly in a warming climate (Arrigo *et al.*, 2008).

Moreover, the physical processes occurring on the shelf play a big role in the transformation of water masses in the Arctic. This is shown in the map in Fig. 11.4. Atlantic water, which enters the Arctic Ocean via the Fram Strait, circulates in an anticlockwise flow around the rim of the deep basin. As it does so, the relatively warm Atlantic water is progressively cooled through heat loss to the overlying surface water and through interaction with the waters of the adjacent shelf. Away from the immediate vicinity of the slope, measurements indicate that the upward heat flux from the warm rim current through the stable halocline is due to the process of double diffusive convection and is too small to account for the observed cooling (Lenn *et al.*, 2009). The main contribution to the cooling is postulated to come from convective stirring by heavy saline water produced by brine rejection on the shelf (e.g. Turner, 2010). In this way, the little understood shelf processes in the Arctic combine to influence the properties of the return flow of cold Arctic water through the Fram Strait. This flow is an important part of *global overturning circulation* since it constitutes the primary source for the formation of North Atlantic Deep Water.

11.2.2 Tropical ROFIs

The shelf seas located in or near the tropics also present an interesting scientific challenge. Many shelf seas in tropical regions are of limited area, extending only a small distance (~50 km or less) from the coastline. This is especially true of the

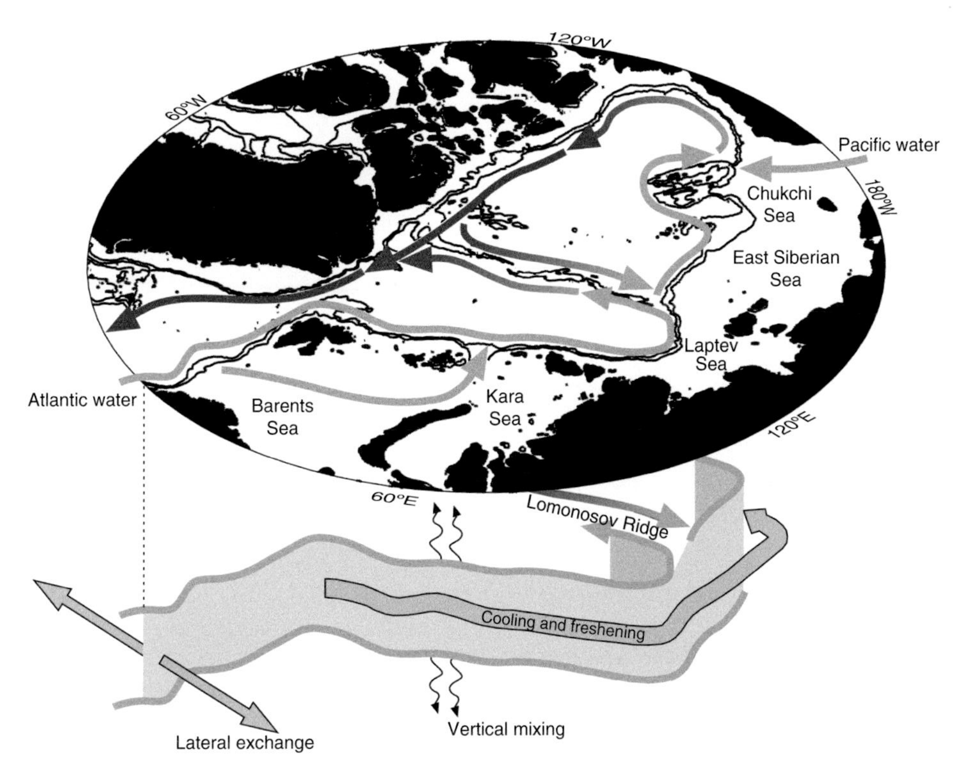

Figure 11.4 Arctic Ocean exchange with the North Atlantic, and the transformation of water masses. Figure courtesy of Yueng-Djern Lenn, Bangor University, UK.

tropical margins of the African and American continents, though in south-east Asia and to the north of Australia there are large areas of broad shelf comparable in extent to the major mid-latitude shelf seas.

As we noted in Chapter 2 (Section 2.3.2), many of these areas are subject to strong seasonal freshwater runoff which dominates over heat exchange in terms of buoyancy input, driven by monsoons, promoting a seasonal cycle which is somewhat analogous to that of the heating-cooling cycle in mid latitudes. They also differ from mid-latitude shelf seas in that, being close to the equator, the Coriolis parameter is close to zero so the influence of rotation is greatly diminished.

To date, these areas have been little studied so that understanding of their physical and biological systems is limited, in spite of their considerable economic importance to the large populations living in the coastal areas bordering these seas. As an example of a tropical ROFI system, consider the Gulf of Thailand, shown in Fig. 11.5. This is a large shelf sea with an average depth of ~50 metres and an area (2.7×10^5 km^2) comparable to that of the North Sea. The Gulf has a relatively energetic tidal regime which will drive considerable tidal mixing, as can be seen from the plot of the *SH* parameter in Fig. 11.5a. Large freshwater inputs to the Gulf occur during the wet season (June to October) from rivers flowing directly into the Gulf and

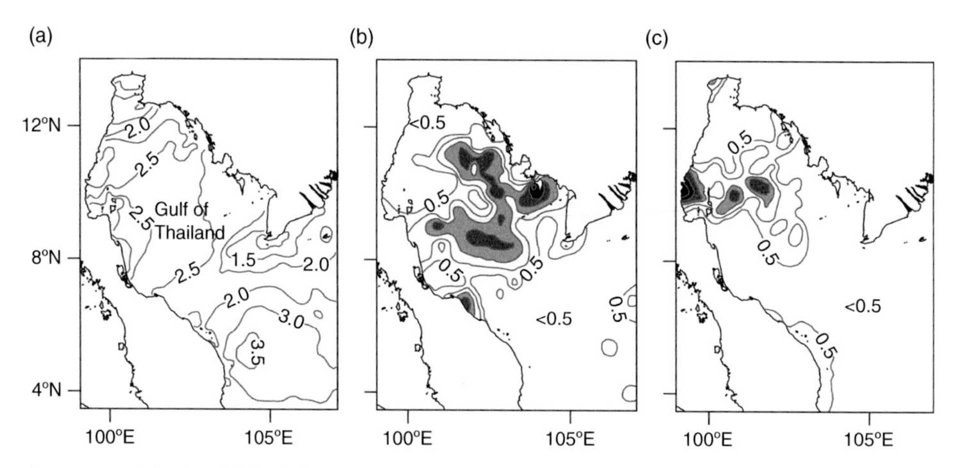

Figure 11.5 (a) The tidal mixing parameter *SH* in the Gulf of Thailand; (b) and (c) Haline stratification in December and in January/February contoured every 0.5. Figures courtesy of Anond Snidvongs, Southeast Asia START Regional Centre, Thailand.

also from the Mekong River whose discharge at the entrance is recruited into the Gulf by the seasonal winds blowing from the northeast during the wet season (Stansfield and Garrett, 1997). Averaged over the area of the Gulf, the freshwater buoyancy input during the wet season exceeds that injected through seasonal heat exchange by a factor of $\sim$3 and leads to the production of pronounced haline stratification (Simpson and Snidvongs, 1998). Figure 11.5b, which is a compilation of the available data from all years, shows strong stratification in December with ΔS up to 3 over a large area of the Gulf. By January and February (Fig. 11.5c), stratification has clearly been eroded by vertical mixing in a spatial pattern which is similar to the tidal stirring parameter of Fig. 11.5a. This suggests an analogy with the heating stirring competition in temperate latitudes but with the monsoonal input of freshwater distributed over the gulf replacing surface heating as the buoyancy source competing with tidal and wind stirring.

As we noted in Chapter 9, modelling the evolution of stratification forced by freshwater from coastal sources is inherently more difficult than the heating-stirring case, where the buoyancy input is spatially uniform, because the spreading of the buoyancy is itself a significant part of the problem. Elucidating the details of the evolution of haline stratification in tropical ROFI systems, therefore, presents a stimulating scientific challenge. It is also a matter of considerable importance for the countries bordering the shelf sea. Areas like the Gulf of Thailand are coming under increasing environmental pressure from nutrient and pollutant inputs from river and coastal discharges as well as from intensive fishing. A pre-requisite for controlling and managing the impact of these inputs is a sound physical understanding of the gulf system and the way it impacts on primary production and higher trophic levels.

11.3 Managing shelf sea resources

There is an increasing demand for sufficient understanding of the biology of the shelf seas to allow robust predictions of how ecosystems might respond to different strategies for managing fisheries, against the back-drop of a changing climate. Oceanographers have responded to this by attempting to predict how the biogeochemistry of the shelf seas might respond to a warming climate, on the basis that the carbon fixed by the primary producers represents a fundamental control on the biomass of fish (e.g. Holt *et al.*, 2009). We might expect large shifts in carbon fixation to have effects through the shelf ecosystem, and so some predictive capability is required. However, we have also seen throughout this book that phytoplankton biomass rarely appears as the principal limiter on the survival of larger organisms. Instead we have seen how physical processes such as frontal jets and convergences, internal tides and solitons, and density-driven flows play more direct roles in the life cycles of important fish species. Developing our understanding, and modelling, of the mechanics of these physical drivers on, for instance, phytoplankton community structure (rather than bulk biomass), larval retention and transport, and prey availability is vital to any meaningful predictions of an ecosystem's response to climate and fishing pressures. Correctly linking the spatial and temporal components of the physics through to the requirements of fish is also needed if we are to design marine protected areas on a scale that both supports the continued existence of fish stocks and provides for sustainable harvesting by fishers (Botsford *et al.*, 2003).

11.4 Shelf seas in the Earth system

The high level of primary production in the shelf seas relative to that of the abyssal ocean is now well documented (e.g. Liu *et al.*, 2010). Important questions remain, however, about the extent to which the resultant particulate carbon is sequestered into shelf sediments, or exported off the shelf and into deep water to be buried in abyssal sediments, or re-mineralised in the water column. There are also interesting questions about which parts of the shelf seas act as sinks for atmospheric CO_2 from the atmosphere while others are net sources (Jahnke, 2010).

11.4.1 Shelf sources and sinks of CO_2

We saw in Chapter 10 how recent studies have suggested that, in mid-latitudes, seasonal stratification exerts a key control on net air-sea fluxes of CO_2. We saw in the North Sea (see Section 10.9.2) that the net balance appears to be a significant net drawdown of carbon for the atmosphere. A recent worldwide synthesis of seasonally averaged fluxes based on measurements of ΔpCO_2 (Chen and Borges,

2009) supports the idea that open shelves in temperate and high latitude regions are net sinks for carbon, while shelf seas in low latitudes appear to act as sources; the net global exchange is estimated to be a transfer of carbon from atmosphere to ocean of ~0.35 GT a^{-1} or ~29% of the total CO_2 absorption by the ocean. The limited sampling of these fluxes, however, means that such estimates are only provisional and subject to large uncertainty. In view of the crucial role of CO_2 absorption in mitigating global warming, there is clearly a pressing need for a future research in this area to improve process understanding and our estimates of fluxes from different shelf regimes.

11.4.2 Cross-slope fluxes at the shelf edge

It might seem obvious that the way to determine the net contribution of the shelf seas to carbon export to the deep ocean would be to measure fluxes at the shelf edge. This would also be the place to observe the transport of nutrients back on to the shelf from deep water sources which are replenished by re-cycling in the deep ocean. As we have seen in Chapter 10, the principal physical processes operating at the shelf edge and over the continental slope have been identified and their importance in relation to the biogeochemistry established. However, quantifying the cross-slope exchange which they bring about at the continental margins of the ocean remains a very challenging and largely unsolved problem. There are major uncertainties in our knowledge of the temporal and spatial variability of physical water fluxes across the shelf edge, and the implications of this exchange on whole-shelf scales. Some of the cross-shelf transport mechanisms, such as the Ekman drain associated with the slope current, are relatively consistent, but others, e.g. cascading and wind-forced transport in the surface layers, are highly intermittent and spatially variable. In the North Sea, for example, the observations of carbon trapped below the seasonal thermocline are clear, but calculating the amount of that carbon which is eventually exported off the shelf edge involves uncertain estimates of the fraction of the lower layer that is exchanged with the North Atlantic before convective overturning re-establishes contact between the bottom layer DIC and the atmosphere.

There is some evidence which indicates that significant quantities of organic matter from the shelf arrive at the shelf break and descend in the bottom layer flow over the slope (e.g. the SEEP studies off eastern North America (Biscaye *et al.*, 1994)). Figure 11.6 shows a schematic summary of measurements of organic carbon made at the Hebridean shelf edge during the UK SES study. While downward fluxes from the surface waters over the slope, determined by sediment traps, showed a settling flux of ~2 g C m^{-2} a^{-1}, the measured rate of respiration in the surficial bed sediments was ~20 g C m^{-2} a^{-1}, while long-term burial was estimated at only ~0.2 g C m^{-2} a^{-1}. Direct observations in the boundary layer over the slope confirmed that large quantities of organic matter were being advected downslope. Much of the material was in the form of 'fresh' phytodetritus which was observed down to at least ~500 metres on the slope, indicating rapid transport from the surface. Interestingly, the remaining organic material contained a significant component of carbon from

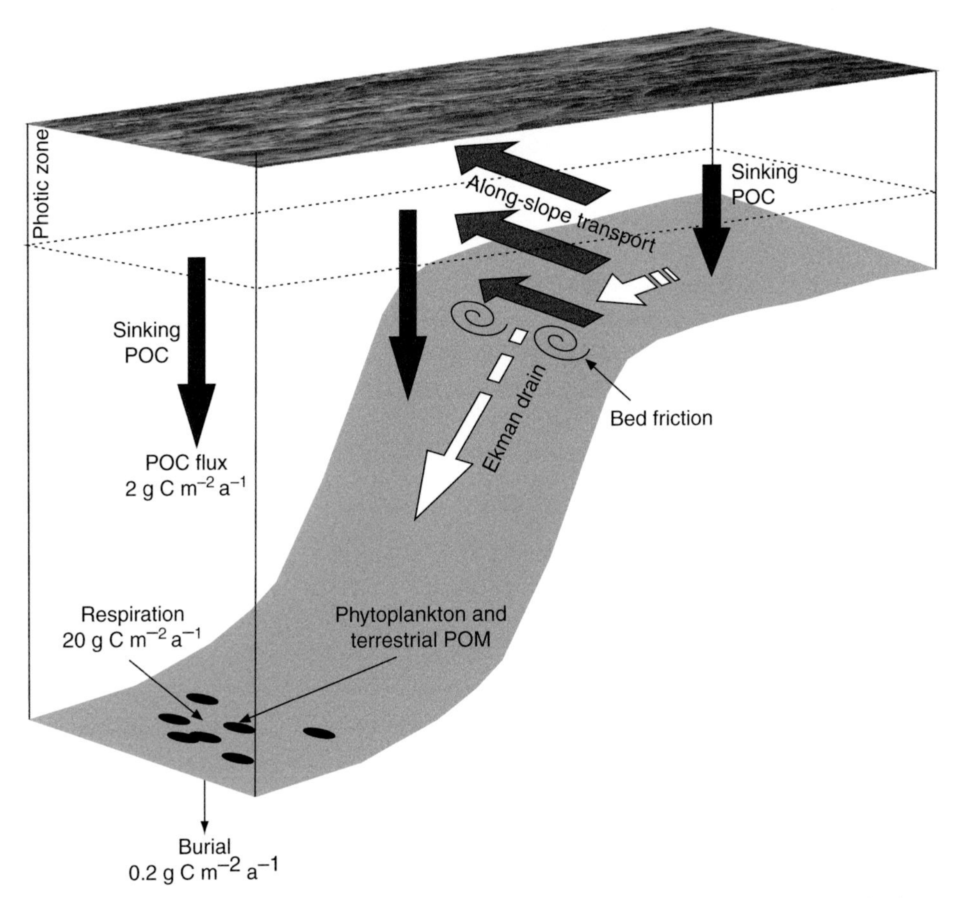

Figure 11.6 Schematic of the particulate organic carbon fluxes off the Hebridean shelf edge. Figure courtesy of George Wolff, University of Liverpool.

terrestrial sources. We can measure the integrated carbon fluxes using traps and seabed sampling, and we have a good understanding of what physical processes are most likely responsible for carrying that carbon. The challenge is to develop techniques that allow us to make compatible and co-located measurements of the biogeochemistry and the physical transport mechanisms if we are to really understand how different processes contribute to carbon export from the shelf.

More studies like SEEP and SES are needed on a representative variety of shelf-slope systems to provide a basis for models of cross-slope transport from which we might hope to obtain globally integrated estimates of carbon export. In addition to the carbon fluxes off the shelf, there is also a need to estimate the reverse flow of nutrients from the deep ocean on to the shelf. It is widely conjectured that the deep ocean is the dominant source of nutrients for new production on the shelf (Liu *et al.*, 2010), but present estimates of these fluxes, even for well-studied shelf systems, are uncertain and often depend on the results of numerical models which have not been validated for the purpose.

11.5 Past perspectives on the shelf seas

Arguably some of the most interesting and important questions for further research in the shelf seas arise in relation to the past and future changes in the physics and the biology of these areas. The dramatic changes in sea level which have occurred since the last glaciation have greatly modified the area and topography of the shelf seas with consequences for the strength and distribution of energy dissipation in the tides. The associated changes in the biology of the shelf seas involves drastic changes in the levels of carbon fixation and hence the shelf seas' role in the Earth system. We shall now consider some of the key issues in the oceanography of the shelf seas looking back to the last ice age.

11.5.1 Changes in tidal dissipation

In previous chapters, we have seen that the tides play a major role in providing much of the energy input to the shelf seas. The magnitude of this power input to shelf seas around the deep ocean can now be determined by numerical models driven only by the lunar and solar tide generating forces. Plots like that of Fig. 2.13 involve the computation of the tidal response of the ocean on a spatial scale which is small enough to resolve the detailed topography of the ocean, a calculation first envisaged by Laplace more than two hundred years ago and only recently made possible by the advent of massive computing power.

As well as determining dissipation in the shelf seas, these model calculations, as we saw in Section 2.5.2, also provide estimates of the dissipation of tidal energy in the deep ocean. In the current situation, they indicate that, of the global total tidal dissipation, approximately 2.5–2.7 TW is dissipated in the shelf with ~1TW being consumed in the deep ocean. These values are consistent with independent estimates of the total tidal dissipation which come from measurements of the acceleration of the moon in its orbit by laser ranging and from satellite altimetry (Egbert and Ray, 2000; Egbert and Ray, 2001). We therefore have a rather good picture of the present-day global tidal dissipation, but questions then arise as to, whether or not, the present pattern of dissipation has always prevailed as sea level has varied, for example through glacial- inter-glacial cycles, and whether total dissipation and its distribution will change as sea level rises in an era of global warming.

There has been a tendency to consider the dissipation rate as fixed; for example, G. H. Darwin, a son of Charles, assumed a dissipation rate constant over long periods of geological time to infer the history of Earth–Moon separation (Darwin, 1899). There seems, however, no fundamental reason why dissipation should be invariant; on the contrary, changes in sea level, in seabed topography and even in the thermal structure of the ocean might all be expected to change the magnitude and distribution of dissipation. Such changes are important, not just in relation to the environment of the shelf seas, but also because it is now recognised that tidal energy makes a major contribution to mixing in the deep

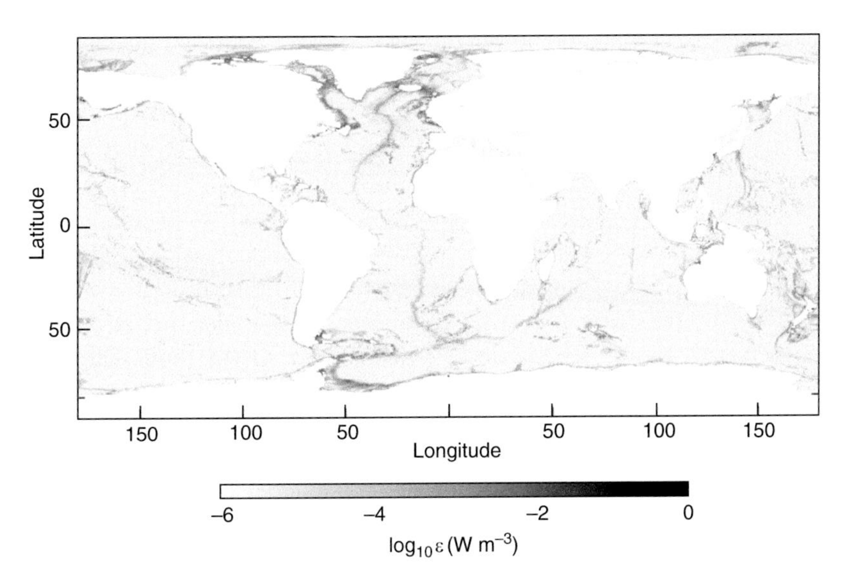

Figure 11.7 See colour plates version. Tidal energy dissipation during the last glacial maximum. From (Green, 2010) with permission from Springer.

ocean which is necessary for the maintenance of the Meridional Overturning Circulation (MOC) (Wunsch and Ferrari, 2004).

Recent model studies indicate that important changes in dissipation have taken place, for example in the period since the last ice age when, at the time of the peak glaciation, sea level was ~130 metres lower than now and much of the present shelf sea area was dry. In this situation, we might expect that the removal of large areas of the shelf, where a considerable proportion of the existing dissipation occurs, would *reduce* the total dissipation. In fact, a number of different models (e.g. Egbert *et al.*, 2004; Griffiths and Peltier, 2008; Arbic *et al.*, 2009) concur that total dissipation was ~33% higher in the Pleistocene with a total dissipation of ~4.7 TW, of which 4.2 TW could be found in the semi-diurnal band (Egbert *et al.*, 2004; Green *et al.*, 2009). At the same time, the rate of energy dissipation in the deep ocean was higher by a factor of ~2.5 relative to the present situation. In consequence, considerably more power was available to produce mixing in the deep ocean with stronger forcing of the MOC during the glacial era (Green *et al.*, 2009). In spite of this increase in forcing, the evidence available suggests that the MOC was not increased but rather decreased during the last ice age, because of the suppression of deepwater formation by the presence of the increased freshwater content of the surface waters of the northern North Atlantic.

How can removal of the damping area (the shelf) increase dissipation? The answer seems to be that the reduction in damping changes the dynamical response of the ocean and moves it closer to resonance in one, or more, of the ocean basins. For example, in Fig. 11.7 the distribution of tidal dissipation at the peak of the last glaciation is seen to be very different from the current situation (Fig. 2.13), most especially in the North Atlantic where the tides are considerably enhanced. The differences are mainly due to the change in sea level, but the changes in topography due to loading of the solid earth by the ice mass have also been taken into account.

11.5.2 Changes in the role of the shelf seas in the carbon cycle

At the time of the last glacial maximum, the total area of the shelf seas was reduced to $\sim 6 \times 10^6$ km^2 which is only ~25% of its present value of 26×10^6 km^2. On this basis alone we would expect the potential of the shelf seas to fix carbon and draw down atmospheric CO_2 (the *continental shelf pump*, see Section 10.9.3) to have increased greatly as sea level rose between 18 000 and 8000 years BP (Before Present). But it is not just a question of the shelf area. As we saw in Chapter 11, the water column structure of a temperate shelf sea exerts an important control on CO_2 recruitment from the atmosphere with stratified areas acting as sinks and permanently mixed regions acting as sources releasing CO_2 back into the atmosphere. Using a numerical model to hindcast the tidal changes since the last glacial maximum, it has been possible to determine the extent of stratified and mixed regimes by applying the *SH* criterion (Chapter 6) with assumptions about the annual cycle of heating and cooling (Rippeth *et al.*, 2008). These hindcasts of tidal conditions are supported by some evidence from a recent study of changes in frontal positions over the last 20 000 years based on bottom temperature estimates derived from proxies (Marret *et al.*, 2004). Making the bold assumption that present-day heat exchange for stratified and mixed regions (Thomas *et al.*, 2004) has remained invariant over the glacial-inter-glacial transition, it is possible to estimate the evolution of the net carbon sink in the temperate shelf seas. The results indicate an increase of the shelf sea sink by a factor of ~3 since 18 000 BP, the effect of which would have been to oppose the increase in atmospheric CO_2 which has taken place since the glacial period.

11.6 Response of the shelf seas to future climate change

Finally, we should look forward in time and consider the important practical questions concerning the changes in the shelf seas on a time scale of decades to centuries which will occur in response to climate change.

A question of immediate interest is, how will the ocean's tidal system react to future increases in sea level due to global warming? Model predictions indicate that tidal dissipation will decrease with rising global sea level (Green, 2010), continuing the trend which occurred as the Earth moved into the present inter-glacial. For realistic sea-level changes (1–2 metres over the next 100–200 years), dissipation on a global scale will decrease rather little (<5%) but there will be significant local effects on the tides in some shelf sea areas (e.g. Hudson Bay, Sea of Japan, the Severn Estuary and Barents Sea). Given the biogeochemical and ecological significance of the internal tide, a challenging question involves determining how changes in the seasonal cycles of heat flux across the sea surface may lead to shifts in the timing and extent of stratification at the shelf edge. Changes in stratification will alter the flux of internal tidal energy, and hence influence the vertical turbulent mixing of nutrients by internal solitons. Recent regional modelling of the northwest European shelf seas has suggested significant strengthening of stratification in a warmer world, as a result of changes in precipitation and in the timing of stratification (Holt, Wakelin, *et al.*, 2010).

The temperate shelf seas are currently warming at a greater rate than the open ocean, with the North Sea, for instance warming at about 0.4 °C decade^{-1} (Sharples *et al.*, 2006) compared to a global average for the ocean surface temperature of 0.16°C decade^{-1} (Solomon *et al.*, 2007). This is largely because the shallower water column provides much less thermal inertia or scope for vertical mixing of surface heat. We are already seeing large shifts in the distribution of animals associated with recent warming (e.g. changes in zooplankton (Beaugrand, 2009) and stocks of commercial fish (O'Brien *et al.*, 2000; Hunt *et al.*, 2002)) which are mainly ascribed to temperature increases and its effects on growth rates (e.g. Stegert *et al.*, 2010; Brander, 1995)). There is also a host of physical changes expected as our climate warms that may affect the ecological environment. Greater heat exchange through the ocean surface could alter the positions of the tidal mixing fronts, or the timing of the onset of stratification in the spring. Shifts in rainfall, either total amounts or the seasonal distribution, could alter the ROFI environment, for instance by changing the density-driven exchange between the estuaries and ROFIs and the adjacent shelf sea, or by modifying the strength of coastal buoyancy currents. Changes in the strength of winds (Kjellstrom *et al.*, 2011) will represent an important alteration of a source of mixing to the shallow seas. Up in the Arctic, the reduction of seasonal ice cover will have enormous impacts on the biogeochemistry of the high latitude shelf regions (Kahru *et al.*, 2011). In deciding how these changes might affect the higher trophic levels of the shelf sea ecosystems, the rate at which the changes become manifest will be a key determinant of whether or not existing species or ecosystems can adapt, or if we should expect wholesale shifts in shelf sea biology.

Conclusion

The shelf seas are intrinsically interesting, complex systems with often remarkable gradients in physical and biological dynamics over short horizontal distances. We have made considerable progress in understanding the fundamental physics of the shallow seas, including recognising the important contrasts with the open oceans, and linking the physics to the biology from microbes up to fish. In this chapter we have suggested several areas where we think further research effort is desirable and necessary, particularly in improving understanding basic processes, combining knowledge from different approaches and in transferring our existing knowledge to try it out on less studied regions.

Interdisciplinary research is now a rapidly developing field. The basic link between the vertical structure of turbulent mixing and the growth of the phytoplankton is well established, but there is work required to understand how the diverse shelf sea phytoplankton communities are controlled by spatially and temporally variable physical processes. In the past there has been an emphasis on taking the physics-biogeochemistry link forwards to understand the responses of higher trophic levels to the perceived patchiness in food supply. Physical processes can, however, influence ecosystems in other ways than exerting control on the growth of the primary producers. We have seen how mean flows in shelf seas can lie at the heart of the life cycles of organisms from phytoplankton to fish, and we are beginning to recognise that neither chlorophyll nor community carbon fixation are sufficient indicators of how an ecosystem reacts to physical processes. Instead we need to understand how turbulence affects the mechanics of nutrient uptake and predator-prey interactions, and how the developmental time scales of different trophic levels lead to different links between the physical environment and the ecosystem.

Bringing together scientists from different fields of oceanography in interdisciplinary studies is intellectually rewarding, but also has an increasing practical urgency. Our exploitation of the resources provided by the shelf seas, including fisheries, fossil fuels and renewable energy sources, has in the past often been pursued with little attention to the broader impacts on the marine system. This approach has been modified in the past decade or so as we begin to see how our use of these resources is combining with anthropogenic climate change to place severe pressures on shallow sea ecosystems. In these circumstances, there can be no doubt that there is a pressing requirement for truly interdisciplinary research, where physicists, biologists, chemists and ecologists work on problems unanswerable by any single discipline.

FURTHER READING

Follows, M. J., *et al.* Emergent biogeography of microbial communities in a model ocean. *Science*, 2007, **315**(5820), 1843–1846. And the Darwin Project: darwinproject.mit.edu/

The ASTroCAT model: faculty.washington.edu/banasn/models/astrocat/index.html

Green, J.A.M. Ocean tides and resonance. *Ocean Dynamics*, 2010, **60**(5), 1243–1253.

Pauly, D. *5 Easy Pieces: the Impact of Fisheries on Marine Ecosystems.* Island Press, Washington, 2010.

GLOSSARY

accessory pigments The main photosynthetic pigment used by photoautotrophs is chlorophyll *a*. Accessory pigments have different light absorption spectra compared to chlorophyll *a*, and so are useful for capturing photons not accessible to chlorophyll *a* and transferring energy to the photosystem. Pigments are often specific to particular phytoplankton groups, as detection of the accessory pigments provides a means to estimate the proportions of different phytoplankton groups within a sampled community.

Acoustic Doppler Current Profiler A device for measuring profiles of water velocity using the Doppler shift of sound reflected back by particles drifting freely with the water.

active fluorescence The active fluorescence technique uses sequences of very short flashes of light to gradually saturate the light-gathering capacity of a phytoplankton sample. It allows assessment of the state of the phytoplankton photophysiology (e.g. number and size of light-gathering units in the cells), and shows some promise in estimating rates of primary production that are compatible with the more standard ^{14}C incubation method.

ageostrophic Not in geostrophic balance. Ageostrophic flows are the result of an imbalance between pressure and Coriolis forces which occurs, for example, due to frictional effects in the surface and bottom boundary layers.

amphidrome A point of zero tidal displacement in the response of a shelf region to forcing by a tidal constituent. The time of high water and other phases of the tide due to the constituent rotate around the amphidrome. *See also* degenerate amphidrome.

autotrophs (autotrophic) Organisms that synthesise organic compounds such as carbohydrates from inorganic matter. In this book we deal exclusively with the photoautotrophs, or the autotrophic phytoplankton, that fix inorganic carbon using sunlight as an energy source.

baroclinic Flow in which surfaces of equal density (isopycnals) are not parallel to isobars (surfaces of equal pressure). In baroclinic regions the horizontal pressure gradient varies with height. A baroclinic pressure gradient might be set up by having a source of freshwater, or by the passage of an internal wave. The flows driven by the pressure gradient will vary through the water column. (See also barotropic.)

baroclinic instability An instability in flows where there are horizontal density gradients; the motion grows through the conversion of potential energy in the density field to kinetic energy of the mean flow leading to the formation of meanders and eddies.

barotropic Flow in which surfaces of equal density (isopycnals) are parallel to isobars, so that the pressure gradient is independent of depth. Barotropic flows tend, therefore, to be depth uniform and the term is often used (rather loosely) to signify depth independence. (See also baroclinic.)

brine rejection In freezing conditions, ice is formed mainly from freshwater. The rejected salt enhances the salinity of the unfrozen seawater and increases its density.

biological pump The export of carbon associated with biological growth from the surface ocean into deep water or the seabed sediments.

buoyancy frequency The natural frequency of oscillation of a water particle when it is displaced vertically in the presence of density stratification and then released. Sometimes referred to as the Brunt-Vaisala frequency.

buoyancy inputs Inputs of heat or freshwater which alter the density of seawater, thus changing its buoyancy.

cell quota The amount of a nutrient, normalized by biomass, within a phytoplankton cell.

cellular respiration Metabolic reactions within a cell that use some of the fixed organic compounds to maintain the cell structures and to synthesise ATP. In the ocean, respiration typically consumes oxygen (aerobic respiration), with CO_2 being released back into the water as a waste product.

coastal currents Flow parallel to a coastline often driven by density gradients arising from freshwater inputs from land.

compensation depth The depth, within the exponentially decaying light profile, at which a phytoplankton cell would need to be held in order that carbon fixation by photosynthesis is just able to balance the energy requirements of cell respiration. Net production by the cell would then be zero.

compensation light intensity The intensity of light (PAR) at the compensation depth.

continental shelf pump The export of carbon from the continental shelves into the adjacent deep ocean.

continuity Conservation principle, which requires that what flows into a defined volume in a defined time, minus what flows out of that volume in the same time, must accumulate in that volume. Continuity is an important constraint on fluid flows which can be expressed in a variety of mathematical statements.

convective chimneys Localised region of sinking of water whose density has been increased, for example by surface cooling.

convective overturning Vertical motions occurring when the water column is unstable due to density increasing towards the surface (for instance, caused by surface cooling). It can lead to very rapid vertical mixing.

convectively unstable Condition in which heavier fluid overlies lighter with the potential to drive overturning motions.

Coriolis force The apparent force which must be included in Newton's laws of motion to describe the movements of a particle in a rotating system. The Coriolis force is a key control on large-scale motions in the ocean and atmosphere. (See also geostrophic flow.)

Coriolis parameter Coriolis force for a particle mass m moving at speed v has a magnitude $m\,f\,v$ where f is the Coriolis parameter $f = 2\Omega \sin\,\phi_L$ where Ω is the angular speed of the Earth's rotation and ϕ_L is latitude.

critical depth If moved continuously by mixing through the exponential light profile between the sea surface and the critical depth, a phytoplankton cell is just able to balance its cell respiration energy requirements with the energy gained from photosynthesis. Net production is just zero. The critical depth is analogous to the compensation depth, but takes into account vertical turbulent mixing in the surface layer; the critical depth is always greater than the compensation depth.

critical mixing In a mixed water column or a surface mixed layer, where the water depth or layer thickness is greater than the critical depth, net phytoplankton production can still be possible. If the rate of vertical mixing is low enough, then the time spent above the critical depth may allow the phytoplankton to generate additional biomass.

cyanobacteria Photosynthetic bacteria (prokaryotes) with cell sizes typically 0.5–2 μm. In the ocean, cyanobacteria tend to be the dominant autotrophs in nutrient-stressed environments (e.g. the oligotrophic open ocean, or the summer surface mixed layer of the shelf seas), They are also referred to as 'blue-green algae'. In the ocean, the most important cyanobacteria are *Synechococcus* and *Prochlorococcus*; they are probably the most abundant photosynthetic organisms on the planet.

deep water waves Waves with a wavelength that is much less than the depth of water through which they are propagating. Such waves have circular particle orbits whose amplitude decreases exponentially from the surface.

degenerate amphidrome An amphidrome which is displaced to a position on land by frictional effects in the tidal regime. Energy lost to friction results in the Kelvin wave reflected at the head of a gulf being weaker than the incident wave, thus producing an asymmetry in the tidal response.

denitrification The conversion of nitrate back to gaseous nitrogen (N_2) by bacteria.

diatom Single-celled autotrophic phytoplankton that use silicate to build their cell wall (called a 'frustule'). They have no ability to swim, but some species achieve vertical movement through changes in cell buoyancy. Diatoms are typically 10–100 μm in diameter, sometimes aggregating into large colonies.

dissolved inorganic carbon (DIC) Dissolved inorganic carbon in the water made up of carbon dioxide, carbonic acid, bicarbonate and carbonate. These different forms of inorganic carbon arise from the transfer of atmospheric CO_2 into the sea, followed by chemical reactions between the CO_2 and the water, About 88% of DIC is bicarbonate, 11% is carbonate and the remaining 1% is CO_2 and carbonic acid.

dinoflagellates Single-celled phytoplankton, typically 10s μm in diameter, and an important part of the marine food chain. About half of the dinoflagellates are autotrophic, the rest being either heterotrophic (part of the microzooplankton) or mixotrophic. Dinoflagellates have two whip-like 'flagella' that allow them to swim and to control their swimming direction. (See also flagellates.)

dissolved organic carbon (DOC) Very tiny particles of organic carbon released by, for instance, the break-up of phytoplankton cells ('lysis') or sloppy feeding by zooplankton. The distinction between 'particulate' and 'dissolved' organic matter is defined operationally, with DOC being able to pass through a filter with pore size 0.45 μm.

dissolved organic matter (DOM) Dissolved organic compounds, such as carbon (DOC), phosphate (DOP) and nitrogen (DON). (See dissolved organic carbon.)

dissolved organic nitrogen (DON) Fixed organic nitrogen released during cell lysis and grazing, and able to pass through a 0.45 μm filter. As with dissolved organic carbon, DON is a component of dissolved organic matter (DOM).

diffuse attenuation coefficient Parameter with units of m^{-1} expressing the rate of exponential decay of the downward flux of diffuse light energy.

DOC See dissolved organic carbon.

DOM See dissolved organic matter.

eddy diffusivity A parameter that describes the transfer of a scalar property by turbulent eddies down the property gradient. It arises from an analogy with the molecular diffusivity in molecular diffusion.

eddy viscosity A parameter that describes the transfer of fluid momentum by turbulent eddies down momentum gradients in the water.

Ekman depth Depth of the water column influenced by a stress imposed at surface or bottom boundaries. The Ekman depth can be related to the magnitude of the imposed stress and the Coriolis parameter.

Ekman drain Downslope transport in the bottom boundary layer induced by friction in an along-slope current moving with the shelf to the right/left of the direction of motion in the northern/southern hemisphere.

Ekman layer Section of the water column influenced by a boundary stress. A stress, imposed, for example, at the surface by the wind, diminishes rapidly with depth so that its influence is confined to a near-surface layer whose thickness is the Ekman depth.

Ekman number Ratio of frictional forces to Coriolis forces in fluid flow.

Ekman spiral Idealised form of velocity profile induced by a steady wind blowing over a homogeneous unbounded ocean. The current, which at the surface is deflected 45° to the right/left of the wind direction, rotates clockwise/anticlockwise in the northern/ southern hemisphere and diminishes exponentially with increasing depth.

Ekman transport The depth integrated transport induced by a steady wind blowing over a homogeneous unbounded ocean. The Ekman transport is at 90° to right/left of the wind direction in the northern/southern hemisphere.

Ekman veering Deflection of a geostrophic current near the seabed as the current begins to slow due to the frictional drag at the seabed boundary.

energy cascade Process of kinetic energy transfer from larger to progressively smaller scales. The motion at the largest scales in a fluid flow generates eddies which lose energy to smaller eddies which, in turn, power even-smaller eddies and so on, through many steps, until the scale is so small that viscosity becomes important and the turbulent energy is dissipated to heat.

equations of motion The rules of fluid motion based on Newton's laws and the principle of continuity.

Equilibrium theory of the tides The simplest theory of the tides which involves a balance between the tide generating force and the surface slope of a uniform ocean covering the whole Earth. This basic model of the tides was used by Isaac Newton to explain many of the qualitative features of the observed tides.

eukaryotes (eukaryotic) Cells, or organisms made up of cells, that contain a cell nucleus enclosed in a membrane.

Eulerian current Velocity measured over time at a fixed point as distinct from the Lagrangian current which is measured following a particle of water.

euphotic zone (see photic zone)

eutrophication (eutrophic) The addition of nutrients, generally nitrates and phosphates, to water leading to excessive growth of phytoplankton. In coastal seas these extra nutrients are largely derived from the runoff of agricultural fertilizers and sewage disposal. Subsequent decay of the large amounts of phytoplankton can reduce dissolved oxygen concentrations in the water. (See also hypoxia.)

export production The amount of carbon fixed by the autotrophs in the photic zone, which sinks, or is otherwise transported, below the thermocline to deeper water or the sediments. If a steady state is assumed, export production is the same as new production.

export ratio The fraction of carbon fixed by the autotrophs that is exported to waters below the winter thermocline, and thus removed from the surface ocean and atmosphere for a long time.

***f*-ratio** The fraction of total autotroph primary production, which is supported by a supply of new nutrients, rather than utilising recycled nutrients. It is the ratio of new production to total production.

Fick's law The diffusive flux of a scalar is proportional to the gradient of the scalar and in the down-gradient direction.

flagellate A single-celled eukaryotic phytoplankton, typically a few to O(10) μm diameter. The term generally refers to any organism that uses one or more whip-like flagella for locomotion (e.g. as in dinoflagellate). In marine studies there is a tendency to use 'flagellate' to refer to cells with one whip-like flagella, usually smaller than dinoflagellates. Flagellates can be autotrophic, heterotrophic or mixotrophic.

flux Richardson number R_f The ratio of the buoyancy flux to the rate of turbulent kinetic energy production in stratified shear flow. R_f can be thought of as an 'efficiency of mixing'.

frontal jet A core of high-velocity flow located in the baroclinic region of an oceanographic front.

gelbstoff Literally 'yellow stuff', also known as Coloured Dissolved Organic Matter (CDOM) and 'gilvin'. Gelbstoff refers to a collection of dissolved organic compounds, mainly tannins arising from the decay of organic material.

geostrophic flow or current In geostrophic flow, there is a balance between Coriolis and pressure forces which requires that the flow direction is at right angles to the pressure gradient. This simple balance is applicable to much of the flow in the deep ocean and the atmosphere and can be applied in many situations in shelf seas.

global overturning circulation Large-scale circulation of the global ocean involving sinking near the poles and slow upwelling in lower latitudes. Also referred to as the 'meridional overturning circulation'.

gradient Richardson number *Ri* A measure of the ratio of water column stability to the kinetic energy available to drive mixing. *Ri* is important in deciding whether turbulence and mixing will occur.

gross community production The gross primary production achieved by all phytoplankton within an autotrophic community.

gross primary production The rate at which the autotrophs fix inorganic carbon. It includes the carbon fixation that is later used in cellular respiration. (See also net primary production.)

group velocity Waves tend to travel in groups which advance at the group velocity. It is important because it is also the speed at which waves transport their energy.

harmonic method Analysis of tidal data in terms of a sum of sine waves (tidal constituents) at different frequencies set by the regular motions of the Sun and Moon. The amplitude and phase of the constituents, determined from the data analysis, are used to construct the sum of constituents for future times and thus predict the tides.

heterotrophs (heterotrophic) Organisms that cannot fix inorganic carbon, but instead need to consume other organisms (autotrophs or heterotrophs) in order to acquire the organic compounds needed for growth.

hydrostatic approximation A balance in the vertical direction between pressure forces and gravity which requires that vertical accelerations can be neglected. The hydrostatic approximation is generally applicable to slowly varying processes in the ocean but is not valid at high frequencies, for example in surface waves and short period internal waves, both of which may involve large vertical accelerations.

hypoxia (hypoxic) Water is hypoxic if the dissolved oxygen concentration drops below 30% of its saturated value, or typically less than 2 mg l^{-1}. Water with a dissolved oxygen concentration of 0% is referred to as anoxic. (See also eutrophication.)

inertial oscillation Circular motion in which there is a balance between Coriolis force acting on a water particle and the particle's radial acceleration. The motion is in the clockwise/anticlockwise sense in the northern/southern hemisphere. Inertial oscillations occur widely in the ocean, especially near the surface in response to changes in wind stress.

inertial period T_I The period of an inertial oscillation is the time required for a particle to travel around a circle and is given by $T_I = 2\pi / f$ where f = *Coriolis parameter*.

inertial subrange The middle section of the turbulence spectrum, between the low-wave-number (large eddy) part and the high-wave number (small scale) part where energy is dissipated to heat by viscous forces.

infinitesimal waves Waves of very small (tending to zero) amplitude. The linear theory of wave motions is based on the assumption of infinitesimal waves but can usually be applied to waves of finite height without serious errors providing the waves are not too steep.

internal Rossby radius Ro' A patch of homogeneous fluid embedded in a stratified water column will spread a distance of order Ro' before coming into geostrophic balance. This length scale is important, for example, in controlling the extent of a region of coastal upwelling.

internal tide Internal waves of tidal frequency involving vertical oscillations of the pycnocline. The internal tide is generated principally by the barotropic tide forcing stratified fluid to move over steep topography.

Kelvin waves Long waves which are influenced by the Earth's rotation. They travel along land boundaries and diminish exponentially in amplitude away from the coast.

Kolmogorov microscales In turbulent motions, energy dissipation to heat becomes increasingly important for eddies whose scale is equal to or less than the Kolmogorov microscale. This length microscale and the corresponding velocity scale depend only on the rate of energy dissipation and the viscosity.

Lagrangian current Velocity measured over time following a particle as opposed to the Eulerian velocity which is measured at a point.

Langmuir circulation Transverse circulation in the surface layer set up by wind-driven waves, often visible as parallel lines of surface foam (called 'windrows') aligned with the wind direction.

level of no motion Level in the water column at which it is known, or can be assumed, that there is no water movement.

limiting nutrient Phytoplankton require a range of macronutrients and micronutrients in order to grow. If a nutrient reaches concentrations low enough for it to be unavailable to the phytoplankton, despite other nutrients being in useable quantities, it is referred to as the limiting nutrient. In temperate shelf seas, nitrate tends to be the limiting nutrient following the spring bloom.

local equilibrium (of turbulence) A simplified version of the turbulent kinetic energy equation in which it is assumed that only the production and dissipation terms are important. There is, therefore, a local equilibrium in which production and dissipation balance at each point in the fluid.

macronutrients The nutrients required by the phytoplankton in relatively large quantities. The key macronutrients are nitrate, phosphate and (for the diatoms) silicate. Ammonium can also be included as a source of nitrogen.

match-mismatch hypothesis The hypothesis, originally proposed by fisheries scientist David Cushing, that higher trophic levels in the marine food chain may need to time their breeding with the flush of phytoplankton at the spring bloom.

maximum quantum yield of photosynthesis Parameter which describes the maximum rate of carbon fixation via photosynthesis at a given light level, normalized by some measure that represents the light-gathering capacity of the phytoplankton (e.g. chlorophyll concentration). It is an important part of the description of the light-growth curve of a phytoplankton species or community, and is a measure of the steepness of the curve at low light.

Michaelis–Menten uptake kinetics A description of phytoplankton nutrient uptake rate as a function of the maximum achievable uptake rate and the ambient nutrient concentration in the water. The uptake rate is initially low at low ambient nutrient, and increases rapidly to reach a plateau approaching the maximum uptake rate at high ambient nutrient concentrations.

microbial loop A vital pathway for returning dissolved organic carbon (DOC) to the marine food web. DOC is unavailable to the micro- and mesozooplankton, but is utilised by heterotrophic bacteria which can then be grazed by the zooplankton.

micronutrients Inorganic compounds required by phytoplankton in small quantities (compared to the macronutrients). Iron, copper, nickel and zinc are recognised as important micronutrients. Iron, in particular, has received much attention because it is now known to limit primary production in the High-Nutrient Low-Chlorophyll (HNLC) regions of the open ocean, despite adequate macronutrients being available.

mixotrophs (mixotrophic) Micro-organisms that can use different sources of energy. In the phytoplankton there are species in all groups, which are able to mix autotrophy and heterotrophy, allowing them to take advantage of shifts in environmental conditions.

mode 1 internal wave Lowest mode of internal wave motion in which the surface and bottom layers oscillate in antiphase with a node located in the interface between the layers.

Monod growth curve A curve describing a functional response in which an organism's growth rate is eventually limited by its capacity to process resources. The photosynthesis-PAR curve describing phytoplankton growth as a function of light is an example of this type of response, with growth rate reaching a plateau at high PAR as the photosystem becomes saturated with light.

neap tides Smallest tides in the fortnightly cycle usually occurring soon after the first and third quarters of the Moon's orbital cycle.

net community production The net primary production achieved by all phytoplankton within an autotrophic community.

net community respiration The total energy, or carbon, requirements of a phytoplankton community to support cellular respiration. Respiration requires dissolved oxygen. In practice, the use of oxygen incubations to measure respiration will also include respiration associated with heterotrophic breakdown of organic material.

net primary production The rate at which the autotrophs fix inorganic carbon minus the carbon fixation that is required for cellular respiration. (See also gross primary production.)

new production (new primary production) The amount of carbon fixed by the autotrophs in the photic zone that is supported by a supply of 'new' nutrients from below the photic zone or thermocline, rather than the result of utilizing nutrients recycled in the photic zone. If a steady state is assumed, new production is the same as export production. (See also *f*-ratio.)

nitracline A vertical region within the water column where nitrate concentrations change rapidly. Typically the nitracline occurs coincident with, or within, the thermocline.

nitrification The biological oxidation of ammonium by bacteria to form nitrite and nitrate.

non-linear (equation or process) Process described by an equation involving the products of the equation variables and their derivatives. Non-linear equations are usually difficult to solve analytically, so appeal is often made to the simplifying assumption that the non-linear terms can be neglected and the equations thus 'linearised'.

non-photochemical quenching A mechanism used by photosynthetic organisms to dissipate excess energy when exposed to high light levels and so avoiding damage to their photosystems.

oligotrophic Literally 'few nutrients'. An oligotrophic region has one or more of the macronutrients in negligible concentrations. The surface layers of the open ocean sub-tropical gyres are oligotrophic. In shelf seas, we see oligotrophic conditions in stratified regions in summer.

ontogenic Relating to, or dependent on, the age or life stage of an organism.

Ozmidov length Scale of the largest overturning eddies in a stratified shear flow. Kinetic energy from the mean flow is required to increase the potential energy of particles displaced upwards in the stratified fluid and this sets a limit to the eddy scale which depends on the energy dissipation rate and the stability frequency.

PAR See Photosynthetically Available Radiation.

particulate organic carbon (POC) Carbon originally fixed from the pool of dissolved inorganic carbon (DIC) by photosynthesis and held within living organisms. POC will not pass through a 0.45 μm filter. (See also dissolved organic carbon (DOC).)

Péclet number The ratio between the advective (e.g. swimming) and diffusive (turbulent) components of transport. It is a useful number when determining the likely use of motility in the ocean; a Péclet number $\gg 1$ implies that swimming is a useful trait, while $\ll 1$ indicates that turbulent mixing overwhelms any swimming capability.

phase speed (phase velocity) The speed of propagation of the waves as determined for example by observing the movement of wave crests. The phase velocity of waves should not to be confused with the group velocity which is the speed at which energy is propagated.

photic zone The region close to the sea surface where there is sufficient light for photosynthesis. Whether or not photosynthesis is possible will depend on the species within the phytoplankton community, but typically the photic zone is taken to reach from the sea surface to the depth at which the surface PAR has dropped to either 1% or 0.1% of its surface value. Also referred to as the 'euphotic zone'.

photoinhibition The reduction of phytoplankton carbon fixation at high light, due to damage of the cells' photosystems.

photosynthetic quotient In phytoplankton, the ratio of oxygen production during photosynthesis to carbon fixation (expressed as moles of oxygen per mol of carbon).

Photosynthetically Available Radiation (PAR) The part of the Sun's spectrum of light that is useful for photosynthesis. It depends on the absorption characteristics of the phytoplankton pigments, mainly chlorophyll *a*. PAR is generally taken to be the light between wavelengths of 400 and 700 nm. A useful 'rule of thumb' is that PAR accounts for about 50% of the total incident radiation at the sea surface.

phytoplankton functional types Collections of phytoplankton that have similar functional, or biogeochemical, traits in a phytoplankton community. For instance, the diatoms are often treated as a single functional type because of their role in cycling silicate; cyanobacteria are a single type because of their role in regenerated production and their lower importance to the export of carbon. Other functional types include the nitrogen fixers (e.g. Trichodesmium), the calcifiers (e.g. coccolithophores) and the flagellates (which might include the dinoflagellates).

potential energy anomaly A measure of the stability of the water column defined as the difference between the potential energy (per unit volume) of the water column after, and before, complete vertical mixing.

practical salinity scale Standard scale for the measurement of salinity based on simultaneous measurements of the conductivity, temperature and pressure of seawater. Salinity is derived from a series of polynomials using the ratio of the observed conductivity to an accurate laboratory standard.

practical salinity unit (psu) Widely used, but strictly inappropriate, unit, since salinity, which is derived from a conductivity ratio, is dimensionless and therefore does not require units.

progressive wave Wave which moves through a medium transferring energy as it travels (as opposed to a standing wave which does not move and does not involve energy transfer).

prokaryotes Organisms, usually single-celled, that lack a cell nucleus. All bacteria are prokaryotes.

Redfield ratio Named after the oceanographer Alfred Redfield, this is the set of molecular ratios between carbon, nitrogen and phosphorus typically found in marine organic matter C:N:P = 106:16:1.

reduced gravity Effective value of the gravitational acceleration in a stratified fluid when buoyancy effects are taken into account. For two layers, differing in density by $\Delta\rho$, the value of the reduced gravity is $g' = g(\Delta\rho/\rho_0)$.

regenerated production Primary production that utilises inorganic carbon and nutrients which have been recycled by bacteria within the surface layer. Regenerated production tends to be carried out by small phytoplankton, particularly the cyanobacteria. It does not result in any net drawdown of CO_2 from the atmosphere.

Region Of Freshwater Influence (ROFI) Coastal region where the input of buoyancy through freshwater discharge from rivers is comparable to buoyancy changes due to surface heating and cooling in the seasonal cycle.

residual circulation The net movement of water as a result of the combined effects of different forces (e.g. wind, tides, density gradients). Residual circulation is generally seen as a long-term averaged flow, filtering out the often much stronger oscillatory tidal flows and episodic wind-driven currents.

resonance Enhanced response of a mechanical system which occurs when the system experiences forcing at one of its natural frequencies of oscillation. Tidal resonance can occur, for example, in a gulf whose natural period is close to that of tide in the adjacent ocean.

Reynolds number *Re* Ratio of inertial forces (mass x accelerations) to viscous forces in fluid flow. *Re* is a key parameter in determining whether a fluid flow will be laminar or turbulent.

Rossby radius of deformation *Ro* Scale of movement involved as a homogeneous fluid, initially at rest, comes into geostrophic balance with, for example, an applied surface slope. *Ro* is the lengthscale which controls the seaward extent of a Kelvin wave travelling along a coastal boundary.

spring bloom The often dramatic, first flush of phytoplankton growth after the low productivity of the winter. The spring bloom is generally most clearly defined in shelf areas, away from coastal sources of fresh water, where the water column becomes thermally stratified in spring.

spring-neap cycle Variation in the range of the tide over a ~14.5 day period as the Moon's position changes relative to the Sun during the lunar orbital cycle.

spring tides Largest tides in the fortnightly cycle usually occurring soon after new Moon and full Moon.

standing wave A wave where the nodes (points of zero amplitude) and anti-nodes (points oscillating with the full wave amplitude) are fixed in position. Standing waves do not involve net energy transfer. Also sometimes called a 'stationary wave'. (See also progressive wave.)

Stokes drift Net motion in the direction of wave travel due to non-linear effects in surface waves of finite amplitude.

subsurface chlorophyll maximum (SCM) In stratified shelf seas, where surface layer nutrients have been removed by an earlier spring bloom, phytoplankton are often found to have their highest biomass in a layer within the thermocline.

Taylor hypothesis An assumption that the advection of a field of turbulence past a fixed point can be taken to be entirely due to the mean flow, thus allowing the determination of the spatial spectrum of turbulence from measurements by a fixed sensor.

thermal wind A term adopted from meteorology for the flow driven by horizontal gradients of density. The thermal wind equation, which relates the vertical shear in the current to the horizontal density gradient, can be integrated to determine the current profile from density observations if the current is known at one level.

Thorpe scale A displacement lengthscale found by re-ordering the particles in the density profile to remove instabilities and then calculating the r.m.s. displacement. The Thorpe scale is closely related to the Ozmidov length, which limits the size of overturning eddies.

tidal constituent A periodic component of the tide in the form of a sine wave at a frequency which is set by the motions of the Moon and the Sun relative to the Earth. The real tide, surface elevation and tidal currents, can be represented as a sum of a number of such constituents.

tidal current ellipse The shape traced out by the tip of the Eulerian tidal current vector over the tidal cycle. For a single tidal constituent, the shape is a true ellipse and can be scaled to show the corresponding displacement of a water particle in the tidal flow.

tidal mixing front Boundary zone with distinct dynamical and biogeochemical properties between seasonally stratified and mixed regimes.

tidal species In a spectrum of tidal energy, the tidal constituents are concentrated into three narrow bands of frequency referred to as 'long period', 'diurnal' and 'semi-diurnal', tidal species.

tidal straining The periodic tilting of isopycnals by shear in an oscillatory tidal current which, in the presence of a horizontal density gradient, induces a periodic component of stratification.

tide generating force The body force producing tidal movements in the ocean. Lunar and solar tide generating forces arise from imbalances between gravitational attraction and centrifugal forces in the orbital motions of the Earth-Moon and the Sun-Earth systems.

top-down control The control of the biomass of an organism, or group of organisms, at one trophic level by the grazing/predation pressure exerted by organisms at a higher trophic level. The opposite is 'bottom-up' control, where biomass is limited by one or more of the resources required to generate growth.

total or material derivative The time derivative of a particle property following the particle as opposed to the 'local derivative' which is that applying at a fixed point.

trophic level The position of an organism within a food web. All organisms at one trophic level will be dependent on the same resources for growth. The primary producers are at trophic level 1, their grazers at level 2, and the predators that eat the grazers at level 3. Humans, and a few other animals, are at trophic level 5: at which predators are not predated upon.

turbulence closure Relationship determining the eddy viscosity and eddy diffusivity in terms of the mean properties of a turbulent flow. Such a relationship is necessary to 'close' a set of equations governing a fluid flow and, hence, allow analytical or numerical solutions to be obtained. Closure schemes vary from simple assumptions of constant eddy viscosities and diffusivities to more elaborate formulations representing more of the relevant physics.

turbulent kinetic energy The total kinetic energy per unit mass present in turbulent motions.

turbulent Prandtl number The ratio of eddy viscosity to eddy diffusivity.

vorticity Flow property depending on velocity gradients in the flow. It can be thought of as the fluid equivalent of the rotation of a solid body to which it is closely analagous.

ANSWERS TO CHAPTER PROBLEMS

Numeric answers are provided here.

Chapter 3

3.1 (i) 7.92×10^{-5} s^{-1}
(ii) 0.142 N; 135°T
(iii) 1.41×10^{-5}, surface slopes upwards towards 135°T

3.2 22.6 hours; 3502 metres

3.3 (i) 4.09 m^2 s^{-1} towards 180°T

3.4 (i) 5 m
(ii) 271 km and 813 km from head of the gulf
(iii) 1.29 metres

3.5 0.58

Chapter 4

4.1 53 kW m^{-1}

4.2 0.36 m s^{-1}; 0.26

4.4 1610 m; 0.263 mg m^{-3}
23.1 hours; 20.9 μg m^{-3}

4.5 0.085 m s^{-1}

Chapter 5

5.2 26 metres in winter
29 metres in summer

5.3 0.009 m^2 s^{-1}

5.4 140 μg l^{-1}

5.5 $>$0.9 mm s^{-1}

Chapter 6

6.2 6.02×10^{8} J m^{-2}; 123 J m^{-3}; 2.7×10^{-3}

6.3 (a) 42.6 metres; (b) 17.5 metres

6.4 (a) 147 J m^{-2}
(b) (i) 24 days; (ii) 7 days; (iii) 3 days

6.5 13 g C m^{-2}

6.6 1.6 g C m^{-2} d^{-1} (assuming the growth phase of the bloom is 7 days)

Chapter 7

7.1 (i) 2; (ii) 600

7.3 (i) 0.3 metres; (ii) 4.5 metres

7.4 22 g C m^{-2} for the bloom; 26 g C m^{-2} for the SCM

7.5 (i) 10 g C m^{-2}; (ii) SCM production is equivalent to about 2.5 storm mixing events.

7.6 0.03 for the Celtic Sea, 58 for Monterey Bay

Chapter 8

8.2 $e_f = 0.0037$; $e = 0.0019$

8.4 1.26; 2.9 km

8.5 (i) 0.26 g C m^{-2} d^{-1}; (ii) 18 g C m^{-2}

8.6 20 km; 1700 tonnes

Chapter 9

9.1 slope = 2.6×10^{-6}, surface current = 0.22 m s^{-1}, transport = 1.98 m^{2} s^{-1}

9.2 14.8 J m^{-3}; 13.7 J m^{-3}

9.4 (i) About 0.1–0.2 mg Chl m^{-3}; (ii) About 5–10 mg Chl m^{-3}

Chapter 10

10.2 0.15 Pa; 1.37 $m^2 s^{-1}$; 0.25 $m s^{-1}$

10.3 (i) 8 g C $m^{-2} d^{-1}$; (ii) 0.8 $m s^{-1}$

10.4 1 mmol $m^{-2} d^{-1}$

10.5 1.5×10^{13} m^3; 7×10^{10} mol C or about 850 kt

REFERENCES

Argo. Retrieved 27th February, 2011, from http://www-argo.ucsd.edu/.

Abraham, E. R. (1998). The generation of plankton patchiness by turbulent stirring. *Nature* **391**(6667): 577–80.

Aguilar-Islas, A. M., M. P. Hurst, K. N. Buck *et al.* (2007). Micro- and macronutrients in the southeastern Bering Sea: Insight into iron-replete and iron-depleted regimes. *Progress in Oceanography* **73**(2): 99–126.

Allen, J. I., J. R. Siddorn, J. C. Blackford and F. J. Gilbert (2004). Turbulence as a control on the microbial loop in a temperate seasonally stratified marine systems model. *Journal of Sea Research* **52**(1): 1–20.

Allen, J. R., D. J. Slinn, T. M. Shammon, R. G. Hartnoll and S. J. Hawkins (1998). Evidence for eutrophication of the Irish Sea over four decades. *Limnology and Oceanography* **43**(8): 1970–74.

Alvarez-Salgado, X. A., J. Aristegui, E. D. Barton and D. A. Hansell (2007). Contribution of upwelling filaments to offshore carbon export in the subtropical Northeast Atlantic Ocean. *Limnology and Oceanography* **52**(3): 1287–92.

Alvarez-Salgado, X. A., M. D. Doval, A. V. Borges *et al.* (2001). Off-shelf fluxes of labile materials by an upwelling filament in the NW Iberian Upwelling System. *Progress in Oceanography* **51**(2–4): 321–37.

Anderson, D. M., P. M. Glibert and J. M. Burkholder (2002). Harmful algal blooms and eutrophication: Nutrient sources, composition, and consequences. *Estuaries* **25**(4B): 704–26.

Arbic, B. K. and C. Garrett (2010). A coupled oscillator model of shelf and ocean tides. *Continental Shelf Research* **30**(6): 564–74.

Arbic, B. K., R. H. Karsten and C. Garrett (2009). On Tidal Resonance in the Global Ocean and the Back-Effect of Coastal Tides upon Open-Ocean Tides. *Atmosphere-Ocean* **47**(4): 239–66.

Armstrong, D. A., H. M. Verheye and A. D. Kemp (1991). Short-term variability during an anchor station study in the southern benguela upwelling system – fecundity estimates of the dominant copepod, calanoides-carinatus. *Progress in Oceanography* **28**(1–2): 167–88.

Armstrong, R. A. (1994). Grazing limitation and nutrient limitation in marine ecosystems – steady-state solutions of an ecosystem model with multiple food-chains. *Limnology and Oceanography* **39**(3): 597–608.

Arrigo, K. R. (2005). Marine microorganisms and global nutrient cycles. *Nature* **437**(7057): 349–55.

Arrigo, K. R., G. van Dijken and S. Pabi (2008). Impact of a shrinking Arctic ice cover on marine primary production. *Geophysical Research Letters* **35**(19): 6.

Azam, F., T. Fenchel, *et al.* (1983). The ecological role of water-column microbes in the sea. *Marine Ecology-Progress Series* **10**(3): 257–63.

Baird, M. E. (2010). Limits to prediction in a size-resolved pelagic ecosystem model. *Journal of Plankton Research* **32**(8): 1131–46.

Balch, W. M., P. M. Holligan, S. G. Ackleson and K. J. Voss (1991). Biological and optical-properties of mesoscale coccolithophore blooms in the Gulf of Maine. *Limnology and Oceanography* **36**(4): 629–43.

Ballestero, D. and S. Boxall (1997). Remote sensing and modelling in upwelling systems. *Earthnet Online*, retrieved July 14, 2010, from http://earth.esa.int/workshops/ers97/papers/ballestero/.

Banas, N. S. (2011). Adding rich trophic interactions to a size-spectral plankton model: Emergent diversity patterns and limits on predictability. *Ecological Modelling* **222**: 2663–75.

Barber, R. T., J. Marra, R. C. Bidigare et al. (2001). Primary productivity and its regulation in the Arabian Sea during 1995. *Deep-Sea Research Part II--Topical Studies in Oceanography* **48**(6–7): 1127–72.

Barkmann, W. and J. D. Woods (1996). On using a Lagrangian model to calibrate primary production determined from *in vitro* incubation measurements. *Journal of Plankton Research* **18**(5): 767–88.

Barlow, R. G., R. F. C. Mantoura, M. A. Gough and T. W. Fileman (1993). Pigment signatures of the phytoplankton composition in the northeastern atlantic during the 1990 spring bloom. *Deep-Sea Research Part II—Topical Studies in Oceanography* **40**(1–2): 459–77.

Batchelor, G. K. (1960). *The Theory of Homogeneous Turbulence*. Cambridge: Cambridge University Press.

Bates, N. R. (2006). Air-sea CO_2 fluxes and the continental shelf pump of carbon in the Chukchi Sea adjacent to the Arctic Ocean. *Journal of Geophysical Research-Oceans* **111**(C10): 21.

Beaugrand, G. (2009). Decadal changes in climate and ecosystems in the North Atlantic Ocean and adjacent seas. *Deep-Sea Research Part II—Topical Studies in Oceanography* **56**(8–10): 656–73.

Begg, G. S. and J. B. Reid (1996). *Spatial variation in seabird density at a shallow sea tidal mixing front in the Irish Sea*. ICES International Symposium on Seabirds in the Marine Environment, Glasgow, Scotland, Academic Press Ltd.

Behrenfeld, M. J., E. Boss, D. A. Siegel and D. M. Shea (2005). Carbon-based ocean productivity and phytoplankton physiology from space. *Global Biogeochemical Cycles* **19**(1).

Behrenfeld, M. J. and P. G. Falkowski (1997). Photosynthetic rates derived from satellite-based chlorophyll concentration. *Limnology and Oceanography* **42**(1): 1–20.

Behrenfeld, M. J., T. K. Westberry, E. S. Boss et al. (2009). Satellite-detected fluorescence reveals global physiology of ocean phytoplankton. *Biogeosciences* **6**(5): 779–94.

Bender, M., K. Grande, K. Johnson *et al.* (1987). A comparison of 4 methods for determining planktonic community production. *Limnology and Oceanography* **32**(5): 1085–98.

Bisagni, J. J. and M. H. Sano (1993). Satellite observations of sea surface temperature variability on southern Georges Bank. *Continental Shelf Research* **13**(10): 1045–64.

Biscaye, P. E. (1994). Shelf edge exchange processes in the southern Middle Atlantic bight – SEEP-ii. *Deep-Sea Research Part II—Topical Studies in Oceanography* **41**(2–3): 229–30.

Biscaye. P. E., C. N. Flagg and P. G. Falkowski (1994). The shelf edge exchange processes experiment, SEEP-ip – an introduction to hypotheses, results and conclusions. *Deep-Sea Research Part II—Topical Studies in Oceanography* **41**(2–3): 231ff.

Blaas, M. and H. E. de Swart (2000). *Vertical structure of residual slope circulation driven by JEBAR and tides: an idealised model. 10th Biennial Conference on the Physics of Estuaries and Coastal Seas (PECS)*, Norfolk, Virginia.

Blackburn, S. I. and G. Cresswell (1993). A coccolithophorid bloom in Jervis Bay, Australia. *Australian Journal of Marine and Freshwater Research* **44**(2): 253–60.

Blaise, S., R. Comblen, V. Legat *et al.* (2010). A discontinuous finite element baroclinic marine model on unstructured prismatic meshes Part I: space discretization. *Ocean Dynamics* **60**(6): 1371–93.

Blondeau-Patissier, D., G. H. Tilstone, V. Martinez-Vicente and G. F. Moore (2004). Comparison of bio-physical marine products from SeaWiFS, MODIS and a bio-optical model with *in situ* measurements from Northern European waters. *Journal of Optics A-Pure and Applied Optics* **6**(9): 875–89.

Botsford, L. W., F. Micheli and A. Hastings (2003). Principles for the design of marine reserves. *Ecological Applications* **13**(1): S25–S31.

Bowden, K. F. (1948). The processes of heating and cooling in a section of the Irish Sea. *Monthly Notices of the Royal Astronomical Society, Geophyscial Supplement* **5**(7): 270–81.

(1950). Processes affecting the salinity of the Irish Sea. *Monthly Notices of the Royal Astronomical Society, Geophyscial Supplement* **6**(2): 63–90.

(1960). The dynamics of flow on a submarine ridge. *Tellus* **12**: 418–26.

(1965). Horizontal mixing in the sea due to a shearing current. *Journal of Fluid Mechanics* **21**: 83–95.

(1983). *Physical Oceanography of Coastal Waters*. Ellis Horward.

Bowden, K. F. and L. A. Fairbairn (1952). A determination of the frictional forces in a tidal current. *Proceedings of the Royal Society* **214A**: 371–92.

(1956). Measurements of turbulent fluctuations and Renolds stresses in a tidal current. *Proceedings of the Royal Society* **A 237**: 422–38.

Bowden, K. F., L. A. Fairbairn and P. Hughes (1959). The distribution of shearing stresses in a tidal current. *Geophysical Journal of the Royal Astronomical Society* **2**: 288–305.

Bowers, D. G., S. Boudjelas and G. E. L. Marker (1998). The distribution of fine suspended sediments in the surface waters of the Irish Sea and its relation to tidal stirring. *International Journal of Remote Sensing* **19**(14): 2789–2805.

Bowers, D. G. and J. H. Simpson (1987). Mean position of tidal fronts in European-shelf seas. *Continental Shelf Research* **7**(1): 35–44.

(1990). Geographical variations in the seasonal heating cycle in northwest european shelf seas. *Continental Shelf Research* **10**(2): 185–99.

Boyd, P. W., A. J. Watson, C. S. Law *et al.* (2000). A mesoscale phytoplankton bloom in the polar Southern Ocean stimulated by iron fertilization. *Nature* **407**(6805): 695–702.

Brander, K. M. (1995). The effect of temperature on growth of atlantic cod (gadus-morhua l). *Ices Journal of Marine Science* **52**(1): 1–10.

Brierley, A. S., P. Ward, J. L. Watkins and C. Goss (1998). Acoustic discrimination of Southern Ocean zooplankton. *Deep-Sea Research II*, **45**(7): 1155–73.

Brink, K. H. (1998). Wind-driven currents over the continental shelf. *The Sea Vol.10: The Global Coastal Ocean: Processes and Methods* A. R. Robinson and K. H. Brink. New York: John Wiley. 3–20.

Broekhuizen, N. (1999). Simulating motile algae using a mixed Eulerian-Lagrangian approach: does motility promote dinoflagellate persistence or co-existence with diatoms? *Journal of Plankton Research* **21**(7): 1191–216.

Brown, J., L. Carrillo, L. Fernand *et al.* (2003). Observations of the physical structure and seasonal jet-like circulation of the Celtic Sea and St. George's Channel of the Irish Sea. *Continental Shelf Research* **23**(6): 533–61.

Brown, J. and E. M. Gmitrowicz (1995). Observations of the transverse structure and dynamics of the low-frequency flow-through the north channel of the Irish Sea. *Continental Shelf Research* **15**(9): 1133–56.

Burchard, H. (2002). *Applied Turbulence Modelling in Marine Waters.* Berlin: Springer.

Burchard, H., O. Petersen and T. P. Rippeth (1998). Comparing the performance of the Mellor-Yamada and the k-epsilon two-equation turbulence models. *Journal of Geophysical Research–Oceans* **103**(C5): 10543–54.

Burchard, H. and T. P. Rippeth (2009). Generation of bulk shear spikes in shallow stratified tidal seas. *Journal of Physical Oceanography* **39**(4): 969–85.

Burrows, M., S. A. Thorpe and D. T. Meldrum (1999). Dispersion over the Hebridean and Shetland shelves and slopes. *Continental Shelf Research* **19**(1): 49–55.

Butman, B. and R. C. Beardsley (1987). Long-term observations on the southern flank of georges bank.1. a description of the seasonal cycle of currents, temperature, stratification, and wind stress. *Journal of Physical Oceanography* **17**(3): 367–84.

Butman, B., J. W. Loder and R. C. Beardsley (1987). The seasonal mean circulation on Georges Bank: observation and theory. *Georges Bank*. R. H. Backus. Cambridge, MA: MIT Press, 125–38.

CalCOFI. *California Cooperative Oceanic Fisheries Investigations.* Retrieved 23rd February, 2011, from http://www.calcofi.org/.

Canals, M., P. Puig, X. D. de Madron *et al.* (2006). Flushing submarine canyons. *Nature* **444** (7117): 354–7.

Canuto, V. M., A. Howard, Y. Cheng and M. S. Dubovikov (2001). Ocean turbulence. Part I: One-point closure model – Momentum and heat vertical diffusivities. *Journal of Physical Oceanography* **31**(6): 1413–26.

Capone, D. G., J. P. Zehr, H. W. Paerl, B. Bergman and E. J. Carpenter (1997). Trichodesmium, a globally significant marine cyanobacterium. *Science* **276**(5316): 1221–9.

Carrillo, L., A. J. Souza, A. E. Hill *et al.* (2005). Detiding ADCP data in a highly variable shelf sea: The Celtic Sea. *Journal of Atmospheric and Oceanic Technology* **22**(1): 84–97.

Cartwright, D. E. (1999). *Tides: a Scientific History*. Cambridge University Press.

Chapman, D. C. and G. Gawarkiewicz (1995). Offshore transport of dense shelf water in the presence of a submarine-canyon. *Journal of Geophysical Research-Oceans* **100**(C7): 13373–87.

Charnock, H., K. R. Dyer, J. M. Huthnance *et al.* (1994). *Understanding the North Sea System.* Chapman and Hall.

Chen, C. S. and R. C. Beardsley (1998). Tidal mixing and cross-frontal particle exchange over a finite amplitude asymmetric bank: A model study with application to Georges Bank. *Journal of Marine Research* **56**(6): 1163–201.

Chen, C. S., Q. C. Xu, R. C. Beardsley and P. J. S. Franks (2003). Model study of the cross-frontal water exchange on Georges Bank: A three-dimensional Lagrangian experiment. *Journal of Geophysical Research-Oceans* **108**(C5): 21.

Chen, C. S., Q. C. Xu, R. Houghton and R. C. Beardsley (2008). A model-dye comparison experiment in the tidal mixing front zone on the southern flank of Georges Bank. *Journal of Geophysical Research–Oceans* **113**(C2).

Chen, C. T. A. and A. V. Borges (2009). Reconciling opposing views on carbon cycling in the coastal ocean: Continental shelves as sinks and near-shore ecosystems as sources of atmospheric CO_2. *Deep-Sea Research Part II–Topical Studies in Oceanography* **56**(8–10): 578–90.

Cheriton, O. M., M. A. McManus, M. T. Stacey and J. V. Steinbuck (2009). Physical and biological controls on the maintenance and dissipation of a thin phytoplankton layer. *Marine Ecology-Progress Series* **378**: 55–69.

Chisholm, S. W. (1992). Phytoplankton size. *Primary Productivity and Biogeochemical Cycles in the Sea*. P. G. Falkowski and A. D. Woodhead. New York, Plenum Press: 213–37.

Choi, B. H. (1980). A tidal model of the Yellow Sea and the eastern China Sea. *KORDI*: 72.

Cloern, J. E. (1991). Tidal stirring and phytoplankton bloom dynamics in an estuary. *Journal of Marine Research* **49**(1): 203–21.

Cloern, J. E. and A. D. Jassby (2008). Complex seasonal patterns of primary producers at the land-sea interface. *Ecology Letters* **11**(12): 1294–1303.

Coachman, L. K. and J. J. Walsh (1981). A diffusion-model of cross-shelf exchange of nutrients in the southeastern bering sea. *Deep-Sea Research Part A–Oceanographic Research Papers* **28**(8): 819–46.

Coale, K. H., K. S. Johnson, S. E. Fitzwater *et al.* (1996). A massive phytoplankton bloom induced by an ecosystem-scale iron fertilization experiment in the equatorial Pacific Ocean. *Nature* **383**(6600): 495–501.

Colloca, F., P. Carpentieri, E. Balestri and G. D. Ardizzone (2004). A critical habitat for Mediterranean fish resources: shelf-break areas with *Leptometra phalangium* (Echinodermata: Crinoidea). *Marine Biology* **145**(6): 1129–42.

Colmenero-Hidalgo, E., J. A. Flores, F. J. Sierro *et al.* (2004). Ocean surface water response to short-term climate changes revealed by coccolithophores from the Gulf of Cadiz (NE Atlantic) and Alboran Sea (W Mediterranean). *Palaeogeography Palaeoclimatology Palaeoecology* **205**(3–4): 317–36.

Condie, S. A. (1995). Interactions between western boundary currents and shelf waters: A mechanism for coastal upwelling. *Journal of Geophysical Research–Oceans* **100**(C12): 24811–18.

Connolly, T. P., B. M. Hickey, S. L. Geier and W. P. Cochlan (2010). Processes influencing seasonal hypoxia in the northern California Current System. *Journal of Geophysical Research–Oceans* **115**: 22.

Conway, D. V. P., S. H. Coombs, J. A. Lindley and C. A. Llewellyn (1999). Diet of mackerel (Scomber scombrus) larvae at the shelf-edge to the south-west of the British Isles and the incidence of piscivory and coprophagy. *Vie et Milieu–Life and Environment* **49**(4): 213–20.

Cooper, L. H. N. and D. Vaux (1949). Cascading over the continental slope of water from the Celtic Sea. *Journal of the Marine Biological Association of the United Kingdom* **28**: 719–50.

Cosper, E. and J. C. Stepien (1984). Phytoplankton-zooplankton coupling in the outer continental-shelf and slope waters of the mid-Atlantic bight, June 1979. *Estuarine Coastal and Shelf Science* **18**(2): 145–55.

Cullen, J. J. and S. G. Horrigan (1981). Effects of nitrate on the diurnal vertical migration, carbon to nitrogen ratio, and the photosynthetic capacity of the dinoflagellate *gymnodinium-splendens*. *Marine Biology* 62(2–3): 81–9.

Cushing, D. H. (1995). *Population Production and Regulation in the Sea: a Fisheries Perspective*. Cambridge: Cambridge University Press.

(1986). The migration of larval and juvenile fish from spawning ground to nursery ground. *Journal Du Conseil* **43**(1): 43–9.

(1990). Plankton production and year-class strength in fish populations – an update of the match mismatch hypothesis. *Advances in Marine Biology* **26**: 249–93.

Darwin, G. H. (1899). *The Tides and Kindred Phenomena in the Solar System*. Boston: Houghton.

Dasaro, E. A., D. M. Farmer, J. T. Osse and G. T. Dairiki (1996). A Lagrangian float. *Journal of Atmospheric and Oceanic Technology* **13**(6): 1230–46.

Deacon, M. (1971). *Scientists and the Sea 1650–1900*. Academic Press.

de Boer, G. J., J. D. Pietrzak and J. C. Winterwerp (2006). On the vertical structure of the Rhine region of freshwater influence. *Ocean Dynamics* **56**(3–4): 198–216.

(2009). SST observations of upwelling induced by tidal straining in the Rhine ROFI. *Continental Shelf Research* **29**(1): 263–77.

Demaster, D. J. (1981). The supply and accumulation of silica in the marine-environment. *Geochimica Et Cosmochimica Acta* **45**(10): 1715–32.

Denman, K. L. and A. E. Gargett (1995). Biological physical interactions in the upper ocean – the role of vertical and small-scale transport processes. *Annual Review of Fluid Mechanics* **27**: 225–55.

Dickey, J. O., P. L. Bender, J. E. Faller *et al.* (1994). Lunar laser ranging – a continuing legacy of the Apollo program. *Science* **265**(5171): 482–90.

Dietrich, G. (1951). Influences of tidal streams on oceanographic and climatic conditions in the sea as exemplified by the English Channel. *Nature, London* **168**: 8–16.

Dillon, T. M. (1982). Vertical overturns: a comparison of Thorpe and Ozmidov scales. *Journal of Physical Oceanography* **87**: 9601–13.

Dittmar, T. and G. Kattner (2003). The biogeochemistry of the river and shelf ecosystem of the Arctic Ocean: a review. *Marine Chemistry* **83**(3–4): 103–20.

Droop, M. R. (1968). Vitamin B12 and marine ecology. IV. The kinetics of uptake, growth and inhibition in monochrysis lutheri. *Journal of the Marine Biological Association of the United Kingdom*. **48**: 689–733.

Dunne, J. P., J. L. Sarmiento and A. Gnanadesikan (2007). A synthesis of global particle export from the surface ocean and cycling through the ocean interior and on the seafloor. *Global Biogeochemical Cycles* **21**(4): 16.

Durazo, R., N. M. Harrison and A. E. Hill (1998). Seabird observations at a tidal mixing front in the Irish Sea. *Estuarine Coastal and Shelf Science* **47**(2): 153–64.

Eady, E. T. (1949). Long waves and cyclone waves. *Tellus* **1**: 33–52.

Egbert, G. D. and R. D. Ray (2000). Significant dissipation of tidal energy in the deep ocean inferred from satellite altimeter data. *Nature* **405**(6788): 775–8.

(2001). Estimates of M-2 tidal energy dissipation from TOPEX/Poseidon altimeter data. *Journal of Geophysical Research-Oceans* **106**(C10): 22475–502.

Egbert, G. D., R. D. Ray and B. G. Bills (2004). Numerical modeling of the global semidiurnal tide in the present day and in the last glacial maximum. *Journal of Geophysical Research–Oceans* **109**(C3): 15.

Egge, J. K. and D. L. Aksnes (1992). Silicate as regulating nutrient in phytoplankton competition. *Marine Ecology Progress Series* **83**(2–3): 281–9.

Einstein, A. (1933). On the method of theoretical physics. The Herbert Spencer Lecture, delivered at Oxford, 10 June. Also published in *Philosophy of Science* **1**, 2(April 1934): 163–9.

Ekman, V. W. (1905). On the influence of the earth's rotation on ocean currents. *Arkiv Mat. Astron. Fysik* **2**(11).

Elliott, A. J. and T. Clarke (1987). Seasonal stratification in the northwest European shelf seas. *Continental Shelf Research* **11**: 467–92.

Emery, W. J. and R. E. Thomson (1997). *Data analysis methods in Physical Oceanography*. Pergamon.

(2001). *Data Analysis Methods in Physical Oceanography*. Amsterdam: Elsevier.

Epifanio, C. E., K. T. Little and P. M. Rowe (1988). Dispersal and recruitment of fiddler crab larvae in the delaware river estuary. *Marine Ecology Progress Series* **43**(1–2): 181–8.

Eppley, R. W., O. Holm-Harisen and J. D. H. Strickland (1968). Some observations of the vertical migration of dinoflagellates. *Journal of Phycology* **4**(4): 333–40.

Eppley, R. W. and B. J. Peterson (1979). Particulate organic-matter flux and planktonic new production in the deep ocean. *Nature* **282**(5740): 677–80.

Escribano, R., G. Daneri, L. Farias *et al.* (2004). Biological and chemical consequences of the 1997–1998 El Nino in the Chilean coastal upwelling system: a synthesis. *Deep-Sea Research Part II— Topical Studies in Oceanography* **51**(20–21): 2389–411.

Fablet, R., R. Lefort, I. Karoui *et al.* (2009). Classifying fish schools and estimating their species proportions in fishery-acoustic surveys. *ICES Journal of Marine Science* **66**(6): 1136–42.

Falkowski, P. G., P. E. Biscaye and C. Sancetta (1994). The lateral flux of biogenic particles from the eastern North-American continental-margin to the north-Atlantic Ocean. *Deep-Sea Research Part II–Topical Studies in Oceanography* **41**(2–3): 583–601.

Falkowski, P. G., T. Fenchel and E. F. Delong (2008). The microbial engines that drive Earth's biogeochemical cycles. *Science* **320**(5879): 1034–9.

Falkowski, P. G. and J. A. Raven (2007). *Aquatic Photosynthesis*. Princeton: Princeton University Press.

Fischer, H. B., E. J. List, R. C. Y. Koh, J. Imberger and N. H. Brooks (1979). *Mixing in Inland and Coastal Waters*. Academic Press.

Flagg, C. N., C. D. Wirick and S. L. Smith (1994). The interaction of phytoplankton, zooplankton and currents from 15 months of continuous data in the mid-atlantic bight. *Deep-Sea Research Part II–Topical Studies in Oceanography* **41**(2–3): 411ff.

Floodgate, G. D., G. E. Fogg, D. A. Jones, L. Lochte and C. M. Turley (1981). Microbiological and zooplankton activity at a front in Liverpool Bay. *Nature* **290**(5802): 133–6.

Fofonoff, N. P. and R. C. Millard (1983). Algorithms for computation of fundamental properties of seawater. *Unesco Technical Papers in Marine Science*, UNESCO.

Fogel, M. L., C. Aguilar, R. Cuhel *et al.* (1999). Biological and isotopic changes in coastal waters induced by Hurricane Gordon. *Limnology and Oceanography* **44**(6): 1359–69.

Fogg, G. E. (1988). The microbiology of sea fronts. *Microbiological Sciences* **5**(3): 74–7.

Follows, M. J., S. Dutkiewicz, S. Grant and S. W. Chisholm (2007). Emergent biogeography of microbial communities in a model ocean. *Science* **315**(5820): 1843–6.

Forget, P. and G. Andre (2007). Can satellite-derived chlorophyll imagery be used to trace surface dynamics in coastal zone? A case study in the northwestern Mediterranean Sea. *Sensors* **7**(6): 884–904.

Franks, P. J. S. and D. M. Anderson (1992). Alongshore transport of a toxic phytoplankton bloom in a buoyancy current – *Alexandrium tamarense* in the Gulf of Maine. *Marine Biology* **112**(1): 153–64.

Franks, P. J. S. and C. S. Chen (1996). Plankton production in tidal fronts: A model of Georges Bank in summer. *Journal of Marine Research* **54**(4): 631–51.

Frederiksen, M., M. Edwards, A. J. Richardson, N. C. Halliday and S. Wanless (2006). From plankton to top predators: bottom-up control of a marine food web across four trophic levels. *Journal of Animal Ecology* **75**(6): 1259–68.

Frost, B. W. and S. M. Bollens (1992). Variability of diel vertical migration in the marine planktonic copepod pseudocalanus-newmani in relation to its predators. *Canadian Journal of Fisheries and Aquatic Sciences* **49**(6): 1137–41.

Fu, L.-L. and B. Holt (1984). Internal waves in the Gulf of California. *Journal of Geophysical Research* **89**(C2): 2053–60.

Gaarder, T. and H. H. Gran (1927). Investigations of the productivity of the plankton of Oslofjord. *Rapp. Cons. Int. Explor. Mer* **42**: 1–48.

Gabric, A. J., L. Garcia, L. Vancamp *et al.* (1993). Offshore export of shelf production in the cape blanc (mauritania) giant filament as derived from coastal zone color scanner imagery. *Journal of Geophysical Research-Oceans* **98**(C3): 4697–712.

Galloway, J. N., F. J. Dentener, D. G. Capone *et al.* (2004). Nitrogen cycles: past, present, and future. *Biogeochemistry* **70**(2): 153–226.

Galperin, B., L. H. Kantha, S. Hassid and A. Rosati (1988). A quasi-equilibrium turbulent energy model for geophysical flows. *Journal of Atmospheric Sciences* **45**: 55–62.

Garabato, A. C. N., K. L. Polzin, B. A. King, K. J. Heywood and M. Visbeck (2004). Widespread intense turbulent mixing in the Southern Ocean. *Science* **303**(5655): 210–13.

Garciasoto, C., E. Fernandez, R. D. Pingree and D. S. Harbour (1995). Evolution and structure of a shelf coccolithophore bloom in the western English Channel. *Journal of Plankton Research* **17**(11): 2011–36.

Garrett, C., J. R. Keeley and D. A. Greenberg (1978). Tidal mixing versus thermal stratification in the gulf of Maine. *Atmospheres and Oceans* **16**(4): 403–23.

Garrett, C. J. R. and J. W. Loder (1981). Dynamical aspects of shallow sea fronts. *Philosphical Transactions of the Royal Society of London* **A302**: 563–81.

Gawarkiewicz, G., and D. C. Chapman (1992). The role of stratification in the formation and maintenance of shelf-break fronts. *Journal of Physical Oceanography* **22**(7): 753–72.

Geider, R. J., H. L. MacIntyre and T. M. Kana (1998). A dynamic regulatory model of phytoplanktonic acclimation to light, nutrients, and temperature. *Limnology and Oceanography* **43**(4): 679–94.

Geider, R. J., C. M. Moore and O. N. Ross (2009). The role of cost-benefit analysis in models of phytoplankton growth and acclimation. *Plant Ecology & Diversity* **2**(2): 165–78.

Geyer, W. R., R. C. Beardsley, S. J. Lentz *et al.* (1996). Physical oceanography of the Amazon shelf. *Continental Shelf Research* **16**(5–6): 575–616.

Gill, A. E. (1982). *Atmosphere-Ocean Dynamics* Academic Press.

Gladman, B., D. D. Quinn, P. Nicholson and R. Rand (1996). Synchronous locking of tidally evolving satellites. *Icarus* **122**(1): 166–92.

Glorioso, P. D. and R. A. Flather (1995). A barotropic model of the currents off SE South America. *Journal of Geophysical Research* **100**(C7): 13427–40.

Graber H. C., E. A. Terray, M. A. Donelan, W. M. Drennan and J. C. van Leer (2000). ASIS – A new Air-Sea Interaction Spar buoy: Design and performance at sea. *Journal of Atmospheric and Oceanic Technology* **17**(5): 708–20.

Graham, G. W. and W. Smith. (2010). The application of holography to the analysis of size and settling velocity of suspended cohesive sediments. *Limnology and Oceanography-Methods* **8**: 1–15.

Grant, H. L., R. W. Stewart and A. Moilliet (1962). Turbulence spectra from a tidal channel. *Journal of Fluid Mechanics* **12**: 241–68.

Green, J. A. M. (2010). Tides and ocean resonance. *Ocean Dynamics* **60**: 1243–53.

Green, J. A. M., C. L. Green, G. R. Bigg *et al.* (2009). Tidal mixing and the Meridional Overturning Circulation from the Last Glacial Maximum. *Geophysical Research Letters* **36**: 5.

Green, J. A. M., J. H. Simpson, S. Legg and M. R. Palmer (2008). Internal waves, baroclinic energy fluxes and mixing at the European shelf edge. *Continental Shelf Research* **28**(7): 937–50.

Green, J. A. M., J. H. Simpson, S. A. Thorpe and T. P. Rippeth (2010). Observations of internal tidal waves in the isolated seasonally stratified region of the western Irish Sea. *Continental Shelf Research* **30**(2): 214–25.

Griffiths, R. W. and P. F. Linden (1981). The stability of buoyancy-driven coastal currents. *Dynamics of Atmospheres and Oceans* **5**(4): 281–306.

Griffiths, S. D. and W. R. Peltier (2008). Megatides in the Arctic Ocean under glacial conditions. *Geophysical Research Letters* **35**(8).

Hadfield, M. G. and J. Sharples (1996). Modelling mixed layer depth and plankton biomass off the west coast of South Island, New Zealand. *Journal of Marine Systems* **8**(1–2): 1–29.

Hales, B., L. Karp-Boss *et al.* (2006). Oxygen production and carbon sequestration in an upwelling coastal margin. *Global Biogeochemical Cycles* **20**(3): 15.

Hales, B., J. N. Moum, P. Covert and A. Perlin (2005). Irreversible nitrate fluxes due to turbulent mixing in a coastal upwelling system. *Journal of Geophysical Research–Oceans* **110**(C10): 19.

Hales, B., T. Takahashi and L. Bandstra (2005). Atmospheric CO_2 uptake by a coastal upwelling system. *Global Biogeochemical Cycles* **19**(1): 11.

Hama, T., T. Miyazaki, Y. Ogawa *et al.* (1983). Measurement of photosynthetic production of a marine-phytoplankton population using a stable C-13 isotope. *Marine Biology* **73**(1): 31–6.

Hamm, C. E., R. Merkel, O. Springer *et al.* (2003). Architecture and material properties of diatom shells provide effective mechanical protection. *Nature* **421**(6925): 841–3.

Hannah, C. G., F. Dupont and M. Dunphy (2009). Polynyas and Tidal Currents in the Canadian Arctic Archipelago. *Arctic* **62**(1): 83–95.

Hansen, B., P. K. Bjornsen and P. J. Hansen (1994). The size ratio between planktonic predators and their prey. *Limnology and Oceanography* **39**(2): 395–403.

Hansen, D. V. and M. Rattray (1966). New dimensions in estuary classification. *Limnology and Oceanography* **11**: 319–26.

Harrison, P. J., P. J. Clifford, W. P. Cochlan *et al.* (1991). Nutrient and plankton dynamics in the Fraser-River plume, Strait of Georgia, British-Columbia. *Marine Ecology Progress Series* **70**(3): 291–304.

Harvey, H. W., L. H. N. Cooper, M. V. Lebour and F. S. Russell (1935). Plankton production and its control. *Journal of the Marine Biological Association of the United Kingdom* **20**: 407–41.

Haynes, R., E. D. Barton and I. Pilling (1993). Development, persistence, and variability of upwelling filaments off the Atlantic coast of the Iberian peninsula. *Journal of Geophysical Research–Oceans* **98**(C12): 22681–92.

Heaps, N. S. (1972). Estimation of density currents in the Liverpool Bay area of the Irish Sea. *Geophysical Journal of the Royal Astronomical Society* **30**: 415–32.

Heil, C. A., M. Revilla, P. M. Glibert and S. Murasko (2007). Nutrient quality drives differential phytoplankton community composition on the southwest Florida shelf. *Limnology and Oceanography* **52**(3): 1067–78.

Hickey, B., S. Geier, N. Kachel and A. F. MacFadyen (2005). A bi-directional river plume: The Columbia in summer. *Continental Shelf Research* **25**(14): 1631–56.

Hickman, A. E., P. M. Holligan, C. M. Moore *et al.* (2009). Distribution and chromatic adaptation of phytoplankton within a shelf sea thermocline. *Limnology and Oceanography* **54**(2): 525–36.

Hill, A. E. (1998). Buoyancy effects in coastal and shelf seas. *The Sea* K. H. Brink and A. R. Robinson. New York, Wiley. **10**: 21–62.

Hill, A. E., J. Brown and L. Fernand (1997). The summer gyre in the western Irish Sea: Shelf sea paradigms and management implications. *Estuarine, Coastal and Shelf Science* **44** (Suppl. A): 83–95.

Hill, A. E., J. Brown, L. Fernand *et al.* (2008). Thermohaline circulation of shallow tidal seas. *Geophysical Research Letters* **35**(11).

Hill, A. E., A. J. Souza, K. Jones *et al.* (1998). The Malin cascade in winter 1996. *Journal of Marine Research* **56**(1): 87–106.

Holliday, D. V., C. F. Greenlaw and P. L. Donaghay (2010). Acoustic scattering in the coastal ocean at Monterey Bay, CA, USA: Fine-scale vertical structures. *Continental Shelf Research* **30**(1): 81–103.

Holligan, P. M., R. P. Harris, R. C. Newell *et al.* (1984). Vertical-distribution and partitioning of organic-carbon in mixed, frontal and stratified waters of the English Channel. *Marine Ecology-Progress Series* **14**(2–3): 111–27.

Holligan, P. M., R. D. Pingree and G. T. Mardell (1985). Oceanic solitions, nutrient pulses and phytoplankton growth. *Nature* **314**(6009): 348–50.

Holt, J., J. Harle, R. Proctor *et al.* (2009). Modelling the global coastal ocean. *Philosophical Transactions of the Royal Society a-Mathematical Physical and Engineering Sciences* **367** (1890): 939–51.

Holt, J. and R. Proctor (2008). The seasonal circulation and volume transport on the northwest European continental shelf: A fine-resolution model study. *Journal of Geophysical Research–Oceans* **113**(C6).

Holt, J., S. Wakelin and J. Huthnance (2009). Down-welling circulation of the northwest European continental shelf: A driving mechanism for the continental shelf carbon pump. *Geophysical Research Letters* **36**: 5.

Holt, J., S. Wakelin, J. Lowe and J. Tinker (2010). The potential impacts of climate change on the hydrography of the northwest European continental shelf. *Progress in Oceanography* **86**(3–4): 361–79.

Hopkinson, C. S., A. E. Giblin and J. Tucker (2001). Benthic metabolism and nutrient regeneration on the continental shelf of Eastern Massachusetts, USA. *Marine Ecology Progress Series* **224**: 1–19.

Horne, E. P. W., J. W. Loder, W. G. Harrison *et al.* (1989). Nitrate supply and demand at the Georges Bank tidal front. *Scientia Marina* **53**(2–3): 145–58.

Horne, E. P. W., J. W. Loder, C. E. Naimie and N. S. Oakey (1996). Turbulence dissipation rates and nitrate supply in the upper water column on Georges Bank. *Deep-Sea Research Part II–Topical Studies in Oceanography* **43**(7–8): 1683–712.

Horsburgh, K. J., A. E. Hill, J. Brown *et al.* (2000). Seasonal evolution of the cold pool gyre in the western Irish Sea. *Progress in Oceanography* **46**(1): 1–58.

Houghton, J. (2002). *The Physics of Atmospheres*. Cambridge University Press.

Houghton, R. W. and C. Ho (2001). Diapycnal flow through the Georges Bank tidal front: A dye tracer study. *Geophysical Research Letters* **28**(1): 33–6.

Howard, L. N. (1961). Note on a paper by John W. Miles. *Journal of Fluid Mechanics* **10**: 509–12.

Hsueh, Y., H. J. Lie and H. Ichikawa (1996). On the branching of the Kuroshio west of Kyushu. *Journal of Geophysical Research–Oceans* **101**(C2): 38517.

Huete-Ortega, M., E. Maranon, M. Varela and A. Bode (2010). General patterns in the size scaling of phytoplankton abundance in coastal waters during a 10-year time series. *Journal of Plankton Research* **32**(1): 1–14.

Huisman, J., P. van Oostveen and F. J. Weissing (1999). Critical depth and critical turbulence: Two different mechanisms for the development of phytoplankton blooms. *Limnology and Oceanography* **44**(7): 1781–7.

Hunt, G. L., P. Stabeno, G. Walters *et al.* (2002). Climate change and control of the southeastern Bering Sea pelagic ecosystem. *Deep-Sea Research Part II–Topical Studies in Oceanography* **49**(26): 5821–53.

Hutchings, J. A. (1996). Spatial and temporal variation in the density of northern cod and a review of hypotheses for the stock's collapse. *Canadian Journal of Fisheries and Aquatic Sciences* **53**(5): 943–62.

(2000). Collapse and recovery of marine fishes. *Nature* **406**(6798): 882–5.

Huthnance, J. M. (1984). Slope currents and jebar. *Journal of Physical Oceanography* **14**(4): 795–810.

Huthnance, J. M., J. T. Holt and S. L. Wakelin (2009). Deep ocean exchange with west-European shelf seas. *Ocean Science* **5**(4): 621–34.

Hydes, D. J., R. J. Gowen, N. P. Holliday, T. Shammon and D. Mills (2004). External and internal control of winter concentrations of nutrients (N, P and Si) in north-west European shelf seas. *Estuarine Coastal and Shelf Science* **59**(1): 151–61.

Ibaibarriaga, L., X. Irigoien, M. Santos *et al.* (2007). Egg and larval distributions of seven fish species in north-east Atlantic waters. *Fisheries Oceanography* **16**(3): 284–93.

Iles, T. D. and M. Sinclair (1982). Atlantic herring – stock discreteness and abundance. *Science* **215**(4533): 627–33.

Inall, M. E., G. I. Shapiro and T. J. Sherwin (2001). Mass transport by non-linear internal waves on the Malin Shelf. *Continental Shelf Research* **21**(13–14): 1449–72.

Irigoien, X., R. P. Harris, H. M. Verheye *et al.* (2002). Copepod hatching success in marine ecosystems with high diatom concentrations. *Nature* **419**(6905): 387–89.

Ivanoff, A. (1977). Oceanic absorption of solar energy. *Modelling and prediction of the upper layers of the ocean*, ed. E. B. Kraus. Oxford: Pergamon: 47–71.

Ivanov, V. V., G. I. Shapiro, J. M. Huthnance, D. L. Aleynik and P. N. Golovin (2004). Cascades of dense water around the world ocean. *Progress in Oceanography* **60**(1): 47–98.

Jahncke, J., K. O. Coyle, S. I. Zeeman, N. B. Kachel and G. L. Hunt (2005). Distribution of foraging shearwaters relative to inner front of SE Bering Sea. *Marine Ecology Progress Series* **305**: 219–33.

Jahnke, R. A. (2010). Global synthesis. *Carbon and Nutrient Fluxes in Continental Margins: A Global Synthesis*. K. K. Liu, L. Atkinson, R. A. Quinones and L. Talaue-McManus. Berlin Heidelberg, Springer-Verlag: 597–615.

James, I. D. (1978). A note on the circulation induced by a shallow sea front. *Estuarine and Coastal Marine Science* **7**(2): 187–202.

(1981). Fronts and shelf circulation models. *Philosophical Transactions of the Royal Society of London* **A302**: 597–604.

(1988). Experiments with a numerical-model of coastal currents and tidal mixing fronts. *Continental Shelf Research* **8**(12): 1275–97.
Jassby, A. D. and T. Platt (1976). Mathematical formulation of relationship between photosynthesis and light for phytoplankton. *Limnology and Oceanography* **21**(4): 540–7.
Jay, D. A. and J. M. Musiak (1996). Internal tidal asymmetry in channel flows: origins and consequences. *Mixing Processes in Estuaries and Coastal Seas*. C. Pattiaratchi. Washington, DC, AGU: 211–49.
Jickells, T. (2006). The role of air-sea exchange in the marine nitrogen cycle. *Biogeosciences* **3**(3): 271–80.
Johnson, J. A. and N. Rock (1986). Shelf break circulation process. *Coastal and Estuarine Sciences* **3**: 33–62.
Johnson, K. S. and L. J. Coletti (2002). In situ ultraviolet spectrophotometry for high resolution and long-term monitoring of nitrate, bromide and bisulfide in the ocean. *Deep-Sea Research I* **49**: 1291–305.
Joint, I., M. Inall, R. Torres *et al.* (2001). Two Lagrangian experiments in the Iberian Upwelling System: tracking an upwelling event and an off-shore filament. *Progress in Oceanography* **51**(2–4): 221–48.
Joint, I. R., N. J. P. Owens and A. J. Pomroy (1986). Seasonal production of photosynthetic picoplankton and nanoplankton in the Celtic Sea. *Marine Ecology Progress Series* **28**(3): 251–8.
Joint, I., R. Wollast, L. Chou *et al.* (2001). Pelagic production at the Celtic Sea shelf break. *Deep-Sea Research Part II–Topical Studies in Oceanography* **48**(14–15): 3049–81.
Kahru, M., V. Brotas, M. Manzano-Sarabia and B. G. Mitchell (2011). Are phytoplankton blooms occurring earlier in the Arctic? *Global Change Biology* **17**(4): 1733–9.
Kamykowski, D. and S. A. McCollum (1986). The temperature acclimatized swimming speed of selected marine dinoflagellates. *Journal of Plankton Research* **8**(2): 275–87.
Kantha, L. H. and C. A. Clayson (1994). An improved mixed-layer model for geophysical applications. *Journal of Geophysical Research-Oceans* **99**(C12): 25235–66.
Karl, D., A. Michaels, B. Bergman *et al.* (2002). Dinitrogen fixation in the world's oceans. *Biogeochemistry* **57**(1): 47–98.
Karp-Boss, L., E. Boss and P. A. Jumars (1996). Nutrient fluxes to planktonic osmotrophs in the presence of fluid motion. *Oceanography and Marine Biology*, Vol **34**. London, U C L Press Ltd. **34**: 71–107.
Katechakis, A. and H. Stibor (2006). The mixotroph Ochromonas tuberculata may invade and suppress specialist phago- and phototroph plankton communities depending on nutrient conditions. *Oecologia* **148**(4): 692–701.
Kelly-Gerreyn, B. A., T. R. Anderson, J. T. Holt *et al.* (2004). Phytoplankton community structure at contrasting sites in the Irish Sea: a modelling investigation. *Estuarine Coastal and Shelf Science* **59**(3): 363–83.
Kiefer, D. A. and J. N. Kremer (1981). Origins of vertical patterns of phytoplankton and nutrients in the temperate, open ocean – a stratigraphic hypothesis. *Deep-Sea Research Part A–Oceanographic Research Papers* **28**(10): 1087–105.
Kingsford, M. J., E. Wolanski and J. H. Choat (1991). Influence of tidally induced fronts and langmuir circulations on distribution and movements of presettlement fishes around a coral-reef. *Marine Biology* **109**(1): 167–80.

Kinsman, B. (1965). *Wind Waves: Their Generation and Propagation on the Ocean Surface*. Englewood Cliffs, NJ: Prentice Hall.

Kiørboe, T. (1993). Turbulence, phytoplankton cell-size, and the structure of pelagic food webs. *Advances in Marine Biology*, 29. London, Academic Press Ltd. **29**: 1–72.

(2008). *A Mechanistic Approach to Plankton Ecology*. Princeton: Princeton University Press.

Kiørboe, T. and K. Johansen (1986). Studies of a larval herring (*Clupea harengus* I) patch in the Buchan area.IV. Zooplankton distribution and productivity in relation to hydrographic features. *Dana-A Journal of Fisheries and Marine Research* **6**: 37–51.

Kiørboe, T., P. Munk, K. Richardson, V. Christensen and H. Paulsen (1988). Plankton dynamics and larval herring growth, drift and survival in a frontal area. *Marine Ecology Progress Series* **44**(3): 205–19.

Kiørboe, T. and E. Saiz (1995). Planktivorous feeding in calm and turbulent environments, with emphasis on copepods. *Marine Ecology-Progress Series* **122**(1–3): 135–45.

Kiørboe, T., P. Tiselius, B. Mitchell-Innes *et al.* (1998). Intensive aggregate formation with low vertical flux during an upwelling-induced diatom bloom. *Limnology and Oceanography* **43**(1): 104–16.

Kiriakoulakis, K., B. J. Bett, M. White and G. A. Wolff (2004). Organic biogeochemistry of the Darwin Mounds, a deep-water coral ecosystem, of the NE Atlantic. *Deep-Sea Research Part I–Oceanographic Research Papers* **51**(12): 1937–54.

Kiriakoulakis, K., E. Stutt, S. J. Rowland *et al.* (2001). Controls on the organic chemical composition of settling particles in the Northeast Atlantic Ocean. *Progress in Oceanography* **50**(1–4): 65–87.

Kirk, J. T. O. (2010). *Light and Photosynthesis in Aquatic Ecosystems*. Cambridge: Cambridge University Press.

Kiszka, J., K. Macleod, O. van Canneyt, D. Walker and V. Ridoux (2007). Distribution, encounter rates, and habitat characteristics of toothed cetaceans in the Bay of Biscay and adjacent waters from platform-of-opportunity data. *Ices Journal of Marine Science* **64**(5): 1033–43.

Kjellstrom, E., G. Nikulin, U. Hansson, G. Strandberg and A. Ullerstig (2011). 21st century changes in the European climate: uncertainties derived from an ensemble of regional climate model simulations. *Tellus Series A–Dynamic Meteorology and Oceanography* **63**(1): 24–40.

Klink, J. (1999). Dynmodes: Matlab code for Normal Modes Analysis.

Kobayashi, S., J. H. Simpson, T. Fujiwara and K. J. Horsburgh (2006). Tidal stirring and its impact on water column stability and property distributions in a semi-enclosed shelf sea (Seto Inland Sea, Japan). *Continental Shelf Research* **26**(11): 1295–306.

Kolber, Z. and P. G. Falkowski (1993). Use of active fluorescence to estimate phytoplankton photosynthesis in-situ. *Limnology and Oceanography* **38**(8): 1646–65.

Kolmogorov, A. N. (1941). The local structure of turbulence in incompressible viscous fluid for very large Reynolds numbers. *Doklady. Akad. Nauk. SSSR* **30**: 301–5.

Kundu, P. K. (1976). Ekman veering pbserved near the ocean bottom. *Journal of Physical Oceanography* **6**(2): 238–42.

(1990). *Fluid Mechanics*. San Diego: Academic Press.

Kundu, P. K. and I. M. Cohen (2008). *Fluid Mechanics*. Amsterdam: Academic Press-Elsevier.

Lai, Z. G., C. S. Chen, G. W. Cowles and R. C. Beardsley (2010). A nonhydrostatic version of FVCOM: 2. Mechanistic study of tidally generated nonlinear internal waves in Massachusetts Bay. *Journal of Geophysical Research-Oceans* **115**: 21.

Lamb, H. (1932). *Hydrodynamics*. Cambridge: Cambridge University Press.

Lamb, K. G. (1994). Numerical experiments of internal wave generation by strong tidal flow across a finite-amplitude bank edge. *Journal of Geophysical Research-Oceans* **99**(C1): 843–64.

Lampert, W. (1989). The adaptive significance of diel vertical migration of zooplankton. *Functional Ecology* **3**(1): 21–7.

Landry, M. R., J. R. Postel, W. K. Peterson and J. Newman (1989). Broad-scale distributional patterns of hydrographic variables on the Washington/Oregon shelf. *Coastal Oceanography of Washington and Oregon*. M. R. Landry and B. M. Hickey, Elsevier. **47**: 1–40.

Langlois, V. J., A. Andersen, T. Bohr, A. W. Visser and T. Kiørboe (2009). Significance of swimming and feeding currents for nutrient uptake in osmotrophic and interception-feeding flagellates. *Aquatic Microbial Ecology* **54**(1): 35–44.

Lasker, R. (1975). Field criteria for survival of anchovy larvae – relation between inshore chlorophyll maximum layers and successful 1st feeding. *Fishery Bulletin* **73**(3): 453–62.

Lavrentyev, P. J., H. A. Bootsma, T. H. Johengen, J. F. Cavaletto and W. S. Gardner (1998). Microbial plankton response to resource limitation: insights from the community structure and seston stoichiometry in Florida Bay, USA. *Marine Ecology-Progress Series* **165**: 45–57.

Laws, E. A., P. G. Falkowski, W. O. Smith, H. Ducklow and J. J. McCarthy (2000). Temperature effects on export production in the open ocean. *Global Biogeochemical Cycles* **14**(4): 1231–46.

Le Corre, P., S. L'Helguen and M. Wafar (1993). Nitrogen-source for uptake by *gyrodinium* cf *aureolum* in a tidal front. *Limnology and Oceanography* **38**(2): 446–51.

Ledwell, J. R., E. T. Montgomery, K. L. Polzin *et al.* (2000). Evidence for enhanced mixing over rough topography in the abyssal ocean. *Nature* **403**(6766): 179–82.

Lee, J, Y., P. Tett, K. Jones *et al.* (2002). The PROWQM physical-biological model with benthic-pelagic coupling applied to the northern North Sea. *Journal of Sea Research* **48**(4): 287–331.

Legendre, L. and R. B. Rivkin (2002). Fluxes of carbon in the upper ocean: regulation by food-web control nodes. *Marine Ecology Progress Series* **242**: 95–109.

Leibovich, S. (1983) The form and dynamics of Langmuir circulation. *Annual Review of Fluid Mechanics*, **15**: 391–427.

Lenn, Y. D., P. J. Wiles, S. Torres-Valdes *et al.* (2009). Vertical mixing at intermediate depths in the Arctic boundary current. *Geophysical Research Letters* **36**.

Lennon, G. W., D. G. Bowers, R. A. Nunes *et al.* (1987). Gravity currents and the release of salt from an inverse estuary. *Nature* **327**(6124): 695–7.

Lentz, S. J. and J. Largier (2006). The influence of wind forcing on the Chesapeake Bay buoyant coastal current. *Journal of Physical Oceanography* **36**(7): 1305–16.

Lewis, M. R., W. G. Harrison, N. S. Oakey, D. Herbert and T. Platt (1986). Vertical nitrate fluxes in the oligotrophic ocean. *Science* **234**(4778): 870–3.

Lie, H. J. (1989). Tidal fronts in the south-eastern Hwanghae (Yellow Sea). *Continental Shelf Research* **9**(6): 527–46.

Linden, P. F. (1979). Mixing in stratified fluids. *Geophysical and Astrophysical Fluid Dynamics* **13**: 3–24.

Linden, P. F. and J. E. Simpson (1988). Modulated mixing and frontogenesis in shallow seas and estuaries. *Continental Shelf Research* **8**(10): 1107–27.

Liu, K. K., L. Atkinson, R. A. Quiñones and L. Talaue-McManus (2010). Biogeochemistry of Continental Margins in a Global Context. *Carbon and Nutrient Fluxes in Continental Margins: A Global Synthesis*.
K. K. Liu, L. Atkinson, R. A. Quinones and L. Talaue-McManus. Berlin Heidelberg: Springer-Verlag, 3–24.

Li, W. K. W., B. D. Irwin and P. M. Dickie (1993). Dark fixation of C-14 – Variations related to biomass and productivity of phytoplankton and bacteria. *Limnology and Oceanography* **38**(3): 483–94.

Loder, J. W., D. Brickman and E. P. W. Horne (1992). Detailed structure of currents and hydrography on the northern side of Georges Nank. *Journal of Geophysical Research–Oceans* **97**(C9): 14331–51.

Loder, J. W., K. F. Drinkwater, N. S. Oakey and E. P. W. Horne (1993). Circulation, hydrographic structure and mixing at tidal fronts - the view from Georges Bank. *Philosophical Transactions of the Royal Society London, Series A Mathematical Physical and Engineering Sciences*, **343**: 447–60.

Loder, J. W. and T. Platt (1985). *Physical controls on phytoplankton production at tidal fronts*. Proceedings of the 19th European Marine Biology Symposium, Cambridge University Press.

Loder, J. W. and D. G. Wright (1985). Tidal rectification and frontal circulation on the sides of Georges Bank. *Journal of Marine Research* **43**(3): 581–604.

Lohrenz, S. E., C. L. Carroll, A. D. Weidemann and M. Tuel (2003). Variations in phytoplankton pigments, size structure and community composition related to wind forcing and water mass properties on the North Carolina inner shelf. *Continental Shelf Research* **23**(14–15): 1447–64.

Long, R. R. (2003). Do tidal channel measurements support k^-5/3. *Environmental Fluid Mechanics* **3**: 109–27.

Longhurst, A., S. Sathyendranath, T. Platt and C. Caverhill (1995). An estimate of global primary production in the ocean from satellite radiometer data. *Journal of Plankton Research* **17**(6): 1245–71.

Lough, R. G. and G. R. Bolz (1989). *The movement of cod and haddock larvae onto the shoals of Georges Bank*. Symp of the Fisheries Society of the British Isles: Fish Population Biology, Aberdeen, Scotland, Academic Press Ltd.

Lough, R. G. and J. P. Manning (2001). Tidal-front entrainment and retention of fish larvae on the southern flank of Georges Bank. *Deep-Sea Research Part II–Topical Studies in Oceanography* **48**(1–3): 631–44.

Lucas, L. V., J. E. Cloern, J. R. Koseff, S. G. Monismith and J. K. Thompson (1998). Does the Sverdrup critical depth model explain bloom dynamics in estuaries? *Journal of Marine Research* **56**(2): 375–415.

Lucas, L. V., J. R. Koseff, J. E. Cloern, S. G. Monismith and J. K. Thompson (1999). Processes governing phytoplankton blooms in estuaries. I: The local production-loss balance. *Marine Ecology Progress Series* **187**: 1–15.

Maar, M., T. G. Nielsen, K. Richardson *et al.* (2002). Spatial and temporal variability of food web structure during the spring bloom in the Skagerrak. *Marine Ecology Progress Series* **239**: 11–29.

(2003). Microscale distribution of zooplankton in relation to turbulent diffusion. *Limnology and Oceanography* **48**(3): 1312–25.

MacIsaac, J. J., R. C. Dugdale, R. T. Barber, D. Blasco and T. T. Packard (1985). Primary production cycle in an upwelling center. *Deep-Sea Research Part A–Oceanographic Research Papers* **32**(5): 503–29.

Mackas, D. L., R. Kieser, M. Saunders *et al.* (1997). Aggregation of euphausiids and Pacific Lake (*Merluccius productus*) along the outer continental shelf off Vancouver Island. *Canadian Journal of Fisheries and Aquatic Sciences* **54**(9): 2080–96.

Mackenzie, B. R., T. J. Miller, S. Cyr and W. C. Leggett (1994). Evidence for a dome-shaped relationship between turbulence and larval fish ingestion rates. *Limnology and Oceanography* **39**(8): 1790–9.

Maddock, L. and R. D. Pingree (1982). Mean heat and salt budgets for the eastern English Channel and southern bight of the North Sea. *Journal of the Marine Biological Association of the United Kingdom* **62**(3): 559–75.

Mamas, C. J. V., L. G. Earwaker, R. S. Sokhi *et al.* (1995). An estimation of sedimentation-rates along the ribble estuary, Lancashire, UK, based on radiocesium profiles preserved in intertidal sediments. *Environment International* **21**(2): 151–65.

Margalef, R. (1978). Life-forms of phytoplankton as survival alternatives in an unstable environment. *Oceanologica Acta* **1**(4): 493–509.

Marra, J. and R. T. Barber (2004). Phytoplankton and heterotrophic respiration in the surface layer of the ocean. *Geophysical Research Letters* **31**(9): 4.

Marra, J., R. W. Houghton and C. Garside (1990). Phytoplankton growth at the shelf-break front in the middle Atlantic bight. *Journal of Marine Research* **48**(4): 851–68.

Marret, F., J. Scourse and W. Austin (2004). Holocene shelf-sea seasonal stratification dynamics: a dinoflagellate cyst record from the Celtic Sea, NW European shelf. *Holocene* **14**(5): 689–96.

Martin, J. H. and S. E. Fitzwater (1988). Iron-deficiency limits phytoplankton growth in the northeast Pacific subarctic. *Nature* **331**(6154): 341–3.

Masson, D. G., V. A. I. Huvenne, H. C. de Stigter *et al.* (2010). Efficient burial of carbon in a submarine canyon. *Geology* **38**(9): 831–4.

Matthews, D. J. (1913). The salinity and temperature of the Irish Channel and the waters south of Ireland. *Scientific Investigations Fisheries Branch Ireland* **4**: 1–26.

Mayer, M. and D. Piepenburg (1996). Epibenthic community patterns on the continental slope off East Greenland at 75 degrees N. *Marine Ecology Progress Series* **146**(1–3): 151–64.

McGlade, J. M. (2002). The North Sea Large Marine Ecosystem. *Large Marine Ecosystems of the North Atlantic*. K. Sherman and G. Hempel. Amsterdam: Elsevier, 339–412.

McHatton, S. C., J. P. Barry, H. W. Jannasch and D. C. Nelson (1996). High nitrate concentrations in vacuolate, autotrophic marine *Beggiatoa* spp., *Applied and Environmental Microbiology* **62**(3): 954–8.

McManus, M. A., R. M. Kudela, M. W. Silver *et al.* (2008). Cryptic blooms: Are thin layers the missing connection?, *Estuaries and Coasts* **31**(2): 396–401.

Meise, C. J. and J. E. Oreilly (1996). Spatial and seasonal patterns in abundance and age-composition of Calanus finmarchicus in the Gulf of Maine and on Georges Bank: 1977–1987. *Deep-Sea Research Part II-Topical Studies in Oceanography* **43**(7–8): 1473–501.

Melville, W. K. and R. J. Rapp (1985). Momentum flux in breaking waves. *Nature* **317**(6037): 514–16.

Miles, J. (1961). On the stability of heterogeneous shear flows. *Journal of Fluid Mechanics* **104**: 496–508.

Miller, J. L. (1994). Fluctuations of gulf-stream frontal position between Cape Hatteras and the Straits of Florida. *Journal of Geophysical Research-Oceans* **99**(C3): 5057–64.

Millero, F. J. (2010) History of the equation of state of seawater. *Oceanography* **23**(3): 18–33.

Milligan, A. J. and F. M. M. Morel (2002). A proton buffering role for silica in diatoms. *Science* **297**(5588): 1848–50.

Milliman, J. D. and K. L. Farnsworth (2011). *River Discharge to the Global Ocean*. Cambridge: Cambridge University Press.

Millot, C. and M. Crepon (1981). Inertial oscillations on the continental shelf of the Gulf of Lions. *Journal of Physical Oceanography* **11**: 639–57.

Mills, D. K., P. B. Tett and G. Novarino (1994). The spring bloom in the south western North Sea in 1989. *Netherlands Journal of Sea Research* **33**(1): 65–80.

Mitchellinnes, B. A. and D. R. Walker (1991). Short-term variability during an anchor station study in the Southern Benguela upwelling system – phytoplankton production and biomass in relation to species changes. *Progress in Oceanography* **28**(1–2): 65–89.

Moore, C. M., D. Suggett, P. M. Holligan *et al.* (2003). Physical controls on phytoplankton physiology and production at a shelf sea front: a fast repetition-rate fluorometer based field study. *Marine Ecology-Progress Series* **259**: 29–45.

Moore, C. M., D. J. Suggett, A. E. Hickman *et al.* (2006). Phytoplankton photoacclimation and photoadaptation in response to environmental gradients in a shelf sea. *Limnology and Oceanography* **51**(2): 936–49.

Morel, F. M. M., R. J. M. Hudson and N. M. Price (1991). Limitation of productivity by trace-metals in the sea. *Limnology and Oceanography* **36**(8): 1742–55.

Morin, P., M. V. M. Wafar and P. Le Corre (1993). Estimation of nitrate flux in a tidal front from satellite-derived temperature data. *Journal of Geophysical Research-Oceans* **98**(C3): 4689–95.

Mork, M. (1981). Circulation phenomena and frontal dynamics of the Norwegian coastal current. *Philosophical Transactions of the Royal Society of London Series a–Mathematical Physical and Engineering Sciences* **302**(1472): 635–47.

Moum, J. N. and J. D. Nash (2000). Topographically induced drag and mixing at a small bank on the continental shelf. *Journal of Physical Oceanography* **30**(8): 2049–54.

Mudge, S. M., D. J. Assinder and G. S. Bourne (1997). Radiological assessment of the Ribble Estuary.3. Redistribution of radionuclides. *Journal of Environmental Radioactivity* **36**(1): 43–67.

Munchow, A. and R. W. Garvine (1993). Buoyancy and wind forcing of a coastal current. *Journal of Marine Research* **51**: 293–322.

Munk, W. (1966). Abyssal recipes. *Deep-Sea Research* **13**: 707–30.

Munk, W. and C. Wunsch (1998). Abyssal recipes II: energetics of tidal and wind mixing. *Deep-Sea Research I* **45**(12): 1977–2010.

Nansen, F. (1906). Northern waters: Captain Roald Amundsen's oceanographic observations in the Arctic Seas in 1901. *Vidensk.Selsk. Skrift., Math.-Naturv. Klasse* **3**: 1–145.

Nealson, K. H. and P. G. Conrad (1999). Life: past, present and future. *Philosophical Transactions of the Royal Society of London Series B-Biological Sciences* **354**(1392): 1923–39.

Nelson, G. and L. Hutchings (1983). The Benguela upwelling area. *Progress in Oceanography* **12**(3): 333–56.

Nixon, S. W., J. W. Ammerman, L. P. Atkinson *et al.* (1996). The fate of nitrogen and phosphorus at the land sea margin of the North Atlantic Ocean. *Biogeochemistry* **35**(1): 141–80.

Nunes, R. A. and G. W. Lennon (1987). Episodic stratification and gravity currents in a marine-environment of modulated turbulence. *Journal of Geophysical Research–Oceans* **92**(C5): 5465–80.

Nunes Vaz, R. A., G. W. Lennon and D. G. Bowers (1990). Physical behavior of a large, negative or inverse estuary. *Continental Shelf Research* **10**(3): 277–304.
O'Brien, C. M., C. J. Fox, B. Planque and J. Casey (2000). Fisheries – Climate variability and North Sea cod. *Nature* **404**(6774): 142.
Oey, L. Y., T. Ezer, G. L. Mellor and P. Chen (1992). A model study of bump induced western boundary current variabilities. *Journal of Marine Systems* **3**(4–5): 321–42.
Officer, C. B. (1976). *Physical Oceanogrphy of Estuaries.* New York: Wiley and Sons.
Okubo, A. (1971). Oceanic diffusion diagrams. *Deep Sea Research and Oceanographic Abstracts* **18**(8): 789–802.
(1987). The fantastic voyage into the deep: marine biofluid mechanics. *Mathematical topics in population biology, morphogenesis, and neurosciences.* E. Teramoto and M. Yamaguchi. Berlin, Springer-Verlag. **71**: 32–47.
Osborn, T. R. (1980). Estimates of the local rate of vertical diffusion from dissipation measurements. *Journal of Physical Oceanography* **10**: 83–9.
Osborn, T. R. and C. W. R. Crawford (1980). An airfoil probe for measuring turbulent velocity fluctuations in water. *Air-Sea Interaction: Instruments and Methods.* F. Dobson, L. Hasse and R. Davis. New York: Plenum, 369–86.
Osborn, T. R. and T. Siddon (1975). *Oceanic shear measurements using the airfoil probe.* Third Biennial Symposium on Turbulence in Liquids Rolla MI.
Ouellet, P., C. Fuentes-Yaco, L. Savard *et al.* (2011). Ocean surface characteristics influence recruitment variability of populations of northern shrimp (*Pandalus borealis*) in the Northwest Atlantic. *Ices Journal of Marine Science* **68**(4): 737–44.
Ozmidov, R. V. (1965). On some features of energy spectra of oceanic turbulence. *Doklady Ak. Nauk. SSSR* **161**: 828–31.
Pacanowski, R. C. and S. G. H. Philander (1981). Parameterisation of vertical mixing in numerical models of the tropical ocean. *Journal of Physical Oceanography* **11**: 1443–51.
Palmer, M. R., T. P. Rippeth and J. H. Simpson (2008). An investigation of internal mixing in a seasonally stratified shelf sea. *Journal of Geophysical Research – Oceans* **113** (C12): 14.
Pauly, D., V. Christensen, S. Guenette *et al.* (2002). Towards sustainability in world fisheries. *Nature* **418**(6898): 689–95.
Perry, R. I., G. C. Harding, J. W. Loder *et al.* (1993). Zooplankton distributions at the Georges Nank frontal system – retention or dispersion. *Continental Shelf Research* **13**(4): 357–83.
Peterson, R. G. and L. Stramma (1991). Upper-level circulation in the south Atlantic Ocean. *Progress in Oceanography* **26**(1): 1–73.
Peterson, W. T. and J. E. Keister (2003). Interannual variability in copepod community composition at a coastal station in the northern California Current: a multivariate approach. *Deep-Sea Research Part II–Topical Studies in Oceanography* **50**(14–16): 2499–517.
Phillips, O. M. (1966). *The Dynamics of the Upper Ocean.* Cambridge University Press.
Pihl, L., S. P. Baden and R. J. Diaz (1991). Effects of periodic hypoxia on distribution of demersal fish and crustaceans. *Marine Biology* **108**(3): 349–60.
Pilskaln, C. H., J. B. Paduan, F. P. Chavez, R. Y. Anderson and W. M. Berelson (1996). Carbon export and regeneration in the coastal upwelling system of Monterey Bay, central California. *Journal of Marine Research* **54**(6): 1149–78.
Pineda, J. (1999). Circulation and larval distribution in internal tidal bore warm fronts. *Limnology and Oceanography* **44**(6): 1400–14.

(1975). The advance and retreat of the thermocline on the continental shelf. *Journal of the Marine Biological Association of the United Kingdom*, **55**(4): 965–74.

(1978). Cyclonic eddies and cross-frontal mixing. *Journal of the Marine Biological Association of the United Kingdom*, **58**(4): 955–63.

(1990). Structure, strength and seasonality of the slope currents in the Bay of Biscay region. *Journal of the Marine Biological Association of the United Kingdom* **70**(4): 857–85.

Pingree, R. D. and B. Le Cann (1989). Celtic and Armorican slope and shelf residual currents. *Progress in Oceanography*, **23**(4): 303–38.

Pingree, R. D. and D. K. Griffiths (1978). Tidal fronts on the shelf seas around the British Isles. *Journal of Geophysical Research* **83**: 4615–22.

Pingree, R. D., P. M. Holligan, G. T. Mardell and R. N. Head (1976). Influence of physical stability on spring, summer and autumn phytoplankton blooms in Celtic Sea. *Journal of the Marine Biological Association of the United Kingdom* **56**(4): 845–73.

Pingree, R. D., P. M. Holligan and G. T. Mardell (1979). *Phytoplankton growth and cyclonic eddies*. Nature, **278**: 245–7.

Pingree, R. D., L. Maddock and E. I. Butler (1977). Influence of biological-activity and physical stability in determining chemical distributions of inorganic-phosphate, silicate and nitrate. *Journal of the Marine Biological Association of the United Kingdom* **57**(4): 1065–73.

Pingree, R. D. and G. T. Mardell (1981). Slope turbulence, internal waves and phytoplankton growth at the celtic sea shelf-break. *Philosophical Transactions of the Royal Society of London Series A-Mathematical Physical and Engineering Sciences* **302**(1472): 663–82.

(1985). Solitary internal waves in the Celtic Sea. *Progress in Oceanography* **14**: 431–41.

Pingree, R. D., G. T. Mardell, P. M. Holligan *et al.* (1982). Celtic sea and armorican current structure and the vertical distributions of temperature and chlorophyll. *Continental Shelf Research* **1**(1): 99–116.

Pingree, R. D., P. R. Pugh, P. M. Holligan and G. R. Forster (1975). Summer phytoplankton blooms and red tides along tidal fronts in approaches to english-channel. *Nature* **258**(5537): 672–7.

PISCO. *PISCO: Partnership for Interdisciplinary Syudies of Coastal Oceans*. Retrieved February 25, 2011, from http://www.piscoweb.org/.

Planas, D., S. Agusti, C. M. Duarte, T. C. Granata and M. Merino (1999). Nitrate uptake and diffusive nitrate supply in the Central Atlantic. *Limnology and Oceanography* **44**(1): 116–26.

Platt, T., C. Fuentes-Yaco and K. T. Frank (2003). Spring algal bloom and larval fish survival. *Nature* **423**(6938): 398–9.

Pomeroy, L. R., P. J. leB. Williams, F. Azam and J. E. Hobbie (2007). The microbial loop. *Oceanography* **20**(2): 28–33.

Postlethwaite, C. F., M. A. M. Maqueda, V. le Fouest *et al.* (2011). The effect of tides on dense water formation in Arctic shelf seas. *Ocean Science*, **7**(2): 203–17.

Prandle, D. (1984). A modeling study of the mixing of cs-137 in the seas of the European continental shelf. *Philosophical Transactions of the Royal Society of London Series A–Mathematical Physical and Engineering Sciences* **310**(1513): 407–36.

(2004). Saline intrusion in partially mixed estuaries. *Estuarine Coastal and Shelf Science* **59**(3): 385–97.

Pugh, D. T. (1987). *Tides, Surges and Mean Sea Level*. Wiley-Blackwell.

Qian, Y. R., A. E. Jochens, M. C. Kennicutt and D. C. Biggs (2003). Spatial and temporal variability of phytoplankton biomass and community structure over the continental margin of the northeast Gulf of Mexico based on pigment analysis. *Continental Shelf Research* **23**(1): 1–17.

Rabalais, N. N., R. E. Turner and W. J. Wiseman (2002). Gulf of Mexico hypoxia, aka 'The dead zone'. *Annual Review of Ecology and Systematics* **33**: 235–63.

Radach, G. and A. Moll (1993). Estimation of the variability of production by simulating annual cycles of phytoplankton in the central North Sea. *Progress in Oceanography* **31**(4): 339–419.

Raven, J. A. (1983). The transport and function of silicon in plants. *Biological Reviews of the Cambridge Philosophical Society* **58**(2): 179–207.

Redfield, A. C. (1934). On the proportions of organic derivations in sea water and their relation to the composition of plankton. *James Johnstone Memorial Volume*. R. J. Daniel. Liverpool, Liverpool University Press: 176–92.

Redfield, A. C., B. H. Ketchum and F. A. Richards (1963). The influence of organisms on the composition of seawater. *The Sea*. ed. M. N. Hill. **2**: 26–77.

Rees, A. P., J. A. Gilbert and B. A. Kelly-Gerreyn (2009). Nitrogen fixation in the western English Channel (NE Atlantic Ocean). *Marine Ecology Progress Series* **374**: 7–12.

Rees, A. P., I. Joint and K. M. Donald (1999). Early spring bloom phytoplankton-nutrient dynamics at the Celtic Sea Shelf Edge. *Deep-Sea Research Part I-Oceanographic Research Papers* **46**(3): 483–510.

Reffray, G., P. Fraunie and P. Marsaleix (2004). Secondary flows induced by wind forcing in the Rhone region of freshwater influence. *Ocean Dynamics*, **54**(2): 179–96.

Reid, D. G. (2001). SEFOS – shelf edge fisheries and oceanography studies: an overview. *Fisheries Research* **50**(1–2): 1–15.

Ribeiro, C. H. A., J. J. Waniek and J. Sharples (2004). Observations of the spring-neap modulation of the gravitational circulation in a partially mixed estuary. *Ocean Dynamics* **54**(3–4): 299–306.

Richardson, K., M. F. Lavin-Peregrina, E. G. Mitchelson and J. H. Simpson (1985). Seasonal distribution of chlorophyll-a in relation to physical structure in the western Irish Sea. *Oceanologica Acta* **8**(1): 77–86.

Richardson, K., A. W. Visser and F. B. Pedersen (2000). Subsurface phytoplankton blooms fuel pelagic production in the North Sea. *Journal of Plankton Research* **22**(9): 1663–71.

Richardson, L. F. and H. Stommel (1948). Note on eddy diffusion in the sea. *Journal of Meteorology* **5**: 238–40.

Richardson, T. L. and J. J. Cullen (1995). Changes in buoyancy and chemical composition during growth of a coastal marine diatom: Ecological and biogeochemical consequences. *Marine Ecology Progress Series* **128**(1–3): 77–90.

Riley, G. A. (1942). The relationship of vertical turbulence and spring diatom flowerings. *Journal of Marine Research* **5**: 66–87.

(1946). Factors controlling phytoplankton populations on Georges Bank. *Journal of Marine Research* **6**: 54–73.

Rippeth, T. P., N. R. Fisher and J. H. Simpson (2001). The cycle of turbulent dissipation in the presence of tidal straining. *Journal of Physical Oceanography* **31**(8): 2458–71.

Rippeth, T. P., J. Scourse, K. Uehara and S. McKeown (2008). Impact of sea-level rise over the last deglacial transition on the strength of the continental shelf CO_2 pump. *Geophysical Research Letters* **35**(24).

Rippeth, T. P. and J. H. Simpson (1998). Diurnal signals in vertical motions on the Hebridean Shelf. *Limnology and Oceanography* **43**(7): 1690–6.

Rippeth, T. P., J. H. Simpson, R. J. Player and M. Garcia (2002). Current oscillations in the diurnal-inertial band on the Catalonian shelf in spring. *Continental Shelf Research* **22**(2): 247–65.

Rippeth, T. P., P. Wiles, M. R. Palmer, J. Sharples and J. Tweddle (2009). The diapcynal nutrient flux and shear-induced diapcynal mixing in the seasonally stratified western Irish Sea. *Continental Shelf Research* **29**(13): 1580–7.

Robinson, C. and P. J. L. Williams (1999). Plankton net community production and dark respiration in the Arabian Sea during September 1994. *Deep-Sea Research Part II-Topical Studies in Oceanography* **46**(3–4): 745–65.

Rodhe, J. (1998). The Baltic and North Seas: a process-oriented review of the Physical Oceanography. *The Sea*. A. R. Robinson and K. H. Brink. **11**: 699–732.

Ross, O. N. and J. Sharples (2004). Recipe for 1-D Lagrangian particle tracking models in space-varying diffusivity. *Limnology and Oceanography-Methods* **2**: 289–302.

(2007). Phytoplankton motility and the competition for nutrients in the thermocline. *Marine Ecology Progress Series* **347**: 21–38.

Ross, O. N. and J. Sharples (2008). Swimming for survival: A role of phytoplankton motility in a stratified turbulent environment. *Journal of Marine Systems* **70**(3–4): 248–62.

Rowe, G. T. and W. C. Phoel (1992). Nutrient regeneration and oxygen-demand in Bering Sea continental-shelf sediments. *Continental Shelf Research* **12**(4): 439–49.

Rudnick, D. L., R. E. Davis, C. C. Eriksen, D. M. Fratantoni and M. J. Perry (2004). Underwater gliders for ocean research. *Marine Technology Society Journal* **38**(2): 73–84.

Runge, J. A., S. Plourde, P. Joly, B. Niehoff and E. Durbin (2006). Characteristics of egg production of the planktonic copepod, Calanus finmarchicus, on Georges Bank: 1994–1999. *Deep-Sea Research Part II—Topical Studies in Oceanography* **53**(23–24): 2618–31.

Ryan, J. P., M. A. McManus, J. D. Paduan and F. P. Chavez (2008). Phytoplankton thin layers caused by shear in frontal zones of a coastal upwelling system. *Marine Ecology Progress Series* **354**: 21–34.

Sabatini, M. E. and G. L. A. Colombo (2001). Seasonal pattern of zooplankton biomass in the Argentinian shelf off Southern Patagonia (45 degrees-55 degrees S). *Scientia Marina* **65**(1): 21–31.

Sagan, C. (1997). *The Demon-haunted World: Science As a Candle in the Dark*. New York: Random House.

Sarmiento, J. L., R. Slater, R. Barber *et al.* (2004). Response of ocean ecosystems to climate warming. *Global Biogeochemical Cycles* **18**(3): 35.

Schahinger, R. B. (1988). Near inertial motions on the south Australian shelf. *Journal of Physical Oceanography* **18**: 492–504.

Schluessel, P., W. J. Emery, H. Grassi and T. Mammen (1990). On the bulk-skin temperature difference and its impact on satellite remote-sensing of sea-surface temperature. *Journal of Geophysical Research–Oceans* **95**(C8): 13341–56.

Schneider, B., R. Schlitzer, G. Fischer and E. M. Nothig (2003). Depth-dependent elemental compositions of particulate organic matter (POM) in the ocean. *Global Biogeochemical Cycles* **17**(2): 16.

Scott, B. E., J. Sharples, O. N. Ross *et al.* (2010). Sub-surface hotspots in shallow seas: fine-scale limited locations of top predator foraging habitat indicated by tidal mixing and sub-surface chlorophyll. *Marine Ecology Progress Series* **408**: 207–26.

Scrope-Howe, S. and D. A. Jones (1985). Biological studies in the vicinity of a shallow-sea tidal mixing front. Composition, abundance and distribution of zooplankton in the western

Irish Sea, April 1980 to November 1981. *Philosophical Transactions of the Royal Society of London Series B–Biological Sciences* **310**(1146): 501–19.

Scully, M. E., C. T. Friedrichs and J. M. Brubaker (2005). Control of estuarine stratification and mixing by wind-induced straining of the estuarine density field. *Estuaries* **28**(3): 321–6.

Scully, M. E. and C. T. Friedrichs (2007). Sediment pumping by tidal asymmetry in a partially mixed estuary. *Journal of Geophysical Research–Oceans* **112**(C7): 12.

Sharples, J. (1999). Investigating the seasonal vertical structure of phytoplankton in shelf seas. *Progress in Oceanography*: 3–38.

(2008). Potential impacts of the spring-neap tidal cycle on shelf sea primary production. *Journal of Plankton Research* **30**(2): 183–97.

Sharples, J., C. M. Moore and E. R. Abraham (2001a). Internal tide dissipation, mixing, and vertical nitrate flux at the shelf edge of NE New Zealand. *Journal of Geophysical Research–Oceans* **106**(C7): 14069–81.

Sharples, J., C. M. Moore, T. P. Rippeth *et al.* (2001b). Phytoplankton distribution and survival in the thermocline. *Limnology and Oceanography* **46**(3): 486–96.

Sharples, J., C. M. Moore, A. E. Hickman *et al.* (2009). Internal tidal mixing as a control on continental margin ecosystems. *Geophysical Research Letters* **36**: 5.

Sharples, J. and J. H. Simpson (1993). Periodic frontogenesis in a region of fresh-water influence. *Estuaries* **16**(1): 74–82.

Sharples, J. and J. H. Simpson (1995). Semi-diurnal and longer period stability cycles in the Liverpool Bay region of fresh-water influence. *Continental Shelf Research* **15**(2–3): 295–313.

(1996). The influence of the springs-neaps cycle on the position of shelf sea fronts. *Buoyancy Effects on Coastal and Estuarine Dynamics*: D. G. Aubrey and C. T. Friedricks, AGU, Washington 71–82.

Sharples, J., O. N. Ross, B. E. Scott, S. P. R. Greenstreet and H. Fraser (2006). Inter-annual variability in the timing of stratification and the spring bloom in the North-western North Sea. *Continental Shelf Research* **26**(6): 733–51.

Sharples, J. and P. Tett (1994). Modeling the effect of physical variability on the midwater chlorophyll maximum. *Journal of Marine Research* **52**(2): 219–38.

Sharples, J., J. F. Tweddle, J. A. M. Green *et al.* (2007). Spring-neap modulation of internal tide mixing and vertical nitrate fluxes at a shelf edge in summer. *Limnology and Oceanography* **52**(5): 1735–47.

Sheldon, R. W., W. H. Sutcliffe and M. A. Paranjape (1977). Structure of pelagic food-chain and relationship between plankton and fish production. *Journal of the Fisheries Research Board of Canada* **34**(12): 2344–53.

Shi, W. and M. Wang (2007). Observations of a Hurricane Katrina-induced phytoplankton bloom in the Gulf of Mexico. *Geophysical Research Letters* **34**(11): 5.

Simpson, J. H. (1981). The shelf-sea fronts – implications of their existence and behavior. *Philosophical Transactions of the Royal Society of London Series A–Mathematical Physical and Engineering Sciences* **302**(1472): 531–46.

(1997). Physical processes in the ROFI regime. *Journal of Marine Systems* **12**(1–4): 3–15.

Simpson, J. H., C. M. Allen and N. C. G. Norris (1978). Fronts on continental-shelf. *Journal of Geophysical Research–Oceans and Atmospheres* **83**(NC9): 4607–14.

Simpson, J. H. and D. Bowers (1979). Shelf sea fronts adjustments revealed by satellite IR imagery. *Nature* **280**(5724): 648–51.

(1981). Models of stratification and frontal movement in shelf seas. *Deep-Sea Research Part A–Oceanographic Research Papers* **28**(7): 727–38.

Simpson, J. H. and D. G. Bowers (1984). The role of tidal stirring in controlling the seasonal heat cycle in shelf seas. *Annales Geophysicae* **2**(4): 411–16.

(1990). Data Bank Oceanography: testing models of the seasonal heat cycle in shelf seas. *Physics of Shallow Seas*. H. Wang, J. Wang and H. Dai. Beijing: China Ocean Press, 291–304.

Simpson, J. H., J. Brown, J. Matthews and G. Allen (1990). Tidal straining, density currents, and stirring in the control of estuarine stratification. *Estuaries* **13**(2): 125–32.

Simpson, J. H., H. Burchard, N. R. Fisher and T. P. Rippeth (2002). The semi-diurnal cycle of dissipation in a ROFI: model-measurement comparisons. *Continental Shelf Research* **22**(11–13): 1615–28.

Simpson, J. H., W. R. Crawford, T. P. Rippeth, A. R. Campbell and J. V. S. Cheok (1996). The vertical structure of turbulent dissipation in shelf seas. *Journal of Physical Oceanography* **26**(8): 1579–90.

Simpson, J. H., D. J. Edelsten, A. Edwards, N. C. G. Norris and P. B. Tett (1979). Islay front – physical structure and phytoplankton distribution. *Estuarine and Coastal Marine Science* **9**(6): 713ff.

Simpson, J. H., J. A. M. Green, T. P. Rippeth, T. R. Osborn and W. A. M. Nimmo-Smith (2009). The structure of dissipation in the western Irish Sea front. *Journal of Marine Systems* **77**(4): 428–40.

Simpson, J. H. and J. R. Hunter (1974). Fronts in Irish Sea. *Nature* **250**(5465): 404–6.

Simpson, J. H., P. Hyder, T. P. Rippeth and I. M. Lucas (2002). Forced oscillations near the critical latitude for diurnal-inertial resonance. *Journal of Physical Oceanography* **32**(1): 177–87.

Simpson, J. H. and I. D. James (1986). Coastal and estuarine fronts. *Baroclinic Processes on Continental Shelves* .C. N. K. Mooers. Washington, DC: Americal Geophysical Union 63–93.

Simpson, J. H., E. G. Mitchelson-Jacob and A. E. Hill (1990). Flow structure in a channel from an acoustic doppler current profiler. *Continental Shelf Research* **10**(6): 589–603.

Simpson, J. H. and R. D. Pingree (1978). Shallow sea fronts produced by tidal stirring. *Oceanic Fronts in Coastal Processes*. M. J. Bowman and W. E. Esaias. New York: Springer-Verlag: 29–42.

Simpson, J. H., T. P. Rippeth and A. R. Campbell (2000). The phase lag of dissipation in tidal flow. *Interaction between Estuaries, Coastal and Shelf Seas*. T. Yanagi. Tokyo: Terra Scientific, 57–67.

Simpson, J. H., J. Sharples and T. P. Rippeth (1991). A prescriptive model of stratification induced by fresh-water runoff. *Estuarine Coastal and Shelf Science* **33**(1): 23–35.

Simpson, J. H. and A. Snidvongs (1998). *The influence of Monsoonal discharge on tropical shelf seas: the Gulf of Thailand as a case for study*. International Workshop on the Mekong delta, Chiang Rai, Thailand.

Simpson, J. H. and A. J. Souza (1995). Semidiurnal switching of stratification in the region of fresh-water influence of the Rhine. *Journal of Geophysical Research–Oceans* **100**(C4): 7037–44.

Simpson, J. H., P. B. Tett, N. L. Argote-Espinoza *et al.* (1982). Mixing and phytoplankton growth around an island in a stratified sea. *Continental Shelf Research* **1**(1): 15–31.

Sims, D. W. and V. A. Quayle (1998). Selective foraging behaviour of basking sharks on zooplankton in a small-scale front. *Nature* **393**(6684): 460–4.

Sims, D. W., E. J. Southall, A. J. Richardson, P. C. Reid and J. D. Metcalfe (2003). Seasonal movements and behaviour of basking sharks from archival tagging: no evidence of winter hibernation. *Marine Ecology Progress Series* **248**: 187–96.

Sin, Y. S., R. L. Wetzel and I. C. Anderson (1999). Spatial and temporal characteristics of nutrient and phytoplankton dynamics in the York River Estuary, Virginia: Analyses of long-term data. *Estuaries* **22**(2A): 260–75.

Sinclair, M. and T. D. Iles (1989). Population regulation and speciation in the oceans. *Journal Du Conseil* **45**(2): 165–75.

Small, J., T. C. Sawyer and J. C. Scott (1999). The evolution of an internal bore at the Malin shelf break. *Annales Geophysicae* **17**: 547–65.

Smayda, T. J. (1997). Harmful algal blooms: their ecophysiology and general relevance to phytoplankton blooms in the sea. *Limnology and Oceanography* **42**(5): 1137–53.

Smayda, T. J. and V. L. Trainer (2010). Dinoflagellate blooms in upwelling systems: seeding, variability, and contrasts with diatom bloom behaviour. *Progress in Oceanography* **85**(1–2): 92–107.

Smetacek, V. (2001). A watery arms race. *Nature* **411**(6839): 745.

Smith, S. D. and E. G. Banke (1975). Variation of sea-surface drag coefficient with wind speed. *Quarterly Journal of the Royal Meteorological Society* **101**(429): 665–73.

Solomon, S., D. Qin, M. Manning *et al.* (2007). *Contribution of Working Group I to the Fourth Assessment Report of the Intergovernmental Panel on Climate Change, 2007.*

Sournia, A., M. J. Chretiennot-Dinet and M. Ricard (1991). Marine-phytoplankton – how many species in the world ocean. *Journal of Plankton Research* **13**(5): 1093–99.

Souza, A. J. and J. H. Simpson (1996). The modification of tidal ellipses by stratification in the Rhine ROFI. *Continental Shelf Research* **16**(8): 997–1007.

(1997). Controls on stratification in the Rhine ROFI system. *Journal of Marine Systems* **12**(1–4): 311–23.

Souza, A. J., J. H. Simpson and F. Schirmer (1997). Current structure in the Rhine region of freshwater influence. *Journal of Marine Research* **55**: 277–92.

Souza, A. J., J. H. Simpson, M. Harikrishnan and J. Malarkey (2001). Flow structure and seasonality in the Hebridean slope current. *Oceanologica Acta* **24**: S63–S76.

Springer, A. M., C. P. McRoy and M. V. Flint (1996). The Bering Sea Green Belt: Shelf-edge processes and ecosystem production. *Fisheries Oceanography* **5**(3–4): 205–23.

Stacey, M. T., J. P. Fram and F. K. Chow (2008). Role of tidally periodic density stratification in the creation of estuarine subtidal circulation. *Journal of Geophysical Research – Oceans* **113**(C8).

Stansfield, K. and C. Garrett (1997). Implications of the salt and heat budgets of the Gulf of Thailand. *Journal of Marine Research* **55**(5): 935–63.

Steele, J. H. and C. S. Yentsch (1960). The vertical distribution of chlorophyll. *Journal of the Marine Biological Association of the U.K.* **39**: 217–26.

Steeman-Nielsen, E. (1951). Measurement of production of organic matter in sea by means of carbon-14. *Nature* **267**: 684–5.

Stegert, C., R. B. Ji and C. S. Davis (2010). Influence of projected ocean warming on population growth potential in two North Atlantic copepod species. *Progress in Oceanography* **87**(1–4): 264–76.

Steinbuck, J. V., M. T. Stacey, M. A. McManus, O. M. Cheriton and J. P. Ryan (2009). Observations of turbulent mixing in a phytoplankton thin layer: implications for formation, maintenance, and breakdown. *Limnology and Oceanography* **54**(4): 1353–68.

Stevick, P. T., L. S. Incze, S. D. Kraus *et al.* (2008). Trophic relationships and oceanography on and around a small offshore bank. *Marine Ecology Progress Series* **363**: 15–28.

Stoecker, D. K., A. Taniguchi and A. E. Michaels (1989). Abundance of autotrophic, mixotrophic and heterotrophic planktonic ciliates in shelf and slope waters. *Marine Ecology Progress Series* **50**(3): 241–54.

Stolte, W. and R. Riegman (1995). Effect of phytoplankton cell-size on transient-state nitrate and ammonium uptake kinetics. *Microbiology* **141**: 1221–9.

Stommel, H. (1948). The westward intensification of wind-driven ocean currents. *Transactions of the American Geophysical Union* **29**: 202–6.

Sugget, D. J., C. M. Moore, A. E. Hickman and R. J. Geider (2009). Interpretation of fast repetition rate (FRR) fluorescence: signatures of phytoplankton community structure versus physiological state. *Marine Ecology–Progress Series* **376**: 1–19.

Sullivan, J. M., P. L. Donaghay and J. E. B. Rines (2010). Coastal thin layer dynamics: Consequences to biology and optics. *Continental Shelf Research* **30**(1): 50–65.

Sundermann, J. (1997). The PRISMA project: an investigation of processes controlling contaminant fluxes in the German Bight. *Marine Ecology Progress Series* **156**: 239–43.

Sverdrup, H. U. (1953). On conditions for the vernal blooming of phytoplankton. *Journal du Conseil* **18**: 287–95.

Swallow, J. C. (1955). A neutral-buoyancy float for measuring deep currents. *Deep-Sea Research* **3**(1): 74–81.

Taylor, G. I. (1922). Tidal oscillations in gulfs and rectangular basins. *Proceedings of the London Mathematical Society* **20**: 148–81.

(1953). Dispersion of soluable matter in solvent flowing slowly through a tube. *Proceedings of the Royal Society of London* **A219**: 186–203.

(1954). The dispersion of matter in turbulent flow through a pipe. *Proceedings of the Royal Society of London* **A223**:446–68.

Tett, P., A. Edwards and K. Jones (1986). A model for the growth of shelf-sea phytoplankton in summer. *Estuarine Coastal and Shelf Science* **23**(5): 641–72.

Tett, P., D. Hydes and R. Sanders (2003). Influence of nutrient biogeochemistry on the ecology of North-West European shelf seas. *Biogeochemistry of Marine systems*. K. D. Black and G. B. Shimmield. Oxford, Blackwell: 293–363.

Thomas, A. C., D. W. Townsend and R. Weatherbee (2003). Satellite-measured phytoplankton variability in the Gulf of Maine. *Continental Shelf Research* **23**(10): 971–89.

Thomas, H., Y. Bozec, K. Elkalay and H. J. W. de Baar (2004). Enhanced open ocean storage of CO_2 from shelf sea pumping. *Science* **304**(5673): 1005–8.

Thomas, P. J. and P. F. Linden (1996). A laboratory simulation of mixing across tidal fronts. *Journal of Fluid Mechanics* **309**: 321–44.

Thompson, S. M., and J. S. Turner (1975). Mixing across an interface due to turbulence generated by an oscillating grid. *Journal of Fluid Mechanics* **67**: 349–368.

Thorpe, S. A. (1971). Experiments on the instability of stratified shear flows: miscible fluids. *Journal of Fluid Mechanics* **46**: 299–319.

(2005). *The Turbulent Ocean*. Cambridge.

(2007). *An Introduction to Ocean Turbulence*. Cambridge University Press.

Torres-Valdes, S., V. M. Roussenov, R. Sanders *et al.* (2009). Distribution of dissolved organic nutrients and their effect on export production over the Atlantic Ocean. *Global Biogeochemical Cycles* **23**.

Townsend, D. W., L. M. Cammen, P. M. Holligan, D. E. Campbell and N. R. Pettigrew (1994). Causes and consequences of variability in the timing of spring phytoplankton blooms. *Deep-Sea Research Part I–Oceanographic Research Papers* **41**(5–6): 747–65.

Townsend, D. W., M. D. Keller, M. E. Sieracki and S. G. Ackleson (1992). Spring phytoplankton blooms in the absence of vertical water column stratification. *Nature* **360**(6399): 59–62.

Townsend, D. W. and N. R. Pettigrew (1996). The role of frontal currents in larval fish transport on Georges Bank. *Deep-Sea Research Part II–Topical Studies in Oceanography* **43**(7–8): 1773–92.

Townsend, D. W. and M. Thomas (2002). Springtime nutrient and phytoplankton dynamics on Georges Bank. *Marine Ecology Progress Series* **228**: 57–74.

Treguer, P., D. M. Nelson, A. J. Van Bennekom *et al.* (1995). The silica balance in the world ocean – a reestimate. *Science* **268**(5209): 375–9.

Trimmer, M., R. J. Gowen and B. M. Stewart (2003). Changes in sediment processes across the western Irish Sea Front. *Estuarine Coastal and Shelf Science* **56**(5–6): 1011–19.

Tsunogai, S., S. Watanabe and T. sato (1999). Is there a 'continental shelf pump' for the absorption of atmospheric CO_2? *Tellus Series B–Chemical and Physical Meteorology* **51**(3): 701–12.

Turner, J. S. (1973). *Buoyancy Effects in Fluids*. Cambridge: Cambridge University Press.

(2010). The melting of ice in the Arctic Ocean: the influence of double-diffusive transport of heat from below. *Journal of Physical Oceanography* **40**(1): 249–56.

Turner, J. S. and E. B. Kraus (1967). A one-dimensional model of the seasonal thermocline. I. A laboratory experiment andd its interpretation. *Tellus* **19**: 88–97.

Tyler, M. A. and H. H. Seliger (1978). Annual subsurface transport of a red tide dinoflagellate to its bloom area – water circulation patterns and organism distributions in chesapeake bay. *Limnology and Oceanography* **23**(2): 227–46.

UNESCO (1981). *Tenth report of the joint panel on oceanographic tables and standards*. UNESCO Technical Papers in marine Science Paris, UNESCO. **36**.

USAF (1960). *Handbook of Geophysics*. New York: Macmillan.

Uye, S., and N. Shibuno (1992). Reproductive-biology of the planktonic copepod paracalanus sp in the inland sea of Japan. *Journal of Plankton Research* **14**(3): 343–58.

Uye, S., T. Yamaoka and T. Fujisawa (1992). Are tidal fronts good recruitment areas for herbivourous copepods? *Fisheries Oceanography* **1**(3): 216–26.

van Duren, L. A. and J. J. Videler (2003). Escape from viscosity: the kinematics and hydrodynamics of copepod foraging and escape swimming. *Journal of Experimental Biology* **206**(2): 269–79.

Veldhuis, M. J. W. and G. W. Kraay (2004). Phytoplankton in the subtropical Atlantic Ocean: towards a better assessment of biomass and composition. *Deep-Sea Research Part I–Oceanographic Research Papers* **51**(4): 507–30.

Verspecht, F., T. P. Rippeth, J. H. Simpson *et al.* (2009). Residual circulation and stratification in the Liverpool Bay region of freshwater influence. *Ocean Dynamics* **59**(5): 765–79.

Villareal, T. A. (1988). Positive buoyancy in the oceanic diatom *rhizosolenia debyana* peragallo. *Deep-Sea Research Part A-Oceanographic Research Papers* **35**(6): 1037–045.

(1992). Buoyancy properties of the giant diatom ethmodiscus. *Journal of Plankton Research* **14**(3): 459–63.

Visser, A. W. and T. Kiørboe (2006). Plankton motility patterns and encounter rates. *Oecologia* **148**(3): 538–46.

Visser, A. W., A. J. Souza, K. Hessner and J. H. Simpson (1994). The effect of stratification on tidal current profiles in a region of fresh-water influence. *Oceanologica Acta* **17**(4): 369–81.

Walsh, J. J. (1991). Importance of continental margins in the marine biogeochemical cycling of carbon and nitrogen. *Nature* **350**(6313): 53–5.

Waniek, J. J. (2003). The role of physical forcing in initiation of spring blooms in the northeast Atlantic. *Journal of Marine Systems* **39**(1–2): 57–82.

Wassmann, P. (1998). Retention versus export food chains: processes controlling sinking loss from marine pelagic systems. *Hydrobiologia* **363**: 29–57.

Weatherly, G. L. and P. J. Martin (1978). Structure and dynamics of the oceanic bottom boundary layer. *Journal of Physical Oceanography* **8**(4): 557–70.

Weir, C. R., C. Pollock, C. Cronin and S. Taylor (2001). Cetaceans of the Atlantic Frontier, north and west of Scotland. *Continental Shelf Research* **21**(8–10): 1047–71.

Werner, F. E., R. I. Perry, R. G. Lough and C. E. Naimie (1996). Trophodynamic and advective influences on Georges Bank larval cod and haddock. *Deep-Sea Research Part II–Topical Studies in Oceanography* **43**(7–8): 1793–822.

Wessel, P. and W. H. F. Smith (1996). A global, self-consistent, hierarchical, high-resolution shoreline database. *Journal of Geophysical Research-Solid Earth* **101**(B4): 8741–3.

Weston, K., L. Fernand, D. K. Mills, R. Delahunty and J. Brown (2005). Primary production in the deep chlorophyll maximum of the central North Sea. *Journal of Plankton Research* **27**(9): 909–22.

Weston, K., L. Fernand, J. Nicholls *et al.* (2008). Sedimentary and water column processes in the Oyster Grounds: A potentially hypoxic region of the North Sea. *Marine Environmental Research* **65**(3): 235–49.

White, R. G., A. E. Hill and D. A. Jones (1988). Distribution of *nephrops norvegicus* (I) larvae in the western Irish sea – an example of advective control on recruitment. *Journal of Plankton Research* **10**(4): 735–47.

Whitney, M. M. and R. W. Garvine (2006). Simulating the Delaware Bay buoyant plume: Comparison with observations. *Journal of Physical Oceanography* **36**: 3–21.

Wiles, P. J., L. A. van Duren, C. Hase, J. Larsen and J. H. Simpson (2006). Stratification and mixing in the Limfjorden in relation to mussel culture. *Journal of Marine Systems* **60**(1–2): 129–43.

Wilkerson, F. P., A. M. Lassiter, R. C. Dugdale, A. Marchi and V. E. Hogue (2006). The phytoplankton bloom response to wind events and upwelled nutrients during the CoOP WEST study. *Deep-Sea Research Part II–Topical Studies in Oceanography* **53**(25–26): 3023–48.

Williams, P. J. leB. and J. E. Robertson (1991). Overall planktonic oxygen and carbon-dioxide metabolisms – the problem of reconciling observations and calculations of photosynthetic quotients. *Journal of Plankton Research* **13**: S153–69.

Williams, R. G. and M. J. Follows (2011). *Ocean Dynamics and the Carbon Cycle: Principles and Mechanisms*. Cambridge: Cambridge University Press.

Winant, C. D., R. C. Beardsley and R. E. Davis (1987). Moored wind, temperature, and current observations made during coastal ocean dynamics experiments-1 and experiment-2 over the northern california continental-shelf and upper slope. *Journal of Geophysical Research–Oceans* **92**(C2): 1569–604.

Winter, D. F., K. Banse and G. C. Anderson (1975). Dynamics of phytoplankton blooms in Puget Sound, a fjord in northwestern United States. *Marine Biology* **29**(2): 139–76.

Wishner, K. F., D. M. Outram and D. S. Ullman (2006). Zooplankton distributions and transport across the northeastern tidal front of Georges Bank. *Deep-Sea Research Part II–Topical Studies in Oceanography* **53**(23–24): 2570–96.

Wollast, R. and L. Chou (2001). Ocean Margin EXchange in the northern Gulf of Biscay: OMEX I. An introduction. *Deep-Sea Research Part II–Topical Studies in Oceanography* **48**(14–15): 2971–8.

Wong, G. T. F., S. Y. Chao, Y. H. Li and F. K. Shiah (2000). The Kuroshio edge exchange processes (KEEP) study – an introduction to hypotheses and highlights. *Continental Shelf Research* **20**(4–5): 335–47.

Woods, J. D. (1968). Wave-induced shear instability in the summer thermocline. *Journal of Fluid Mechanics* **32**: 791–800.

Wunsch, C. and R. Ferrari (2004). Vertical mixing, energy and the general circulation of the oceans. *Annual Review of Fluid Mechanics* **36**: 281–314.

Xing, J. X. and A. M. Davies (1998). A three-dimensional model of internal tides on the Malin-Hebrides shelf and shelf edge. *Journal of Geophysical Research-Oceans* **103**(C12): 27821–47.

Yanagi, T. and H. Tamaru (1990). Temporal and spatial variations in a tidal front. *Continental Shelf Research* **10**: 615–27.

Yool, A. and M. J. R. Fasham (2001). An examination of the 'continental shelf pump' in an open ocean general circulation model. *Global Biogeochemical Cycles* **15**(4): 831–44.

Young, J. W., R. Bradford, T. D. Lamb *et al.* (2001). Yellowfin tuna (*Thunnus albacares*) aggregations along the shelf break off south-eastern Australia: links between inshore and offshore processes. *Marine and Freshwater Research* **52**(4): 463–74.

Yu, J. and P. Gloersen (2005). Interannual variations in global SST deviations from SMMR from 1978–1987. *International Journal of Remote Sensing* **26**(24): 5419–31.

INDEX

Printed in the United States
By Bookmasters